STAPLE TO SUPERFOOD

Columbia Studies in International and Global History

COLUMBIA STUDIES IN INTERNATIONAL AND GLOBAL HISTORY
Cemil Aydin, Timothy Nunan, and Dominic Sachsenmaier, Series Editors

This series presents some of the finest and most innovative work coming out of the current landscapes of international and global historical scholarship. Grounded in empirical research, these titles transcend the usual area boundaries and address how history can help us understand contemporary problems, including poverty, inequality, power, political violence, and accountability beyond the nation-state. The series covers processes of flows, exchanges, and entanglements—and moments of blockage, friction, and fracture—not only between "the West" and "the Rest" but also among parts of what has variously been dubbed the "Third World" or the "Global South." Scholarship in international and global history remains indispensable for a better sense of current complex regional and global economic transformations. Such approaches are vital in understanding the making of our present world.

Pierre Singaravélou, trans. Stephen W. Sawyer, *Tianjin Cosmopolis: An Alternative History of Globalization*

James De Lorenzi, *Feasting on History: Ethiopia and the Orientalists*

Jie-Hyun Lim, trans. Megan Sungyoon, *Victimhood Nationalism: History and Memory in a Global Age*

Hale Eroğlu, *Muslim Transnationalism in Modern China: Debates on Hui Identity and Islamic Reform*

Sandrine Kott, *A World More Equal: An Internationalist Perspective on the Cold War*

Julia Hauser, *A Taste for Purity: An Entangled History of Vegetarianism*

Hayrettin Yücesoy, *Disenchanting the Caliphate: The Secular Discipline of Power in Abbasid Political Thought*

Anne Irfan, *Refuge and Resistance: Palestinians and the International Refugee System*

Michael Francis Laffan, *Under Empire: Muslim Lives and Loyalties Across the Indian Ocean World, 1775–1945*

Eva-Maria Muschik, *Building States: The United Nations, Development, and Decolonization, 1945–1965*

Jessica Namakkal, *Unsettling Utopia: The Making and Unmaking of French India*

Michael Christopher Low, *Imperial Mecca: Ottoman Arabia and the Indian Ocean Hajj*

Nicole CuUnjieng Aboitiz, *Asian Place, Filipino Nation: A Global Intellectual History of the Philippine Revolution, 1887–1912*

Mona L. Siegel, *Peace on Our Terms: The Global Battle for Women's Rights After the First World War*

Raja Adal, *Beauty in the Age of Empire: Japan, Egypt, and the Global History of Aesthetic Education*

For a complete list of books in the series, please see the Columbia University Press website.

Staple to Superfood

A GLOBAL HISTORY OF THE SWEET POTATO

Q. Edward Wang

Columbia University Press
New York

Columbia University Press
Publishers Since 1893
New York   Chichester, West Sussex

Copyright © 2025 Columbia University Press
All rights reserved

Library of Congress Cataloging-in-Publication Data
Names: Wang, Q. Edward, 1958– author
Title: Staple to superfood : a global history of the sweet potato /
Q. Edward Wang.
Description: New York, NY : Columbia University Press, [2025] |
Series: Columbia studies in international and global history |
Includes bibliographical references and index.
Identifiers: LCCN 2025009685 (print) | LCCN 2025009686 (ebook) |
ISBN 9780231217354 hardback | ISBN 9780231217361 trade paperback |
ISBN 9780231561709 ebook
Subjects: LCSH: Sweet potatoes—History | Sweet potatoes—Health aspects |
Sweet potatoes—Harvesting
Classification: LCC SB211.S9 W36 2025 (print) | LCC SB211.S9 (ebook)

Cover design: Chang Jae Lee
Cover image: © Shutterstock

GPSR Authorized Representative: Easy Access System Europe, Mustamäe tee 50,
10621 Tallinn, Estonia, gpsr.requests@easproject.com

Staple to Superfood

A GLOBAL HISTORY OF THE SWEET POTATO

Q. Edward Wang

Columbia University Press
New York

Columbia University Press
Publishers Since 1893
New York   Chichester, West Sussex

Copyright © 2025 Columbia University Press
All rights reserved

Library of Congress Cataloging-in-Publication Data
Names: Wang, Q. Edward, 1958– author
Title: Staple to superfood : a global history of the sweet potato / Q. Edward Wang.
Description: New York, NY : Columbia University Press, [2025] | Series: Columbia studies in international and global history | Includes bibliographical references and index.
Identifiers: LCCN 2025009685 (print) | LCCN 2025009686 (ebook) | ISBN 9780231217354 hardback | ISBN 9780231217361 trade paperback | ISBN 9780231561709 ebook
Subjects: LCSH: Sweet potatoes—History | Sweet potatoes—Health aspects | Sweet potatoes—Harvesting
Classification: LCC SB211.S9 W36 2025 (print) | LCC SB211.S9 (ebook)

Cover design: Chang Jae Lee
Cover image: © Shutterstock

GPSR Authorized Representative: Easy Access System Europe, Mustamäe tee 50, 10621 Tallinn, Estonia, gpsr.requests@easproject.com

Contents

List of Illustrations vii
Acknowledgments ix

INTRODUCTION 1

I American Origin and Oceanian Diffusion 33

II A Sweet Connection Between Europe and America 71

III Mundane or Miracle? Feeding and Fueling China's Population 153

IV Hunger Food? Daily Meals? Sweet Potato in Japan and Korea 220

V Sweet Potato Islands, Sweet Potato Peoples 271

VI From "Asian Crop" to "African Crop": Recent Global Expansions 329

EPILOGUE 371

Notes 387
Selected Bibliography 457
Index 469

Illustrations

Figure 2.1	The illustration of *batatas* (sweet potatoes) in Carolus Clusius's *Rariorum aliquot stirpium per Hispaniam observatorum Historia*. 81
Figure 2.2	The illustration of sweet potatoes in John Gerard's *Herball* (1597). 106
Figure 2.3	The illustration of skirrets in John Gerard's *Herball* (1597), another popular root plant in the British Isles. 125
Figure 3.1	The illustration of the sweet potato, referred to as 地瓜 (*digua*, a regional name for the plant in China), in Xu Guangqi's 徐光啓 *Nongzheng quanshu* 農政全書 (Complete treatise on agriculture). 162
Figure 5.1	Map of Taiwan, which resembles the shape of a sweet potato. 273
Figure 5.2	A display of various purple sweet potato products in Okinawa 298
Figure 7.1	Sweet potato ice cream for sale in Kawagoe, Japan 378

Acknowledgments

As the production of this book begins, I take great pleasure in reflecting on the research and writing journey I have undertaken over the past five years. Though I cannot pinpoint exactly when the idea of writing a global history of the sweet potato first occurred to me, I know it is closely tied to my work on *Chopsticks: A Cultural and Culinary History* (2015). In that project, I sought to reconstruct the history of chopsticks, tracing their spread across the Sinic world and beyond, while also examining their gradual replacement of the spoon, the oldest eating utensil used by Asians in ancient times. One reason for this shift was population growth. China, where both chopsticks and the spoon were invented, experienced an impressive population increase from the fifteenth century onward.

To explain this demographic boom, the venerated historian Ping-ti Ho proposed in the 1950s that the introduction of American crops such as sweet potatoes, maize, peanuts, and, later, white potatoes played a pivotal role. As Ho's theory resonated with me, it also led me to Alfred Crosby's seminal work, *The Columbian Exchange: Biological and Cultural Consequences of 1492* (1972). Crosby argued that the worldwide spread of American crops significantly contributed to global population growth—except, ironically, in the Americas, where the native population faced a devastating decline due to diseases brought by Europeans. Crosby's research has inspired countless studies, but as I reviewed them, I noticed a curious gap: While many scholars had explored the far-reaching impact of these crops, none had comprehensively examined the sweet potato's unique role in shaping global history.

The sweet potato's high yield and adaptability made it a staple food across Asia and Africa, where its influence was as transformative as that of the white or Irish potato in Euro-America. Ping-ti Ho's thesis, in this regard, holds merit. However, my findings show that China's population growth didn't directly align with the introduction of American crops; rather, these crops helped sustain and accelerate this growth from the mid-eighteenth century onward. A similar pattern unfolded in Japan, where sweet potatoes were introduced in the eighteenth century to alleviate famine. In both cases, the sweet potato became an "insurance crop" or "famine food"—in stark contrast to its image in Euro-America, where it was embraced as a sweet delicacy from its earliest introduction.

This dichotomy in perception—a root plant seen as a critical survival food in one context and as a treat in another—barely scratches the surface of the sweet potato's fascinating global journey. Its diverse receptions in different regions highlight the plant's adaptability and importance as it spread out from the Americas at the turn of the sixteenth and seventeenth centuries. Through this project, I hope to uncover and analyze the many ways this humble crop has shaped history, nourished populations, and woven itself into the cultural fabric of societies worldwide.

My research journey began in China, a country where the sweet potato held a vital place in the diet of its people for centuries, up until the 1980s. Along the way, I have been fortunate to receive support from a number of dear friends and students. I owe particular thanks to Li Longguo, Zhang Yibo, Deng Rui, Sun Weiguo, Tan Ying, Wang Fuchang, Xu Cheng, Yang Jingjing, Xu Yue, Li Leibo, and Li Yuetong, all of whom contributed to shaping this project in its various stages. Special gratitude is reserved for Xiao Dan, then a graduate student at Peking University studying Spanish literature, who patiently helped me locate, read, and translate critical Spanish sources that documented the sweet potato's early transfer from the Americas to Europe.

My first serious attempt at writing this book began in North America, where I have the privilege of teaching. I am grateful to my colleagues in the History Department of Rowan University, where I first presented a portion of my research at a work-in-progress session. The insightful feedback and constructive criticism I received from my colleagues, especially Debbie

Sharnak, a historian of Latin America; Kelly Duke-Bryant, a historian of Africa; and Stephen Hague, a specialist in British imperial history, were invaluable. Laurie Kaplis-Hohwald, who taught in the World Languages Department at Rowan, beautifully rendered the precious passages from a sixteenth-century Spanish recipe book, which was a great help in my research. I also wish to thank my Rowan students Kendra Hahn, Joshua Saldan, and Julia Destra for their diligence in locating sources in the library and using its interlibrary services. A special thank-you goes to Julia Destra for compiling the initial list of index entries for the book, which is of course a tremendous help. Additionally, Rowan University provided me with a seed grant in the summer of 2018, which allowed me to travel to Japan, particularly Okinawa. This visit offered me the opportunity to observe the rich tradition of sweet potato cultivation and consumption in Japan, both past and present. My research and writing on the sweet potato's culture in Japan also received helpful comments and valuable assistance from Aida Y. Wong, Chen Liwei, Liu Qunyi, Liu Shilong, Liu Gang, and Yang Li.

My deep appreciation goes to Rowan University for granting me a year-long sabbatical during the fall of 2021 and spring of 2022, which allowed me to complete the bulk of my writing during the challenging period of the COVID-19 pandemic. Reflecting on this time evokes mixed emotions. Writing about a global phenomenon while the world was in the grip of a pandemic underscored the interconnectedness of our histories. Despite the limitations of library access, I was fortunate to have friends like Pan Kuang-che in Taiwan, who tirelessly located and scanned books for me, as did the dedicated staff at the Rowan University library. Lai Guolong, who was then teaching at the University of Florida, also found and copied an MA thesis held in their library for my research. I have also enjoyed sharing my interest in global history and the role Asia/China has played in it with specialists such as Nicola Di Cosmo, Robert B. Marks, Kenneth Pomeranz, and Fan Xin.

One of the most memorable periods of my sabbatical was my stay at the University of Göttingen, made possible through the generous funding of Dominic Sachsenmaier. My four months there provided me with an ideal environment to work closely on this project. I had the honor of delivering a formal lecture on my research at the East Asian Seminar organized by Professor Sachsenmaier, and I am thankful for his invitation. My time in Germany also brought me to the University of Bochum, where I shared my work with faculty and students, thanks to the warm hospitality of Stefan Berger and Christina Moll-Murata. It was through Professor

Moll-Murata that I first encountered Felix Siegmund's master's thesis, as well as his other writings, which greatly informed my research on sweet potato cultivation and consumption on the Korean Peninsula.

In the summer of 2023, I was fortunate to return to Göttingen for an additional three months, thanks again to the generous fellowships provided by Professors Sachsenmaier and Berger. During this visit, I also had the pleasure of returning to Bochum, where I met Dr. Felix Siegmund in person, alongside Professors Moll-Murata and Berger and a host of other welcoming colleagues. The intellectual exchanges and friendships I forged with my German colleagues during these stays constitute a *sweet* memory of my time writing on the sweet potato. In addition, my stay in Germany allowed me to witness the sweet potato's resurgence as a highly regarded "superfood" among Europeans, a development that beautifully illustrated the ongoing relevance of this remarkable crop. The warmth and hospitality of Jin Yan and her family also made my two stays in Göttingen particularly memorable, especially the first visit, which took place during the height of the pandemic. Their friendship added a comforting personal touch to an otherwise challenging time.

Most of all, I wish to express my heartfelt gratitude to Dominic Sachsenmaier. From the moment he first heard of my project in 2017, his unwavering encouragement has been a source of immense support. It brings me great joy to know that this book is being published as part of Columbia University Press's Studies in International and Global History series, of which Professor Sachsenmaier is a coeditor. His belief in this project has been a cornerstone of my efforts, and for that, I am profoundly grateful. I would also like to thank Zhao Xiaoyang and Tang Zhengyu, Professor Sachsenmaier's two graduate assistants, who helped me in various ways during my research and stay in Germany.

In the course of my writing, I have had the opportunity to present my research in China, where the project first took root. Though the sweet potato is no longer a staple food in that country, it remains a significant part of Chinese culinary culture, which made my work resonate with audiences. I am grateful for the invitations from Zhan Xiaobai of Beijing Normal University, Hu Xiaobai at Nanjing University, Li Longguo and Ouyang Zhesheng at Peking University, and Sun Jinghao (Howard) at Zhejiang University. The insightful feedback I received from Wei Xiaoji, Li Huaiyin, Hou Shen, and Zhang Fan, among others during these occasions was beneficial. In 2022, through the generous arrangements of Victor Xiong,

I had the honor of delivering a lecture entitled "Was 1492 a Turning Point in World History?" for the Burnham Macmillan Lecture Series hosted by Western Michigan University's History Department. I am thankful to Professor Linda J. Borish, chairperson of the department, and her colleagues and students for graciously hosting my talk and offering their thoughtful feedback. In the summer of 2023, I presented my research at the annual conference of the International Council for Philosophy and Humanistic Studies, held at Tokyo and Keio Universities in Japan. I am grateful to Hsiung Ping-chen, the secretary-general of the council, for her invitation and to Torbjorn Loden for chairing the session.

Lastly, I would like to extend my appreciation to the two anonymous experts who meticulously reviewed my manuscript for Columbia University Press. Their thoughtful critiques and suggestions greatly enhanced the quality of my work. I am also immensely grateful to Caelyn Cobb and Emily Elizabeth Simon at Columbia University Press for their exceptional editorial support. Caelyn Cobb devoted several months to carefully reviewing the manuscript and offering invaluable recommendations for improvement, while Emily Simon skillfully oversaw the production process, providing steady guidance throughout. Ryan Perks, my copy editor, also deserves special thanks for his diligence in going through the entire manuscript, including meticulously checking the enormous number of footnotes. Their professionalism and dedication have undoubtedly improved this book. I am equally indebted to my family, whose support has been essential to the completion of this project. Of course, any remaining imperfections are solely my responsibility.

STAPLE TO SUPERFOOD

Introduction

My Doe with the black Scut? Let the Sky rain Potatoes, let it thunder to the Tune of Green Sleeves, hail kissing-Comfits, and snow Eringoes; let there come a Tempest of Provocation, I will shelter me here.

William Shakespeare's Falstaff, in *The Merry Wives of Windsor*, uttered these words in description of the potato.[1] This scene was one of the first European references to the American plant. Shakespeare was far from the only European who had been influenced by the newly discovered tuber from the Americas during his lifetime. Shakespeare's contemporary and fellow playwright John Fletcher expressed his affection for potatoes in his verse:

> I have fine potatoes,
> Ripe potatoes....
> Will your Lordship please taste a fine potato?
> 'Twill advance your wither'd State.
> Fill your Honour full of most noble itches,
> And make Jack dance in your Lordships Breeches.[2]

In the early seventeenth century, but half a world away, Xu Guangqi—a high-ranking official and devout Christian in Ming China—also praised the potato for its remarkable value. "Potato spuds can be sown in the field from early spring through summer, though the sizes of the roots may differ. The roots usually grow in a couple of months and can then be dug out to eat, sell, or leave in the ground to grow more.... The drought-resistant root can be planted in high hills with just a little irrigated water.

It is also less prone to flooding because it grows on hills. Furthermore, because it has underground roots, it is resistant to locusts. Every peasant family should farm it every year for these reasons. In short, the root is the best of all non-grain plants and a key famine-relief crop."[3]

What is more, despite writing on opposite sides of the globe, the English playwrights and the Chinese scholar-official were describing the same plant—the sweet potato (*Ipomoea batatas*), not the white or Irish potato. According to Redcliffe Salaman, a trained biologist and author of the authoritative *History and Social Influence of the Potato*, "there is no reason to suppose that Shakespeare knew of the existence of our [white] potato (*Solanum tuberosum*)." The same could be said of John Fletcher because most Europeans confused the two potatoes during the sixteenth and seventeenth centuries.[4] In retrospect, the confusion was caused by Spaniards, who were the first to plant both potatoes in Europe. In his 1992 book *España, encrucijada de culturas alimentarias: Su papel en la difusión de los cultivos americanos* (Spain, crossroads of food cultures: Its role in the diffusion of American crops), Spanish food historian Eloy Terrón writes, "In the first half of the eighteenth century, the newly created Royal Spanish Academy, when compiling the Dictionary of the Authorities, defined the word potato as 'the same as the sweet potato.'" According to other research, the confusion between sweet potato and white potato may have persisted in parts of Europe until the nineteenth century.[5]

Sweet and white potatoes were exotic plants to both Europeans and Chinese in the sixteenth century. They were spread from the Americas to the rest of the world, along with maize, peanuts, tobacco, chile peppers, tomatoes, and cocoa. Alfred Crosby, an American geographer and historian, published *The Columbian Exchange: Biological and Cultural Consequences of 1492* in 1972. As implied by the title, Crosby intended for the book to draw attention to previously understudied aspects of the European conquest of the American continents, or the New World. He said that following the conquest, Europeans brought animals and food crops from the Old World to the Americas and brought back the aforementioned New World flora; the diseases they carried also brought about an apocalyptic tragedy for the Amerindian population. Although Crosby initially had trouble finding a publisher, the book "became one of the foundational texts for the field of environmental history, . . . and by the 1990s the notion of the Columbian Exchange had worked its way into several textbooks on American and world history," states John R. McNeill, a prominent environmental historian who wrote the foreword to the book's thirtieth anniversary edition.⁶

In my opinion, Alfred Crosby's *The Columbian Exchange* made a significant contribution by igniting a strong and enduring interest among academics from other fields in analyzing the significance that New World food crops, and New World root plants in particular, played in forming the post-Columbian world. According to Crosby, American crops such as maize, white potatoes, sweet potatoes, and cassava (manioc) contributed to population growth in regions outside the New World. In *The Columbian Exchange*, each of these root crops receives some attention, though sweet potatoes are notably discussed far less than white potatoes and cassava, which are treated with roughly equal weight. Addressing the historical confusion between the two types of potatoes in sixteenth-century Europe, Crosby remarks—somewhat dismissively—that the distinction was "of no great relevance because neither had much value except as novelties and aphrodisiacs!"⁷

However, since the publication of *The Columbian Exchange*, and particularly in recent years, research on the white or Irish potato has received significantly more attention than that of its distant "cousin." Although the two potatoes are linguistically close in most European languages, they are actually unrelated botanically. Both potatoes were once despised by many Europeans because they were root plants. For instance, the historian Rebecca Earle, in her study on the Spanish response to Amerindian foods, notes that the conquistadors had a contemptuous attitude toward the native diet because they thought that "bread made from roots" was not proper bread. Instead, while the plants themselves entered European menus, the Amerindian cuisine featuring them was ignored, a fact that the historian Rachel Laudan says complicates the typical understanding of the "Columbian Exchange."⁸ The titles of certain books on the white potato also suggest there were cultural and gastronomic biases against the tuber. For instance, the title of George George's book, written in the middle of the nineteenth century, was *Potatoes: The Poor Man's Own Crop*. More recently, Larry Zuckerman released *The Potato: How the Humble Spud Rescued the Western World*, emphasizing its lowly beginning.⁹

But there have also been a number of others who have recognized the significance of the white potato in the history of the contemporary world. If it is indicating a new tendency in Western academia, it may have started with William H. McNeill's groundbreaking article "How the Potato Changed the World's History," which was published in 1999. McNeill's strong argument for the historical significance of the potato is based on the fact that after 1750, the pace of population growth in the world reached an all-time high. But because it was so simple to enhance food sources by introducing potatoes to fallow fields, the growth of the human population in northern Europe "far outstripped the rate of population growth in other anciently civilized places," he said.[10] White potatoes may have contributed to Europe's fight for democracy, argues Linda Civitello, author of *Cuisine and Culture*, which is an early attempt to offer an overview of the history of food in the world. Similarly, Rebecca Earle writes that potato production and consumption in Europe were "entangled with the emergence of liberal nations and the notion that the individual, rather than the collective, should form the primary building block of society." The author of a global history of the white potato, Andrew F. Smith, then draws readers' attention to more contemporary instances of what he calls "potato politics" in both the U.S. and global political spheres.[11] In short, there are many stories to be told about the transfer of plants between the New and Old Worlds in the post-Columbian eras.

The search for the sweet potato in history

What about the sweet potato? After all, it was one of the first American food plants Christopher Columbus (1451–1506) encountered after crossing the Atlantic Ocean to reach America. It is also widely assumed that he brought the root back to Europe on either his first or second return trip. In contrast, not only did Columbus and other early Europeans who arrived in the New World have little to do with the introduction of the white potato, but earlier assumptions about who was responsible for transferring it, such as Francis Drake and Sir Walter Raleigh in the sixteenth century, have now been proven to be incorrect. Indeed, prior to the first half of the twentieth century, when Redcliffe Salaman and Berthold Laufer, a German-American anthropologist and historical geographer, published their studies on the history of the white potato, information on its spread from the Americas to Europe and around the world was rather hazy. In fact, when it came to the early history of the two potatoes' dispersal, the sweet potato was far better documented than the white one, not only in European but also in Asian languages. Xu Guangqi's earlier description was a telling example, and he was not the only one in Asia with knowledge of the American plant. Salaman conducted valuable research on the sweet potato while accounting for the origin and early spread of the white potato in Europe, as the two shared not only the same linguistic root but also a tangled history.[12]

Besides the white potato, maize, another important food crop from the Americas, has also received adequate scholarly attention over the years. *Corn and Capitalism* by Arturo Warman is just one example. According to Warman, despite its poor reception as a food crop in Europe, maize played a role in shaping the modern world. There was, he writes, "a clear link between the expansion of corn and the expansion of world capitalism."[13] In addition, monographic studies on peanuts and chile peppers, two other food plants from the Americas, have been conducted in recent years.[14] By contrast, no book-length study of the sweet potato and its diverse impact on the course of world history has been published in any language. Among the major American food plants that spread around the world beginning in the sixteenth century, the sweet potato is one of the few that has not received the attention it deserves from historians. When compared to the aforementioned works and much more on the white potato, which I will mention below, the gap becomes obvious, because, as I will explain and expound on throughout this book, the sweet potato occupied a comparably significant position to the white potato in shaping the modern world or worlds. The various ways in which the sweet potato has been embraced as a food crop around the world exemplifies and explicates the diverse paths that the modern world has taken. That is, the modern era did not emerge in a linear fashion; rather, there has been an overlooked historical plurality embodied by the many different receptions of the sweet potato during the post-Columbian eras.

This book is thus the first historical attempt to describe and analyze the sweet potato's importance as a valuable food, as well as the various cultural, political, and religious meanings associated with its cultivation and consumption for peoples all over the world. As a historian specializing in the history of Asia and its role as a global nexus in modern times, I feel compelled to take on the task because, following its export from the Americas, the sweet potato enjoyed its greatest flourishing as a food crop in Asia, particularly in China. Today, Asia still leads the world in both area (60.75 percent) and production (86.89 percent) of sweet potatoes. Meanwhile, the root has been a staple food in South and Central America, as well as Sub-Saharan Africa and Oceania. While sweet potato has mostly been used as an ingredient in sweet/dessert or side dishes in Europe and North America, it plays an important role in African American cooking.

George Washington Carver, an accomplished African American botanist of the early twentieth century, is credited with discovering over 100 uses for the sweet potato in a variety of products. In sum, sweet potato is a high-energy food crop (higher than white potato) that is rich in vitamins (particularly vitamin A) and minerals. It is currently grown in approximately 120 countries in a variety of soils ranging from sea level to altitudes of nearly 3,000 meters. Its position as a food source is unrivaled because it can feed more people per hectare than any other food crop.[15]

Ping-ti Ho, an eminent twentieth-century Chinese historian based in the United States, was among the first to recognize the importance of American plants in Chinese agronomy. Ho asserted in his influential 1959 book *Studies on the Population of China, 1368–1953* that the introduction and subsequent cultivation of sweet potatoes, peanuts, and maize was a major reason for the country's population explosion during the late imperial period. Alfred Crosby based his *Columbian Exchange* primarily on Ho's study, which he described as "splendid," and stated the following: "However important maize is to China, the sweet potato is an even greater boon. It arrived at least as early as the 1560s and was adopted rapidly because it did not compete with rice and other traditional crops, but prospered in previously unutilized soils, such as the rocky Shantung coast, the rice-deficient southeast provinces and the drought-ridden highlands."[16] In other words, while his research was not limited to the American root plant, Ping-ti Ho assisted Crosby and us in recognizing, perhaps for the first time, the sweet potato's historical significance in Asia.

How far-reaching was the impact? As a starting point, consider William McNeill's previously quoted claim that, if the white potato, which McNeill referred to simply as the "potato," helped the European population grow at an exponential rate in the eighteenth and nineteenth centuries, then the sweet potato's role in China and Asia was equally significant. In fact, McNeill's finding that population growth in Europe "outstripped" the size of demographic expansion in other parts of the world may need to be qualified. Though not universally agreed upon, the consensus is that China's population increased roughly fourfold between the thirteenth and eighteenth centuries. More specifically, according to Angus Maddison, a renowned macroeconomist, China's population increased from 103 million in 1500 to 130 million in 1700 and more than doubled in a century to 381 million in 1820. In comparison, Europe's population grew from 71 million in 1500 to 100 million in 1700, then steadily, but not as rapidly as in

China, to 170 million in 1820. Furthermore, between 1500 and 1820, China surpassed India (inclusive of the modern nations of Bangladesh and Pakistan) to become the world's most populous country.[17]

Ping-ti Ho's thesis on the role of American food plants in driving Chinese population growth has sparked debate among academics. Suffice it to say that, while acknowledging the contribution of American plants to increasing food supply during the period, Ho appeared to be somewhat unsure about the specific period during which the root plant had exerted its impact as a food resource for the Chinese. Overall, it appears that Ho's conclusion came across as Malthusian; he believed that the population explosion had left China in a vulnerable position as a nation in the nineteenth century. All the same, Ho was the first person writing in English to point out that the root had had a remarkable run as a food crop in China since Xu Guangqi's promotion of sweet potato farming in the early seventeenth century. By the early twentieth century, the country had easily become the world's largest sweet potato producer, and the root was the third most important source of food for its people, trailing only rice and wheat.[18] Sweet potato farming and consumption in China and other parts of Asia have declined in recent decades compared to the previous century. Nonetheless, China remains the world's largest producer of the root plant, producing 46.6 million tonnes in 2022, with Malawi and Tanzania trailing far behind at 6.9 and 4.4 million tonnes, respectively.[19]

Indeed, after leaving the Americas in the early sixteenth century, the sweet potato became an essential food source for many peoples in Asia and Africa. And in parts of Oceania, such as Polynesia and New Zealand, as discussed in chapter 1, the root had been a staple food for Indigenous peoples long before European contact. Ruben L. Villareal, a Filipino horticulturist, described the root plant as "an Asian crop" in 1982, while organizing and coediting the first international conference on the sweet potato. Other conference attendees agreed with him, calling the root "the most important root crop" in Asia.[20] The Philippines was one of the first Asian destinations where the root plant arrived before continuing its journey onto the Asian mainland. True, since the second half of the twentieth century, sweet potato farming in Asia has not been as robust as it once was. However, production of the crop has increased in Africa. Aside from Malawi and Tanzania, sweet potato has become a major food plant throughout Sub-Saharan Africa. For centuries, the root plant has also been the primary source of food in Papua New Guinea, Tonga, and other Pacific islands.[21] Indeed, states Jennifer Woolfe, author of *Sweet Potato: An Untapped Food Resource*, while the sweet potato is ranked seventh in total food production in the world, "it is fifth on the list of the developing countries' most valuable food crops."[22] In other words, I would like to argue that paying more attention to the sweet potato's varying role in history, which predated that of the white potato in both Europe and Asia, will allow us to see the importance of New World crops beyond the West and gain a better understanding of their global impact in the modern world.

What we know about the sweet potato?

Jennifer Woolfe's *Sweet Potato: An Untapped Food Resource* was published in 1992 in collaboration with the International Potato Center in Peru, which includes a section on sweet potato research. It is a scientific study of the root that covers its cultivation, consumption, and nutritional value as human food and animal feed, as well as its processing to become a material for the food and pharmaceutical industries. The publication of Woolfe's book indicates that sweet potato has received ample attention in the scientific community outside of the field of history. The establishment of the International Potato Center (Centro Internacional de la Papa) in 1971 is but one example. Indeed, botanists, agriculturalists, and ethnographers have conducted substantial research on the root plant over the century, greatly expanding our understanding of the sweet potato. For instance, in contrast to the common parlance by which the sweet potato is often referred to as a tuber, like the white potato, Woolfe enjoins instead that "It should be noted from the outset that the storage organ of the sweet potato is a root not a tuber. This fact seems to be ignored by many specialists who write about the sweet potato. Storage roots are true roots whereas tubers are modified stems, and as such, differ from roots in anatomy and physiology." Woolfe's injunction is echoed by others.[23] In other words, white potato, or *Solanum tuberosum*, is a tuber by scientific definition, whereas sweet potato is a root; the two are classified as belonging to different botanical families. According to scientists, the term "sweet potato" is "really a misnomer," because the root differs from white potatoes in the same way that carrots do. As a result, the National Sweetpotato Collaborators Group and the United States Sweetpotato Council in 1989 endorsed renaming the crop "sweetpotato" rather than "sweet potato." Sweet potatoes are not potatoes that are sweet, according to science.[24] However, because this book is about the history of the sweet potato, I will continue to use the term "sweet potato" because the term itself recorded the root plant's historic journey around the world. Nonetheless, I will refer to the sweet potato as a root or root plant throughout to distinguish it from the white potato, whose edible part is its swollen stems, not its roots.

The sweet potato, or *Ipomoea batatas*, is a dicotyledonous plant. Botanically speaking, it belongs to the morning-glory family, or Convolvulaceae, consisting of over a thousand species. Other than the *Ipomoea aquatica*, which is known in English as water/river spinach, a popular vegetable in East and Southeast Asia, *Ipomoea batatas* is the only other species in the family that can be consumed as food. As a food resource, sweet potato has been domesticated and cultivated over several millennia. Indeed, as Felipe Fernández-Armesto, a noted global historian of food, observes, the plant was "almost certainly cultivated first" among all other food crops in the world.[25] During the course of its history, especially in more recent centuries when it was cultivated around the world, the sweet potato has evolved and diversified into a large number of cultivars, more than those of cassava and yam.[26] Some of these cultivars were results of natural hybridization and mutation, whereas many more were products of human intervention.

Compared with other major food crops in the world, sweet potato possesses several distinguishing characteristics. One of them is that both its roots and foliage are edible. Of the countries where it is cultivated, most consume both, but some countries and regions choose to breed only one of them for consumption. Due to the preference, different varieties have been cultivated over time because farmers and scientists find that sweet potato's vines can grow more at the expense of its roots and vice versa, depending on the changing farming techniques and climatic condition.[27] All the same, both the roots and vines are valued food resources for humans and animals, a quality that is also found in maize and white potato.

Sweet potato's other distinct property, setting it apart from most other food crops, including the white potato, is that its swollen roots can be consumed both raw and cooked. In Bangladesh, mainland China, and Taiwan, for instance, yellow-fleshed sweet potatoes are occasionally eaten raw by people, especially among children. Elsewhere, such as in Indonesia, Japan, and parts of the United States, raw or slightly cooked sweet potatoes can be a salad ingredient served in restaurants. Yet in most cases, humans eat cooked and/or processed sweet potatoes. There are indeed a variety of ways of turning the root of the sweet potato into different food products, such as grounding it into flour to make—mixing with wheat or corn flours—noodles and bread, sweet desserts, snacks like fried chips and flakes, purees or mashes, and jams or candies. But around the world, the most common ways to eat the roots are simply to roast them whole or boil them cubed. In addition, sweet potato is often processed into starch, which can be used in cooking like cornstarch to thicken soups or stews. The starch is also an important material in the preparation of both alcoholic and nonalcoholic beverages; for the former, it is indispensable for making distilled liquors such as *shōchū* in Japan, *soju* in Korea, as well as over a dozen well-known *baijiu* (lit. white spirits) in China. When sweet potato is used to make cellophane noodles (or "glass noodles" in some usages), or *fensi/fentiao*, a very popular starch food originating from China but consumed also in Korea, Vietnam, Japan, and Thailand, it is said that its quality is superior to cornstarch but inferior to mung bean starch, a more traditional type of which the foodstuff is made. Sweet potato flour can also totally replace wheat flour in soy sauce production. Then the sweet potato easily surpasses wheat, mung beans, and soybeans in availability and quantity as it is a high-yield crop.

When humans consume the green tops of the sweet potato, they mostly treat them as a vegetable, picking usually between ten and fifteen centimeters of the apical tip of the vine, including the leaves, petioles, and stem. There are also regional differences, with some consumers preferring petioles, others the leaves. In the Philippines, the green tips of the sweet potato are sold in markets and eaten as one of the most common leafy vegetables in the Filipino diet. Interestingly, another popular leafy vegetable there is known in Filipino as *Kangkong*, or *Ipomoea aquatica*, which belongs to the same botanical family as mentioned earlier.[28] In China, as stir-frying is a popular method for preparing vegetables, sweet potato's green vines are usually cooked in a wok with hot oil, by which they are often ready in just several minutes as a vegetable dish. Stir-fried sweet potato leaves is also a popular dish in the South Pacific, where coconut cream is often added for flavoring. There are, of course, other ways to prepare the vines for human consumption, such as boiling or mixing them into a flour dough to make pancakes. And, if they are fresh enough, it is also possible to eat the green vines raw as a salad, such as in certain African countries, where sweet potatoes are gaining in popularity as a food.[29]

While an important food in various forms for humans, the sweet potato has also been a common form of livestock feed in many regions where it is grown.[30] In the case of China, its largest producer, more than half of the harvest was traditionally used for feeding the animals, especially for raising pigs. And its use as animal feed is increasing as human consumption has declined there since the early 1980s.[31] Yet in Papua New Guinea, raising pigs on sweet potatoes has remained a prosperous industry, one that the country is increasingly dependent on.[32] As an animal feed, sweet potatoes, both their roots and tops, are usually eaten raw with some cutting and cubing. In some countries, it is also possible to let pigs graze on sweet potato fields. Both the roots and tops can also feed animals in dried forms, in addition to being used for silage. For their different digestibility, there have also been experiments in mixing the roots and vines and turning the mixture into a sweet potato meal for animals in Korea.[33] Compared with human consumption, sweet potatoes as a livestock feed by and large are much easier to prepare, which can often include cull roots and hard stems that may be otherwise plowed back into fields, as well as the leftovers, or waste, from the process when the crop is used to make alcohol, starch, flour, and other food or nonfood products.

When the sweet potato is processed into starch, it also becomes a valuable raw material for industry, which is another interesting characteristic of the plant. And compared with other crops, the starch yield in sweet potato is notably higher. More specifically, it is 30 percent higher than that of rice and corn and 49 percent higher than that of wheat. As mentioned earlier, sweet potato starch is often essential to make alcoholic and nonalcoholic beverages. But given its high yield, its starch is also commonly used in making industrial alcohol for medical use, among other applications. Moreover, sweet potato starch can be produced in a variety of ways. One can use sun-dried sweet potatoes, cutting them first into chips and/or slices and then processing them into starch as needed all year round. If sweet potato fields are to be used for planting other crops, as intercropping is a common method for growing the root crop, then the other popular way is by grinding them immediately after harvest and making wet starch instead, which can either be stored for several months or go directly into the process of purification, concentration, and dehydration, through which starch powder is produced.[34]

In addition to its versatile forms for consumption, the sweet potato displays a number of other remarkable qualities. The first and perhaps most notable advantage of the sweet potato is its high yield per hectare, mentioned earlier. Indeed, its yield of 18–36 tonnes per hectare (achieved by Asian farmers in modern times) is higher than that of the white potato, which has averaged about 15 tonnes per hectare throughout the world. If compared with other popular food crops, this advantage becomes much more impressive, for the per hectare yield of wheat and rice, two of the most widespread grain staples in the world, is usually no more than 10 tonnes. The seeming exception is observed in China and India, where in recent years successful technological experiments have significantly raised the rice yield to about 20 tonnes per hectare—but that remains only comparable, or still inferior in some cases, to that of the sweet potato. Second, as it is a root crop, the sweet potato plant is less affected by weather conditions than other grain crops. For instance, many farmers find that it is drought tolerant. And, over the centuries, it has grown in both temperate and tropical or subtropical regions, even though it originated from the latter. In fact, research has found that the sweet potato yields better in temperate than in tropical climate zones. But farmers in the latter can usually compensate for the lower yield by growing the crop twice or more per year. Third, the sweet potato is not only an easy-growing but also a fast-ripening plant. For example, compared with the white potato, which possesses some of similar qualities, the sweet potato, as originally a tropical plant, can ripen in as short as in three months, whereas the white potato, a temperate plant, ripens in about four months. As it is quick-ripening, sweet potato can have two or even three seasons a year in humid tropics. In fact, in those warm regions, sweet potatoes can be left in the fields, ready for harvest whenever is necessary. The main reason is that the plant is perennial, though it can also be planted annually, as it usually is in temperate zones. And fourth, sweet potato can grow well in a wide range of soils. What makes it somewhat extraordinary is that it grows better in sandy or sandy loam soil. As such, it is possible to grow sweet potato on hills and mounds as well as in ridges and furrows, which are often ill-suited for other food crops.[35] For all these reasons, the sweet potato has been called "the survival crop," "the insurance crop," "the golden spud" (*jinshu* 金薯), as well as "the famine relief food," attesting to many of its advantages as a food plant. However, as I will explain in the following chapters, the crop has also been a victim of its own success, so to speak; its hardiness and high yield have seen it be described as "a poor person's food," making it less desirable for some who detest the association.[36]

There are, admittedly, a few seeming downsides to the sweet potato's status as a food resource for humans, which may explain its being ranked

only seventh in terms of worldwide production relative to other crops, and down from the late 1970s, when the root plant was ranked sixth.[37] At present, the worldwide production of sweet potatoes is below that of wheat, rice, maize, white potatoes, barley, and cassava, but above that of soybean, sorghum, and bananas. To most researchers, this relatively recent decline, apparently caused by the decrease in production in China, the crop's largest producer, since the 1980s, is somewhat unjust, for as a food crop, the sweet potato's root alone produces more edible energy per hectare and per day than that of rice and wheat.[38] In fact, scientists find, the energy content in sweet potato roots is roughly one and a half of that of the white potato, its unrelated "cousin." Indeed, if we compare the nutritional values of all the food crops around the world, then the sweet potato not only contains multiple vitamins and minerals but, despite the "sweet" epithet, also helps stabilize blood sugar level and lower insulin resistance. All this makes it a healthier choice than other more popular food staples.[39]

Jennifer Woolfe, who hopes to "rescue" the sweet potato from its current position as an underrated food crop, believes that the main difference between the sweet potato and other more popular food crops, which may have affected its acceptance as the main food resource, is that the latter are usually bland in flavor or taste.[40] That is, rice, wheat, and white potatoes are "flavor carriers," which absorb the tastes of non-grain foods, or dishes, in a meal. By contrast, as the name indicates, sweet potatoes are usually sweet, which seems to have made them appeal to some, such as people in Africa and Papua New Guinea, while other consumers, such as those in China, where the crop was once ranked number two in production, found that the sweetness tended to cloy if they had to eat sweet potatoes as an everyday staple, as some of them did prior to the 1980s.[41] In Taiwan, where the sweet potato was the main food for many islanders over several centuries, the most popular form of consumption involves cooking cubed sweet potato with rice, sometimes into porridge, in which the cubes become softened and even dissolve into the mixture. In a word, sweet potato has been consumed as a staple or a co-staple food—the Taiwanese case above is but one example—in some regions. But in other places, perhaps due to its sweet taste (though some of its cultivars are not necessarily so), it is eaten more as a vegetable, a snack, a side dish, and a dessert. Jennifer Woolfe's explanation, which others support, seems plausible: As a result of its sweetness, the sweet potato has not been readily accepted as a daily staple by many people; it thus remains "an untapped food resource." Meanwhile, remedies for this concern certainly exist. Treating the root crop as a co-staple food, cooking it with other grains, has certainly been a widely accepted method. Another, which has also been experimented with in some regions for centuries, is to continue exploring ways of breeding new cultivars that contain less or little sugar. A prime example is the dry, starchy, and bland cultivar—which still bears the familiar orange flesh—that is popular in Sub-Saharan Africa.[42]

However, a look at the history of the sweet potato reveals that the root has at times appealed to people precisely because of its sweetness. The spread of the plant in early modern Europe was a case in point, though far from the only one. Europeans, particularly Spaniards, turned to the American plant primarily to satisfy their sweet tooth, for which they chose to bring back its sugary variety rather than the starchy variety from the Caribbean. This preference in sweet potato consumption is still prevalent in both Europe and North America today. Sweet potato has been a fixture on dinner menus as a side dish or sweet dish for holidays such as Thanksgiving in the United States, for example. In other words, the root plant has produced a large number of cultivars during its evolution and spread. Scientists have classified storage root varieties into four types based on flesh color, dry matter, total sugar, and taste. The first is the dry, low-sweet starchy type in white, yellow, or cream color. The second is the soft, moist, and sweet type in orange color, ideal for desserts. In the United States, orange-fleshed sweet potato is often mistakenly called yam, even though it is different from the latter in terms of both botany and physiology. The third is the dry and starchy type, bland in taste, while the fourth is also dry and low in sweetness with a purple-colored flesh.[43] The first two types perhaps are the most common and the last two represent newer developments, which suggest that the current trend in sweet potato consumption, especially among those who eat the root as a staple, is to prefer starchy and less sugary varieties to moist and sweet ones. Moreover, since the root plant is a valued source for animal feed and industrial use, including the latest attempt to investigate its use in bioethanol production, agriculturalists and horticulturalists are experimenting with even newer varieties for meeting these needs as well.[44]

Sweet potato's entrée into Europe

I was on the poop deck at ten o'clock in the evening when I saw a light. It was so indistinct that I could not be sure it was land, but I

> call Pedro Gutiérrez, the Butler of the King's Table, and told him to look at what I thought was a light. He looked, and saw it. I also told Rodrigo Sánchez de Segovia, Your Majesties' observer on board, but he saw nothing because he was standing in the wrong place. After I had told them, the light appeared once or twice more, like a wax candle rising and falling. Only a few people thought it was a sign of land but I was sure we were close to a landfall.[45]

Christopher Columbus wrote this in his journal on October 11, 1492, when his *Santa María* crossed the Atlantic Ocean and arrived in the Bahamas. The next day, after a three-month voyage, Columbus and his crewmen finally made landfall, which later became Columbus Day in the Americas and elsewhere.

It is well-known that Columbus's motivation for crossing the Atlantic was his belief that, because the earth is round, he could reach Asia by going west. Despite the fact that his ship landed on a continent previously unknown to Europeans, he never doubted that he had arrived in Asia. Why was Asia so appealing to Columbus? One answer can be found in his fascination with the thirteenth-century explorer Marco Polo's *Book of the Marvels of the World*, which he had carefully read before his trip, jotting down his notes in the margins. One can easily see from Columbus's own journal how he hoped to find Cathay, or China, and its famous cities, Quinsay and Zayton (Zaitun, Zaiton), which Marco Polo described in detail. Quinsay, or Jingshi in Chinese, was the capital of China's southern Song dynasty, and Zayton, or Quanzhou, was/is a coastal city in southeastern China where Marco Polo allegedly met a group of Christians. But, perhaps more importantly, Columbus was drawn to Zaiton/Quanzhou because, according to Marco Polo, it was a port where goods from India arrived. In other words, Columbus saw it as a link between China and India.[46] On October 21, for example, after landing on what he thought was an island, Columbus, for example, wrote, "I am still determined to continue to the mainland, to the city of Quinsay, and to give Your Majesties' letters to the Great Khan and return with his reply." Ten days later, he again showed his determination after finding a big river: "I am sure that this is the mainland, and that Zayton and Quinsay lie ahead of us, each about hundred leagues from here."[47]

Columbus mentioned the Great Khan a total of six times in his journal, and he was convinced that the latter ruled China/Asia and lived in

Quinsay. His references show that he was desperate to meet the mighty Mongol ruler. Despite never abandoning his longing, Columbus was unable to reach Asia, his dreamland, in the several voyages he attempted later in his life.

However, it appears that Columbus did make a connection with Asia in the end, as his voyages led to the European discovery of the Americas, which eventually occasioned the "Columbian Exchange" between the New and Old Worlds. The term "Columbian Exchange," coined by Alfred Crosby in his eponymous book, refers to the transfer of New World crops such as maize, sweet potato, white potato, peanuts, and chile peppers to the rest of the world following Columbus's voyages. All of the plants mentioned above arrived in Asia in the sixteenth century and have since played an important role in the foodways of the continent. Indeed, the introduction of these American food plants, particularly sweet potatoes and maize, had a significant impact on the Asian economy and agricultural system. To write a history of the sweet potato, we must first examine the impact of Columbus' transatlantic voyages on the modern world.

In retrospect, Columbus's success in crossing the Atlantic ushered in the "Age of Discovery" for Europeans, as the transatlantic voyages not only "discovered" America but also sparked a series of maritime expeditions led by other Europeans such as Vasco da Gama and Ferdinand Magellan in the decades that followed. While da Gama fulfilled Columbus's dream of reaching Asia by sailing westward from Europe in 1498, Magellan and his assistants completed the circumnavigation of the globe in 1522. Columbus's actions at the end of the fifteenth century thus became a watershed moment in modern world history. Following these maritime explorations, Europeans gained access to the Western Hemisphere as the New World, which included the Americas and Oceania, while Eurasia and Africa became the Old World. Maize, sweet potato, white potato, peanuts, tobacco, and other crops were referred to as New World crops, and their subsequent transfer to the Old World was an important part of the Columbian Exchange. The transfer of flora (wheat, olive trees, cabbage, etc.) and fauna (horses, dogs, pigs, etc.) as well as diseases (smallpox, measles, syphilis, etc.) from the Old World to the New was also part of the exchange.

In other words, what occurred following the European "discovery" of the Americas, or the New World, constituted a defining moment in global history. This assessment, however, reflected a Eurocentric view of historical development, because Columbus, though regarded as a "discoverer" by modern Europeans, did *not* discover an uninhabited "new world," nor was he the first European explorer to land in America. It is established fact that, before Columbus, Leif Erikson of Iceland landed in North America and founded a settlement in Vinland, or today's Newfoundland, off the northeastern coast of North America. In his best-selling book *1421: China Discovered the World*, Gavin Menzies, a retired naval officer from the United Kingdom, also hypothesized that a Chinese fleet from Zheng He's legendary maritime expeditions in Ming China reached America by sailing around Africa, roughly seventy years before Columbus's "discovery."

Indeed, Columbus's historical role has been contested and contentious. And the conflict appears to have begun during his lifetime. Columbus quickly rose to fame in Spain and Europe following his first trip to America. However, his failure as an administrator in managing the colonies soured his relationship with the Spanish court in his later years, and he died in relative obscurity. Columbus made his third voyage to America in 1497, arriving in Mexico, where another Italian explorer, Amerigo Vespucci, had arrived several months earlier. In contrast to Columbus, who believed he had arrived in Asia (though some doubts did arise during his fourth and final voyage), Vespucci was clearly aware that he had landed on a new continent. This distinction was one of the reasons that, beginning in the early sixteenth century, Vespucci's first name, rather than Columbus's, was used to refer to the continent. However, it appears that over the next two and a half centuries, Columbus gradually redeemed himself as the great European explorer who discovered the New World. During the American Revolution, Columbus was fashioned as an American hero who had played a pivotal role in the formation of the American nation. "Columbus figures," Heike Paul observes, "as a patron and ancestor of those Americans who were demanding their independence from England and who later became citizens of the new republic."[48] From the nineteenth to the twentieth centuries, this "Americanization" of Columbus was accelerated by the publication of Washington Irving's exhaustive biography, *The Life and Voyages of Christopher Columbus*, and George Bancroft's *History of the United States*. Columbus thus acquired a very positive image in the United States. In their biography published in 1992, the quincentennial of Columbus's transatlantic voyage, William D. Phillips and Carla Rahn Phillips explain the reason for Columbus's "Americanization" in the following terms: "Americans took Columbus, a Genoese merchant mariner sailing for Spain, as one of their national heroes. When the United States expanded westward in the nineteenth-century, his reputation rose. Scores of cities, counties, and institutions from east to west were named after Columbus, or his poetic counterpart Columbia, all attesting to the special role that Columbus had assumed in the self-definition of the United States."[49] It should not be surprising, then, that October 12 became known as Columbus Day around the turn of the twentieth century. It became a national and local holiday in the United States, as well as in several Latin American and European countries.

However, disagreements about Columbus's historical role persisted. Columbus did not arrive in a completely new world; rather, his voyages brought Europeans into contact with Native Americans, whom he referred to as Indians. Columbus's mistreatment and enslavement of Amerindians as governor of the colonies he established in America was one of the reasons for his tainted relationship with the Spanish court near the end of his life. More importantly, if the exchanges that occurred following his voyages resulted in significant demographic and economic changes in the Old World, one must logically consider how they impacted the New World. Without a doubt, what happened to Indigenous peoples in the aftermath of the European "Age of Exploration" was disastrous. Exposed to diseases brought by European settlers to which they had no resistance, the Amerindian population was drastically reduced while Europeans built their New World. In other words, when assessing Columbus's historical significance, one must adopt a global perspective, examining how the European "Age of Discovery" inaugurated by Columbus changed not only Afro-Eurasia, but also the Americas.

At the end of the nineteenth century, as the fourth centennial of Columbus's initial voyage lionized his status in the United States and Europe, Justin Winsor, a recognized authority on the early history of North America, published a well-researched biography of Columbus in which he presented indisputable evidence of how Columbus not only enslaved Amerindians but also initiated the transatlantic slave trade. Winsor's controversial work on Columbus both altered and tarnished Washington Irving's celebrated portrayal of this European explorer. Scholars compared Columbus's treatment of Amerindians to that of one of his contemporaries, Bartolomé de Las Casas, a Spanish colonist who sympathetically recorded the misfortunes of the Amerindians under Spain's colonial empire in America.

Nonetheless, it was not until after World War II that serious criticisms of Columbus's voyages and their consequences emerged in Western academia.

In fact, Alfred Crosby's writing of *The Columbian Exchange* was a prime example, even though it appears that he went about expanding on Columbus's historical contribution. In his 2003 preface to the thirtieth anniversary edition of his book, Crosby recalls how the events of the 1960s (civil rights and Black Power, the anti–Vietnam War movement, etc.) prompted him to question some of his long-held convictions. "The sixties 'globalized' my mind a quarter century before that word entered journalistic jargon," he wrote. The events of the decade opened his eyes to a common historical phenomenon: When one people is mistreated and oppressed by another, they always mount a strong resistance. Crosby maintained that the success of the European conquest of the Americas was attributable as much to their advanced weaponry as to the epidemic diseases they carried, which ravaged the native populations.[50] In "The Contrasts," the first chapter of his *Columbian Exchange*, Crosby describes the differences between Eurasia and the Americas prior to Columbus's exploratory voyages. It lays the groundwork for subsequent chapters, which examine the processes and consequences of the "Columbian Exchange," or how the exchange brought together the Old and New Worlds in botany and zoology—how the European "Age of Exploration" gradually unified the flora and fauna of the two worlds. However, after finishing the book, one realizes that such unification was not in fact the main reason Crosby wrote the book. What he meant to do was to present the other "contrast" in the post-Columbian era: While the transfer of New World crops contributed to population growth in the Old World, the spread of epidemics—caused by diseases brought from the Old World to the New—devastated Indigenous populations across the Americas. Crosby's book concludes on a somber note, with the following reflection: "The Columbian exchange has left us with not a richer but a more impoverished genetic pool. We, all of the life on this planet, are the less for Columbus, and the impoverishment will continue."[51]

Indeed, following the publication of Crosby's book in the early 1970s, many critical voices were raised about the European "Age of Discovery" initiated by Columbus's transatlantic voyages at the end of the fifteenth century. For example, in 1976, Hans Koning, a prolific Dutch author, published *Columbus: His Enterprise: Exploding the Myth*. Tzvetan Todorov, a Bulgarian-born French scholar, published *Conquest of America: The Question of the Other* in 1984, arguing for the need to transcend Eurocentric historical perspectives. Todorov argued in a subsequent book, *The Morals of History*, that the term "discovery" is incorrect and that the "Age of Discovery" should be renamed the "Age of Invasion." Todorov explained that the term "discovery" is "legitimate only if one has decided that the history of humanity is identical to the history of Europe, and as a result, the history of the other continents begins the moment the Europeans arrive."[52]

As the quincentenary of Columbus's voyages was approaching in 1992, the criticisms of his "discovery" of America came to a head. In 1991, Hans Koning published *The Conquest of America: How the Indian Nations Lost Their Continent*, in which he criticized Columbus's legacy. For the occasion, a number of anticolonial films and videos were also shown, with titles like *The Columbian Invasion: Colonialism*, *The Indian Resistance*, and *Columbus Didn't Discover Us*.[53] William and Carla Phillips's *The Worlds of Christopher Columbus*, one of the year's written works and visual presentations, may offer one of the most positive assessments of Columbus's role in history. According to the Phillipses, Columbus's action had a global impact on peoples on both sides of the Atlantic by opening the route from Europe to the Americas: Columbus's role, in their words, was to have "placed the world on the path leading toward global interdependence, with enormous consequences—both good and ill—for the peoples of the world."[54] By comparison, the 1992 celebration for the five hundredth anniversary of Columbus's voyage, which took place in the United States, Spain, Italy, Mexico, and the Dominican Republic, was deemed a "failure," according to Stephen J. Summerhill and John Alexander Williams, the latter of whom served as director of the Christopher Columbus Quincentenary Jubilee Commission from 1986 to 1988. Meanwhile, they believe that this "failure" could become a "success" only if it helps people see the flaws of Eurocentrism and reconsider Columbus's historical legacy. Their conclusion is as follows: "Planners set out to celebrate an imperial past but found themselves confronting difficult questions about the rise of colonialism, the destruction of native American societies, and the disruption of biological habitats throughout the globe. In this way, 1992 contributed to an increased public recognition of the importance of nature, native peoples, and human rights in contemporary society."[55]

The Columbian Exchange beyond Europe

Writing this book on the history of sweet potatoes, I believe, contributes to the aforementioned efforts to examine multifaceted global consequences of Columbus's voyages and the Columbian Exchange. It will attempt to provide a different perspective on the formation of the modern world by transcending the Eurocentric view of history. To be sure, efforts have been made to assess Columbus's historical impact from a global perspective. Felipe Fernández-Armesto's *1492: The Year Our World Began* was published in 2009, and it includes chapters on the histories of contemporary China, Japan, Korea, India, and the Americas. Despite his book's global scope, Fernández-Armesto remains convinced that Columbus's historic voyages were a one-of-a-kind Western phenomenon. His biography of Columbus differs from Alfred Crosby's in focus; he, for example, barely mentions the transfer of animals and plants between the Old and New Worlds.[56] In comparison, Charles C. Mann's *1493: Uncovering the New World Columbus Created*, published in 2011, expands on Crosby's approach by providing detailed descriptions of various post-Columbian exchanges around the world. Mann's book, due to its breadth, draws attention to the "accidental" yet "devastating consequences" maize and sweet potatoes have had on ecosystems in modern East Asia. His main goal in writing the book is to illustrate how the Columbian Exchange eliminated botanical, zoological, and biological distinctions that existed between the world's continents prior to the fifteenth century. Mann's critical assessment of Columbus's exploration, or "exploitation," of a world previously unknown to Europeans is also demonstrated in his 2005 book *1491: New Revelations of the Americas Before Columbus*.[57]

In recent decades, a number of monographs have been published by contemporary experts that examine the importance and impact of American crops over the *longue durée* and from a global perspective, with maize receiving much attention. Michael Blake's *Maize for the Gods*, for example, describes the Amerindian domestication of the crop over the past nine millennia, while Duccio Bonavia's *Maize: Origin, Domestication, and Its Role in the Development of Culture* compares its cultivation in the Americas and around the world during the post-Columbian era. In particular, Bonavia points out that while maize arrived in Europe earlier, it blossomed as a staple in North China in the nineteenth century, just as the sweet potato did in South China.[58] Indeed, despite its poor reception as a food crop in Europe, corn has played a significant role in shaping and transforming the modern world. In addition to flourishing in China, corn played a key role in the United States, both economic and social, as well as a central position in North American culinary culture: "American cooking was rooted in corn," says Betty Fussell in *The Story of Corn*.[59] To examine the global significance of maize, Arturo Warman linked its diffusion to the development of capitalism around the world. However, he argues that while capitalism promised a uniform economic model, global modernization was not a homogeneous process, as evidenced by the diverse routes and impacts that maize traveled and produced around the world. To illustrate this diversity, Warman turns to the cultivation of maize in China, while Elizabeth Fitting presents her analysis of the transformative role of the crop in Mexican agronomy.[60]

The study of the white potato has also taken a "global turn" since the early twenty-first century, with expanding attention to its role in Asian food culture. In 2011, Reaktion Books, as part of its Edible food history series, published *Potato: A Global History* by Andrew F. Smith, a prolific food scholar in the United States. A concise survey of the potato's importance as a food crop, it devotes several pages to discussing its impact in Asia, both past and present. Smith points out that despite a global decline in potato consumption as food, the tuber is still a staple of Asian diets today, from China to Japan to India.[61] While not claiming to be a global history of the potato, John Reader's *Potato: A History of the Propitious Esculent* (2008) provides adequate coverage of the potato's rapid growth outside the West and around the world. Indeed, he devotes an entire chapter to the potato's growing appeal to the Chinese, who became the world's largest producer in 1993. Unlike Andrew Smith, Reader attributes the rise in Chinese potato consumption to Deng Xiaoping's (1904–1997) Reform and Opening-Up policy in the post-Mao era. He discovers that "French fries have accounted for most of the rise in potato consumption." Since the 1980s, when global fast-food chains became available throughout the country, Chinese consumers, mostly of the younger generation, have turned to McDonald's and KFC. In other words, China and India lead the world in white potato consumption not because the tuber is central to their dietary traditions, but because certain segments of the two countries' enormous populations have grown to enjoy it, and they consume the tuberous crop in the same form that it has traditionally appealed to people in the Western world.[62]

Thus, food consumption reflects the effects of political policy and cultural change. Rebecca Earle's two recent books offer a comparative study of the tuber's new appeal in Asia from the standpoint of political influences. Earle investigates the modern-day global cultivation of the white potato and its relationship to the formation of nation-states. According to her, the increase in potato production in China was caused by state intervention. She observes that since the 1960s, China has established a large number of research institutes aimed at developing new white potato cultivars. Meanwhile, the government ran campaigns in the country, such as "Happy Potato Family," to promote potato cultivation and consumption. All of this, she believes, is reminiscent of similar state-led initiatives and programs to promote the tuber as a staple food in eighteenth- and nineteenth-century Europe and the United States.[63]

Scholars across academic disciplines in mainland China have produced valuable studies that analyze the country's marked change from the late imperial to modern period from the perspectives of food history and demography, inspired by Ping-ti Ho's seminal work on the impact of American food plants. Zhang Jian's *Xindalu nongzuowu de chuanbo he yiyi* (The spread and significance of New World agricultural plants) provides a comprehensive account of the global spread of American food crops. It discusses not only food crops like maize, sweet potato, potato, and cassava, but also plants like cocoa, coffee, sunflower, rubber, tobacco, fruits, vegetables, and herbs. While European exploration of the Americas resulted in colonialism and imperialism, Zhang acknowledges that it also created a powerful force unprecedented in history, leading to a more connected world. Without mentioning Alfred Crosby, Zhang observes that in the post-Columbian era, the "great exchange and great dissemination" (*da jiaoliu* 大交流, *da chuanbo* 大交換) of agricultural products pitted Europe against Africa, Asia, the Americas, and Australia. The rise of capitalism and colonialism in the former caused the people in the latter to fall into the abyss while giving rise to globalization.[64]

Zhang, a world historian by training, avoids covering only the American plants that had a significant impact in China or Asia—sweet potato is just one of the food plants he describes in the book. The Columbian Exchange and its botanical impact on the world have received considerable attention across the Sea of Japan. Not only have Crosby's books long been available in Japanese, but so have some related works, such as Mann's *1493*. Meanwhile, Japanese scholars from various disciplines have begun their own research on the subject, producing a number of notable studies in recent years. One of them is Sakai Nobuo's *Bunmei o kaeta shokubutsu-tachi: Koronbusu ga nokoshita shushi* (Plants that changed civilization: The seeds of Columbus), which identifies six American plants (potatoes, corn, cacao, pepper, rubber, and tobacco) that the author believes have helped reshape modern culture and society. Sakai, like many of his Western counterparts, argues that potato farming helped to strengthen European nation-states. Another is Inagaki Hidehiro's *Sekaishi o ōkiku ugokashita shokubutsu* (World history influenced by plants). Inagaki, a prolific author with a background in botany, adds a plant-history perspective to the world's historical development. He claims that plants generated food and wealth, giving rise not only to nation-states but also to conflicts and wars between great powers. If Inagaki takes a critical stance on the Columbian Exchange's consequences, as Crosby did in his writing, he is not alone. Another compelling example is Yamamoto Norio's *Koronbusu no fubyōdō kōkan sakumotsu dorei ekibyō no sekaishi* (Columbus's unequal exchange: A world history of crops, slaves, and plagues). Yamamoto, who is also trained in botany and anthropology, has previously written a book on the role of chile pepper in history, and here he describes critically the misfortunes and miseries of Africans and Amerindians during the Columbian Exchange. He also concludes his book with a full-fledged critique of Columbus's role in world history, as does Sakai Nobuo.[65]

Of all the American crops that have influenced history, the sweet potato has probably received the most attention in Asia. Indeed, if sweet potato is an "Asian crop," it is well documented in East Asian history and historiography. In contrast to the anglophone world, a number of studies on the root crop have appeared in Asian languages over the last century, including a few historical texts. From the mid-twentieth century, when the Chinese government promoted sweet potato as a beneficial crop, Chinese agriculturalists penned several pamphlets on its farming, harvesting, storage, and utilization as food, feed, and industrial product, which were widely distributed throughout the country. Historians have also compiled source books that trace its spread in the country beginning in the sixteenth century.[66] In Japan, Miyamoto Tsuneichi, the historian and ethnographer, wrote *Kansho no rekishi* (History of the sweet potato), which provides a concise history of the root's spread in Japan between the sixteenth and the nineteenth centuries after tracing its origins in the Americas and transmission to the Old World. Miyamoto states in his preface that the cultivation of the American crop reminded him of the struggle his people, or "sweet potato eaters" (*imo kui*), faced during the country's difficult postwar years.[67]

At the turn of the twenty-first century, agriculturalists in Japan established the history and cultivation of the root plant in Japan and around the world, how Japanese farmers improved its farming and utilization, as well as how it became woven into the fabric of Japanese culture and society. In 2010, Itō Shōji published *Satsuma imo to Nihon jin: Wasure rareta shoku no ashiato* (Sweet potato and the Japanese: Forgotten footprints of food), which is a cultural, if also nostalgic, account of the sweet potato's reception in Japan's recent past. Incidentally, Itō has also written and edited books on the white potato.[68]

The aforementioned books, as indicated by their titles, tend to take a national and regional approach to sweet potato cultivation and consumption as a food. Another example is Cai Chenghao and Yang Yunping's writing of *Taiwan fanshu wenhuazhi* (Cultural annals of the sweet potato in Taiwan). Taiwan saw the cultivation of the American plant spread throughout the island beginning in the sixteenth century, owing primarily to immigrants from mainland China. Taiwan emerged as a major producer of the American crop in Asia during the period of Japanese rule between 1895 and 1945. From the postwar years through the 1980s, the trend continued and intensified. According to Cai and Yang, sweet potato has been so intertwined with Taiwan's history that many Taiwanese refer to themselves as "sweet potato folks" (*fanshulao* 蕃薯佬).[69] Indeed, the first international symposium on the sweet potato was held in Taiwan in 1982, which was somewhat predictable given the crop's importance to the islanders at the time. Its proceedings were later published, and they discussed the sweet potato's importance as an "Asian crop," its production, harvesting, processing, and utilization, as well as its benefits and drawbacks as a common food across the continent. Another conference proceedings from the same decade, *Sweet Potato Research and Development for Small Farmers*, echoed the same sentiment: "Sweet potato is an Asian crop in spite of its New World origin."[70] Both volumes included studies of how the American root crop played a central role in the agronomies of western Oceania, including Papua New Guinea, and West and South Africa, in addition to their Asian focus.

The sweet potato has a longer planting history in Southeast Asia and Oceania than it does in East Asia. Evidence suggests that the root arrived in Papua New Guinea around the beginning of the second millennium, while Europeans brought it to Maritime Southeast Asia several decades later in the same millennium. The sea route used by the Spanish "Manila galleons" between Mexico and the Philippines played a significant role in the transmission. Meanwhile, Portuguese merchants were credited with opening the Europe–Africa–Asia trade route, which led to the cultivation of the sweet potato in West Africa in the early sixteenth century, and in India and Indonesia in the following decades. In recent years, the growing importance of the sweet potato in Africa has elevated the continent to the world's second-largest producer, trailing only East Asia, but ahead of both Southeast Asia and Oceania. This dramatic rise is reflected, inter alia, in *Potato and Sweetpotato in Africa: Transforming Value Chains for Food and Nutrition Security*, an up-to-date and comprehensive account of recent scientific advances in dissecting the crop's botanical propensity, improving breeding and harvesting, and facilitating food, feed, and industrial uses.[71]

In sum, I must state that all of the valuable studies mentioned above attest to the need for a global history of the sweet potato. The root has long been a valued food crop for many peoples around the world since its dispersal from the Americas in the early sixteenth century. Given that it is a tropical and subtropical plant, sweet potato is a food for the "Global South" rather than the North. Though Europe was the first continent where the root was grown outside the Americas, it has remained a secondary food in European gastronomy. The same could be said for its status in North American culinary practice, where it was later introduced via Europe and Africa. White or Irish potatoes, on the other hand, a temperate crop that was initially ignored and even despised as poisonous, eventually flourished as a popular source of food for Europeans and now around the world. This historical fortuity, however, should not be used to justify a historiographical bias against the sweet potato.

Reconstructing the history of the sweet potato on a global scale, I believe, will allow us to see the modern world in a new light. In fact, it appears to me that the white potato received a better reception than its unrelated "cousin" in Western food history and historical scholarship in general because the tuber played a significant role in Europe's industrialization and urbanization. However, this was another case of historical contingency, because the sweet potato, maize, and cassava are all easy-growing, high-yielding food crops that support population growth. Moreover, the latter three have all contributed to past and present demographic expansions, albeit in different parts of the world. Sweet potato first made a notable contribution to supporting Qing China's impressive population explosion in the eighteenth century, which continued through the nineteenth century and well into the twentieth century, aided by extensive maize farming in the country. However, in both Chinese and Western historical scholarship, this demographic expansion has generally been characterized as one of the country's modern woes.[72]

Indeed, it appears that in modern history, sweet potatoes and white potatoes met very different fates, which have been associated with the contrasting images of "the fall of Asia" and "the rise of Europe" in previous scholarship. William McNeill's enthusiastic praise for the white potato's historic role in supporting European population growth recalls the esteem the tuber enjoys in influential works by both Adam Smith and Thomas Malthus. In other words, if China was perceived as Europe's "other," the perceived dichotomy had already emerged in the eighteenth century. Smith saw white potatoes as a potentially very valuable food, superior to wheat but not inferior to rice, and believed that widespread cultivation in Britain and Europe could increase the population's food supply. Despite agreeing with Smith and being unaware that both potatoes had already entered the Chinese agronomic system during his time, Malthus implied that China's unchecked population growth would have disastrous consequences.[73]

Recent scholarship has called into question the entrenched China-Europe dichotomy in economic development. If, beginning around 1750, when McNeill says the white potato began propelling Europe's population expansion, and many Chinese demographers believe marks the beginning of Qing China's demographic explosion, Europe and China embarked on two different economic paths, this "great divergence" was by no means predetermined, nor destined to lead to a "fall of Asia" and "rise of Europe." While R. Bin Wong criticizes the limitations of European historical experience in explaining China's path to development, James Z. Lee and Wang Feng question the Malthusian thesis on China's uncontrollable population, and Kenneth Pomeranz highlights changes in the "global junctures" of the eighteenth and nineteenth centuries that aided England's path to industrialization. Pomeranz directs our attention to the English colonization of North America, recognizing that both England and China's Jiangnan (Lower Yangzi Delta) were experiencing an ecological bottleneck hampering further economic growth. That is, England's "luck" in "plucking" the energy resources, particularly coal, of the New World enabled it to break free from the bottleneck and begin its industrialization. "If we accept the idea that population growth and its ecological effects made China 'fall,'" he argues forcefully, "then we would have to say that Europe's internal processes had brought it very close to the same precipice—rather than the verge of 'take-off'—when it was rescued by a combination of overseas resources and England's breakthrough (partly conditioned by geographic good luck) in the use of subterranean stores of energy."[74]

If the role of the white potato in assisting European population growth is celebrated, then it is necessary to have an equally appreciative attitude, if not more so, toward the role of the sweet potato, a high-energy and high-yield food superior to the white potato, in assisting China's population expansion beginning in the eighteenth century. We must move beyond Adam Smith and Thomas Malthus and stop viewing China and Asia as Europe's contrasting "other." As alluded to above, demographers and historians in China have identified 1750 as the start of the Chinese population explosion,[75] which coincided with the expansion of sweet potato farming in the country. This phenomenon represented Chinese farmers' innovative response to the population growth of preceding centuries. Because of the response, China was able to maintain its position as the world's largest economy, contributing roughly one-third of global GDP through the first half of the nineteenth century. Recent studies also show that, despite its growing population, China's standard of living remained roughly constant during the period. Some economists argue that China's success in labor absorption to support its growing population was part of the "East Asian miracle," which was characterized by a labor-intensive approach that absorbed its massive population and was inextricably linked to the "European miracle" of industrialization.[76]

As sweet potato helped China sustain its population in the eighteenth century, it became even more important as a food crop in the nineteenth and twentieth centuries, not just in China, but throughout East Asia. During those centuries, the continent saw Western incursions as well as several intra-Asian conflicts and upheavals. In the midst of famines, rebellions, wars, and revolutions, Asian farmers turned to high-yield crops such as sweet potato and maize to make a living. By the early twentieth century, China had become the world's largest producer of sweet potatoes. Its population also recovered from the previous century's losses and resumed its ascent to new heights. In addition to China, Japan, Vietnam, and Taiwan increased sweet potato farming during and after World War II. As a result, there are reasons to believe that sweet potato played a significant role in East Asia's remarkable postwar recovery. It was an essential food source for the labor force behind the "economic miracles" achieved by Japan beginning in the 1960s, by Taiwan and the other "four little dragons" beginning in the 1970s, and mainland China beginning in the 1980s, just as it is now in supporting the populations in Sub-Saharan Africa, Southeast Asia, and South and Central America. Indeed, if sweet potato became an "Asian crop" in the first four centuries after leaving the Americas, its growing importance in Africa today is making it increasingly an "African crop."

Ultimately, a global history of the sweet potato allows us not only to discover and describe its significant role as a food crop in the modern world, but also to discern and appreciate the various images and perceptions of the American root around the world. While its association with Asian and African agronomies has given people the impression that it is a "subsistence crop" and "poor person's food," its initial reception in Europe painted a rather different picture—Shakespeare and his contemporaries saw it as an aphrodisiac rather than a food. Today, the root retains its status as a luxury or treat—it is an ingredient for making dessert to top off the dining experience in both daily meals and holiday celebrations (e.g., Thanksgiving dinner).[77] Granted, as mentioned above, global sweet potato production has declined—the root, which was the sixth most cultivated food crop through the 1970s, is now the seventh, replaced by cassava, owing to reduced production in China and Asia as the country and continent shifted more toward white potato cultivation. However, in developed countries, a new, distinct trend is emerging to value sweet potato as a healthy food option, while consumption of white potatoes has declined.[78] Sweet potato fries, for instance, are becoming a popular alternative to french fries in many fast-food restaurants around the world. All of this implies that if the sweet potato's journey around the world has become entangled with that of the white potato, this entanglement may continue in the coming years. Both potatoes, along with other plants like maize, cassava, and peanuts, will remain valuable foods into the foreseeable future as Amerindian gifts to the world in the aftermath of the Columbian Exchange.

Now let me turn to brief descriptions of the book's chapters. Beginning with Thor Heyerdahl's 1947 expedition to cross the Pacific and reach Polynesia from South America, chapter 1 describes how the sweet potato originated in the Americas and spread to the eastern part of Oceania in pre-Columbian times. Drawing on anthropological, linguistic, and ethnobotanic research, it presents different theses advanced by scholars of various disciplines related to the three routes of transfer for the root's diffusion in Oceania before and after European contact. Sweet potato, a relatively new food crop in the region, was quickly adopted as a staple food due to islanders' familiarity with taro and yam cultivation. Despite its varied importance, the root plant has been taken as a valuable source of both food and feed throughout the vast Pacific.

Chapter 2 centers on sweet potatoes' global journey outside the Americas following Columbus's transatlantic voyages. Columbus brought the root back to Spain, where it was quickly embraced as a novel food, not only because it was easily cultivated in a similar climate, but also because its sweetness appealed to Europeans. Indeed, the *batata* variety, from which the English word "potato" was derived, was preferred by Spanish and Portuguese conquistadors for its sweet flesh compared to other available varieties. Some even thought it had aphrodisiac properties. All the same, sweet potato became a popular ingredient in both meat and sweet dishes, as evidenced by cookbooks from the seventeenth and eighteenth centuries. Having traveled from Europe back to North America, the root played an important role in shaping the food culture of the American South, particularly the African American foodway. Due to the southern influence, it eventually became a staple in both Canadian and American Thanksgiving celebrations. Meanwhile, despite notable fluctuations, sweet potato has remained an essential component of modern Latin American cuisine.

The importance of the sweet potato in China from the late sixteenth century to the present is examined in chapter 3. Though the question of when the American root was first cultivated in the country remains open, no one has doubted its rapid acceptance as a famine food. Sweet potato spread from south to north thanks to the promotion of notable officials, which was also supported by imperial fiat, overcoming the initial challenge of preserving the root through the winter. As a result, the majority of the population adopted the root as a staple. The cultivation of sweet potatoes in China increased along with the country's population during the eighteenth and nineteenth centuries, and as a result, China became the world's most populous country and the largest producer of sweet potatoes. And despite, or perhaps because of, the change in government in 1949, the parallel development continued well into the late twentieth century—sweet potato production and consumption both reached record highs in Mao's China. The sweet potato's success as a food crop in China calls into question the Malthusian thesis about food resources and population growth. It also prompts us to reconsider the divergent paths of development between Europe and China. This reconsideration could include both demographic and economic factors.

Chapter 4 covers the history of sweet potato cultivation and consumption in Japan and Korea. The root plant met a different fate in each country, despite their being close neighbors. Introduced to the archipelago via

Ryukyu or Okinawa, it was embraced as a valuable food resource in Tokugawa Japan, where its role in dealing with several famines was highly appreciated. On the other hand, though it was introduced to the Korean Peninsula around the same time, sweet potato did not become a staple there, except in some southern regions. When Japan began its wars of expansion at the turn of the twentieth century, sweet potato became an essential source of food to support the country's imperial ambition. It also helped the Japanese survive the defeat and starvation that followed. It is no surprise, then, that sweet potato evokes nostalgia in many Japanese, whereas in Korea, its popularity can be seen in the preparation of glass noodles for *japchae*, an iconic dish once served to royalty, retaining the flavor of rare and exotic food. These disparities in dietary experiences have influenced the recent development of sweet potato culture in both countries.

Chapter 5 looks at four "sweet potato islands," or places where sweet potatoes were/are a staple food: Taiwan, Ryukyu/Okinawa, Papua New Guinea, and Hawaii. The reason for focusing on them is not only because sweet potato has been a particularly important food for islanders, but also because I believe that the study of food history cannot be limited by national borders. In terms of political development, the four islands have taken very different paths. But they have one thing in common: the dominant role the American root plant has played in their agronomies and societies, both past and present. Taiwan's sweet potato cultivation, for example, was inextricably linked to its colonial experience with Japan, as was Okinawa's, albeit earlier. In comparison, the histories of both Papua New Guinea and Hawaii bear signs of European colonialism. The root plant played an important role in shaping gender roles in Papua New Guinea, both socially and economically. Similarly, the root's significance, which has seen it overtake taro as a food plant in parts of the Hawaiian archipelago, is well evidenced in the islands' religious culture.

The sweet potato's growing popularity that extends from Asia, particularly South and Southeast Asia, to the Middle East and Africa is discussed in chapter 6. Because of the differences between these regions, sweet potato has been received in different ways as a food. The plant's roots and vines are common foods in Southeast Asia, in such places as the Philippines, Vietnam, and Indonesia, whereas in the Middle East and South Asia, sweet potato is seen and consumed primarily as a vegetable, similar to the white potato. Despite mixed reactions throughout history, sweet potato has unquestionably become a global food, reaching every corner of the globe today. And the most significant progress has been made in Sub-Saharan Africa. Since the turn of the twenty-first century, Africa has surpassed Southeast Asia to become the world's second-largest producer of sweet potatoes. Given the enormous potential for African farmers to increase per hectare yield, it is not an exaggeration to say that the cultivation of the root in Africa will likely determine its global status as an important food plant in the future.

In the epilogue, I look at two interesting developments in the sweet potato world. One is the Chinese government's recent campaign to make white potatoes, rather than sweet potatoes, the country's fourth staple food, after rice, wheat, and corn. The other is the emerging global trend, particularly in the West, of representing the root plant as a healthy food, or a so-called superfood. Needless to say, given that China is still the world's largest producer of sweet potatoes, the government's decision to shift its focus to white potatoes would have a significant impact on the future of both plants. However, despite official efforts, the campaign to make white potato a staple food in China has yet to bear fruit. Sweet potato, on the other hand, has been growing in popularity among the younger generation in the country. Likewise, the practice of consuming sweet potato for its health benefits is gaining tremendous traction around the world, evinced by the surge in sales of sweet potato fries as an alternative to white potato french fries and a series of experiments with new sweet potato products ranging from sweet potato bread, sweet potato pizza, and sweet potato sponge to sweet potato beverages, sweet potato doughnuts, sweet potato ice cream, and sweet potato yogurt. In a nutshell, thanks to the renewed interest in consuming the root for health reasons, sweet potato is taking on new life in our times and into the future.

CHAPTER I

American Origin and Oceanian Diffusion

A Queer Craft—Out in the Dinghy
Unhindered Progress—Absence of Sea Signs
At Sea in a Bamboo Hut
On the Longitude of Easter Island
The Mystery of Easter Island
The Stone Giants—Red-Stone Wigs
The "Long-Ears"—Tiki Builds a Bridge
Suggestive Place Names
Catching Sharks with Our Hands
The Parrot
LI 2 B Calling-Sailing by the Stars
Three Seas—A Storm
Blood Bath in the Sea, Blood Bath on Board
Man Overboard—Another Storm
The Kon-Tiki Becomes Rickety
Messengers from Polynesia

Thor Heyerdahl, a Norwegian explorer, wrote this poem to describe his daring raft crossing of the Pacific Ocean from South America to Polynesia in 1947. The expedition was named "Kon Tiki" after the Inca god Viracocha, the great creator in the Inca mythology of South America's Andes region—Kon Tiki or Con Tici were the deity's old names. The raft was made of balsa wood, a lightweight material native to the Americas that Amerindians commonly used. Heyerdahl and five of his crewmen sailed the raft west in the ocean on April 28, 1947, using private loans and donated equipment from the U.S. Navy. After 101 days and 4,300 miles, their raft collided with a reef in the Tuamotu Islands and successfully landed on August 7, 1947. Heyerdahl and his companions were received as heroes upon their safe return, including an invitation to the White House by U.S. President Henry Truman (1884–1972), to whom Heyerdahl presented the U.S. flag he and his companions had brought on the raft. A year later, Heyerdahl published *Kon-Tiki: Across the Pacific in a Raft*, a book about his successful expedition that became an instant best seller and was translated into several languages. The poem with which this chapter opens is an excerpt from his book.[1]

Why did Thor Heyerdahl embark on the Kon-Tiki expedition? Why did he try to recreate the technique that prehistoric people in South America might, allegedly, have used to sail a balsa raft across the Pacific? What was he hoping to demonstrate? His book contains some of the answers to these questions. At the outset, Heyerdahl tells his readers unequivocally that the expedition was motivated by his desire to answer the "Polynesian question"—namely, where the islanders of Polynesia came from, whether and how they were related to other islanders in the Pacific, and, perhaps most importantly, whether or not the Polynesians were related to the Amerindians of the Americas. As some have noted, a major motivation for Heyerdahl to embark on the expedition was his belief in "the Nordic cult of a Great White Race" that, endowed with their adventurous spirit, became "the originator of the civilisations of the Americas and Polynesia."[2]

Heyerdahl's racist intention aside, his adventure highlights something important. That is, if the origins of the Polynesians, the so-called Polynesian question, or the issue of prehistoric trans-Pacific contact in general has piqued people's interest, it has been largely due to the sweet potato.[3] Historical accounts and scientific research have shown that it had been cultivated in America and Oceania as a main food resource for both Amerindians and some Pacific Islanders before the European arrival. Besides the eyewitness accounts recorded by European explorers, including that of Capitan James Cook, there is also linguistic evidence demonstrating that the sweet potato provided the key connection between the Americas and Oceania. Many scholars have noted that while the plant is referred to by a variety of names, it was frequently called *kumar/cumar* or *kumara* in Polynesia, New Zealand, and other islands in Oceania, which is also a name in the Quechuan language, spoken among the Amerindians in South America. The first person who recorded the identical pronunciation of a word for sweet potato in both Oceania and South America might be John Crawfurd, a Scottish physician and diplomat who authored *History of the Indian Archipelago* in 1820. Crawfurd's observation was subsequently noted by the German botanist Berthold Seemann, who boasted extensive knowledge of the plants in both regions in the *Journal of Botany* in 1866.[4] Regardless of whether people from the Americas went to Polynesia or vice versa, that the sweet potato was grown in both regions clearly piqued the curiosity of scientists and the public about the possibility of intercontinental communication before the European "Age of Discovery." It is worth mentioning that in carrying out the Kon-Tiki expedition, Thor Heyerdahl and his crew brought with them sweet potato roots, which accompanied them to Polynesia. "Like our prehistoric forerunners," Heyerdahl wrote, "we also had with us sweet potatoes and gourds from Peru."[5] And in his extensive book detailing the Kon-Tiki adventure, Heyerdahl devoted an entire section to discussing the sweet potato. Unlike breadfruit and bananas, Heyerdahl stressed, there were other plants that could not "have reached Polynesia merely by the usual marginal diffusion from Melanesia, but speak for direct relations with early America. The principal of these species is undoubtedly the sweet potato."[6]

Indeed, besides the sweet potato, few other species were seen in both the Americas and elsewhere around the globe prior to the Columbian Exchange. One can broadly compare the flora and fauna of the New World with that of the rest of the world to see this dearth of connections. Sweet potato, the protagonist of this book, was indeed a quite extraordinary, if unique, species in this sense. For with respect to fauna, as Alfred Crosby aptly described in his seminal book *The Columbian Exchange*, "The contrast between the Old and New World fauna has impressed everyone who has ever crossed the Atlantic or Pacific. Some species are common to both worlds, especially in the northern latitudes, but sometimes this only serves to point up other contrasts."[7] Jared Diamond is more specific about the contrast in his more recent but equally well-known book *Guns, Germs, and Steel*. He notes that the Americas had only one wild ancestor that gave rise to the "llamas and alpacas," compared to the fourteen common animals in Asia, Africa, and Europe. This is because, Diamond explains, "the Americas may formerly have had almost as many candidates as Africa, but most of America's big wild mammals (including its horses, most of its camels, and other species likely to have been domesticated had they survived) became extinct about 13,000 years ago. Australia, the smallest and most isolated continent, has always had far fewer species of big wild mammals than has Eurasia, Africa, or the Americas." He also points out that while Polynesians and Aztecs raised dogs, theirs were a very different breed.[8]

The plants in the Old and New Worlds were as distinct as the animals. Indeed, because the floras of the Old and New Worlds are so dissimilar, sweet potatoes stand out from other plants. In his 1939 study, Donald D. Brand observed that

> Among the pre-Columbian cultivated species of the New World there was only one of definite Old World origin. This was the bottle gourd (*Laganaria vulgaris Ser.*), native in Africa, which probably was drifted by ocean currents to South America and thence spread from Brazil to Mexico. The coconut (*Cocos nucifera L.*) was the only other cultivated species common to the two hemispheres prehistorically, and this was raised in the Old World but was known only wild in the New World along the Pacific coast of Central America and Columbia. This species also was probably brought across by ocean currents. No other species of plant cultivated in pre-Columbian days was common to both the Old World and the New.[9]

Grand's observation explains why the Kon-Tiki expedition brought both sweet potatoes and bottle gourd to Polynesia. In addition to the linguistic evidence already mentioned, archaeologists have recently discovered sites in Polynesia proving that the plant was unquestionably cultivated in the region prior to European contact.[10]

Sweet potato: Its origin and mysterious spread in Oceania

Polynesia, and most of Oceania in general, of course, is not considered part of the Old World. Nonetheless, given the stark differences in fauna and flora between the Old and New Worlds, the fact that, aside from bottle gourd, sweet potato was the only known plant domesticated outside the Americas has helped arouse the interest of scientists, historians, and adventurers like Thor Heyerdahl. In fact, Heyerdahl's Kon-Tiki expedition was launched on two assumptions. The first was that sweet potato was a native plant of the Americas, and the second was that Amerindians brought it from South America to Polynesia and the rest of Oceania.

Was Heyerdahl correct in basing his decision on these two assumptions? Let us first consider the origin of the sweet potato. As it is one of the world's most important food crops, much research has been conducted to determine sweet potato's birthplace or birthplaces. Most scientists since the late nineteenth century have believed that the root plant originated in the Americas; one of the first scientists to advocate this theory was French botanist Alphonso de Candolle in 1884.[11] Today, the majority of scientists believe it was grown in the vast area between Mexico's Yucatan Peninsula and Venezuela's Orinoco River. Sweet potato, in other words, was domesticated in both Central and South America. However, scientific research appears to be somewhat inconclusive in terms of the crop's exact birthplace or which side of the Isthmus of Panama it was first grown on. The earliest archaeological discovery of dried sweet potato remains, estimated to be ten thousand years old, was made in the caves of Peru's Chilca Canyon.[12] Then, also in Peru, sweet potato roots dating between from 1785 and 1120 BCE were discovered in the Casma Valley, in the country's coastal region. However, according to research on molecular markers relevant to the natural distribution of *Ipomoea batatas* and related species, the first domestication of the sweet potato should have occurred in Central America, followed by the Peru-Ecuador region in South America.[13]

Central America's status as one of the sweet potato's birthplaces also reflects the plant's early diffusion throughout neighboring regions. During his first voyage to the Americas in 1492, Christopher Columbus observed Indigenous peoples cultivating and consuming the root as a staple food in the Bahamas and the Caribbean, indicating its widespread significance across diverse parts of the Americas. Columbus and his crewmen appeared to have also grown accustomed to eating sweet potatoes on their return voyage.[14] These islands are located east of the Yucatan Peninsula, which was identified as one of the first areas where sweet potato was grown. Did the root plant travel westward to the Pacific islands over the millennia as a source of food for people? For over a century, many scientists and historians had been intrigued by this question. The distance between the Americas and Polynesia, or the eastern side of the Pacific islands, is much farther than the short distance between the Yucatan Peninsula in southeastern Mexico and the Caribbean islands. In fact, the distance between Easter Island, Polynesia's most eastern point, and the coast of South America is approximately 3,510 kilometers (2,180 miles). And Hawaii, the most well-known Polynesian island, is about 4,000 kilometers or nearly 2,500 miles from America's coast.

Despite the long distance, it is clear that sweet potatoes are a source of food for islanders not only on Easter Island, but also in Hawaii and other Polynesian islands. In fact, sweet potatoes are eaten throughout Oceania, including almost all Pacific regions such as Micronesia, Melanesia, and

Australasia. There are, of course, differences in the crop's importance as a food source in the various regions of Oceania. The sweet potato is more central in the local diet on the eastern islands of triangular Polynesia, such as Hawaii, Easter Island, and New Zealand, than on the other islands. The sweet potato is the most planted species in the morning-glory family in Hawaii, and according to recent scientific research using GIS analysis, the crop played an even larger role in the past, as it was the main plant to be grown well in dryland areas. Its agricultural significance was also attested by the Hawaiians' development of a social ritual that corresponded to its growth cycle, which, incidentally, could be as short as three months. In other words, most scientists today believe that the sweet potato originated in Polynesia prior to Captain Cook's arrival in the late eighteenth century.[15] Similarly, when Europeans arrived in New Zealand around the same time, they discovered that sweet potatoes were already a staple food for the Indigenous Maori people.[16]

In western Oceania, and particularly in Papua New Guinea, sweet potato is also a staple food, as it is for Tongans and Fijians, whose islands are in the center of Oceania, connecting Melanesia and Polynesia. In a recent study of food and agriculture in Papua New Guinea, the authors observe that "Sweet potato is by far the most important staple food in PNG [Papua New Guinea]. It provides around two-thirds of food energy from locally grown food crops and is an important food for 65% of rural villagers."[17] A similar situation can be found in Tonga and the neighboring Solomon Islands, where the root plant is also the main source of food. Indeed, Tonga and the Solomon Islands have the highest per capita sweet potato production in the world, surpassing Papua New Guinea and China, among others.[18]

If the sweet potato is widely available in the food systems of Oceania today, the questions that follow are how and when it arrived from the Americas. Scholars have been engaged in a lengthy debate in the hopes of finding convincing answers to these questions since the late nineteenth century. To some extent, the debate is still ongoing, with new hypotheses and theories emerging from international scholarly communities.

In the main, most scientists believe that there are only two possible ways for a plant like the sweet potato to cross the vast ocean in prehistoric times. The first is through natural diffusion, and the second is through human transfer. Concerning the former, scientists have conducted extensive research to determine (1) whether sweet potato roots or seeds could be carried by ocean currents from the Americas to Polynesia and (2) if sweet potato seeds could be carried by birds flying over to Polynesia. As for the first question, scientists generally hold that after drifting long distances in the ocean water, sweet potato's capsules and seeds might lose their hard skin and consequently decay, though more experiment remains to be done. However, they are quite certain that, after considerable time drifting, sweet potato roots would decay and lose germination power.[19] Moreover, unless floating on a tree trunk, both sweet potato seed and roots would sink in the ocean, as shown by a floatation experiment as early as 1906.[20]

Findings related to the second question have so far been less straightforward. There are many cases in nature where birds eat fruits and excrete seeds, which fall to the ground and germinate. And when sweet potato seeds are excreted from birds, streaky cracks in the hard skin are observable, which may actually help germination. The challenge again is the long distance: South American birds hoping to reach Oceania have thousands of kilometers of nonstop flight distance to cover. Granted, a migratory bird could fly thousands of miles in autumn from Alaska to the South Pacific islands, including Hawaii, and spend the winter. For instance, the golden plover, which has been seen in both Polynesia and the coast of South America, has the capability for a sufficient flight distance. But its role in carrying seeds is unknown. The question of whether sweet potato seeds could be transported such distances by birds is also complicated because seeds or capsules can be transmitted in some cases through bird feathers and claws. Even though this is a possibility, the bird's long flight distance is likely to be a barrier, reducing the likelihood of such a transfer.[21]

Consider the evidence that Polynesians and the Quechua people in Peru, Ecuador, and other parts of South America name the sweet potato in a similar way; that is, if not transmitted by humans, how could it be pronounced the same? Nonetheless, there are still difficult and complex issues. For one thing, if we agree that sweet potato was native to Central and South America and that Indigenous peoples such as the Quechua referred to it as *kumar/cumar* or their cognates, then the fact that Polynesians adopted this terminology would lend support to Thor Heyerdahl and others who believe the Polynesian population originated in America.

Polynesia aside, D. E. Yen, the Australian ethnobotanist, paints a more detailed picture of the sweet potato's diffusion in the entire Pacific region. In his comprehensive study of the sweet potato and its propagation in Oceania, *The Sweet Potato and Oceania*, published in 1974 (though some of the views expressed in the book have been updated in his contribution to a similarly entitled anthology in 2005),[22] Yen first gives a succinct summary of the various theories regarding the plant's transfer to Oceania. Unlike other scholars on the subject, Yen considers not only linguistic and historical but also archaeological and botanical evidence. He believes that the worldwide cultivation of the sweet potato was most likely due to human agency.[23] Yet, interestingly, if this were the case, then it would also support other hypotheses that the plant was native to places other than the Americas. That is, if the dispersal of the sweet potato resulted from human action, then it might well have taken a westward direction in its distribution.

In his book, Yen carefully goes over the proposed theories and offers a thoughtful discussion. There had previously been various hypotheses about how the sweet potato spread to Oceania: For those who believe it originated in America, the direction of its diffusion was from east to west, or from the coastal regions of the Americas to the Pacific islands. There had also been theories about its spread from west to east, incidentally. Some believed it traveled from Asia to Oceania, while others argued that it originated in Africa and described how it traveled to Asia and then to the Pacific. However, by the time Yen wrote his book, most of the theories about sweet potato's eastward spread from Asia to the Americas in the Pacific had been debunked, as explained in more detail in chapter 6.[24]

More specifically, four theories were advanced regarding the origin of the sweet potato and its early spread in Oceania. The first was that the plant migrated westward from Asia to the Pacific, following the path of human migration. However, "This view is not widely held now," states Yen, "for unlike many of the other crop plants in the area, the evidence for Asiatic origin has not stood up to critical scrutiny."[25] The second was that the root crop originated in Africa and was transferred first to Asia before its spread to Oceania and America. Captain Cook's explorations in the Pacific formed the basis of this theory, developed chiefly by Elmer D. Merrill, an American botanist and taxonomist, in the mid-twentieth century.[26] The third was a position advanced, among others, by George Murdock in his *Africa: Its People and Their Culture History*, published in 1959. Murdock argued for an east-to-west direction of the dispersal, that the plant was native in the Americas and transferred first to Polynesia and then to Asia and Africa. Yet he was not so specific about the question of how the sweet potato spread from America to Oceania; that is, whether it was transferred by Polynesians or by Amerindians. This remains an open question to this day. While many—including Japanese scientist Uchibayashi Masao, whose point of view will be discussed more below—agree that given their superior seafaring ability, Polynesians should have been the carriers who brought the sweet potato from the Americas to Oceania, others argue against this view. Aside from Thor Heyerdahl's conviction that Polynesia was a demographic extension of South America, another example is a 2001 study by Robert Langdon. Extending Heyerdahl's theory, Langdon claims that in prehistoric times, Amerindians learned to use bamboo rafts from Southeast Asians and took long-distance voyages around the American continent, possibly leading to the discovery of Easter Island.[27] A fourth theory was proposed in 1963 by Jeremy W. Purseglove, who offered an alternative solution to the debate on its origin. Purseglove's research led him to speculate that the root crop might have multiple origins instead of only one. In particular, he believed that the sweet potato could be grown in both the Americas and Polynesia. If his hypothesis is accepted, then it would solve the debate about the plant's transfer from the former to the latter once and for all.[28]

Few, however, have accepted Purseglove's proposal since the mid-twentieth century; the debate on the transfer and transferor of the plant from America to Polynesia and the rest of Oceania has continued more or less to this day.[29] In addition to the work of Uchibayashi and Langdon, Roger C. Green, a New Zealand anthropologist, presented his research on a different route of the America–Polynesia transfer in 2005, in which he reiterated the theory that Polynesians, rather than Amerindians from South America, were the transferors. Besides the sweet potato, Green speculated that the Polynesian voyager also brought back bottle gourd and soapberry plants from the American continent. And after the initial transfer, he stated, the sweet potato's further diffusion around the entire Pacific region was by the hands of both the islanders and Europeans.[30]

Yet, beginning in 2018, new research conducted by Tom Carruthers, Pablo Muñoz-Rodríguez, and their team at Oxford University seems to have revived Purseglove's speculation on the natural dispersal of the sweet potato between the Americas and Oceania. A plant geneticist, Carruthers has collected and analyzed 199 specimens of the sweet potato and 38 of its wild relatives. Their finding is that as *Ipomoea batatas*, sweet potato's closest relative is *Ipomoea trifida*, also belonging to the morning-glory family. More importantly, about 800,000 years ago, sweet potato originated from

Ipomoea trifida before diverging from it about 100,000 years ago. The specimens collected by European explorers like Captain Cook in Polynesia, according to the analysis of Carruthers and Muñoz-Rodríguez, showed the divergence, as they were distinct from those in Central and South America. In other words, sweet potatoes spread to Polynesia thousands of years before human colonization. This new research, which is still ongoing, would lend support to Purseglove's earlier belief that sweet potatoes were cultivated independently in both Polynesia and the Americas. It also helps the conjecture on its natural diffusion, which is indeed favored by Carruthers, Muñoz-Rodríguez, and their team. They emphasize that their research finding provides "strong support for its presence there [Polynesia] as a result of naturally occurring long-distance dispersal," which means that "the evidence against human-mediated transport of the sweet potato to Polynesia is, therefore, extremely strong."[31] What is unknown is whether the national diffusion was by wind, water, or birds.

However, there are problems with Carruthers and Muñoz-Rodríguez's research, readily noted by their critics. One mystery is that if sweet potatoes were long present in Polynesia in prehuman times, then how could it be named *kumar/cumar* or *kumara*, as it is in South America? The other, posed by Caroline Roullier, an evolutionary biologist in France, is that if sweet potato's gene had mutated long ago, then how could it look physically the same after it was domesticated in different places?[32]

Sweet potato and the debate on Polynesians

This question brings us back to Thor Heyerdahl's second assumption when embarking on the Kon-Tiki expedition. That is, he wanted to show that Amerindians, at some point in prehistory, introduced the sweet potato to Polynesia and Oceania in general. By loading sweet potatoes onto his boat for the Kon-Tiki voyage, Heyerdahl intended to seek evidence of both pre-Columbian contact between South America and Oceania as well as the possibility that Amerindians, or members of his own race, whom he referred to as "White Gods," may have traveled to Polynesia and brought the American root plant with them. Heyerdahl's belief was that "our own race first reached the Pacific islands after the discovery of South America."[33] Given the space limitations, we will disregard his thesis about the earlier

European discovery of South America and instead concentrate on how he argued about American colonization of Polynesia.

Heyerdahl first presented his position that the Polynesians were not Aborigines, but migrants, in response to the "Polynesian question" with which he began his *American Indians in the Pacific: The Theory Behind the Kon-Tiki Expedition*. "The zoologist can tell us that the Polynesians are immigrants and not autochthonous to the Pacific Islands," he says. He then examines the various theories about the origins of Polynesians advanced by previous scholars since the mid-nineteenth century. If they were migrants, they would have come to Polynesia either from the east, the Americas, or the west, Maritime Southeast Asia. By describing and analyzing the physical appearance of the Polynesians and comparing it with the Malays, who were native to the Malay Peninsula, Heyerdahl concludes that the "Malayo-Polynesian theory" was inconsistent, nor, drawing on other evidence and for that matter, were Polynesians related to Indonesians.[34] In other words, Polynesians and Amerindians shared many commonalities, which led to the hypothesis supported by Heyerdahl that the islanders might *not* have come from the West, or Asia, but instead from the East, or the Americas.

Heyerdahl also examines different possibilities with respect to the relationship between Amerindians and Polynesians, including whether Pacific Islanders might have populated the American continents. His conclusion, however, is that the former were more likely the ancestors of the latter, rather than vice versa. He explains in the following terms:

> If we suppose the American aborigines to be the ancient inventors and givers, and their young and unidentified East Pacific islander neighbours to be subsequent emigrants who have carried ancient culture elements abroad with them in rather recent centuries A.D., we find at once that the American seniority of these culture elements embodies no problem at all, but offers instead the necessary foundation for just such a working hypothesis.

However, this remains a hypothesis that requires further investigation. Heyerdahl therefore continues to elaborate on his mission plans: "When we turn our attention from the west of Polynesia to the east, this is not primarily done to argue diffusion or to look for a reason for all the similarities and parallels between America and Polynesia that have so far not been adequately explained; it is done to carry the search for Polynesian migratory routes into a marginal territory that has not previously been explained for such possibilities."[35]

In other words, if Amerindians were the cultural ancestors of Polynesians, the next logical question was how they got there from the American coast. Polynesia is made up of a number of islands, with Easter Island being the closest to the coast of South America. The distance between the two is approximately 3,510 kilometers (2,180 miles). And the distance from South America to Hawaii, the largest island in Polynesia, is approximately 4,000 kilometers or nearly 2,500 miles. Without modern means of transportation, how could the Amerindians cross the seemingly insurmountable distance in the Pacific Ocean to have arrived in Polynesia? This appeared to be how the Kon-Tiki expedition gained prominence, as Thor Heyerdahl hoped to show the world that Amerindians could cross the vast Pacific Ocean on a wooden raft to reach Polynesia. Indeed, as depicted at the beginning of this chapter, Heyerdahl and his crew passed by Easter Island and arrived at one of the islands of the Tuamotus Archipelago after traveling 4,300 miles.

Having successfully completed their experiment, as mentioned before, Heyerdahl and his crew were hailed as heroes. Their expedition also inspired others to follow in their footsteps. From the mid-twentieth century to as recently as 2015, a half-dozen attempts have been made by people of various nationalities, with varying degrees of success.[36] After the Kon-Tiki expedition, Heyerdahl himself also launched several other seafaring attempts, and later in life, he moved to Peru and organized an archaeological project aimed at discovering the Americas' largest pyramid complex. At the time of his death in 2002, Heyerdahl was regarded as a national hero in Norway, despite his controversial scholarly claims based on unfounded racial propositions.[37]

Without a doubt, the Kon-Tiki expedition was a huge success. However, despite the media and public interest generated by it, Heyerdahl and his colleagues were unable to persuade the scientific community that Polynesia was colonized by Amerindians from South America during the prehistoric period. And many of his detractors opposed Heyerdahl's experiment, and not because they condemned his racist motivation.[38] In terms of the origins of Pacific Islanders, or Austronesian peoples, including Polynesians, standard scholarship today holds that they migrated eastward from Asia all the way to Polynesia and New Zealand at various historical times. Polynesians, like their counterparts in Micronesia and Melanesia, were Asian in origin. The aforementioned Japanese scholar Uchibayashi Masao provides a brief overview of this scholarship in his 2006 study. He states that perhaps due to the climate change between 5,000 and 3,500 years ago, the inhabitants in the Asian mainland drifted from north to south, causing the people in the southeast to migrate to the islands in the Pacific, such as to today's New Guinea, in Melanesia. This constituted the first wave of eastward migration of the Asians to Oceania. The second wave of migration began around 3,500 years ago, resulting in the colonization of Fiji and through Tonga to Samoa approximately 3,100 to 3,000 years ago. For an unknown reason, the colonization of Oceania then took a long break. Yet, beginning around the fourth century, the tide resumed again and finally reached Polynesia. It took place first in the Marquesas Islands and then, around the fifth century, in Easter Island and in the sixth century in Hawaii. Then it moved southwest to the Society Islands and New Zealand around the ninth century. All these dates are of course estimates, but, Uchibayashi stresses, they are more or less accurate because the calculation is based on the ages of the archaeological remains found on each island, combined with local traditions.[39] However, while most scholars agree that there was a notable hiatus in Austronesians' migration to Polynesia, the pause actually occurred within the region rather than without: Having reached Samoa and Tonga in the ninth and tenth centuries BCE, they did not continue to colonize the Cook Islands, Hawaii, and the rest of Polynesia until a millennium later, or from the eighth century CE forward.[40]

Besides archaeological evidence, more recent research has used DNA samples to help ascertain the colonization of Oceania, which generally lends support to the theory that the Pacific Islanders were related to the ancestors of the Malays, in today's Maritime Southeast Asia. Specifically, the commonly received knowledge is that the formation of the Austronesian peoples began around what is now Taiwan, whence they spread around Oceania over the span of a few millennia. However, scholars differ in their opinions on the migratory routes and pace. Uchibayashi Masao's theory, summarized above, is one of them; in terms of the colonization of Polynesia, this represents what is known as the "Entangled Bank Model," emphasizing that the region was populated by several waves of migration. In contrast to the "Entangled Bank Model," others have presented the "Express Train Model" and the "Slow Boat Model," both of which agree that

Melanesia was where the Polynesians migrated from while differing on the pace of the movement in the process. The "Express Train Model" earns its name because, in the words of Erik Thorsby—a Norwegian, like Thor Heyerdahl—the theory "proposes a rapid migration of the ancestors of the Polynesians from the vicinity of Taiwan, without extensive contacts with near Oceanic populations along the way."[41] In other words, while Melanesia was a stepping-stone, it did not form a close connection with Polynesia with respect to the latter's population.

Erik Thorsby wrote his essay in 2012. He reviewed the theories about the colonization of Polynesia because he intended to revisit the thesis advanced by Thor Heyerdahl when he launched his Kon-Tiki expedition a half century earlier. While he supported the notion that Austronesians shared a common origin in the vicinity of Taiwan, Thorsby also compared DNA samples of Polynesians, Easter Islanders in particular, and South Americans. He concluded that Heyerdahl's hypothesis, while generally unsupported by the scientific community, might be at least partially right, for Amerindians had indeed contributed to the gene pool of Polynesians. Thorsby's cautious conclusion is as follows:

> The results of our molecular genetic studies of some highly selected Easter Islanders are fully compatible with the notion that the first inhabitants of the island were Polynesians arriving from the west. Our investigations of HLA alleles in the studies population, however, also demonstrate for the first time early genetic traces of some Amerindians on the island.
>
> We cannot establish by our investigations when the first Amerindians reached Easter Island. The data provide strong evidence, however, for their introduction prior to the Peruvian slave trades in the 1860s, and suggest that it occurred in prehistoric time, but probably after the island was inhabited by Polynesians.[42]

In a word, Thorsby's research has demonstrated that though Polynesians migrated from the west, or Asia, they had contacted and mixed with Amerindians before the European arrival during the sixteenth century. As such, it lends credence to the notion that pre-Columbian contact existed between Amerindians and Polynesians, and that the *kumara* line of sweet potato transfer from America to Oceania was both the result and proof of that contact.

The European exploration controversy

However, in Oceania, *kumar/cumar* and their cognates are far from the only vernacular names used by the locals to refer to the root crop. Scholars discovered that the closer the islands were to Asia, the more names the islanders used to refer to the sweet potato, implying that there were other routes by which the root plant could have been transferred to Oceania. In tropical America, which includes the West Indies and the Caribbean, the sweet potato was known as *camote* or *batata*, with some variations. And both *camote* and *batata*, as well as their cognates, are used by islanders to refer to the sweet potato in Micronesia, Melanesia, and the Pacific regions near Asia. As a result, D. E. Yen and others believe that there were possibly two additional lines of transfer from the Americas to the Pacific, one called the *camote* line and the other the *batata* line. The former ran between Mesoamerica and Southeast Asia, while the latter ran between the Caribbean and Europe, then from Europe to the Pacific via Asia (India?). The root crop's multilinear diffusion was also confirmed by comparing its varieties in the Americas and Oceania.[43] For example, in Papua New Guinea, it is widely assumed that the root crop was introduced in the 1700s and gradually replaced taro to become the country's primary staple food.[44] In other words, the *camote* and *batata* lines of transmission began in the seventeenth century, at the height of European colonial expansion around the world.

As previously stated, sweet potato became an important source of food for all islanders after it was introduced to Oceania. Indeed, it appeared to be a major American food plant widely accepted by the inhabitants of the vast Pacific (along with, though more popular than, maize, peanuts, and lima beans). Its popularity rivaled that of the native plant, the banana, to varying degrees. Meanwhile, it is worth noting that the importance of sweet potatoes as a type of food varied (and continues to vary) throughout Oceania. Specifically, sweet potato was/is a major food crop in the four extremes of Oceania: Easter Island in eastern Polynesia, Hawaii in northern Polynesia, New Zealand in southern Polynesia, and Papua New Guinea in western Melanesia, with its significance waning in the middle, or the islands (e.g., Tonga and Fiji) of eastern Micronesia, Melanesia, and western Polynesia. This varying importance of sweet potato as a food plant lends support to the idea that it arrived in Oceania via multiple routes at various historical times.

While the issue of the sweet potato's historical diffusion in Oceania is still being debated, most academics tend to concur with D. E. Yen's assessment of the four hypotheses concerning the root plant's dispersal in the Pacific (covered in the previous section): all four hypotheses have imperfections, despite being valuable in their own right. Yen put forth the following three new ideas in his study from 1974 and his contribution to a 2005 anthology. First is that the American origin of the sweet potato ought to be established without much dispute, as it has generally been recognized by most experts. Second, the transfer was more likely the result of human agency rather than natural diffusion, such as by ocean currents or by birds. And third, the plant's spread around Oceania took three routes rather than one single path from South America to Polynesia and the rest of Oceania. In a nutshell, "The American origin is maintained," Yen writes, "but the sweet potatoes of the western Pacific are treated as separate introductions from those of Polynesia." He believes that there were three routes for the plant to enter Oceania.[45]

In other words, if the Polynesians, not Amerindians from South America, as Thor Heyerdahl suggested, brought the sweet potato back to Oceania through the *kumara* line of transfer, it was actually Europeans who opened the *camote* and *batata* lines that dispersed the root plant throughout the rest of the Pacific. The transatlantic voyage of Christopher Columbus in 1492 marked the beginning of the Europeans' discovery of the American plant, as will be discussed in the following chapter. Columbus encountered the sweet potato on his first trip to the Caribbean, when he described the root as looking like yams and tasting like chestnuts. Columbus returned with not only ten Amerindians, but also a number of animals and fruits, which may or may not have included sweet potatoes, however; as will be discussed further in the following chapter, opinions differ as to which return trip saw Columbus introduce the root to his fellow Europeans. When Columbus and his companions encountered the sweet potato, they noted that Amerindians called the root various names, including *ajes, ajies, ajes, ages,* and *asses*.[46] According to modern research, these names are still used to refer to the sweet potato by islanders (e.g., Haitians) in the region.[47]

Yet *ajes/ages* was *not* the only vernacular name for the root used by the locals in the Caribbean. Within a couple decades of Columbus's voyages, more Europeans followed to various parts of the Americas, where they took note not only that sweet potato was commonly grown in the region, but that it was called by a variety of names, including *patata* or *botato* and their cognates. Peter Martyr d'Angleria, a prolific writer who detailed European exploration of the Americas in his well-known *De Orbe Novo* (The new world) in 1530, described the sweet potato not only as *ages* but also as *patatas* in Jamaica. The natives, Martyr wrote, "eat roots which in size and form resemble our turnips, but which in taste are similar to our tender chestnut." In another place, he noted that the root had many species with different names, and that their outside skin and inside color also differed—the outside can be violet, red, or purplish, while the inside can be white or yellowish, or even vice versa.[48] The above description was later confirmed by the explorer turned historian Gonzalo Fernández Oviedo y Valdés in 1535. Like Peter Martyr, Oviedo recorded the Spanish colonization of the Caribbean. In his account, *Historia de las Indias*, he noted that the words *batatas* and *ajes* had both been used by the locals to describe the root plant in Panama, though he thought the two were referring to two different species.[49] Over time, the terms *ages* or *ajes* used by Columbus and his contemporaries were dropped. "It was as the 'batatas,'" wrote Redcliffe Salaman, author of a comprehensive study of potatoes, "that the *Ipomoea* became generally known."[50] Consequently, the route of sweet potato's transfer from the Caribbean to Europe was referred to as the *batata* line, whereas the aforementioned lexicon parallel of *kumar/cumar* or *kumara* between the Americas and Polynesia became the basis for another line of transfer known as the *kumara* line. It is worth pointing out that in a later time, the *batata* line also extended from Europe to Africa and Asia, thanks mainly to the Portuguese, who, following Vasco da Gama's voyage of 1497–1499, opened and controlled the sea circuit, or the Portuguese India Armadas, from Europe via Africa to South Asia.[51] Another interesting point is that the word *batata*—or *patata*; both were variations of "potato"—was later used by Europeans to refer mainly to the white or Irish potato when it landed in Europe over a half century after the sweet potato. In order to distinguish the two, they gradually added the prefix "sweet" to the latter, becoming the "sweet potato" to refer to *Ipomoea batatas*.[52] Indeed, as Donald Brand noted, " 'Potato' is an example of a word that has changed meaning, as it was used by Shakespeare and other Englishmen of his time to mean what we now designate as sweet potato." In other words, before the seventeenth century, the word "potato" had been associated with the sweet potato in many parts of Europe.[53]

Around the time when more navigations were attempted to explore new ocean routes from Europe to Asia, Africa, and elsewhere, the conquistadors consisting of Spaniards, Portuguese, and other Europeans, colonized the Americas, decimating Amerindian populations in while establishing their empire. In the process, the colonists continued their exploration around the New World, which led to the Spanish explorer Vasco de Balboa's sighting of the Pacific Ocean in 1513. Knowing the existence of the Pacific, which was then called the "South Seas" by the Europeans, Ferdinand Magellan hoped to chart a new, westward path to reach Asia, allowing the Spanish to circumvent the Portuguese monopoly of the eastward Europe–Asia sea route established through the Treaty of Tordesillas in 1494. After finding and passing through what we now know as the Strait of Magellan, at the southern end of the continent, Magellan and his fleet entered the Pacific, navigating through the Tuamotu Archipelago in Polynesia and the Mariana Islands in Micronesia before arriving in Maritime Southeast Asia. Magellan was consequently killed in the Philippine Islands in a clash with the locals, but his admiral Juan Sebastián Elcano (1476–1526) managed to take the remaining fleet back to Spain and complete the circumnavigation of the earth in 1522.

The significance of Magellan's voyage in the spread of the sweet potato in Oceania was that it helped open up, in addition to the *kumara* and *batata* lines, the *camote* line of transfer for the plant to reach western Oceania. Few, however, followed him around South America through the Strait of Magellan before entering the Pacific. Instead, many people traveled directly from Mesoamerica to the Philippines. The sweet potato was given the name *camote*, which came from *camotli* in the Nahuatl language spoken by the Indigenous peoples of central Mexico. In places throughout Micronesia, such as Guam, Mariana, and Palau, as well as Mindanao and Luzon, in Maritime Southeast Asia, *camote* and its variations were adopted by the locals, suggesting a direct line of transfer for the sweet potato to reach western Oceania.[54] This *camote* line was operated along the well-known trading route plied by the "Manila galleons," linking Spain's two major colonies in Mexico and the Philippines between the mid-sixteenth and the early eighteenth centuries. It was also along this route that sweet potatoes spread from Southeast Asia to coastal regions in South China, Taiwan, and Ryukyu around the turn of the sixteenth and seventeenth centuries.

The discovery of the South Seas, or the Pacific Ocean, revived for the Europeans of the time the antiquated idea of the existence of the "Southern Land," or Terra Australis, an idea that dated back to Romans like Ptolemy and Cicero. In search of that land (i.e., today's Australia), further explorations were attempted, which led some navigators to locate more Pacific islands from the mid-sixteenth century. Álvaro Mendaña, a Spanish navigator, was one of them. Having first settled in Peru, Mendaña launched two voyages from America into the vast South Pacific, first between 1567 and 1569 and then from 1595 to 1596. Both trips, however, failed to find Australia. But Mendaña and his crew, especially Pedro Fernandes de Queirós, a Portuguese serving the Spanish Empire in South America, described both voyages in detail in one of the earliest written records of the European navigation of Oceania. These records later helped Captain Cook in his search for Australia in the eighteenth century, though some suspected that Queirós might have landed in Queensland without realizing it.

Mendaña's first attempt to find Australia reached the Solomon Islands in Melanesia and Tuvalu in Polynesia, according to Queirós's accounts, which were based on the reports he sent back to the Spanish court on his voyages. Mendaña was unable to establish a colony due to the native islanders' resistance, prompting him to embark on his second voyage thirty years later. He prepared a much larger fleet for his second attempt but fared much worse. After arriving in the Marquesas Islands in Polynesia, he failed to return to the Solomon Islands; instead, after searching for a couple of months, he eventually landed instead on Santa Cruz, where, according to Queirós's records, they saw sweet potatoes growing on the island: The islanders ate them "roast or boiled."[55] Suffering from illness, Mendaña's plan to establish a colony failed as well; he himself died on the island. Afterwards, Queirós led the surviving ships westward to the Philippines.

Queirós's mention of the sweet potato in his account was one of the first by a European of the root's consumption in Oceania. However, as valuable as the account was, it also sparked debate about the spread of the sweet potato in the region. For example, if Mendaña and Queirós visited islands in both Melanesia and Polynesia on their voyages, why did Queirós only record seeing the root in Santa Cruz and nowhere else? As mentioned before, both of them had lived in Peru before embarking on their voyages, where the root had been grown and consumed by Amerindians; this experience would have enabled them to identify the sweet potato if they saw it in other places.

The subject is complicated by the fact that, almost two centuries after Mendaña and Queirós's initial explorations of Oceania at the end of the sixteenth century, European navigators made successful "discoveries" of the Pacific islands and Australia. While the Spaniards effectively ran the Manila galleons and the Portuguese traveled across the Indian Ocean and around Africa's Cape of Good Hope to Asia, neither made further significant explorations of Oceania during the sixteenth and eighteenth centuries. Of course, by extending the *batata* line, the Portuguese could have brought the sweet potato all the way to western Oceania, including New Guinea, a possibility to which we will return later in the book. And, by operating Manila galleons after Álvaro Mendaña and Pedro Queirós, the Spaniards may have discovered Hawaii and other Pacific islands, such as Pitcairn and Vanuatu, during the period. However, as British historian Henry Kamen has pointed out, Spain's imperial power began to decline in the mid-seventeenth century, making further exploration of the region at latitude 20° north impossible.[56] In its place rose the Dutch, followed by the British. For instance, Tonga, New Zealand, and Easter Island were "discovered" by the Dutch. And from the mid-eighteenth century, as it were, it was the British navigators who took over the exploration, Captain Cook being the most famous. In his three voyages to the Pacific, Cook not only "discovered" and/or "rediscovered" Hawaii, New Zealand, and other islands but also described them and their inhabitants in detail. The Dutch and British navigators, too, were credited for "finding" Australia: While the former called it "New Holland," Cook named its eastern coast New South Wales.

While their exploration was significant, the existing records of Dutch and British explorers concerning the sweet potato in Oceania were uneven and inconsistent, which gave rise to debates about its origin and routes of diffusion among the Pacific islands. After Mendaña and Queirós, Dutch navigator Abel Tasman went to the South Pacific in 1642, where he located New Zealand; and in the early eighteenth century, his countryman Jacob Roggeveen sailed to Easter Island and Samoa in Polynesia. Tasman did not report seeing the sweet potato in his voyages, whereas Roggeveen did. According to Roggeveen, sweet potato was grown widely on Easter Island as a main source of food.[57] Then, in the late eighteenth century, Captain Cook sailed to Hawaii and New Zealand, where he recognized that sweet potatoes were an important food for the Maori. In his journal, Cook wrote that in his first voyage landing in New Zealand, "they [the Indigenous people] bringing us fish and now and then a few sweet potatoes and several trifles which we deem'd curiosities for these we gave them cloth, beeds nails, etc."[58] And then in his second voyage, landing on

Easter Island, Cook again described his encounter with the natives that "after distributing among them some Medals and other trifles, they brought us sweet potatoes, Plantains and some Sugar cane which they exchanged for Nails etc."[59] Importantly, European explorers found that the Maori people in New Zealand also called sweet potato *kumara*, as did Polynesians in Easter Island and Hawaii. However, regarding the root in other parts of Polynesia and Melanesia, these European explorers seemed short of providing an unequivocal account of its being an essential food plant for the islanders.

Nonetheless, given the robust existence of the *camote* line, which successfully delivered sweet potatoes to the Philippines and the remainder of Southeast Asia on the Manila galleons, scholars (except those who advocated America–Polynesia transfer in pre-Columbian times, as we saw earlier) have long held that Spaniards were instrumental in helping spread the plant throughout the entirety of Oceania from the sixteenth century, as they had spread it across Europe in the same period. As early as 1884, when the French botanist Alphonse de Candolle described the sweet potato in his sweeping survey of all cultivated plants, he was quoting Columbus's contemporaries that "the cultivation of this plant was already common in Spain from the beginning of the sixteenth century." "After Columbus brought the sweet potato to Spain following his first voyage to the Caribbean in 1492–93," commented Robert Langdon in this century, "it was dispersed to other parts of the world so quickly that its place of origin was soon forgotten. By the second half of the eighteenth century, it was so widespread that Cook and his associates were not surprised to find it cultivated at far corners of Polynesia and at other islands in between." John Reader, author of a recent global history of the white potato who believes that it found its way to Europe a half century later than did the sweet potato, concurs by saying that "By the second or third decade of the sixteenth century the sweet potato was already widely cultivated in southern Europe wherever conditions were suitable."[60] In other words, as will be demonstrated in chapter 2 of this book, the sweet potato's introduction into Europe was quite successful, giving rise to the belief that Europeans were responsible for spreading the root crop to other parts of the world, including Oceania.

In 1929, Georg Friederici, a noted German ethologist specializing in cultural and ethnographic relations between the Americas and Oceania, wrote an extensive article that, applying the method of philology, offered a detailed account of how Europeans introduced the sweet potato to the Pacific region. He provided linguistic evidence about the three lines of transfer (i.e., *camote*, *batata*, and *kumara*) from the Americas identified by Alphonse de Candolle, which became a foundation for later scholars to expound the theory. Meanwhile, Friederici hypothesized that Álvaro Mendaña and Pedro Queirós were the pioneers in bringing the sweet potato to Oceania, despite the scant evidence—besides Queirós's brief mention of the root in Santa Cruz, Mendaña actually did not offer much proof. But according to Friederici,

> The potatoes were considered a contemptuous food by the Spaniards of Peru, while the *batatas* were valued and popular throughout the Spanish colonial area. Therefore, while sowing maize and planting out potatoes was reported to be of particular benefit in the context of the general instructions and the customary practice of Spaniards' spread of useful plants and domestic animals, the attempts made with *kumara* were, of course, not mentioned.

In other words, Friederici believed that given their preference for the sweet potato over the white potato, it was undoubtedly the Europeans who had brought the former on their voyages to Oceania, beginning with Mendaña and Queirós. More specifically, he stated the following:

> So the *Ipomoea Batatas* came from America to Polynesia and brought its old native name *kumara* from the areas around Páyta. Mendaña brought the plant to the Solomon Islands on his first voyage across the Pacific Ocean and, thirty years later, also to the Santa Cruz Islands on his second voyage. Finally, nine years later, Queirós followed suit on his third major Spanish southward voyage who, having previously sailed through the Tuamotu archipelago, brought the *kumara* to the New Hebrides. When the Europeans visited these islands and archipelagos again in the second age of great discoveries, thus they found the *kumara* there, as well as in Fiji, New Caledonia and all Polynesian islands and archipelagos.[61]

Georg Friederici's viewpoint, which smacked of the Eurocentric take on the sweet potato's diffusion in Oceania, echoed Berthold Laufer's comprehensive overview of the plants from the New World published in the same year. A prominent sinologist of his time, Laufer declared that "none of the American cultivated plants occurs in Europe, Asia or Africa prior to the age of discovery."[62] In 1932, however, Ronald B. Dixon, a Harvard anthropologist, challenged their proposition. Having combed through almost all the travelogues by European explorers in the region from the sixteenth to the nineteenth century, Dixon wrote "The Problem of the Sweet Potato in Oceania," in which he raised questions about Friederici's theory that after the voyages of Mendaña and Queirós in the sixteenth century, Pacific Islanders learned about the sweet potato and planted it from Polynesia to Melanesia during the seventeenth and eighteenth centuries, before Cook and others' explorations. Dixon pointed out that the accounts of European explorers pertaining to the sweet potato were far from consistent and sweeping. For instance, while Cook and others saw the importance of the root in Easter Island, Hawaii, and New Zealand, they failed to mention it in their records about other Pacific islands. Dixon's conclusion was as follows:

> In the face of all the evidence it thus seems clear that a diffusion of the sweet potato during the seventeenth and eighteenth centuries from the Marquesas to the outlying groups in Polynesia would have been practically impossible, and we must admit that certainly in New Zealand, and with greater probability in Easter Island and Hawaii, the sweet potato had been in use as an important food product long before the earliest Spanish contacts with Polynesia took place.[63]

Having enlisted ample evidence against Georg Friederici's hypothesis of the Spaniards as the sole transferors of the sweet potato in Oceania, Dixon also piqued a new interest among his readers: If the diffusion of the root crop had occurred in pre-Columbian times, then when did it happen and—if it was by the hands of humans—by whom? Dixon himself failed to conduct further research because he passed away a few years after publishing this article. But his work was influential. In 1938, Peter Buck (a.k.a. Te Rangi Hīroa), an anthropologist and medical doctor in New Zealand whose mother was Maori, published his popular book *Vikings of the Sunrise*, in which he told a romance about the superb seafaring ability of Polynesians in exploring the Pacific Ocean before Europeans. According to Buck, who later served as a longtime director of the Bishop Museum in Hawaii, sometime before the thirteenth century, a Polynesian voyager sailed east from Marquesas Islands to search for a new land. Entering the vast and empty ocean, he and his companions had no other choice but to continue their voyage. Eventually they landed in Peru by accident, bringing the sweet potato back to Polynesia. In concluding the story, Buck offered the following reflection:

> The unknown Polynesian voyager who brought back the sweet potato from South America, made the greatest individual contribution to the records of the Polynesians. He completed the series of voyages across the widest part of the great Pacific Ocean between Asia and South America. Tradition is strangely silent. We know not his name or the name of his ship, but the unknown hero ranks among the greatest of the Polynesian navigators for he it was who completed the great adventure.[64]

In hindsight, it seems Buck's lament has been a bit misplaced, for more recent scholarship seems to hold that if there were contacts between South America and Polynesia before the European "Age of Discovery," then Polynesians should be the ones who went to the continent rather than vice versa. For instance, impressed as much as Buck was by the Polynesians' seafaring excellence, Japanese scientist Uchibayashi Masao echoed Buck by stating in 2006 that "there is currently a large number of theories that the Polynesian's bold return trip from Peru brought sweet potatoes to Polynesia, which I also support."[65]

However, it is worth noting that in the 1930s and 1940s, just before Thor Heyerdahl attempted his adventure in 1947, there were some who seemed to believe that Amerindians sailed westward to reach Polynesia before Europeans. James Hornell (1865–1949), an English expert on seafaring ethnography, published two articles in 1945 and 1946 on the pre-Columbian contact between South America and Polynesia and Oceania. Expanding on Dixon's criticism of Friederici, Hornell offered more evidence that the sweet potato's diffusion was quite uneven in the Pacific and that before European contact, the root had been cultivated not only in Polynesia but also in Micronesia and Melanesia. Drawing on his expertise in seafaring knowledge, Hornell carefully studied the currents, drifts, and winds to assess the likelihood of a pre-Columbian communication between Oceania and the Americas. He agreed that Polynesians' seafaring ability was unquestionable: "With the knowledge obtained after careful analysis of the many traditional accounts of long voyages successfully undertaken by Polynesians between the tenth and fourteenth centuries, there can be no doubt that this people produced a succession of hardy adventurers during this period who sailed to and from over the length and breadth of the Pacific." That is, according to Hornell, it was "well within the power" of Polynesian adventurers to reach the Peruvian coast on one of their voyages during the period. However, he appeared much less certain about whether they could make an equally successful return trip from South America, let alone bring back the sweet potato to either Easter Island or Marquesas Islands in Polynesia, for there were "great and well-nigh insuperable difficulties" in terms of ocean currents and wind directions. In the end, Hornell cautiously offered some concluding thoughts: (1) Amerindians and Polynesians had made at least one contact without European intermediation; (2) "the sweet potato's introduction into Polynesia was due to this contact arising from an isolated incident"; and (3) whether this introduction was made by Polynesians during their return trip from America, or whether it was the result of "an involuntary drift voyage from Peru, must remain uncertain; I incline to favour the second explanation as being much the more likely."[66]

Over the last few decades, few have done more research that better explains whether or not Polynesian seafarers were able to travel to and from South America in prehistoric times, carrying back the sweet potato. However, advanced technology have recently enabled scientists to provide more convincing evidence establishing the three routes of sweet potato diffusion in Oceania. In their 2013 study, Caroline Roullier and her colleagues carefully examined the genetic samples of the plant throughout Oceania. Like Yen and other previous scholars, Roullier's study also considers the linguistic evidence, for the sweet potato, as we have seen, is called by a variety of names in the Pacific region and Americas. Specifically, the tripartite routes of the plant's diffusion in Oceania consisted of the *kumara* line from South America to Polynesia, the *camote* line from Mesoamerica to Island Southeast Asia and the *batata* line, originating also from Mesoamerica but going eastward first to Europe and Africa and then to western Oceania. These lines of dispersal were so named because of the linguistic variety by which peoples in those regions referred to the sweet potato. Roullier and her associates have focused their research on whether the genes and names of the sweet potato dispersed together in the Pacific region. What they find is that their research provides "strong genetic support, previously lacking, for the tripartite hypothesis, notably concerning the Kumara line, the pre-Columbian diffusion of sweet potatoes from South America into Polynesia."⁶⁷ This conclusion thus offers strong credence to Yen's thesis of multilinear routes for the sweet potato's transfer from the Americas to Oceania.

Sweet potato's reaches in Oceania

It is now time to look more closely at sweet potato as a food plant in Oceania. Prior to the European explorers who traversed the vast Pacific Ocean beginning in the sixteenth century, others had visited the region. A Chinese explorer named Wang Dayuan left us his *Daoyi zhilue* (Brief descriptions of island barbarians) in the mid-fourteenth century, which later served as a guide for the explorer Zheng He's maritime expedition in the early fifteenth century. As indicated by its title, Wang's account was based on his seafaring trips from the South China Sea to the Pacific and Indian Oceans, reaching as far as India, North Africa, Australia, and many places in Maritime Southeast Asia. Indeed, according to Barbara Watson Andaya and Leonard Y. Andaya, authors of *A History of Early Modern Southeast Asia*, Wang's *Daoyi zhilue* provides "far more information on areas to the east of Malay Peninsula than any other source and affirms growing maritime connections with the Philippines, Maluku, and Timor."⁶⁸ While brief, Wang's account gives information on food crops and records that in places like Malacca, Manila, and Langkawi, it was common for the natives to plant "yams and taro."⁶⁹ The Chinese terms for yams and taro (*Colocasia esculenta*) he used were *shu* 薯 and *yu* 芋, which might also include *Dioscorea esculenta*, or lesser yam, in addition to *Dioscorea alata*, or greater yam. Regardless their varieties, both yams and taro were native to the Asian continent and Asian tropics.

If Wang Dayuan had ventured to the eastern part of Oceania, then he might have seen that yams and taro were also important crops in Melanesia and Polynesia. The broad appeal of root crops like yams and taro to Pacific Islanders, or Austronesians, provided another proof of their eastward migration from East Asia to Polynesia. Meanwhile, it also helped explain why the sweet potato, another root plant, was adopted by them as a food source once it reached Oceania from the Americas. In other words, thanks to their established dietary tradition, Pacific Islanders were accustomed to growing root plants as food, which facilitated the diffusion of the sweet potato in Oceania. Harold M. Ross finds that in the Solomons

Islands, sweet potato and taro are both important food plants and that the former rivals the latter because it is immune to blights and other diseases that affect taro.[70] Besides yams and taro, the *ti* plant, or *Cordyline fruticosa*, an evergreen flowering plant widely available from the Philippines to northern Australia, is also essential to the animistic religion practiced among Austronesians. In Hawaii, its root was consumed as a sort of confection, whereas elsewhere the plant was regarded as too sacred to be treated as food. In addition, turmeric and Polynesian arrowroot were two other root plants traditionally processed and consumed as starchy foods in the region.

Sweet potatoes were a newcomer to the Pacific Islanders' root diet. However, while scholars disagree about who brought sweet potatoes to Polynesia, they all agree that it happened a few centuries before European contact in the sixteenth century. There have been various methods for dating the arrival of the sweet potato in the region, just as there are still disagreements about the Austronesian settlement chronology of its islands. In 2005, Roger Green proposed that so far as its spread in Hawaii was concerned, it occurred via human agency in the eleventh and twelfth centuries, "along with the bottle gourd and perhaps the soapberry plant." He also maintained, agreeing with Peter Buck, the Maori scientist at the Bishop Museum in Hawaii, that it was the Polynesians who "went to South America and returned to the central regions of east Polynesia with the sweet potato." According to Green, eastern Polynesia was not populated until the ninth century, later than the fifth and sixth centuries proposed by Uchibayashi Masao (cited above). After their settlement in the Marquesas Islands, for instance, Polynesians, with their excellent seafaring skill, continued to sail eastward to South America, from where they brought back the sweet potato and bottle gourd. They chose the two plants, but not maize, a major food crop in the Americas, because "the sweet potato was probably treated as another kind of yam. It was also a plant that could be transferred in the form of rootstock and could be propagated vegetatively." They chose bottle gourd because "it was almost identical to the wax gourd (*Benincasa hispida*)," which "was already known to the Polynesians." Green's explanation is radically different from the one offered by J. W. Macnab of Victoria University of Wellington, who argues that sweet potato had been transferred to Polynesia from America before 750 BCE, or before maize became a major food crop for Amerindians.[71]

Roger Green draws on Helen Leach's research to make his observation about the acceptability of sweet potatoes in Polynesia. Leach, a fellow anthropologist in New Zealand, argues that sweet potato was adopted in eastern and southern Polynesia because the root was similar in many ways to *ufi* yam, a specific variety grown by the locals after their settlement on the islands. According to Leach, in New Zealand, *ufi* yam is even referred to as *ufi kumara*, implying that the Maori people treated the two crops equally. In her study, she enumerates five reasons for the adoption of the sweet potato in the region, ranging from its physical appearance, planting, storage, seasonality, and importance as a food crop compared to that of the *ufi* yam. Needless to say, as she observes, the two roots are not identical in all aspects. But Leach's general argument runs as follows: "Successful introduction required not only survival of the original planting material, but an appropriate classification within existing frameworks of knowledge, a vital act of analogy. Successful spread of sweet potato within east Polynesia then depended on the plant's acceptance as a new and improved variety of something they were already familiar with." That is, whoever brought the sweet potato back from the Americas to Polynesia did not equip themselves with sufficient knowledge of how best to plant it. This has been widely agreed upon by all researchers. As a result, Leach's thesis holds water: The root was more or less readily adopted by Polynesian islanders because it resembled the *ufi* yam with which they were familiar. Polynesians eventually recognized the differences between the two roots and adjusted their sweet potato farming practices accordingly. Leach concludes that sweet potato surpassed yam due to its "fast-growing and less-demanding" qualities.[72]

Helen Leach and Roger Green's thesis on the acceptability of the sweet potato in Oceania has broad implications to the extent that it helps to explain why, of all the New World crops, the sweet potato became almost the only one to be adopted as a food crop in the region. In Helen Leach's words, "East Polynesians already had a good variety of tree crops and a handful of root crops. Why did they bother with a new and unfamiliar one? And why only one of the South American crops?"[73] In fact, this is one of the key issues that D. E. Yen considered in his authoritative *The Sweet Potato and Oceania*. Besides the sweet potato's botanical proximity to other root crops, which to him accounts for why maize failed to take root in Oceania,[74] Yen examined a variety of factors that paved the way for the sweet potato's acceptance. He began by looking at the dates of its diffusion. Having carefully reviewed various proposals about the advent of the sweet potato in Oceania, ranging from a couple millennia BCE to a few centuries before European contact, Yen believed that the latter would probably be a better possibility. That is, the plant was transferred to Oceania prior to the European "Age of Discovery," but not many centuries prior. For instance, he thought the sweet potato perhaps arrived in New Zealand in 1350, brought by settlers in Hawaii, the Marquesas Islands, and other islands from the north. According to Maori mythology, their ancestors were from Hawaii and reached New Zealand in the Great Fleet. Incidentally, this later date, later than Roger Green's estimate about the introduction of the root in Polynesia in 1000–1100 CE, mentioned above, seems to align more closely with other estimates. James Coil and Patrick V. Kirch of University of California, Berkeley, conducted archaeological digs in the Hawaiian Islands and concluded that sweet potatoes were not planted there until approximately the fourteenth century. In their study of the sweet potato in Easter Island (Rapa Nui), Paul Wallin, Christopher Stevenson, and Thegn Ladefoged also maintained that the root was not present until 1200–1300 CE.[75] Thus viewed, it is reasonable for Yen to propose that the root plant reached New Zealand in approximately 1350.

The above dates for the sweet potato's spread in Polynesia suggest that, if we accept its American origin, not only did Polynesians bring it to Oceania, as Buck and Green argue, but they also played a key role in dispersing it in eastern Oceania. Linguistic evidence suggests both possibilities. As previously stated, *kumara* was a Quechuan word used by South American Amerindians. And the fact that the sweet potato was known as *kumara* and its various cognates throughout Polynesia proves the plant's origin and spread. Furthermore, the cognates indicate the routes of its dispersal within and between Polynesia and Melanesia. D. E. Yen explains it in more detail:

> The dropping of one or two consonants, with or without the substitution of glottal stops, seems to have been the main modification, if it is accepted that the Quechua form arrived in Polynesia with the plant. The substitution of /r/ with /l/ occurs in the western islands Tonga, Futuna, and Uvea, and was also recorded by Solander for New Zealand. It also occurs in Samoa and Hawaii, but with accompanying modifications in the first two consonants.

Examples of the above observation can be seen in the fact that sweet potato is called *kumara, kumala, kakau,* and *oomara* in New Zealand; *kumara* in

Easter Island and Tuamoto; *kuma'a* in the Marquesas; *umara* and *umaa* in Tahiti, *kumala* in Tonga; and *uala* and *uwala* in Hawaii. By comparison, the root is referred to as *kumar, kumara, cumar', umar'*, and *kumal* in Peru and Ecuador. That sweet potato is pronounced *uala* in Hawaii, different from that of the rest of Polynesia, was noticed by John F. G. Stokes, a pioneer of Hawaiian archaeology, who suspected that there might be a different route for the sweet potato to land on the island.[76]

James Hornell, the English ethnographer specializing in the Pacific region, postulated a theory that the above linguistic variations suggest the routes of the sweet potato's diffusion in eastern Oceania. The presence or absence of the consonant k/ could be a sign of the intra-Polynesian diffusion, according to Hornell's essay "How Did the Sweet Potato Reach Oceania?," which was previously mentioned. Specifically, he believed that "Once introduced, the plant, together with its Quichuan name in some modified form, became widely distributed throughout the Polynesian islands by local seafaring activity." And over time, the diffusion also reached many islands in Melanesia and some places in Micronesia by the fifteenth century, before European contact. One of the values of Hornell's research was that he described, quite convincingly, the Polynesians' seafaring routes in the Pacific:

> Guided in the first instance by the seasonal flights of migratory birds, voyages between Hawaii and Tahiti, probably via the Marquesas and Fanning Island, were accomplished, involving a distance of more than 2400 miles. So, too, were voyages made in far-off days between Tahiti and New Zealand; even the names of many of the leaders are known, together with the fanciful names of the double-canoes in which they sailed. Distant Easter Island was occasionally visited—a far more hazardous journey because of the small size and lonely situation of the island, hundreds of miles beyond the eastern limit of the closely set island groups of central and eastern Polynesia.[77]

Here Hornell departs from Thor Heyerdahl by arguing that the Marquesas Islands were the sweet potato's point of entry to the Pacific Ocean. Roger Green supports Hornell's position in his more recent and extensive study. Green proposes, based on historical, linguistic, and botanical evidence, that an "ellipse region" existed for the introduction of the sweet potato into Oceania, consisting of the Galapagos Island near the coast of South America on one end and the Marquesas Islands and Tuamotu Archipelago in eastern Polynesia on the other. That is, after their settlement in Polynesia, Polynesians "proceeded further eastwards and then south-east to Mangareva thence to Rapa Nui. Therefore, some Polynesians went to South America and returned to the central region of east Polynesia with the sweet potato, the white flower bottle gourd, and (arguably) the soapberry plant. Further voyages took them, or their descendants, back to Rapa Nui, up to Hawaii, and down to New Zealand."[78] Green's study, therefore, enriches our understanding of the *kumara* line of diffusion for the sweet potato's spread from the eastern to the central Pacific regions.

This transfer, while not as ancient as some believe, occurred prior to the arrival of the Europeans. When European explorers sailed to the Pacific region, they came across the sweet potato on various islands, where it had already been grown as a food crop of varying importance. As previously stated, Pedro Queirós saw the plant in the sixteenth century in Santa Cruz, Melanesia. Sweet potatoes were planted in the Marianas, in the upper west corner of Micronesia, in the same century, according to Antonio Pigafetta, a Venetian scholar and explorer who accompanied Ferdinand Magellan on his voyage across the Pacific Ocean between 1519 and 1522. "Their provisions are certain fruits," he wrote of the islanders, "called Cochi [coconuts] and Battate." As sweet potato had already been introduced to southern Europe in the previous century and was known as *batate* in Spain, it is likely that Pigafetta made an accurate identification.[79] Then Fray Juan de Torquemada, a Spanish missionary, saw the sweet potato in Vanuatu, south of Santa Cruz in eastern Melanesia, in the early seventeenth century. As for the importance of the sweet potato in Polynesia, Jacob Roggeveen discussed it during his travels to Easter Island in the early part of the eighteenth century and Capitan Cook confirmed it in Hawaii and New Zealand later in the same century, as we heard earlier. On his third voyage to the Pacific Ocean, Cook, a careful observer, described the sweet potatoes he saw in Hawaii "as big as a man's head," suggesting that the natives had mastered the technique for growing the plant.[80] William Ellis, an English missionary who resided in Hawaii and other parts of Polynesia for several years in the early nineteenth century, also wrote that the sweet potato, which the natives called *umara*, is "large," "has been long cultivated," and "is one of the principal means of subsistence" for the Indigenous people.[81]

Sweet potato's intervention as a food crop

These records by European explorers and missionaries, needless to say, have contributed to the likelihood of the sweet potato's pre-European presence in Oceania. Given the late colonization of Polynesia between the tenth and fourteenth centuries, their accounts also revealed that, despite being a latecomer, sweet potato as a food crop was relatively quickly accepted by the islanders. D. E. Yen explains this success as follows:

> If the sweet potato had arrived in eastern Polynesia as a pioneer cultigen, a hypothetical result would have been that its agronomic advantages would have transcended those of the Melanesoid species. Its yielding capacities under a wider range of ecological and soil preparation conditions, its speed of production, lack of seasonality, and simple preparation methods, together with its adaptability to the pounding processes, endow the species with the competitive properties over the other available plants with could have resulted in a building of a tradition of agriculture considerably more contrastive with western Polynesia than was the case.[82]

The Melanesoid species to which he refers are taro and yams in the main. Compared with them, which were much more entrenched in the region's agronomy, the sweet potato offers three principal advantages as a food crop. The first and most important is its higher fecundity relative to that of taro and yams; the second is its high adaptability to climate and soil; and the third is its short growing period. For its abundance, sweet potato was widely adopted in the swidden economy by the people in the region stretching from Papua New Guinea to Easter Island. And its tolerance of cold weather made it a more ideal crop than yam in New Zealand, for the island has a relatively harsher weather system than that of the rest of Polynesia. This also explains its importance in highland Papua New Guinea, where sweet potato supplanted taro as the most important food crop. Sweet potato, on the other hand, did not fare as well in the lowlands of Hawaii or Papua New Guinea. And, throughout the Pacific, the sweet potato's fast growth made it an excellent ground cover, as well as the best animal feed.

While its botanical advantages are obvious, the sweet potato remained a later introduction to Oceania, much later than the Melanesoid species the Austronesians carried with them when colonizing the Pacific. The recency of its "intrusion" means that its acceptance has varied considerably from place to place in the region. Specifically, sweet potato is a major, or the most important, staple food in Easter Island, Hawaii, New Zealand, and Papua New Guinea, whereas elsewhere (e.g., Melanesia and the islands of the central Pacific) it may trail taro and/or yams in terms of its significance.[83] Nonetheless, given its fecundity, sweet potato has without question been the most popular source of feed and forage throughout the Pacific.

Local folklore, religious beliefs, and rituals have reflected the sweet potato's recent intervention in Oceania's economy. Take, for example, Hawaii, where sweet potato is a major crop. As a result, the root, along with taro, is offered to the rain god Lono on the islands. Chants about the sweet potato were also developed and often sung on certain occasions as part of the ceremonies. For instance, Craighill Handy and Elizabeth Green Handy in their ethnographical writing note that in Kona, Hawaii, where the rain god was worshipped, "food, including hogs and sweet potato (the staple in this southern dry area), were laid on altars (*ahu*) dedicated to Lono" at the annual ritual of reenactment during the harvest festival of Makahiki.[84] However, because sweet potato is a newly introduced crop, whereas taro is a historical food crop that is ubiquitous throughout Hawaii in both wild and cultivated forms, it is secondary to taro in some regions. This is especially the case in shady areas and damp forests because taro likes moisture. There was also a gender-related difference regarding the ceremonial value of the two plants to Hawaiians. In *The Hawaiian Planter*, Craighill Handy observed that "the sweet potato did not enter into the ritual of the kahunas as did the taro. Despite the fact that there was more worship associated with the planting of the potato than the taro, because of the pressing need for rain in the areas where it was planted, the sweet potato was not a sacred plant. This is shown by the fact that women as well as men cultivated the humble potato, but only men the noble taro."[85]

A similar situation could be observed in Easter Island and New Zealand, where taro also initially occupied a dominant position in the local agronomy. Like elsewhere in Polynesia, taro and yams were native plants in Easter Island and the sweet potato was not. But thanks to its edaphic flexibility and climate tolerance, sweet potato was quickly adopted and integrated into the existing agronomical system. Despite the dry environment, which made taro and yams more difficult to cultivate in large quantities, sweet potato was effectively grown with large outputs in vast inland

and upland areas. The sweet potato's importance was reflected in folklore. According to legend in Easter Island, early settlers planted yam as the first crop and clashed with other people on their reconnaissance tour. And then they met Hotu Matua, a chief from elsewhere (probably the Marquesas) who brought sweet potato, yam, and gourd on his two canoes to the island. Over time, Hotu Matua was worshipped as the ancestor of the Rapa Nui people on the island.[86] The legends of the Maori in New Zealand, as alluded to above, tell of their ancestors from Hawaii. As the sweet potato successfully replaced taro as the most popular plant on the island before the European arrival, the root was associated quite closely with the Maori rain god Rongo, or Lono. The Maori held festivals equivalent to Makahiki for praying to Rongo before the sweet potato planting.[87] By contrast, taro "appears to have been considerably less important than the sweet potato in the cropping pattern, and little or no ceremonial was attached to its cultivation." Again, this replacement seemed to be more recent in that taro retained some prestige as a food to greet the visitors and offer to the dead in mourning.[88]

Though the sweet potato figured notably in local mythology and folklore in eastern Polynesia, it occupied a much vaguer position in the rest of Oceania. Corresponding to the root's lesser importance as a food crop, the sweet potato made a rare presence in the ethnological literature of the islands that comprise central Polynesia. Of course, rainmaking rituals and harvest festivals were commonplace among the natives, but taro and yams, along with breadfruit, coconuts, and banana, rather than sweet potato, usually played a part on such occasions. Further west in the same direction, one encountered a comparable phenomenon on such islands as Samoa, Tonga, Fiji, Vanuatu, and New Caledonia in western Polynesia and eastern Melanesia. In other words, with the exception of the Solomon and Santa Cruz Islands, where the sweet potato is a staple crop, there is a clear trend of diminishing importance of the sweet potato as a food plant along the route from Polynesia to Melanesia. Meanwhile, the lexical term *kumara* was commonly used by the locals to refer to the root plant, suggesting yet again that once sweet potato had reached Polynesia, Polynesians spread it further on their east–west voyages to Melanesia and probably beyond as well. This connection between Polynesia and Melanesia is shown in the plant's varieties in both regions, according to the scientific research of D. E. Yen and others.[89] All the above, perhaps, may lead us to think of the likelihood that, given time, the sweet potato, as a new botanical intrusion following the east-west direction, could have assumed a more important place in the agronomy of not only Polynesia but also of Oceania as a whole.

In Papua New Guinea and western Oceania, sweet potato has indeed occupied a dominant position in the local economy as a food crop. In fact, scientists hypothesized that the root plant had been cultivated in parts of Melanesia before European contact as early as the sixteenth century. James Hornell argued in his aforementioned article "How Did the Sweet Potato Reach Oceania?," which was based on the accounts of early European explorers in the area, that the presence of both *camote* and *batata*, Amerindian words for the sweet potato, in the regional languages were "definite identifications [that] appear adequate as furnishing presumptive evidence that the sweet potato was under cultivation in Melanesia before the first known contact with Spanish voyages from South America." More recently, Christopher Ehret has proposed, supported also by linguistic evidence, that the sweet potato variety grown in western Oceania crossed the Indian Ocean and landed in East Africa during the pre-Columbian era. In fact, Wolfgang Grüneberg, Maria Andrade, and their colleagues at the International Potato Center in Peru make a more compelling case that "The crop crossed the Pacific in pre-Columbian times and became a staple in the relatively cool tropical highlands of Papua New Guinea and adjacent Irian Java/Indonesia, where it developed a secondary genetic center of diversity." They provide evidence that the sweet potato variety that thrived in Tanzania's highlands has its roots in Papua New Guinea.[90]

On the other hand, some scholars have disagreed with the hypothesis that the sweet potato spread throughout Oceania and even reached Africa in pre-Columbian times. As early as 1963, Harold C. Conklin of Columbia University questioned the possibility of an Oceanian–African route of root plant transfer predating European contact. His claim was supported by three kinds of evidence: ethnoecological, historical, and lexical. Conklin noted that sweet potatoes were referred to as "Spanish yam" or "white man's yam" in both Malaysia (where such a route would have passed through) and Africa. He also expressed his disbelief in the historical accounts offered by Hornell and others, calling them "inconsistent" and lacking in botanical and linguistic evidence. Moreover, he contended that if the Oceanian–African route existed, some non-European- and non-Amerindian-derived sweet potato names should have been used in both places. But he could not find any such lexical proof.[91]

Conklin's argument primarily contested the hypothesis that sweet potato arrived in Africa from Oceania via the Indian Ocean prior to European colonialism. It did not rule out the possibility of pre-Columbian sweet potato cultivation in western Oceania or New Guinea. After carefully weighing the botanical evidence, D. E. Yen in his *The Sweet Potato and Oceania* notes the presence of one "common and indistinguishable" variety in both "Polynesian and Melanesian populations of sweet potato varieties," which might suggest the extent to which the root plant spread from Polynesia to Melanesia. But Yen quickly adds that the "history of the interchange is obscure," meaning it could have occurred before or after European contact. His general observation is that the sweet potato dispersed in Oceania via multiple, rather than unilineal, routes.[92]

Notwithstanding the possibility of pre-Columbian sweet potato cultivation in New Guinea, economists who study food and agriculture in Papua New Guinea believe that its introduction there had more to do (indirectly) with Europeans than with Polynesians.[93] If we accept this observation, the range of Polynesian voyagers might have ended somewhere in central/western Oceania, attesting to the relative recency of the root's historical presence in the Pacific. Its multiple vernacular names also provided linguistic evidence for the breadth of its spread via various forms of human agency. While in the sixteenth century, the Italian explorer Antonio Pigafetta had found *batate* on his trip to the Mariana Islands, the locals also refer to the sweet potato as *kamote*, *kamuti*, or *kamute*, as do the people of Guam and other nearby islands. And this multiplicity in nomenclature was not unique to eastern Micronesia but is also seen in Melanesia (e.g., Papua New Guinea) and the Philippines. All of this may suggest, as some scholars have expounded, that insofar as sweet potato farming in central and western Oceania is concerned, there might be multiple times during which the sweet potato entered the region. And the Europeans may well be the main agents. Indeed, if the vernacular name *kamote* and its cognates in Micronesia was evidence for the Spanish-induced *camote* line of transfer, then the fact that the sweet potato is called *butete* on the Pacific island of Santa Cruz could suggest the farthest reach of the *batata* line facilitated by the Portuguese.[94] In a nutshell, outside of eastern Polynesia, sweet potato is known by a variety of vernacular names, indicating that its dispersal may have occurred various times throughout Oceania.

The case of Santa Cruz, in eastern Melanesia, is interesting as it allows us to see the possible role played by European explorers, whalers, sailors, and missionaries in dispersing the sweet potato. As we know, Pedro Queirós had seen the root as food in Santa Cruz in the sixteenth century. Yet Queirós's mention was so brief that it is difficult to know for sure if it was planted as an important crop on the island. More recently, Matthew G. Allen, a geographer at the Australian National University, conducted a detailed study of the plant in Melanesia. Contrary to other studies about the pre-European diffusion of the sweet potato, Allen, drawing on careful document analysis, proposed that if it were Polynesians who spread the crop to Melanesia, then its dissemination has been rather limited in some isolated islands in eastern Melanesia. In western Melanesia, such as the Bismarck Archipelago and New Guinea, the sweet potato was not present until Europeans made frequent contact with the region's inhabitants in the eighteenth and nineteenth centuries. One of Allen's key findings was that on the Solomon Islands, Santa Cruz's western neighbor, the sweet potato was only introduced in the mid- or late nineteenth century and, due to the outbreak of a taro blight in the mid-twentieth century, it rapidly assumed the dominant position, a status it maintains in the local agronomy today.[95]

In Papua New Guinea, the sweet potato was introduced in a similar manner. Scholars believe that the three routes for the sweet potato into Oceania actually converged in the island nation. Indeed, the root has been given a variety of names by the locals, including variations of *batata*, *camote*, and *kumara*.[96] Of course, this fact alone is not enough to explain the dominance of the root crop in Papua New Guinea's economy. But without doubt, while still debating the dates of its introduction, experts generally agree that sweet potato has played a crucial role in transforming agriculture in Papua New Guinea, a topic discussed in detail in chapter 5. Of the eighteen chapters comprising *The Sweet Potato in Oceania: A Reappraisal*, edited by Chris Ballard, Paula Brown, R. Michael Bourke, and Tracy Harwood in 2005, no less than eight deal with the root crop in Papua New Guinea. Although not a part of Asia, Asian scholars paid ample attention to sweet potato farming in western Oceania when they organized symposiums on the American root's value as a food crop back in the 1980s.[97] Michael Bourke and Tracy Harwood write in *Food and Agriculture in Papua New Guinea* (2009) that there are twenty myths about the country's economy, one of which is that rice is the most important food for its people. Instead, they state, "locally grown staples provide an estimated 68% of food energy, with sweet potato by far the most important of those crops."[98] The economic impact of the sweet potato's rise in Papua

New Guinea was comparable to, if not greater than, it was in the Solomon Islands. The traditional root crop, taro, was displaced because of sweet potato's intervention, which is not surprising given that the country has many highland areas where sweet potato is a more suitable plant than taro and yam. Sweet potato has a prominent place in the ritual repertoire of highland societies, much more visible than elsewhere in Melanesia, as a reflection of its high status. All of this suggests that Papua New Guinea, like Taiwan, is a "sweet potato island" in Oceania, as shown in chapter 5.

In 2005, after helping organize an international symposium on the sweet potato's origin, spread, and importance in Oceania at the Australian National University, Chris Ballard, then a young scholar at that institution, offered his own thoughts in his contribution to the conference proceedings, in which he compared the history of the sweet potato in the Pacific to that of the white potato in Europe. "Much as the potato has had the broader impact historically upon European societies than it did in the location of its original domestication," Ballard writes, "so too the sweet potato has proved more influential in Oceania than in Central and northern South America, enabling substantial population expansion and increase, and fuelling social transformations. Sweet potato supplies historians with a powerful tool, a thread drawing together a range of significant questions about the Oceanic societies."[99] What Ballard said is undoubtedly correct, as the content of this chapter hopefully demonstrates. Yet the sweet potato's transformative role has not been limited to Oceania, as vast as that region is, but has been seen throughout the world.

CHAPTER II

A Sweet Connection Between Europe and America

> This people is very gentle and timid,
> naked as I have said,
> without arms or law;
> these lands are very fertile,
> they are full of *niames*
> which are like carrots,
> and have the flavor of chestnuts.

On November 4, 1492, Christopher Columbus wrote the above words in his journal, describing his very first trip to the Bahamas in the Caribbean Sea and his first encounter with the sweet potato.[1] This may also be the earliest description of the local root crop provided by a European. Columbus may have had a suspicion that it was something else based on his description even though he identified the sweet potato as *niame*, or *ñame*, which was an African name for yam (*Dioscorea*), a native plant of Africa, Asia, and the Americas. Columbus named the sweet potato *niame/ñame* because he had previously visited Guinea in West Africa and seen the root plant there.[2] In modern Spanish, yam is still known as *ñame*, which, along with the Portuguese word *inhame*, became the etymological origin of the English word "yam." Because sweet potatoes and yams both come in a variety of shapes, colors, and characters, it was easy for people to confuse the two back then and now. To this day, many Americans in the United States refer to sweet potatoes, particularly those with orange flesh, as yams. As a root crop, yam, too, can be confused with taro (*Colocasia esculenta*); in Portuguese, for example, the term *inhame* can refer to both yam and taro. Similar cases also exist in Asia and Oceania, as discussed in the previous chapter.

However, there is a significant difference between yam and sweet potato: The latter can be eaten raw and has a sweet taste, whereas the former must usually be boiled or roasted before consumption because it may contain toxins. There is insufficient information to determine whether Columbus tasted the sweet potato raw or cooked on his first encounter—his description of it tasting like chestnut may lead one to believe he ate it roasted. Because sweet potato was common in the Caribbean,[3] Columbus encountered it several times during his voyage. On November 6, for example, two days after he offered the above-quoted observation on the sweet potato, he recorded that the "land is very fertile and much cultivated with yams (sweet potatoes?) and beans."[4] The more time he spent on the island, the more knowledge Columbus seemed to have gained about the plant. On December 16, he wrote not only about how the natives planted it but also how they made bread out of it. He wrote that the roots "are certain little branches that they set out and at the foot of them grow some roots like carrots which serve as bread, and they rasp and knead and make bread of them; and afterwards they proceed to plant the same little branch in another place, and again it gives four or five of the same roots, that are very palatable, like the taste of chestnuts." According to the description, Columbus could have tried the sweet potato in a slightly cooked form by then. And his use of the word "palatable" to describe the root confirmed that what Columbus tasted was not yam, which is usually flavorless or slightly bitter.

A week later, on December 21, Columbus learned from the islanders their name for the plant. "Some [villagers] ran here and others there," he wrote, "to bring us bread which they make of *niames* [*ñames*] which they call *ajes*, which is very white and good."[5] In other words, while the Spanish explorers still called the sweet potato *ñames* (yams), they knew by then it was called *ajes* by the Indigenous peoples of the Caribbean. Then, on Columbus's second voyage, one of his companions, Guillermo Coma of Aragon, wrote letters to his Italian friend Nicolò Syllacio, who translated them into Latin and gave them to the Duke of Milan. The sweet potato was described in detail in the letters:

> The island is therefore covered with foliage and trees, and is decked with plants of many hues; it is neither niggardly nor unbountiful but is fertile and accessible throughout. The most notable fruit of the island is called *asses*. They are very similar to smooth turnips except that they grow somewhat larger, like pumpkins. It should be noted also that they have a variety of tastes. When eaten raw, as in salads,

they taste like parsnips; when roasted, like chestnuts. When cooked with pork, you would think you were eating squash. You will never eat anything more delicious or with more appetite than *asses* soaked in the milk of almonds. It is a dish which lends itself to all the culinary arts and the requirements of gourmets. It has such a pleasing variety of uses and is so gratifying to the palate that you would think it was the manna of the Jews, i.e., the Syrian dew. Moreover, since they are not injurious to the stomach or to the digestive system, they are prescribed for the sick and the diseased with good results, on the advice of the doctors on the royal payroll who came out with the expedition. Seeds have been sent to Spain, so that our world might not lack this beneficial plant and its great variety of gustatory sensations.[6]

The preceding description demonstrated not only that *asses* was another local name for the sweet potato, but also that the Spaniards were so taken with its taste that they decided to bring the plant back to Spain. Dr. Diego Álvarez Chanca, who accompanied Columbus on his second voyage and treated his malaria, also confirmed that the Spaniards ate sweet potatoes, which were the main food in the Caribbean at the time:

They [the natives] all come loaded with *ages*, which are like turnips, very excellent for food, which we dressed in various ways. This food was so nutritious as to prove a great support for all of us after the privations we endured when at sea, which were more severe than ever were suffered by man; for as we could not tell what weather it would please God to send on our voyage, we were obliged to limit ourselves most rigorously with regard to food, in order that, we might at least have the means of supporting life. This *age* the Caribbees called *napi*, and the Indians [Tainos] *hage*.[7]

What Dr. Chanca recorded here is extremely valuable, particularly in terms of the vernacular names of the sweet potato spoken by Caribbean natives, because all of these names, some of which are still in use in the region, appeared in other contemporary accounts discussed below.

While early European explorers were generally enamored of the sweet potato, historians disagree on when Columbus first brought the root plant back to Europe. According to a recent essay on the sweet potato, while Columbus and his crew saw the root crop on their first trip, it was not until Columbus's fourth trip that he "brought the plant back to his homeland."[8] This claim appears to contradict what Guillermo Coma of Aragon wrote to his friend Nicolò Syllacio. No information was actually given about the sweet potato in the existing records pertaining to Columbus's fourth voyage; what they saw and ate was cassava, another popular root crop in Central America. In his account of the voyage, Ferdinand Columbus (1488–1539), the second of Columbus's sons, for instance, wrote that "The Indians were pleased, and for trifles brought us whatever we needed.... If they brought rounds of the bread which they call cassava, made with grated roots of a plant, we gave them two or three green or yellow rosary beads."[9] This was not the Spaniards' first encounter with cassava. In fact, Columbus saw and tasted cassava bread made by Amerindians around the same time he saw the sweet potato on his first voyage.[10]

Francisco López de Gómara, the sixteenth-century Spanish historian who wrote about the Spanish conquest of the Americas, believed that Columbus brought back the sweet potato, together with other American plants such as maize, after his first voyage. "This was the first house or village," wrote Gómara, "that the Spaniards made in the Indies. He took ten Indians, forty parrots, many gallipavos, rabbits (which they call hutias), *batatas*, *ajíes*, maize, from which they make bread, and other things strange and different from ours, for testimony of what he had discovered."[11] Gómara's statement was cited by such scholars as Redcliffe Salaman, who wrote that on Columbus's return, "he exhibited his ten Indian natives and a variety of animals, fruits, and ornaments of gold." And among the fruits Columbus brought back were *ajíes* and maize.[12] Despite the fact that Gómara's account of the Spanish conquest of the Americas was detailed, including additional descriptions of the sweet potato that we will discuss below, he had never visited America. Gómara's proclamation that Columbus brought back the sweet potato on his first return is not entirely trustworthy, as he relied on others to construct his work, which has been criticized by many since the sixteenth century for its inaccuracies. Consequently, it remains unclear when Columbus first brought the plant back to Europe.[13]

The advent of *ajes*, *batatas*, and *camotli* in Europe

What was interesting about Gómara's writing, however, is that when describing the sweet potato, he interchangeably used the words *batatas* and *ajíes*. Were *ajíes*, or *asses*, which appeared in Nicolò Syllacio's letters, the same sweet potato as the *ajes* and *ages* mentioned by Columbus and Dr. Chanca? According to the descriptions above, it appears that these terms are all symbols of the root plant used by Indigenous peoples in the Caribbean, or the West Indies as the Spanish explorers called it. Why, out of all these references, did *batatas* become the most popular Spanish and Latin term for sweet potato?

Francisco López de Gómara, writing approximately half a century after Columbus's first voyage, was far from the first European to use both *ajes* and *batatas* to refer to the sweet potato. According to Isaac H. Burkill, an English botanist who provided a detailed analysis of the two terms in a 1954 article, Peter Martyr d'Anghiera (1457–1526), an Italian historian then serving the Spanish court as its chronicler, had earlier recorded both references in his *De Orbe Novo* (On the new world), in the second of the book's ten big volumes, known as *Decades*. A comprehensive account of the Spanish conquest of the New World, Peter Martyr's *De Orbe Novo* was published between 1511 and 1530, and the second volume appeared in 1516.[14] But Burkill might be mistaken, for in the version of his book prevalent today, Martyr actually recorded the sweet potato in the first *Decade* of his *De Orbe Novo*, which was published in 1511. Martyr's narrative goes like this: "They [Spaniards] eat roots which in size and form resemble our turnips, but which in taste are similar to our tender chesnuts [sic]. These they call *ages*. Another root they eat they call *yucca*; and of this they make bread. They eat the *ages* either roasted or boiled, or made into bread." Without question, the *ages* recorded by Peter Martyr was a variation of *ajíes*.[15]

Then in the Third Decade, Martyr mentioned sweet potato again, together with maize and *yucca*, regarding them as staple foods among Amerindians. Yet this time, he used both *ages* and *batatas* in the same sentence: "We have sufficiently explained how maize, *ages*, *yucca*, *batatas*, and other edible roots are sown, cultivated and used."[16] A few pages later, Martyr offered a more detailed description of the *ages* and their varieties:

> There are numerous varieties of *ages*, distinguishable by their leaves and flowers. One of these species is called *guanagax*; both inside and out, it is of a whitish colour. The *guaragua* is violet inside and white outside; another species of ages is *zazaveios*, red outside and white inside. *Quinetes* are white inside and red outside. The *turma* is purplish, the hobos yellowish and the *atibunieix* has a violet skin and

white pulp. The *aniguamar* is likewise violet outside and white inside and the *guaccaracca* is just the reverse; white outside and violet inside. There are many other varieties; upon which we have not received any report.[17]

The botanic characteristics of the sweet potato and its various cultivars are consistent with Martyr's description, which was based on reports sent to him by Spanish explorers. However, aside from mentioning *batatas* once, he provided no additional information about the plant or discussed its varieties. What is the relationship between batatas and *ajes/ages*?

The answer to this question might be found in the writings of Bartolomé de Las Casas. Las Casas, a priest who arrived in Hispaniola, Columbus's first European settlement in the New World, in 1502, witnessed the conquistadors' atrocities against Indigenous people. His empathy for the latter prompted him to record the Spanish conquest in detail, resulting in such works as *Brevísima relación de la destrucción de las Indias* (A short account of the destruction of the Indies) and *Apologética historia summaria de las gentes destas Indias* (Apologetic history of the Indies). Meanwhile, Las Casas began writing his *Historia de las Indias* (History of the Indies) as early as 1527, completing the three-volume book in 1561. He, too, copied Columbus's journals and made annotations to them. On November 4, 1492, when Columbus entered his first record of the sweet potato, Las Casas paraphrased the entry by stating that the *niames* that Columbus identified as tasting like chestnuts were "called *ajes* and *batatas* which are very tasty."[18]

Las Casas recorded more than once in his own writings that *ages*—or *ajes* in his own hand, following the usage established in Columbus's journals—and *batatas* were actually two different species:

> There are other roots Indians called *ajes* and *batatas*, which are two species of them; these latter are more delicate and of a more delicate and of more noble nature; they are grown in heaps of plants in the same way as yucca, which is different. The plant of these roots is in the manner of the gourds of our land, but is much more beautiful and delicate; it does not have those like little thorns that the pumpkin plant has, but it is softer, thinner, cleaner and more soft, thin, clean or smooth, and the leaves are the same size, and so smooth, soft and beautiful, like those of the vines of Castile.[19]

Las Casas continued to provide information on how *ajes* and *batatas* were prepared and consumed after describing the botanical characteristics of the two:

> These roots of *ajes* and *batatas*, the syllable in the middle is long, are not poisonous and can be eaten raw, roasted and cooked, but roasted they are better. And for them to be very good, especially the *batatas*, which are of a more delicate nature, they have to be put in the sun for eight or ten days, first sprinkled and even washed with a small amount of brine, more water than salt, and covered on top with rare grass so that the sun does not shine on them, which in fact, those that are to be eaten roasted, placed in the embers of the fire until they are tender, come out covered as if they were taken out of a canning jar; and if you want them cooked, fill a pot with them and put inside a small bowl of water, not to cook them, but so that the pot, being dry at the beginning, does not break, and cover the pot with leaves of the plant, or vines or other good leaves, so that the water does not come out of the pot, and cooking in this way for one, two, or three hours, or whatever it takes, because they do not need much time, until water is soaked and so much honey or syrup comes out, and they are all packed as if they were a preserve, but they are more tasty and better than anything else.[20]

In other words, while *ajes* and *batatas* were two distinct species—according to Las Casas, *batatas* were more delicate than *ajes*—the two could be eaten raw, roasted, and cooked in the same manner. Nonetheless, *ajes/ages/ajíes* and *batatas* represented two sweet potato varieties.

In the sixteenth century, Bartolomé de Las Casas was not the only one who noticed that *ajes* and *batatas*, or sweet potatoes, were common root plants consumed as the primary food by the Indigenous peoples of America. Francisco López de Gómara, the previously mentioned Spanish historian, includes both *ajes* and *batatas* in his *Historia general de las Indias* (General history of the Indies). "There is amber, jasper, chalcidonias, sapphires, emeralds, and pearls," Gómara wrote. "The land is fertile and irrigated; corn, yucca, *batatas*, and *ajes* multiply a lot. The yucca, which in Cuba, Haiti and the other islands is deadly when raw, is healthy here; eat it raw, roasted, cooked, in casseroles or stews, and however it is good tasting." He also noticed that *ajes* and *batatas* were virtually the same plant except the latter was more delicate and sweeter: "The *ajes* and *batatas* are almost the same thing in size and flavor, although *batatas* are sweeter and more delicate. The *batatas* are planted like the yucca, but they do not grow like that, because the branch does not rise from the ground more than the blond one, and it throws the leaf like ivy; they take half a year to be seasoned to be good; they taste like chestnuts with sugar or marzipan."[21]

Gonzalo Fernández de Oviedo, an official historiographer of the West Indies appointed by the Spanish court in 1523, provided another important contemporary source about the New World. Unlike Gómara, Oviedo visited the Caribbean only a few years after Columbus and meticulously recorded what he saw. In his *Historia natural de las Indias* (Natural history of the West Indies), which he began in 1526, Oviedo observed "*batatas* and *ajes* and the other things they [the Indigenous people] use for their sustenance." And, in another place, he also described their botanical characteristics: "There are other plants that are called *ajes* and others that are called *batatas*. They are sown from the slips, and the stem and leaves are more like lesser bindweed or ivy, lying on the ground, and not as thick as ivy leaves, and under the ground they produce roots like turnips or carrots. The *ajes* have a color between blue and purple and *batatas* are browner. Roasted they are both excellent and hearty eating, but *batatas* are better."[22] In other words, when Gómara and Oviedo, the two Spanish historians, encountered sweet potato, the root "sparked immediate enthusiasm," to quote the words of Sophie D. Coe, author of *America's First Cuisines*.[23]

Interestingly, though *ajes* and *batatas* were mentioned in parallel by these Spaniards, it was the latter variety that prevailed en route to Spain, whereas the former gradually fell into oblivion over time. It was also *batatas* that later became the linguistic root of *Ipomoea batatas*, the officially designated term for the plant.[24] Moreover, if this was the case, it appeared to have occurred as early as 1568, when Bernal Díaz del Castillo, a sixteenth-century Spanish conquistador who recalled his experience in the Americas, wrote about it. Following Hernán Cortés, Díaz went to Cuba in 1516 and later joined the Spanish conquest of Mexico and the destruction of the Aztec Empire. A major critic of Gómara for his biography of Cortés, Díaz decided in his later years to offer an eyewitness account of the Spanish conquest, exposing some of the inaccuracies found in Gómara's writing. Díaz's book, entitled *Historia verdadera de la conquista de la Nueva España* (The true history of the conquest of New Spain), centered more on the military aspect, but in one place he does mention sweet potato: "We named this town Santa Cruz, because four or five days before Santa Cruz we saw it; there were in it good apiaries of honey and many *boniatos* and *batatas* and herds of pigs of the earth, which have on their backbone the navel; there were three small villages in it, and this one where we disembarked was larger, and the other two were smaller, each one on a point of the island; it will have about two leagues in length."[25] Instead of *ajes* or *ajíes*, used by his contemporaries in the early sixteenth century, Díaz chose *boniatos*, another Spanish term for both sweet potato and yam still prevalent today, together with *batatas*, to name the root plant.

    The first attempt to use *batatas* as the rubric term to refer to the sweet potato was likely made by Nicolás Monardes in his *Historia medicinal de las cosas que se traen de nuestras Indias Occidentales* (Medical study of the products imported from our West Indian possessions), which first appeared in 1565 and was enlarged and finalized through the subsequent decades. Monardes, a physician and botanist, diligently gathered information about plants from the New World, relying on firsthand accounts from soldiers, merchants, Franciscans, colonial officials, and women. His work was one of the first scientific and comprehensive studies of the American continents' flora. As a result, it was translated into many European languages. In addition to the Latin translation, *Historia medicinal de las cosas que se traen de nuestras Indias Occidentales* was translated into English by John Frampton, an English merchant in Spain, in 1577, bearing the title *Joyfull Newes Out of the Newe Founde Worlde*. Monardes also chose the term *batatas* to describe the sweet potato:

> The Batatas, which is a common fruit in those countries, I do take them for a victual of much substance, and that they are in the midst between meat and fruit, truth it is that they be windy, but that is taken from them by roasting of them, chiefly if they be put into wine being fine: there is made of them conserve very excellent, as of Marmalade, and small morsels, and they make potages and broths, and cakes of them very excellent: they are subject to be made upon them any manner of conserve, and any manner of meat: there be so many in Spain, that they bring from Velez Malaga every year to Seville, ten

or twelve caravels laden with them: They be sown of the same plants that are set, the smallest of them, or pieces of the greatest in the earth that is well tilled, and they grow very well, and in eight months the roots are very gross, that you may eat of them: They be temperate, and roasted, or otherwise dressed, they do soften the belly, and being raw, they are not good to be eaten, because they are windy, and hard of digestion.[26]

Though brief, Monardes's description covered many aspects of sweet potatoes as food, including how and where they were planted, how they were prepared, and how they were consumed in various ways, including noting their flatulent effect. Perhaps the most important information from the book was that just a half century after the plant was brought back to Spain, the sweet potato had become a popular food in several regions of the country, including Monardes's home province of Seville.

The Latin translation of Nicolás Monardes's *Historia medicinal de las cosas que se traen de nuestras Indias Occidentales* was rendered by Carolus Clusius (also known as Charles l'Écluse), who, hailing from France, was another pioneering botanist in the late sixteenth century. Clusius also authored a Latin text, *Rariorum aliquot stirpium per Hispaniam observatorum Historia* (History of the observations of rare plants in Spain) in 1576, which followed Monardes to substitute *batatas* for *ajes* and its cognates in reference to the sweet potato. *Rariorum aliquot stirpium per Hispaniam observatorum Historia* had a chapter on the root plant, in which Clusius said that it had originated in the New World and that the Spaniards called it either *batatas*, *camotes/amotes*, or even *ajes*. And if the three terms represented three cultivars of the sweet potato, like his predecessors, Clusius wrote that *batatas* had longer roots and were softer and sweeter. And Clusius lumped the three under one heading—"De Batatas"—and described how the sweet potato was grown in the field: "It stretches out on the ground like a wild cucumber, with fat enough stems, full of juice and soft, with succulent leaves, green with white, rather like arum, or a bit like spinach leaves. I don't recognize any of the flowers or seeds it produces, but the roots are strong, usually three and four, sometimes bigger than a radish and more blunt on the sides." He also observed that as a root plant, it spread from America and nearby islands to Andalusia—Seville, Cádiz, and Málaga in particular—and other regions of southwestern Spain, for it could not be planted in colder regions.[27]

Figure 2.1 The illustration of *batatas* (sweet potatoes) in Carolus Clusius's *Rariorum aliquot stirpium per Hispaniam observatorum Historia* (Antwerp, 1576), arguably the earliest depiction of the American plant in European literature.

The growing preference for *batatas* to *ajes* was also shown in a contemporary Spanish text *Historia de las plantas de Nueva España* (History of the plants of New Spain), written by Francisco Hernández, a Spanish court physician under Philip II in the sixteenth century who, inspired by Nicolás Monardes, went to Mexico, or New Spain, where he developed a strong interest in Nahuatl medicine and the therapeutic uses of American flora and fauna. Published posthumously in 1615, in *Historia de las plantas de Nueva España* Hernández wrote about the sweet potato under the heading "Camotli or Batata," in which he began by saying that "The herb that the Haitians call *batata*, the Mexicans call *camotli* by the shape of the root, which is the main and most useful part; and although its genera have long since begun to be known to our compatriots, I do not want to omit in this place what refers to its nutritional properties and how to sow and cultivate it." Omitting the name *ajes* altogether, Hernández continued to offer a detailed description of the variety, cultivation, and nutritional value of the root plant as food:

> There are some varieties of this plant, different only by the color of the root (for they all have voluble stems, angular and round leaves, and flowers with white calyxes with purple), which is sometimes red on the outside and white on the inside, and is called *acamotli*; other times the outer membrane is purple and the inside white, and is called *yhoicamotli*; when the exterior is white and the interior is yellow or reddish, it is called *xochicamotli*; there are times when both the interior and exterior are red or completely white, and it is then called *camopalcainotli* or *poxcauhcamotli*, names imposed for many centuries according to the variety of colors. All genera have an oblong root, sometimes bulky according to the nature of the soil, and of the different colors I mentioned before. This root is good to eat raw or cooked, and in various viands that are prepared with it; it tastes very similar to chestnuts, and provides a similar food, good, though crude and suitable to produce flatulence. The stem is, as we said, voluble, cylindrical and thin, drags along the ground, has leaves similar to those of berengena or wild apple tree and color that pulls to purple, and purple flowers, small and oblong. It is sown in the month of August, putting its branches in dug soil, and the root is pulled up and used during the fall, winter and spring. It grows, cultivated, in temperate climate, or also in somewhat cold or

somewhat hot weather, but with more exuberance in cultivated and humid soil.²⁸

Instead of treating *ajes* and *batatas* as two different species, as did Bartolomé de Las Casas and Gonzalo Fernández de Oviedo, botanists Nicolás Monardes, Carolus Clusius, and Francisco Hernández considered them more as two cultivars of the sweet potato. Moreover, Hernández recorded that among the Aztecs in Mexico, the root plant was called *camotli* or *camote*, which, as discussed in chapter 1, has been more commonly used by later scholars to identify its route of transfer to the Pacific and Maritime Southeast Asia by the Spaniards.

The term *camotli* also appeared in Bernardino de Sahagún's stupendous *Historia general de las cosas de la Nueva España* (General history of the things of New Spain), written in the second half of the sixteenth century. A Franciscan friar, Sahagún went to New Spain in 1529 and studied the Aztec language Nahuatl, becoming arguably the most learned European ethnographer of Nahuatl culture and language. Sahagún's work, which is over 2,400 pages in the extant version, was bilingual with Spanish and Nahuatl on opposing folios, in addition to pictorial illustrations. In the volume entitled *Earthly Things*, Sahagún recorded in detail a variety of plants he observed in Mesoamerica. One of them was sweet potato, which he twice associated with the name *camotli*: "It is a root, cylindrical, round, ball-like, twisted. The *camoxalli* is the small *camotli*. Its foliage just creeps like the bean, like the *caxtlatlapan* herb. And as for transplanting, to be propagated, only its foliage, its vine is transplanted. It can be cooked in an olla; it is edible raw. The *camotli* is planted here. Here the foliage is transplanted; here it is grubbed up, here it is cooked in an olla, here it is baked. I eat it raw." In another place, Sahagún's description focused more on the plant's appearance above the ground, which he identified as *Ayauhtona* or *Aiauhtona* in Nahuatl. Then he said that its root was *camotli*, and that "Some are white; their name is *iztac camotli* [or] *poxcauhcamotli*. Some are yellow; their name is *xochicamotli*. Some are blue; their name is *tlapalcamotli*. And the name of those mentioned above is *camoxalli*. They are cookable in an olla; they are edible uncooked."²⁹

These names, which referred to various characteristics of the sweet potato, were similar to those recorded by Francisco Hernández, implying that by the seventeenth century, following the establishment of New Spain, the Spaniards had gained good knowledge of the sweet potato as a food crop in Mexico and regions of Central America, in addition to their early contact with the plant in the Caribbean.

Why *batata*? The sweet potato's appeal to Europeans

From the early sixteenth century, sweet potato was introduced to southern Europe and planted in southwestern Spain and Portugal. "As the sweet-potato needs a warm climate, it never spread to Northern Europe," wrote biologist-cum-historian Redcliffe Salaman, "though it was grown for a short time in some parts of southern France; it was, however, a common import to London, where it was regarded as a delicacy."[30] Indeed, as Spanish scholar Eloy Terrón points out more recently in his *España, encrucijada de culturas alimentarias: Su papel en la difusión de los cultivos americanos* (Spain, crossroads of food cultures: Its role in the diffusion of American crops), the Iberian Peninsula seems to be a suitable place where many New World crops could be successfully transplanted. And the sweet potato was one of the earliest: "Among the first plants to reach the Peninsula were the sweet potato, bell pepper, tomato, corn, prickly pear cactus and various other ornamental or curious plants." The reason, Terrón explains, is that its climate is similar to that of Mesoamerica:

> The climatic diversity of Spain, which ranges from subtropical climate zones, temperate and dry climate zones, mild and humid climate zones, has made possible the gradual acclimatization of tropical crops such as sweet potatoes, prickly pear, peppers, tomatoes, corn, beans and some trees such as cherimoya, avocado, pineapple, papaya and some tinctorial plants, although several of these plants were reduced to the Mediterranean coast and the Ebro delta.[31]

The following pages will discuss the early appeal of the sweet potato to Europeans, especially how and why it was treated as a delicacy when it first arrived in the Old World. This treatment was reflected in both nomenclature and taxonomy. In the process of its transfer from the Americas to Europe, the root plant became identified more as *batatas* and *camote* than *ajes*, the pronunciation Christopher Columbus had learned from the natives when he first saw the plant in the Caribbean. In Spain, sweet potato was also called *patata*, which combined *papa*, the Quechua word for the white potato. (In Quechua, a language spoken by the Incas in what is now Peru, the term for the sweet potato was *kumara* or *khumara*, as discussed in previous chapters.)[32] From the expression *patata* (*batata* plus *papa*) in Spanish came the English derivative "potato," which referred to the sweet potato rather than the white potato in the sixteenth century and most of the seventeenth. Indeed, as explained in the introduction, nomenclatural confusion between white potato and sweet potato persisted in some parts of Europe until the nineteenth century. Nonetheless, *batata* remains the most common term for sweet potato in both Spanish and Portuguese.

According to historian Rachel Laudan, members of the Jesuit order played a significant role in the transfer of knowledge about the flora of the New World to the rest of the world.[33] José de Acosta (ca. 1539–1600), a Jesuit missionary from Spain who traversed North and South America in the late sixteenth century, uses both *batata* and *camote* for the sweet potato in his comprehensive *Historia natural y moral de las Indias* (Natural and moral history of the Indies). Acosta wrote that while the Old World has had carrots, turnip, radishes, and garlic, "in those countries [of the New World] they have so many divers sortes, as I cannot reckon them; those of which I remember besides Papas, which is the principall, there is Ocas, Yanaocas, Camotes, Batatas, Xiquimas, Yuca, Cochuchu, Cavi, Totora, Mani, and infinite number of other kindes."[34] His account shows that it was *batata* and the Nahuatl term *camote* (but not *aji*, as recorded originally in Columbus's journal) that had entered the Spanish language as names for the sweet potato; as is widely known, *batata* also became the basis for the English word "potato."

However, despite its later popularity, the linguistic origin of *batata* is more obscure than that of *ají* and its cognates such as *aha*, *ahi*, *aje*, *age*, *hage*, *haje*, and *napi*. It is generally agreed that *ají* is a Taino word that, according to the ethnobotanist D. E. Yen, mentioned in the previous chapter, remains more or less current in the circum-Caribbean region. Botanists and linguists, on the other hand, disagree on whether *batata* is a Taino or a Chibchan word. Indeed, citing a study by British botanist Isaac H. Burkill, Yen notes that "the origin of the word *batata* is the most obscure of the many names for the plant."[35]

In modern Spanish, *ají* refers instead to chile pepper (*Capsicum*), another New World plant that Spaniards fell in love with when they tasted it on Columbus's first voyage. "Also there is much *axí*," Columbus wrote in his journal, "which is their pepper, and it is stronger than pepper." Bartolomé de Las Casas noted that the Spaniards quickly became "addicts of chile."[36] Francisco López de Gómara, too, wrote about the Spaniards' first encounter with chile pepper: "They tasted *ají*, a spice of the Indians, which burned their tongues, and batatas."[37] In the Taino expression, the word *ají/axí* also had an aspirate, or *haxí*, which was lost when it was borrowed into Spanish.[38] Of course, one could speculate that the Spaniards chose *batata* and *camote* over *aji* to name the sweet potato in their language over time in order to avoid confusion with references to the chile pepper. However, no evidence has been presented to support such speculation.

Though a plausible linguistic explanation for why Spaniards preferred the term *batata* to refer to the sweet potato has been lacking, Isaac Burkill offered his botanic analysis of why *batata* as a food crop appealed to the Spaniards. After reviewing European explorers' early accounts of the Americas, he followed the descriptions of Bartolomé de Las Casas and Gonzalo Fernández de Oviedo, agreeing that *ajes* and *batatas* were two "races" of the root crop. By drawing on other contemporary sources, Burkill stated that if *batatas* were sweeter and more delicate, then *ajes* were more starchy and less sugary. As the "starchy field crop was easily raised in the Indies," according to him, *ajes* were "planted abundantly on all farms and tenements" in Hispaniola, for Columbus and his fellow colonists would like to supply "a basic and cheap food" for the enslaved Indians and Africans—the latter having been brought to the colonies in the early sixteenth century. Burkill then postulated that as a root plant, *ajes* were an "inferior race" to *batatas*; the latter was what the Spanish colonists chose to bring back home. "Writer after writer," he argued, "popularized the name *batata* as belonging to the better races that the Spaniards in the New World took care to have available for their tables while they saw to an abundance of *aji* for the slaves. Having returned home they might be compelled to accept something short of the best, but they were not willing to accept it under the name of *aji*; consequently, every race when raised abroad for the masters was a *batata* and this name soon spread beyond Spain."[39]

In other words, Isaac Burkill believed that the Spaniards preferred the *batata* variety of the sweet potato over the *aji* variety because it was more sugary. That is, the sweet potato spread to Europe beginning in the sixteenth century due to its sweetness. This preference appears to run counter to the experiences of other regions where sweet potato was accepted as a food in later centuries. In her *Sweet Potato: An Untapped Food Resource*, Jennifer Woolfe points out that while a valuable food resource, the consumption of sweet potato has declined in many centuries (China being a conspicuous example, as mentioned in the introduction to this book). One of the causes for this decline, Woolfe observes, is "the high levels of sweetness and the strong flavour associated with many cultivars," which "have reduced its popularity as a staple food."[40] Woolfe's observation appears credible in that the most popular food crops around the world today are corn, rice, wheat, and the white potato—all of which have a plain taste that allows them to become flavor carriers when consumed with non-grain foods like meat and vegetables. Having lost popularity to the white potato a century or so after arriving in Europe in the mid-sixteenth century, the sweet potato has also fallen to seventh place in world production as a food crop in recent decades, overtaken by cassava, another New World crop of which only certain varieties are sweet in taste.[41]

The question is, if *aji* was more starchy and less sugary than *batata*, why did the Spanish colonists not prefer it? In fact, in his article, Burkill offered the following comparison of *aji* and the white potato: "The *aji* is starchy and was suitable for the basic place that the Irish potato now occupies in temperate Europe and other parts of the world; it could be used for bread; the *batata* is sugary." And, he went on to say, *aji* was indeed preferred by Amerindians as a staple food. But unfortunately, Spanish conquistadors "destroyed Hispaniola by 1630 and left it the desert that they had created. With the disappearance of the Indians the local language disappeared, save words useful to the Spaniards who had adopted '*aji*' at first, but let it slip apparently because as they spread from Hispaniola they found its use by the Indians too local."[42]

Burkill's hypothesis is thus as follows: Comparing *aji* and *batata*, the two varieties of the sweet potato, we can see that the former was a low-class food, more suitable than the latter to meeting the basic nutritional needs of the Amerindians and enslaved Africans working on Spanish-owned plantations in Hispaniola. In the wake of the colony's destruction, the variety and its name were forgotten since *aji* had not been favored by the Spaniards in the first place. In his classic study from 1982, *Cooking, Cuisine and Class*, Jack Goody, the Cambridge anthropologist, drew our attention to the difference between cooking and cuisine in class societies. According to his research, which was conducted on a comparative scale covering major societies (Chinese, Indian, Russian, Muslim, English, etc.) across

Eurasia, food preparation and consumption invariably reflect and extend an existing socioeconomic structure, which can either expedite or resist change in the corresponding gastronomy.[43] Since the publication of his work in 1982, many food scholars have extended and expanded on Goody's thesis with additional case studies. In terms of cultural and botanic exchanges following European exploration and colonization of the Americas, the previous section's discussion of the early writings of Spanish navigators, botanists, and historians demonstrates that Europeans were interested in learning about the New World, its flora and fauna, and its native populations. This interest, one could say, contributed to the initial phase of the "Columbian Exchange" described by Alfred Crosby: Early European explorers, beginning with Columbus, brought back samples of seeds, fruits, and plants from the New World. But, as food historian Jodi Campbell states in a recent study,

> Curiously, in spite of the potential transformation inherent in the "Columbian exchange" of new plants, animals, and cooking styles across the Atlantic, and in spite of Spaniards' eagerness to acquire unusual and expensive foodstuffs, they were remarkably resistant to New World cuisines. As their empire expanded into the Americas, they refused to adopt most of the native foodstuffs, even the fruits that grew in such abundance in the tropics.[44]

Campbell's curiosity stemmed from the fact that, following the Columbian Exchange, New World crops such as maize, potatoes, and chile peppers became widely adopted as main foods and essential ingredients all over the world. That is, if Spanish explorers and colonists were "resistant" to the foodstuffs discovered in the Americas, how could the aforementioned foodstuffs, and indeed many others, eventually become accepted and adopted by peoples outside of the Americas? In my opinion, our investigation into the global spread of the sweet potato offers a compelling illustration of this phenomenon. Indeed, as Jack Goody observed in his study of the West African food system, whether a new food could be adopted into a region's and civilization's gastronomy was a class issue. The Spanish preference for the *batata* variety of sweet potato over the *aji* variety, as suggested by Isaac H. Burkill, appeared to be a case in point. *Aji* was too inferior as a food for the Spaniards because it was more of a starch reserved for feeding the Amerindians and Africans on the plantations. In other words, the European selection of New World foods reflected their class consciousness as colonizers vis-à-vis the colonized Amerindians and Africans.

The actual processes of the Columbian Exchange, particularly in its early stages, thus were far more complex than one might imagine. Historian Rachel Laudan observes in her 2013 book *Cuisine and Empire*, a global history of major cooking traditions and their modern permutations, that class awareness was an integral part of Christian cuisine in Europe, as evidenced by the received hierarchical order of foodstuffs. For instance, "Root vegetables such as turnips," she writes, "were by nature earthly (dry and cold) and thus better left to peasants." By comparison, sugar cookery, indebted to the Islamic influence from the twelfth century, "was good." After making contacts with Amerindians, many Europeans regarded the Indigenous peoples as less human and/or "natural slaves described by Aristotle." Consequently, "Europeans recoiled from their foods, including unfamiliar grains such as maize; starchy roots (although they did eat the cassava bread); the viscous, mildly alcoholic pulque prepared from the sap of the agave; and cactus paddles, eaten as a vegetable."[45] Of course, as previously discussed in this chapter, Europeans from Columbus onward tried and ate sweet potatoes while exploring the New World.

As a result of European suspicion of and vigilance against foodstuffs in the Americas, the traditional notion of class difference in food was extended. Granted, the early setbacks these European explorers had experienced, such as Columbus's repeated destructions and reconstructions of Hispaniola, may have given them reason to be concerned about their survival in the New World, as it was a completely new environment to which their bodies were unaccustomed. However, Rebecca Earle has argued that a more fundamental reason originated from Europeans' contempt for Indigenous peoples in the Americas and their wish to maintain their supposedly superior racial and class/caste identities. Most Europeans of the time, states Earle, were genuinely concerned that if they consumed the local foods, then their bodies would also take on the humors of Amerindians and became indistinguishable from the Indigenous people: "You will become like them if you eat their food." That is, for those early settlers, "diet was believed to help *create* the physical differences that separated Europeans from Amerindians and Africans." To overcome the unfamiliarity of the new environment, Europeans (beginning with Columbus) attempted to bring foods from the Old World to the New World, such as wheat and wine. Despite their apparent enthusiasm for learning about the New World, some European historians and botanists, such as Gonzalo Fernández de Oviedo, Francisco Hernández, and José de Acosta, warned others about the dangers of eating native foods. They reasoned that if the environment, including the air, soil, and temperature, shaped one's physical appearance and temperament, then food could help preserve their European identity—"good food trumped climate." To them, wheat bread was the only proper bread, whereas root bread was not. However, their efforts were not always fruitful. Spanish attempts to grow wheat and barley on Caribbean islands, for instance, were unsuccessful.[46]

As many have pointed out, while Europeans appeared fascinated by the flora and fauna of the Americas, they were initially hesitant to incorporate New World crops into their diet. Although maize was one of the earliest New World arrivals, it took nearly a century for the plant to be grudgingly accepted in Europe. Indeed, according to David Gentilcore, the path maize took to enter the European agricultural system was anything but smooth. It went through several stages in which maize's naturalization progressed "from animal fodder, to famine food, food for the poor and, eventually, for the well-off as well." And the white potato, another widely planted food crop in the world today, did not fare much better in its journey to Europe. Gentilcore writes that "the European reaction to the potato" was "characterized by mistrust and suspicion" from the late sixteenth century to the early eighteenth century.[47] Andrew F. Smith, author of *Potato: A Global History*, has gone farther; he observes that the negative views of the white potato "survived in some of parts of Europe almost until the end of the eighteenth century," for several European botanists in the period "were convinced that potatoes were poisonous and caused leprosy, dysentery and other diseases."[48] Larry Zuckerman, author of *The Potato: How the Humble Spud Rescued the Western World*, has gone the farthest; he says that "Over the past four centuries, Western societies have feared, mistrusted, disdained, and laughed at the potato. Even today, we deride it, as in the phrases *couch potato* or *potato head*."[49]

However, there were two notable exceptions to European reluctance to accept New World plants: chocolate and sweet potatoes. It appears that the relatively quick adoption of these two items, as well as some New World fruits like the pineapple, into European diets was due to the fact that both of them shared one common feature—their sweetness appealed to Europeans. As I previously stated, sweet potato—namely, the *batata* variety—was popular among Europeans because it was sweet, as "sweet things were scarce in early Modern Europe," according to Rebecca Earle.[50] With respect to chocolate, it was long believed by many that drinkable sweetened chocolate was a Spanish practice, adopted after they learned of the beverage from Amerindians. But Marcy Norton's recent study reveals that "native Mexicans and Mayans already sweetened many cacao beverages with honey."[51] The attraction of Europeans to sweets was no surprise. In ancient Greece, Aristotle already enjoined that "nothing can nourish the human body unless it participates in some sweetness."[52] Modern scientists have also pointed out that human beings have a natural preference for sweet foods. "Prehistoric hunter-gatherers," writes environmental archaeologist Alan K. Outram, "had something of a sweet tooth. It is no accident that the taste buds that sense sweetness are on the tip of our tongues. Their presence there is an adaptive trait in our evolution." However, as sugar-rich foodstuffs were rare, at least in ancient times, humans depended on carbohydrates, which could be "easily metabolized for energy and sugar."[53] Thus, throughout most of human history and across most cultures, carbohydrates or grain foods have been the primary component of a meal, whereas non-grain dishes are a supplement whose primary function is to aid in the consumption and digestion of the grain, whether wheat, rice, or corn.[54]

Needless to say, a quicker way to sate our sweet-inclined taste buds is to eat sweets. But from ancient times through most of the Middle Ages, sugary foods, while desired, were by and large absent from European gastronomy. "In 1000 A.D.," observes the anthropologist Sydney W. Mintz, author of *Sweetness and Power*, "few Europeans knew of the existence of sucrose, or cane sugar."[55] Jeri Quinzio, author of *Dessert: A Tale of Happy Endings*, points out that while "dessert" is a French word referring to sweet dishes that has now being accepted in many cultures, "until the nineteenth century, the word was simply used to mean dishes that were served after the previous ones had been cleared."[56] As such, the distinction between sweet and salty, or between "savory" and "sweet," was not a matter of course to Europeans until the mid-seventeenth century.[57] Of course, cakes, puddings, and tarts, at least in their proto-forms, had existed centuries before and their sweetness was achieved by adding honey and/or fruit juices, a technique developed and perfected by people in the Middle East. The reason was simply that before the colonization of the Americas, "sugar was rare and expensive, [such that] even members of the top tier of European society had little. Sugar was a spice, a medicine, to be kept locked in the spice cabinet, used with discretion, but flaunted whenever possible."[58] Indeed, it took a long journey for sugar to travel from India to Persia, the Middle East, and the Mediterranean before reaching Europe from the eleventh century. It was after the Crusades and the conquest of Jerusalem that Europeans were first introduced to sugar, which was noted as "precious" and "very necessary for the use and health of mankind." From then on, there emerged an ever-growing demand for sugar among monarchs and upper-class nobles in Europe.[59] A study of a fifteenth-century Portuguese cookbook reveals that many recipes, targeted clearly for the rich, used the spice in a variety of dishes, suggesting the robust interest of Europeans in the gastronomical and medicinal efficacy of sugar when it became more available to them.[60] The same level of enthusiasm for cooking dishes with sugar continued in the following century, as shown in Cristoforo di Messisbugo's (?–1548) *Banchetti*, a cookbook printed posthumously in 1549, which recommended seasoning almost everything with sugar.[61] Europeans' interest in sweets was also reflected in their preference for sweetened drinks. In Tudor England, for instance, not only was mead, made solely of water and yeast sweetened by honey, "an extremely popular drink," but wine, usually from France, Spain, Madeira, or Greece, was also "strong and sweet" (with added sugar), most favored by the royal court.[62] Having examined a number of dietary texts in the Renaissance era, food historian Ken Albala concludes by paraphrasing the words of Platina (Bartolomeo Sacchi, 1421–1481), Messisbugo's predecessor who authored *De honesta voluptate et valetudine*, the first ever printed cookbook of 1474, that "sugar was considered a universal condiment, suitable for flavoring all foods."[63] While European colonization of the Americas increased Europeans' access to sugar, Muslim and Asian influence persisted—words like "syrup," "candy," "sherbet," and, indeed, "sugar" itself were Arabic and/or Sanskrit imports.

It was no coincidence, then, that following the European "discovery" of the New World, colonists pursued an obvious interest in both sugarcane and sweet potato: For the former, they established plantations throughout the Caribbean to meet the high demand for sugary foods back in Europe, and for the latter, they exhibited a preference for *batata*, the sugary variety of the sweet potato, as opposed to the starchy *aji*. As previously discussed, early Europeans were skeptical and hesitant to embrace flavorless starches such as white potato and maize. In a nutshell, concludes the historian Rebecca Earle, "a European fondness for sweet things" explains the early "popularity of the sweet potato."[64]

Exotic as aphrodisiac: Chocolate and sweet potato

"All (or at least nearly all) mammals like sweetness." Thus writes Sydney W. Mintz in his influential *Sweetness and Power*, which examines the history of sugar production and consumption in the modern world. Mintz notes that while this natural disposition is scientifically proven—that human milk "is sweet is hardly irrelevant," "sugar (sucrose) from the sugar cane is a late product that spread slowly from the first millennium or so of its existence, and became widespread only during the past five hundred years."[65] He is here referring, of course, to the establishment of sugarcane plantations in the Caribbean and Americas in the wake of European colonization, which made sugar increasingly more available from the sixteenth century on.[66] Indeed, Christopher Columbus was known for his keen interest in sugar. Before his voyages, he had visited Madeira to purchase sugar, and if he can be credited with the "discovery" of the Americas, then one of his "discoveries" was his realization that the Caribbean islands he had visited could grow sugarcane. Therefore, "On his second voyage to the Caribbean in 1493, Columbus stopped in the Canary Islands and picked up seed cane, which he introduced to the Caribbean Island of Hispaniola." Due to his pioneering effort, Hispaniola became "the most important New World sugar producer" of the time.[67] Following the example of Columbus, Spanish and Portuguese explorers established other sugar plantations throughout Central and South America. According to López de Gómara's 1553 *Historia general de las Indias*, Spanish colonists introduced a variety of vegetables to Hispaniola, such as radishes, lettuce, onions, parsley, cabbage, carrots, turnips, and cucumbers. But what they desired most was sugar. "Sugar has increased mightily," wrote López de Gómara, "and there are thirty factories and sugar mills. Pedro de Atienza was the very first Spaniard who planted sugar cane and Miguel Ballestero, a Catalan, the first who harvested it."[68] To meet the high demand for sugar production, European conquistadors not only used Amerindians, but also brought in large numbers of African slaves to work on the plantations.

Many of us believe that without sugar, there would be no chocolate. Chocolate, as many have noted, was one of the early food products from the New World that had been adopted in European gastronomy from the sixteenth century onward.[69] But the way most people consume chocolate today—chewing cacao butter rather than sipping chocolate drink—did not begin until the early nineteenth century.[70] In other words, the consumption of chocolate in Mesoamerica, its birthplace, by Amerindians differed markedly from how it is consumed today. Mayans and Aztecs made beverages out of chocolate mass, prepared by roasting, grinding, and fermenting the seeds of cacao trees, what we call cacao beans. Moreover, chocolate beverages made in Mesoamerica were not necessarily sweet, though some were mixed with honey. Christopher Columbus himself was the first European to record the consumption of this chocolate drink. On December 22, 1492, during his first trip, Columbus wrote in his diary that "they [Amerindians] threw a grain into a porringer of water and drank it, and, said the Indians whom the Admiral took along, this was a most wholesome thing."[71]

José de Acosta, in his *Historia natural y moral de las Indias*, offered a more detailed description:

> The chief use of this cacao is in a drink which they called chocolaté, whereof they make great accompt in that Country, foolishly and without reason, for it is loathsome to such as are not acquainted with it, having a skumme or froth that is very unpleasant to taste.... Yet it is a drink very much esteemed among the Indians, wherewith they feast noble men as they passe through their Country. The Spaniards, both men and women, that are accustomed to the Countrey, are very greedy of this chocolaté. They say they make diverse sortes of it, some hote, some colde, and some temperate, and put therein much of that chili; yea they make paste thereof, the which they say is good for the stomacke and against the catarrh.[72]

Acosta's observations were notable in two ways. One was that he had a negative and suspicious attitude toward the drink, which was shared by the majority of his Jesuit contemporaries at the time, despite the fact that many of his countrymen had grown accustomed to it. The other was that, in accordance with Amerindian custom, he described chocolate being consumed in the form of a drink, which was typically cold, colored, and spiced.[73] The Amerindians often added chile as well as vanilla, another New World spice, to the drink to offset the bitterness of the cocoa beans. Another way to combat bitterness was to add honey, but Amerindians preferred to use spices.

By the end of the sixteenth century, chocolate drink had become common in southern Europe, in parts of France and Italy as well as Spain. During the period, while Europeans kept the New World spices for flavoring, they also learned how to sweeten it so as to satisfy their desire for sweets. They did it "at first with honey, and later with sugar." As noted before, the drinking of sweetened beverages was already an entrenched practice in European high society. Yet the substitution of honey with sugar took place when the chocolate drink spread beyond the elite. According to Andrew F. Smith, author of *Sugar: A Global History*, the first treatise on cacao was written by the Spanish physician Antonio Colmenero in 1631. Colmenero's recipe for the drinkable form blended New World ingredients with sugar for sweetness: "Take a hundred cacao kernels, two heads of Chili or long peppers, a handful of anise or orjevala, and two of mesachusil or vanilla—or, instead, six Alexandria roses, powdered—two drachms of cinnamon, a dozen almonds and as many hazelnuts, a half pound of white sugar, and annotto enough to color it, and you have the king of chocolates." Over time, Smith observes, Europeans gradually dropped many exotic spices (except vanilla) from the New World, but retained the practice of sweetening the chocolate beverage, for which they used sugar.[74] By 1700 or so, states historian David Gentilcore, "drinking chocolate of the sugar-and-vanilla variety" had become standard among European aristocrats, especially in France.[75] This was not surprising given that Europeans also liked sweetened tea and coffee, the other two popular drinks at the time. "What is interesting about tea, coffee, and chocolate," observes Mintz, "is that none had been used exclusively with a sweetener in its primary cultural setting."[76]

If Europeans in the early modern period liked their chocolate drink sweet, their sweet tooth also attracted them to the sweet potato, because *batata*, the variety preferred by Europeans at the time, is naturally sweet. On the other hand, while cocoa beans are typically unsweet, one variety, criollo cocoa, produced in what is now Venezuela, is sweet and was in high demand in Spain at the time.[77] However, the association of sweetened chocolate and the sweet potato, two seemingly unrelated food plants, went beyond that. Thomas Gage (ca. 1603–1656), an English Dominican friar who traveled widely in the Americas, recorded an interesting experience upon his arrival: In a procession to a cathedral, his group's supervisor "entertained us very loving with some sweetmeats, and everyone with a Cup of the Indian drink called chocolate."[78] He did not name the sweetmeat, but it certainly could have been made of sweet potato.

In general, it appeared that sweet potato and chocolate formed a relationship in three particular ways during the European encounter with New World foods. First, when Spaniards encountered the two plants, they were prepared as food and drink by Amerindian women. From the earliest records of both Columbus and Dr. Diego Álvarez Chanca, who accompanied him on his second trip, local food and women were often mentioned together. Both Hernán Cortés (1485–1547) in his letters to the Spanish court and Bernal Díaz del Castillo, his associate, in his book *Historia verdadera de la conquista de la Nueva España*, frequently mentioned that Amerindian women made bread, cooked food, and delivered fruits for the Spaniards.[79] Drawing on those sixteenth-century accounts, Rebecca Earle states that if Spaniards learned how to cook and consume native foods after landing in the Americas, then "the vector of transmission was undoubtedly Meso-American women." And, she continues, it is from Amerindian women that Spanish women began to learn how to prepare delicious dishes from "sweet potatoes and other new-world ingredients."[80] Sergio Ramírez concurs by stating that "indigenous women introduced unknown staples to the Spanish palate."[81] The association between chocolate drinks and Amerindian women was also widely observed. "Across Mesoamerica," write Sarah Moss and Alexander Badenoch in their *Chocolate: A Global History*, "chocolate was usually prepared by women, although often consumed by men." Marcy Norton points out in her research that Spanish colonists learned to adjust themselves to consuming American foods and drinks because they had a material dependence on Amerindians, and that these "cross-cultural contacts flourished in intimate settings, some voluntary, and others coerced. Both a drastic shortage of Spanish women and a conscious and explicit strategy of appropriation and conquest through matrimony led to many marriages, as well as less formal domestic unions, between Indians and Europeans in the early sixteenth century."[82] Yet Spanish men were not the only ones who were acculturated by such contacts. Moss and Badenoch state that after exposure to the drink, Creole and European women were in fact more receptive to chocolate than their male counterparts, a phenomenon that had already been noticed and, in fact, bemoaned by José de Acosta, the Jesuit father, in his *Historia natural y moral de las Indias*.[83] Yet there were others who wrote positively about the women's attraction to the drink. In a report by an envoy sent by the Medici court, it was claimed that "Chocolate is something all the most noble Spanish ladies boast about preparing, spending as much time making it as they do in drinking it, . . . chocolate is sipped in the most elegant homes."[84] Apparently, such enthusiasm also went beyond "elegant homes." Citing a contemporary source, David Gentilcore tells us that in the early seventeenth century, sixty nuns from a convent in Mexico City spent 2,916 pesos on the chocolate drink and only 390 pesos on poultry, eggs, and wine.[85]

Second, after being brought from the Americas to Europe in the sixteenth century, chocolate and sweet potatoes satisfied people's palates in two ways: as a drink and as a delicacy—a kind of confection. And, interestingly, they were both thought to have aphrodisiac properties at the time. Bernal Díaz del Castillo was probably the first to mention this, which he did in his *Historia verdadera de la conquista de la Nueva España*. Describing Hernán Cortés's campaign in Mexico, he wrote, "They [Aztecs] brought him . . . all the fruits in the land, but he ate but very little, and from time to time they brought some cups of fine gold, with a certain drink made of the same cocoa, which they said was to have access with women; and then we did not look at it; but what I saw, that they brought about fifty large jars made of good cocoa with its foam, and of what he drank."[86]

Third, from José de Acosta's complaint against chocolate, one could also detect somewhat of a hint of such propensity, even though it lacks scientific proof. Acosta wrote that chocolate "disgusts those who are not used to it, for it has a foam on top, or a scum-like buddling. . . . And the Spanish men—even more the Spanish women—are addicted to it." Chocolate's reputation as an aphrodisiac might be from Spanish observers contemporaneous with Acosta, who recorded that the Aztec elite consumed the drink "for success with women."[87]

But this reputation remains a legend. In writing her *The True History of Chocolate*, which was completed after her death by her husband, Michael D. Coe, Sophia D. Coe strongly disputed the claim. She pointed out that in conquering the Americas, Spanish conquistadors searched avidly for both aphrodisiacs and laxatives, and that the first one who assigned the aphrodisiac attribute to chocolate was probably Francisco Hernández, the sixteenth-century Spanish court physician mentioned earlier, in his

Historia de las plantas de Nueva España. Though the evidence provided by Hernández, according to Coe, was far from convincing, she acknowledged that "the probably baseless claim that chocolate has aphrodisiac properties was one that was to arise again and again in Europe, and obviously also appeals to modern authors." In other words, while chocolate's reputation as an aphrodisiac has somewhat persisted to this day, Coe states unequivocally that "the reader should stop to consider if there has ever been a consumable substance that has not had this reputation at some time in some place."[88]

Compared with chocolate, few see the sweet potato as an aphrodisiac today. But back in the sixteenth and seventeenth centuries, when the root plant first landed and spread in Europe, it was widely seen as having an aphrodisiac effect, or the ability to "incite to Venus," by many authors. As mentioned at the outset of this book, William Shakespeare was a noted example. Not only does his character Falstaff reference the potato, but he also has Thersites delivers the following praise in *Troilus and Cressida*: "How the devil luxury, with his fat rump and potato-finger, tickles these together! Fry lechery, fry!" Historian Redcliffe Salaman, who quoted these two passages, observed that "In both cases the potato is regarded not as a food but solely as an aphrodisiac."[89] Shakespeare's belief in the sweet potato's effect was shared by John Fletcher, his emulator. The latter described the sweet potato as an aphrodisiac in two of his plays. One of which reads as follows:

> A Banquet!—Well! Potatoes and eringoes
> And, I take it, cantharides! Excellent!
> A priapism follows; and as I'll handle it
> It shall, old lecherous goat in authority.

The other contained the following lines:

> Will your Lordship please taste a fine potato?
> 'Twill advance your wither'd State.
> Fill your Honour full of noble itches.[90]

The sweet potato's reputation as an aphrodisiac, however, could have originated in Spain, where the plant first gained popularity. Citing

Estebanillo González, a popular novel in mid-sixteenth-century Spain, Eloy Terrón observes that "In Spain, as well as in England, as in other European countries, the widely known tuber or root was the sweet potato, and for reasons not exactly nutritional; it was a popular belief that the sweet potato had aphrodisiac virtues and that it provoked sexual desire." Estebanillo González, the novel's eponymous protagonist, was supposed to have said, upon embarking on a trip to Málaga, that " I went to the promontory of the raisin and almond and to the potato sea."[91] The potato mentioned here should be referred to as the sweet potato. Málaga, a major area in Andalusia, was the region where the sweet potato had been grown in large quantities and where the locals, according to contemporary accounts, planted it like turnips from the 1580s.[92]

Over the course of the late sixteenth to early seventeenth century, the sweet potato came to be widely regarded as an aphrodisiac in Europe. In his famous cookbook *The Good Huswife's Jewell*, written in 1585, Thomas Dawson recommended that sweet potato could be used "to make a tart that is a courage to a man or woman."[93] Then in 1577, the *Description of England* by William Harrison, an English clergyman, referred to "the potato, and such venerous roots as are brought out of Spain, Portugal and the Indies to furnish up our banquets."[94] Gaspard Bauhin, a Swiss botanist, was even more specific. In his *The Prodromos*, which appeared in 1619, he wrote that "Our own people sometimes roast them [sweet potatoes] under embers in the manner of Tubers and having taken off the cuticle eat them with pepper: others having roasted them and cleaned them cut them up into slices and pour on fat sauce with pepper and eat them for exciting Venus, increasing semen: others regard them as useful for invalids since they believe them to be good nourishment."[95] Toward the end of the seventeenth century, an English sailor known as J. H. described a sea trip to the Bermudas and the fruits and plants he encountered there. He wrote enthusiastically about such exotic fruits as pomegranate, papayas, and fig trees, as well as sweet potatoes: "Here's roots as well as trees, potatoes good for sustenance of man to make pure blood."[96]

Like chocolate, however, the idea that sweet potatoes have aphrodisiac properties is a myth that lacks scientific evidence. Indeed, according to Felipe Fernández-Armesto, author of a global history of food, "although eating and sex seem to be complementary, mutually lubricating forms of sensuality, every particular aphrodisiac is a kiss in the dark. None has anything that could remotely be called scientific endorsement."[97] The most likely reason for the two New World plants acquiring such a reputation was their comparable exoticness, which made them luxuries only for the upper class at the time of their introduction to Europe. This is the third characteristic that the two different American food plants shared in their initial acceptance among Europeans. With regard to the luxuriousness of chocolate as a drink, it had begun with how it was consumed by the Aztecs in Mesoamerica. As a desirable beverage, it was reserved for warriors and nobility, and in Aztec society it had already been seen as an "exotic, luxurious product."[98] That it was usually prepared by high-ranking Aztec women also added to its magnificence, which was passed on to Spanish settlers in the Americas and then in Europe. Referring to Europeans' attitudes toward New World products, historian David Gentilcore writes that "foreignness was viewed as a sign of luxury, decay and moral weakness,"[99] which perhaps helped enhance the mysterious appeal of chocolate. Indeed, writes Irene Fattacciu, "when cacao arrived to Spain its circulation remained for a long time limited to the court and the nobility, and the incredible success gained was the other main reason for which it was so important to determine its qualities/properties."[100] A similar fate befell the sweet potato's early reception in Europe. Regarding the sweet potato's perceived properties, Redcliffe Salaman provided a compelling analysis that linked its alleged aphrodisiac effect to its status as a foreign and rare food: "The idea would seem more likely to have arisen from the fact that the employment of aphrodisiacs and the circumstances in which they were indulged, were only available to the wealthy; such agents were expensive. A rare and costly luxury, especially when imported from abroad, would for these reasons, if no other, tend to acquire exotic attributes."[101] That is, exoticism elevated both chocolate and sweet potato to the status of luxury item. Not only were tea, chocolate, and coffee newly discovered imports from overseas in early modern Europe, but the root plant as a food and its sweet taste also suggested a foreignness among European consumers. Examining dietary texts from the Renaissance period, as experts like Ken Albala have done, we can see that it was not uncommon for Europeans at the time to attribute aphrodisiac properties to rare (and thus expensive) foods like the sweet potato. Truffles, for example, with which both potatoes and sweet potatoes were compared or even confused in those texts, had long been regarded as an aphrodisiac due to their scarcity. "Renaissance physicians," comments Albala, "recognized the allure of exotic delicacies and often explicitly connected perverse tastes in food with sexual license."[102]

It was also no surprise, from a dietary and nutritional standpoint, that sweet potato became an aphrodisiac during this period. Sweetness, as perceived by dieticians of the time, suggested the hot and moist nature of the food, which was considered nourishing to the heart and body by most European dieticians during the medieval and early modern periods. In addition, nourishment increased sexual drive and desire. "Sex is safe," writes Albala, "only when there is an abundance of heat and moisture in the body and a plethora of blood and sperm." Sweet potatoes, whether boiled or roasted, are naturally hot and moist, in addition to being sweet. Honey, for example, has long been thought to be an aphrodisiac. Aside from honey, there was a long list of foodstuffs that the Europeans of the time believed had aphrodisiac properties. These included common foods like garlic, leeks, capers, chickpeas, quail, pine nuts, mint, eggs, parsnips, oysters, turnips, asparagus, celery, cloves, carrots, as well as unusual ones like truffles, saffron, and galangal. These foods, whether hot or cold—but mostly hot—helped nourish the body, increasing virility. It is worth noting that, in addition to being hot and moist, sweet potato causes flatulence, which was noted almost immediately in the earliest descriptions of the root by European botanists and nutritionists. Windy foods were also considered "useful for sex" at the time, which reflected an old belief dating back to Galen of Pergamon (129–216), the Greek physician who had a profound influence on dietary writings throughout early modern Europe.[103]

Finally, while the exoticism of the sweet potato and chocolate increased their appeal to sixteenth- and seventeenth-century Europeans, it also prevented them from being integrated into the existing gastronomic system. As previously stated, many food experts have claimed that European settlers and consumers generally shied away from the foodstuffs provided by Indigenous peoples in the New World. For the early Spanish colonists, it appeared that they shared a common belief, derived from their sense of racial and cultural superiority, that, in the words of Bernardino de Sahagún, "you will become the same way if you eat their food." After colonizing the Americas, the Spanish and Portuguese settlers made numerous efforts to introduce European food crops and maintain their dietary customs in the New World. Rachel Laudan, revising Alfred Crosby's thesis about the "Columbian Exchange," believes that even if there were exchanges between the Old and New Worlds, Europeans did not show much interest in

Amerindian culinary traditions. As a result, "European and Asian food-processing and cooking technology was transferred wholesale to the New World," she observes, "while essentially no American food-processing technology was moved to the Old World."[104] Rebecca Earle concurs in her detailed analysis of European attitudes toward native foods in the Americas, stating that while European settlers quickly realized that Amerindian bread was made of different food crops such as maize and cassava, they remained firmly convinced that wheat bread was the "proper bread" and that "bread made from roots" was not. European religious faith also shaped and reinforced the belief that "only wheat bread could undergo transubstantiation." And Amerindians "fabricated their breads" from maize, cassava, and other food materials that differed from wheat "in this crucial way," thereby rendering them incapable of becoming "the body of Christ."[105]

All of this helps to explain why chocolate and sweet potato, while adopted relatively quickly as drinks and foods by Europeans, were to serve as supplements rather than staples in European gastronomy. As previously stated, Europeans were drawn to chocolate drinks and sweet potatoes at the time due to their sweetness—the former often thanks to artificial additives, first honey and later sugar, and the latter naturally, particularly in the *batata* variety. Gregorio Saldarriaga, a historian of modern Colombia, writes that Europeans regarded the sweet potato as "a proper food for indigenous people because it is a tuber. Nevertheless, they appreciated its sweetness early on and deemed it suitable for their consumption." And, continuing his analysis, "the sweet potato could be a good food for Spaniards, not only because of its sweetness, but because it was *not* an everyday food, as it was for black people and indigenous people, but a complement to the diet."[106] Indeed, observes Rachel Laudan, there was a "layered cuisine" in New Spain: Europeans and their descendants mostly dined on foods whose ingredients were largely transferred from Europe, whereas the Indigenous population continued their own dietary traditions with foodstuffs from local sources.[107] Of course, foods like the sweet potato were consumed in New Spain and, as a result, were transferred from the colony to Europe. Yet the introduction of the sweet potato, which had been consumed as a staple in regions across the Americas, to Europe was not intended to shake the received traditional and hierarchical foundations of European gastronomy; the root plant, though treated as a luxurious delicacy at times, was to be consumed primarily as a sweet dish that complemented the main meal, or, in William Harrison's words, to "furnish up our banquets." But over time, it assumed a wider use.

Sweet potato: Between meat and fruit

Take two Quinces, and two or three Burre rootes, and a Potaton, and pare your Potaton, and scrape your rootes and put them into a quart of wine, and let them boyle till they be tender, & put in an ounce of Dates, and when they be boyled tender, Drawe them through a strainer, wine and all, and then put in the yolkes of eight Egges, and the braines of thrée or foure cocke Sparrowes, and streine them into the other, and a litle Rose water, and seeth them all with Sugar, Synamom and ginger, and Cloues and mace, and put in a litle sweete butter and set it vpon a chafingdish of Coles betweene two platters, and so let it boyle till it be something bigge.

The above was a recipe offered by Thomas Dawson in *The Good Huswife's Jewell* for a sweet potato tart, which, as mentioned before, would supposedly be "a courage" to both men and women.[108] Written in the late sixteenth century, Dawson's book was one of the first and most well-known cookbooks in early modern Europe. This sweet potato recipe, which refers to the tuberous root by way of the term "potaton," has also been identified as the very first one by a European cook detailing how to prepare the New World food plant—in this case as a type of sweet.[109]

Thomas Dawson lived in England during the reign of Queen Elizabeth I (1558–1603). As discussed previously, the early part of the century had witnessed the initial spread of the sweet potato from New Spain in the Americas to southern Europe. According to Jeffrey L. Forgeng, author of *Daily Life in Elizabethan England*, sweet potato, called simply "potato" at the time, was first introduced to the British Isles under Elizabeth I. And "they were an extremely expensive delicacy!" Resonating with what we discussed above, Forgeng also notes that during the time, people of means in the country liked "to indulge in a variety of sweet foods, including gingerbreads and cakes, candies, marzipan, conserves, and marmalade—ordinary people probably had such foods only on special occasions, such as holydays."[110] Aside from the sweet potato tart, Dawson offered a variety of confections in *The Good Huswife's Jewell*, including cake, trifle, pie, pudding, cheese, cream, and other types of tarts. It should also be noted that, while people desired sweet foods at the time, dessert as a type of food or the custom of eating dessert after a meal did not appear to be common, as Dawson also used animal meat such as hare and veal to make either puddings, pies, or tarts.

The sweet potato recipe found in Thomas Dawson's sixteenth-century cookbook was also extraordinary for another reason: Given the "newness" of the sweet potato to Europeans at the time, many of his countrymen in Britain, as well as most dieticians in continental Europe, were unfamiliar with the root. One such example was William Vaughan, who wrote a popular health book entitled *Directions for Health Both Naturall and Artificial* in 1600. Though a learned person, Vaughan was oblivious to New World foods.[111] Another example is *The Closet of the Eminently Learned Sir Kenelme Digbie Kt. Opened*, attributed to Kenelme Digby, a prominent English courtier of the time. Printed as late as 1669, the text offered hundreds of recipes of traditional English dishes and pastries, including some of the foods Sir Digby might have learned from his travels in continental Europe. However, potato, or sweet potato, failed to enter the text as an ingredient for either a dish or a pastry. The *Closet Opened* did list a pretty long list of "Roots," such as Alexander, angelica, asparagus, beet, betony, bittersweet, eryngo, fennel, fern, galangal, horseradish, marshmallow, parsley, and strawberry, most of which were recommended for use as spices in cooking a dish.[112] The absence of the sweet potato, as well as the white potato—which was also being introduced to Europe during the same period—as an ingredient among the recipes show that both plants remained somewhat untraditional in English gastronomy.

However, people in Elizabethan England were apparently becoming aware of the sweet potato as it began to be imported from southern Europe. Dawson's contemporary William Harrison correctly identified sweet potato as being from Spain, Portugal, and the West Indies in his encyclopedic account of Elizabethan England, referring to it as a "venerous roots" comparable to sweet cicely.[113] Another important English text from the period that revealed Europeans' knowledge about the sweet potato was John Gerard's *The Herball, or Generall Historie of Plantes*, which was first published in 1597 in a massive edition of nearly 1,500 pages, complete with fancy woodcuts illustrations, and soon became the most prevalent botany text in seventeenth-century England. Gerard's work may have synthesized European scholarship on botany as an emerging discipline of the time, as many now believe that it was an unacknowledged translation of the book with the same title by Flemish botanist Rembert Dodoens. In contrast to his predecessor William Harrison, Gerard stated in the *Herball* that there were two types of potatoes, or "potatus." The first was simply named potato, or "common potato," which actually referred to the sweet potato, judging by

Gerard's own description and the work of later food scholars.[114] With regard to its origin, Gerard identified at least three—"India [the West Indies], Barbarie [North Africa] and Spaine and other hot regions," but he also pointed out that some referred to it as "Skyrrets [skirrets] of Peru." His emphasis on sweet potato as a plant from "hot regions" was based on his own experience of planting the sweet potato in his own garden, "where they flourished until winter, at which time they perished and rotted." He also talked about how to propagate the sweet potato: "[its] being divided into divers parts and planted, do make a great increase, especially if the greatest roots be cut into divers goblets, and planted in good and fertile ground."

Gerard offered detailed descriptions of the propensities of the sweet potato, covering not only how the root should be planted but also describing its nutritional value as a food:

> The Potato roots are among the Spaniards, Italians, Indians, and many other nations, ordinarie and common meat; which no doubt are of mighty and nourishing parts, and doe strengthen and comfort nature; whose nutriment is as it were a mean between flesh and fruit, but somewhat windie; yet being rosted in the embers they lose much of their windinesse, especially being eaten sopped in wine.
>
> Of these roots may be made conserves no lesse toothsome, wholesome, and dainty, than of the flesh of Quinces; and likewise those comfortable and delicate meats called in shops, *Morselli, Placentulæ*, and divers other such like.
>
> These roots may serve as a ground or foundation whereon the cunning Confectioner or Sugar-Baker may worke and frame many comfortable delicat Conserves and restorative sweet-meats.[115]

Gerard saw sweet potato as an ingredient in the preparation of a sweet delicacy. Indeed, as historian Alison Sim explains, Gerard's comparison of the sweet potato to quince, a tree fruit of the genus *Cydonia* in the Rosaceae family (e.g., apple and pear), was a high compliment, because "Quince preserves were greatly prized at the time."[116]

Another early English record of the sweet potato was found in John Parkinson's *Paradisi in sole paradisus terrestris* (Park-in-sun's terrestrial paradise) published in 1629. A prominent herbalist of the age who was later appointed royal botanist under King James I (1566–1625), Parkinson examined the potatoes he encountered and categorized them, according

Figure 2.2 The illustration of sweet potatoes in John Gerard's *Herball* (London, 1597), labeled "potatoes." This indicates that sweet potatoes arrived in Europe before white potatoes, leading to a naming confusion that persisted into the late seventeenth century. The descriptor "sweet" was later added to differentiate the two.

to their origins, into three different varieties: that of Spain, Virginia, and Canada. In doing so, however, Parkinson made a few mistakes. The "potato of Canada" was actually Jerusalem artichoke, though he insisted that "potato of Canada" was a better name. (Incidentally, a native root plant in North America, "Jerusalem artichoke" is itself a misnomer because it has nothing to do with Jerusalem, nor is it an artichoke.) Nonetheless, all three "potatoes" received detailed descriptions in his *Paradisi in sole paradisus terrestris*, which was a useful compendium on the cultivation of plants in what Parkinson called the three gardens: the flower garden, the kitchen garden, and the orchard garden. About the sweet potato, he provided the following description:

> The Spanish kinde hath (in the Islands where they growe, either naturally, or planted for increase, profit, and vse of the Spaniards that nourse them) many firme and verie sweete rootes, like in shape and forme vnto Asphodill rootes, but much greater and longer, of a pale browne on the outside, and white within, set together at one head; from whence rise vp many long branches, which by reason of their weight and weaknesse, cannot stand of themselues, but traile on the ground a yard and a halfe in length at the least (I relate it, as it hath growne with vs, but in what other forme, for flower or fruit, we know not) whereon are set at seuerall distances, broad and in a manner three square leaues, somewhat like triangled Iuie leaues, of a darke greene colour, the two sides whereof are broad and round, and the middle pointed at the end, standing reasonable close together: thus much we haue seene growe with vs, and no more: the roote rather decaying then increasing in our country.

From this description, it seems that Parkinson had grown the sweet potato in his garden. The following passage from his book shows that he also prepared and consumed it:

> The Spanish Potato's are roasted vnder the embers, and being pared or peeled and sliced, are put into sacke with a little sugar, or without, and is delicate to be eaten.
>
> They are vsed to be baked with Marrow, Sugar, Spice, and other things in Pyes, which are a daintie and costly dish for the table.

The Comfit-makers preserue them, and candy them as diuers other things, and so ordered, is very delicate, fit to accompany such other banquetting dishes.[117]

Like his predecessor John Gerard, Parkinson identified another potato from Virginia. Let us first look at the description offered by Gerard. In his *Herball*, he argued that the potato shared some similar propensities with the "common potato," or the sweet potato from West Indies and Spain. He wrote, "The vertues be referred to the common Potato's, being likewise a food, as also a meat for pleasure, equall in goodnesse and wholesomnesse to the same, being either rosted in the embers, or boiled and eaten with oile, vineger and pepper, or dressed some other way by the hand of a skilfull Cooke." Gerard also said that "It groweth naturally in America, where it was first discovered, as reporteth *Clusius*, since which time I have received roots hereof from Virginia, otherwise called Norembega, which grow & prosper in my garden as in their owne native country."[118] His experiment in growing the potato successfully in his garden further suggested that it might be the white potato rather than the sweet potato, as the former can endure the winter better than the latter.

John Parkinson, too, compared the "Virginia potato" with the sweet potato from Spain, noting that "The Potatoes of Virginia, which some foolishly call the Apples of youth, is another kinde of plant, differing much from the former, sauing in the colour and taste of the roote. . . . The rootes are rounder and much smaller then the former, and some much greater then others, dispersed vnder ground by many small threads or strings from the rootes, of the same light browne colour on the outside, and white within, as they, and neare of the same taste, but not altogether so pleasant."[119]

Notably, when Gerard and Parkinson introduced their readers to the American food plants, John Forster, a contemporary whose background is unknown, was a strong supporter of their benefits. In 1664, Forster published *Englands Happinesse Increased, Or a Sure and Easie Remedy Against All Succeeding Dear Years by A Plantation of the Roots Called Potatoes*, in which he identified not three, but four varieties. Aside from the potatoes of Spain, Virginia, and Canada (similar to Parkinson, Forster stated that potatoes of Canada were falsely called "Artechocks of Jerusalem" in English), Forster mentioned a fourth type called Irish potatoes, which he said were "a little different from those of Virginia."[120]

According to the writings of John Gerard, John Parkinson, and John Forster, the "Virginia potato" thus was most likely the white potato. Alison Sim writes in *Food and Feast in Tudor England* that the "Virginia potato" featured in Gerard's and Parkinson's books "is the one familiar to us today."[121] However, Virginia was not yet established as an English colony when Gerard's *Herball* was first printed at the end of the sixteenth century. When the colony was founded, in 1607, the settlement was not permanent until a few decades later, or around the time Parkinson's work appeared. Was the potato they described really from Virginia? The author of *A History of English Food*, Clarissa Dickinson Wright, is skeptical. While Sir Walter Raleigh, the two botanists' contemporary in Elizabethan and early Stuart England, was credited with English colonization of Virginia, she writes, "As far as we know, potatoes were not grown in Virginia until long after Raleigh's time." She suspected the botanists made a "genuine mistake" in confusing the potato that had come from South America to Europe via Virginia for one grown there. She also suspected that if Raleigh was the one who introduced the tuber to Europe, he might have started with his Irish estates at Youghal, near Cork.[122] According to Redcliffe Salaman's research, potatoes were accepted by the Irish as early as the sixteenth century, several decades before the English.[123]

Regardless of whether the "Virginia potato" seen and planted by John Gerard and John Parkinson was from Virginia, they both clearly preferred the "Spanish potato," or sweet potato. When the root plant was introduced to Britain, it was prepared as a delicacy to satisfy people's palates by its sweet taste. And, thanks to the introduction of sugar, this interest in sweets was certainly prevalent at the time. Alison Sim and Terry Breverton both describe how sugar consumption remained a luxury for many Europeans in the sixteenth century—only the wealthy could afford it, but it was no doubt coveted by all. As a result, sweet foods in general, including sweet potatoes, were regarded as delicacies. However, as Sim explains, echoing some of the other culinary scholars discussed earlier, the desire to savor sweet foods has long been ingrained in English/European gastronomy. "The use of the [sweet] potato in conserves sounds very odd today," she writes in reference to John Parkinson's recipe, "but made perfect sense at the time. Wealthier people's meals in the Middle Ages often ended with a compost, which was a dish of root vegetables and fruit, such as pears, mixed up in kind of sweet and sour sauce. So the idea of putting root vegetables in sugar was a very old one."[124]

If all these English authors considered the sweet potato to be a Spanish import, how did Spaniards prepare it as a food after it was brought there from the Americas? According to Isaac Burkill's linguistic analysis of the *batata* variety preferred by Spaniards, which contained more sugar than *aji*, the sweet potato appealed to the Spanish palate because of its sweetness. The root plant was described, for example, by the above-mentioned Spanish authors Bartolomé de Las Casas and López de Gómara, not only in terms of how it could be grown, but also how it could be eaten. According to Las Casas, the root should be cooked in a sort of preserve with water for a few hours until the sweet juice comes out. "They are more tasty and better than anything else," he said. Likewise, Gómara described that sweet potatoes as "tast[ing] like chestnuts with sugar or marzipan." And Gonzalo Fernández de Oviedo also said that "a batata well cured and well prepared is just like fine marzipan." All of this supports Burkill's theory that sweet potatoes were consumed as a sweet delicacy by Spaniards. Sophie Coe found evidence that sweet potato could have been presented as a treasured gift to the people of Spain in the late sixteenth century.[125] Sweet potato had already been grown extensively in Seville and other southwestern regions of the country, such as Málaga, according to Nicolás Monardes. Monardes recognized that sweet potato could be windy but praised it as "very nutritious, somewhere between meat and fruit. They make excellent preserves, similar to quince."[126] Indeed, when explaining why the sweet potato entered Spanish gastronomy, Gregorio Saldarriaga confirms that the root's appeal to Europeans was primarily due to the fact that "they could be transformed into various delicacies, depending on their mode of preparation: grilled, cooked with meat, bathed with almond milk, cured, or preserved."[127]

Despite its nutritional value (primarily due to its sweetness according to the aforementioned Iberian dietary tradition), sweet potato was not widely accepted and described in Spanish cookbooks of the time. Indeed, food historians such as Ken Albala argue that while Spain established a global empire beginning in the sixteenth century, its culinary practice was far from cosmopolitan. As a result, before the seventeenth century, New World foods had no impact on the diet of the Spanish court and nobles.[128] In other words, as evidenced by the dominance of Robert de Nola's *Llibre del Coch* (Cookery book) throughout the century, Spanish gastronomy remained medieval and parochial, in contrast to the Spanish Empire's reputation as one "on which the sun never sets" under Charles V (1500–1558) and Philip II (1527–1598). Luis Lobera de Ávila, a Spanish physician who worked for Charles V's royal court, wrote *Banquete de nobles caballeros* (Banquet of noblemen) in 1530. Despite its contemporary status as a comprehensive text on Spanish gastronomy, Ávila did not mention a single plant from the Americas. If this omission was understandable given the recent discovery of the New World, then such parochialism appeared to have undergone a significant shift toward the end of the century. Two cookbooks with the same title, *Libro del arte de cozina* (Book of the art of cooking), were published in 1598 and 1607, respectively, by Diego Granado and Domingo Hernández de Maceras. Little is known about either author, but we do know that they were trained chefs; the former worked for the royal court, while the latter taught at the University of Salamanca, in northwest Spain, for several decades. Both texts included recipes that suggested significant changes in Spanish cuisine. However, given Granado's advantageous position as the "official cook" of the Spanish court, his cookbook was more comprehensive than Maceras's. Granado relied heavily on the cookbook prepared by Bartolomeo Scappi, a famous Renaissance Italian chef whose massive *Opera dell'arte del cucinare* (Art of cooking) contained over one thousand recipes. Granado's book included recipes that incorporated New World foods, such as *batata* and *patata*, which likely referred to sweet potatoes. In contrast, Maceras's slimmer text largely failed to include New World foods. This omission could be explained by the fact that he spent his entire life working in northern Spain, while the introduction of New World crops was then taking place primarily in the southwestern part of the country.

The two sweet potato recipes offered by Diego Granado were both about how to use the root to make preserves. One was called "Lemon and Sweet Potato Jelly" (*Carne de limon y batatas*) and the other "Citron and Sweet Potato" (*Cidra y patata*)—and while the two recipes used two different terms, I believe both were referring to the sweet potato because Eloy Terrón has stated before that *batata* and *patata* were interchangeable in Spanish before the eighteenth century. The first recipe instructs readers to "take ripe lemons and slice them in half, immerse them in lukewarm brine for eight days and then immerse them in boiling water until they are soft." Then "wash the sweet potatoes twice, and cook them in a pot of boiling water until they can be easily peeled." And after mashing the sweet potato, "combine one pound of (mashed) sweet potato, one pound of (mashed) lemon, and two and a half pounds of sugar, and, if desired, two dozen finely pounded almonds. When combining the sugar with the fruit preserve, it should be very clarified and smooth [*a punto*], not so much as for peach [preserves]. Cook the mix over low heat; when the bottom of the cooking pan becomes white, the preserves are done."

The second recipe for making "Citron and Sweet Potato" is as follows:

> cut a ripe citron into quarters and remove the bitter portion. Peel and grate the citron and cook it by bringing it to a boil three times. Remove it from the heat and let it cool, then rinse in lukewarm water. Put it in a sieve and rinse in cold water to remove the bitter taste and drain well. The sweet potatoes should be large and thoroughly rinsed. Boil them and peel them when they are tender, then put them through a ricer/grater [*cedazo despojador*]. Weigh the sweet potato puree and combine it with the grated citron; mix well. Combine in a large pot with clarified sugar and cook over low heat; stir continuously so the mixture does not stick. Cook the mixture over low heat; when the bottom of the cooking pan becomes white, the preserves are done.[129]

Sweet potato thus entered European gastronomy and, initially, was mainly used to make sweet delicacies, as demonstrated by the two recipes in Granado's cookbook. That is, as discussed in the previous section, Spaniards resisted accepting new foods from the Americas as staples because they held a hierarchical attitude toward culinary traditions in the Old World versus those in the New World. To Granado and Maceras's contemporary Francisco Núñez de Oria, a physician, hygienist, and humanist who wrote a dietary text during this period, it was even implausible that some of his compatriots would be interested in food from their newly established American colony. While Núñez de Oria included a number of New World food plants in his book, he cast them in a negative light. Even though it was widely known that Columbus's motivation for seeking a new route to Asia was to obtain not only gold but also spices, Núñez de Oria believed that the search for new foods was totally pointless. "It is a strange thing in human nature," he commented, "to go to the Indies, to sail to the Straits of Magellan and to appease the rage of appetite and gluttony . . . because why put a man's life in so many perils to seek precious food?"[130]

Moreover, even the use of sweet potatoes as an ingredient in desserts was not always universally welcomed. A monumental work on the history of Spanish gastronomy, Francisco Martnez Montiño's *Arte de cocina, pastelera, bizcochera y conservera* (Arts of cookery, pastry, bakery, and preserves), was critical of the culinary impact associated with the import of New World foodstuffs. Montiño worked in the Spanish palace for thirty-four years, from Philip II to Philip IV (1605–1665). His cookbook, which contained over five hundred recipes, was one of the most notable gastronomic compendiums written in Spanish, and it went through several editions after its initial publication in 1611. Montiño was well aware that sweet potato had been used as an ingredient in the preparation of sweets, given his broad knowledge, which ranged from cooking various dishes and desserts to hygienic kitchen use. However, he warned his readers that while cakes and pies were delicious, especially those made with cherries, chestnuts, carrots, figs, and sweet potatoes, "they do great harm to people who eat them."[131]

Francisco Martnez Montiño's warning seemed to coincide with a shift in European attitudes toward sweets beginning in the late sixteenth century. Bucking what had become a traditional notion during the medieval and early Renaissance periods, many dieticians in the seventeenth century and later no longer assumed that a food's good taste was proof of its healthy properties. In fact, they believed that "anything too exciting," as Ken Albala observes, "smacks of sinful indulgence. The most salient change in the details of dietary recommendations is the excision of sweets." Sweetness, in other words, was no longer regarded as a sign of nourishment. It now stood for "difficult to digest, gross, and oppilative." Sugar became increasingly available and affordable to European consumers beginning in the seventeenth century, owing largely to the expansion of sugarcane plantations in the New World. However, it was quickly labeled a "dangerous temptation."[132] To some extent, this changed attitude toward sugary foods, or a sense of guilt associated with indulging in sweets/desserts, has persisted in the Western world to this day.

Despite these initial cautions and criticisms, over time Spaniards became more acquainted with New World plants, including the sweet potato. Spanish historian Eloy Terrón observes aptly in his *España, encrucijada de culturas alimentarias*,

> It is striking how little curiosity and even less rigor intellectuals, that is to say, the people who knew and were able to write (at that time individuals of the nobility, of the judicature, or of the church)

displayed with respect to a question so important and that had to have so much influence on the Spanish people as the introduction of the American crops, which changed the feeding of the working masses, peasants, and artisans; among these crops it is necessary to mention the bell pepper, the tomato, the sweet potato (called by many potato), the beans [*alubias*]; the corn, the potato (or papa, in all America of Castilian speech, and in Andalusia and the Canary Islands).

In other words, Spanish society's attitudes toward new foods from the Americas were shaped by both social class and regional variation. Pedro Cieza de León and Garcilaso de la Vega, two conquistadors who took part in the Spanish conquest of the Americas, are duly acknowledged by Terrón for their more enthusiastic depictions of food crops from the New World. It is important to note that Cieza de León attempted the first history of Peru, the country where early evidence of sweet potato domestication was found, and showed great interest in all the crops grown by Amerindians in his writing.[133] Members of the working class in Spain were consuming New World foods like tomato, sweet potato, and white potato from the sixteenth century onward, according to evidence.[134] Seville, in the southwest, was a hub for botanical exchanges between the Old and New Worlds at the time. Andrea Navagiero, the Venetian ambassador to Spain, provided firsthand information in his writings as early as 1525. He stated that Seville's fertile land produced wheat and wine, which were then transported to the New World. He also wrote that "In Seville I saw many things from the Indies and I tasted the roots they call *batatas*, which taste like chestnuts. I also saw and tasted a beautiful fruit that arrived fresh and is called *ananá* and that tastes somewhere between a melon and a peach, with a strong fragrance, and it was truly very nice."[135]

Though his recipes were mostly aimed at the upper class, Diego Granado's cookbook did leave us with the impression that sweet potatoes were not just eaten as preserves among the rich and famous. The root may have a more versatile use for members of the lower social strata. Sweet potato was after all regarded as an ingredient "between meat and fruit," according to Monardes. It can also be "grilled, cooked with meat, bathed with almond milk, cured, or preserved," as Gregorio Saldarriaga described. *Olla podrida*, a popular Spanish dish, both then and now, is a prime example. Aside from the two recipes for sweet potato preserves mentioned above,

Granado's book also included one for *olla podrida*.[136] *Olla podrida* was a popular Spanish stew from the time of Diego Granado and Domingo Hernández de Maceras, if not earlier. It was eaten by people of all social classes and later spread throughout Europe. Literally translated as "rotten pot," it contained a variety of ingredients. *Olla podrida*, as seen in Granado's and Maceras's recipes, could be cooked with a variety of meats, ranging from lamb, mutton, beef, and hare to sausage, tongue, pigeon, and various pork parts. Its use of vegetables also varied, ranging from chickpeas, peas, beans, and garlic to turnips, chestnuts, and onions. While Granados and Maceras's recipes, as well as those of Bartolomeo Scappi, did not call for the use of potatoes, other types of *olla podrida*, known as cocido or olio in English, did. In 1660, Robert May, an experienced chef trained first in France and later in England, published *The Accomplisht Cook*, a collection of recipes that he himself cooked with during his professional career. On his recommendation, the *olla podrida*, or *olio podrida*, was made with potatoes. And the potato here should be the sweet potato, as May distinguished it from the other potato—the "Virginia potato"—in his book, the latter of which he recommended as an ingredient in "Sallet" (salad).[137]

As such, it is reasonable to assume that when Spaniards made *olla podrida* in the sixteenth and seventeenth centuries, they also used sweet potatoes. At the time, the root appeared to have widespread culinary appeal. Giovanni Domenico Sala, an Italian nutritionist of the seventeenth century, stated in his book *De alimentis et eorum recta administratione* (Foods and their direct management) that "In Spain, they cook them [sweet potatoes] in the coals, peel them and cut them into slices and eat them with a bit of wine, rose water, and sugar."[138] The mention of wine and sugar here may imply that the vegetable was sweet potatoes rather than white potatoes, because, as we can see from the recommendations of Nicolás Monardes, Thomas Dawson, and John Gerard, eating sweet potato in this manner was quite common in early modern Europe.

The sweet potato was popular in Spanish recipe books from the seventeenth to the eighteenth centuries. Meanwhile, these books demonstrated how it was gradually replaced by the white potato as Europeans' fondness for sweets waned. That is, while its sweetness allowed it to be easily transformed into a dessert, sweet potato could also be cooked with meat and other vegetables, a practice that is now uncommon. Indeed, given Europeans' predilection for sweet flavors during the early modern period, sugar was ubiquitous in many dishes, as previously discussed. The sweet potato's sweetness thus made it a popular ingredient at the time. However, it was gradually replaced by the white potato. Indeed, pairing meat with white potatoes is a common culinary practice today, whether cooked together in a stew with the latter cut into cubes or served on a plate with the tuber mashed.

Two mid-eighteenth-century Spanish cookbooks depicted the sweet potato's changing status and diverse use. Included in Juan de la Mata's 1747 *Arte de repostería* (Art of confectionery) were two sweet potato reserves: "Málaga Sweet Potatoes in Dry and Liquid Form" and "Compote of Sweet Potatoes from Málaga." Aside from indicating that Málaga was the region in Spain where sweet potato was grown in abundance, the recipes clearly reflected the influence of Granado's sweet potato delicacies. Little is known about Juan de la Mata, but his work has been regarded as an important reference for "historical studies of recipes of the eighteenth century; many of the traditional preparations of today's Spanish confectionery can be found in his book."[139] The following was the recipe for "Málaga Sweet Potatoes in Dry and Liquid Form": After cleaning, peeling, and cutting the sweet potatoes into pieces, they should be thoroughly boiled with "a pound or five quarts of clarified sugar, leaving everything to cook while the syrup is reduced to the small pearl." Afterwards, "the syrup and the sweet potatoes will be put together in a glazed earthenware pot, leaving them to infuse for twenty-four hours. Then everything will be reheated in the pot until the syrup has become the good pearl." One could eat the sweet potato on its own or with the syrup. A few pages later, Juan de la Mate proposed that if the sweet potatoes in dry form were cooked with another pound of sugar until the syrup formed small feathers, it could be served in a cold compote bowl as the "Compote of Sweet Potatoes from Málaga." In other words, they were both sweet potato compotes, despite being two distinct dishes.[140]

The sweet potato was treated as a fruit in Juan de la Mata's recipe book. In comparison, the root was cooked with meat in Juan de Altamiras's 1745 cookbook *Nuevo arte de cocina: Sacado de la escuela de la experiencia económica* (New art of cooking: From the school of economic experience). Again, information about the author is limited, but most people now believe that Juan de Altamiras was the pseudonym of the eighteenth-century Aragonese Franciscan friar Raimundo Gómez; the subtitle of the book also seemed consistent with its purported purpose—to provide economical recipes for commoners rather than the upper class or the royal court. *Nuevo arte de cocina* was an instant success upon its release. Through the early twentieth century, it was expanded in at least twenty editions, exemplifying the tradition of Spanish gastronomy before it was overrun by the strong influence of French haute cuisine. While the book was written by a Franciscan friar, Vicky Hayward, its annotator, describes *Nuevo arte de cocina* as "an elegy instead to earthy everyday eating . . . [and] revealed friary food as something very different to monastic cooking. Shaped by exchange with neighbors, it was improvised around food gifts and kitchen-garden produce as well as the rhythms of the religious calendar." Moreover, the cookbook reflected influences of the New World foods, for six thousand Franciscans had traveled to the Americas. Altamiras "may have met returning mission cooks as well as travelers from closer lands. Perhaps a visitor from Florence's Spanish friary taught him how to make breadcrumb noodles. Maybe a French friar taught him how to cook ragout or a Portuguese traveler told him about a cookbook called *Arte de Cozinha*, by Lisbon court cook Domingues Rodriguez, a book structured around meat-eating and lean days."[141]

One recipe from the *Nuevo arte de cocina* was "Braised Lamb with Pomegranate Juice." It was suggested that the meat be cooked in a pot with salt, pepper, fat bacon, and onion before "squeez[ing] in the juice of two or three boiled pomegranates. They will give you a very good, tasty sauce." Furthermore, two additional touches could be added to the dish: "a lame stock made with the meat's roasted bones and a puree of wood-roasted sweet potato."

The other recipe is a variation on the famous Spanish omelet known as a "Lamb Tortilla" in the *Nuevo arte de cocina*. Though a meat dish, it demonstrated the possibility of replacing meat with potato, because its ingredient could be either meat (lamb) or potato, according to Hayward. It could also be cooked sweet or unsweet: "Make your meat into small dice or chop it as you would for *gigote*, braise it till it is tender, drain off the cooking liquid, and put the meat in a basin, adding breadcrumbs, sugar, cinnamon and eggs. You can fry these tortillas in soft lard. If you do not want to sweeten them season them with all your spices, but if they are sweetened, then serve them fried with sugar and cinnamon on top."

Moreover, Hayward notes that after potatoes were cultivated in La Mancha in the late eighteenth century, they were frequently used to make the Spanish omelet, known in Spain as a potato tortilla.[142] *Tortilla de patatas* (potato omelet) is listed as a traditional dish in Maria Paz Moreno's *Madrid:*

A Culinary History, implying that the Spanish omelet is perhaps more commonly cooked with potato rather than meat in modern times. Another example, according to Maria Paz Moreno, was *olla podrida*, which Queen Isabella II (1830–1904) favored during her reign. While her *olla podrida* may have used meat, those of the lower classes in Spanish society at the time tended to use potatoes as a substitute for meat in theirs.[143] However, the dish could have been sweet in its original form, and thus the sweet potato could have been included as an ingredient, because locals would refer to the white potato as "Irish sweet potato," according to Hayward.[144]

Indeed, it was quite common for Europeans not only to confuse the sweet potato with the white potato in nomenclature, but also to substitute one for the other in cooking during the sixteenth and seventeenth centuries. In fact, when John Forster praised potatoes in the mid-seventeenth century, he clearly lumped white potatoes and sweet potatoes together in describing their various uses. For instance, while Forster recommended baking potato bread, which he thought was comparable if not superior to corn bread, he also suggested making potato cakes, cheesecakes, custards, and pudding. Needless to say, both sweet potato and white potato could be used as ingredients in the latter.[145] According to Terry Breverton, author of *The Tudor Kitchen: What the Tudors Ate and Drank*, while two first course dishes in Tudor times used sweet potatoes rather than white potatoes, it would also become possible for people in later and modern times to replace the former with the latter. The first was "Breakfast Soup—Bubbly Beer Cheese Breakfast Soup," which included onions, potatoes, and cheese and was cooked in chicken broth. The second was a traditional Welsh dish called "Cawl Llysiau'r Gaeaf," or winter vegetable stew. Breverton believes that this straightforward vegetable soup would have included onions, leeks, swedes, carrots, parsnips, and bay leaves in addition to sweet potatoes, rather than white potatoes, as the main ingredient in Tudor times. Salt and black pepper would also be added to the soup to taste.[146]

It should be noted that the consumption of sweet potatoes (and later white potatoes) in Todor England was somewhat unusual for the time, as the two root crops took much longer to be readily accepted into the gastronomic system in continental Europe. While the sweet potato was grown in many parts of the Iberian Peninsula and was adopted, albeit marginally, into Spanish cooking, its fate in neighboring Portugal appeared to be somewhat different. *Livro de cozinha da Infanta Maria* (The cookbook of Infanta Maria of Portugal), from around the mid-sixteenth century, is arguably the oldest Portuguese cookbook in existence. There was no mention of sweet potato. A century or so later, the root failed to appear in Domingos Rodrigues's *Arte de cozinha* (Art of cooking), the first printed Portuguese cookbook, appearing in 1680, and probably consulted by Juan Altamiras. However, like in Spain, *olla podrida*, also known as *olha podrida* in Portuguese, appeared to be a popular dish in Portugal. Its recipe was included in Rodrigues's *Arte de cozinha*,[147] which may have included (sweet) potatoes as one of the ingredients, as we speculated earlier regarding the preparation of the same dish in Spain.

Going east from Iberia to continental Europe, the procession for sweet potato (and white potato) consumption was equally slow. In 1581, Marx Rumpolt, a head cook for the elector of Mainz, published *Ein New Kochbuch* (A new cookbook) in central Europe. The book, a massive collection of about two thousand recipes, did not include one for preparing the sweet potato—or the white potato—as a delicacy or an ingredient in a meat dish. The sweet potato also failed to catch on in French cuisine during the sixteenth and seventeenth centuries in southwestern Europe. It was said that under Louis XIV (1638–1715), the "Sun King," whose long reign marked the pinnacle of French political, military, and cultural power on the European continent, French culinary practice made a revolutionary breakthrough. François Pierre de la Varenne's *Cookery*, also known as *Le Cuisinier françois*, was a seminal work of the time that documented how French cuisine departed from its Italian influences and established itself as a culinary tradition in its own right. The cookbook, known in English as *La Varenne's Cookery*, was divided into three sections: "The French Cook," "The French Pastry Chef," and "The French Confectioner." Despite its comprehensiveness, *La Varenne's Cookery* did not include sweet potato as an ingredient for either a dish or a confection. In fact, only a few New World foods were mentioned in the book.[148] Martin Lister, an English naturalist who visited Paris at the end of the seventeenth century, observed that "one scarcely finds any potatoes at market—those wholesome and nourishing roots which make up so great a resource for the people of England." Though it is unclear whether Lister was referring to the sweet potato or the white potato, Barbara K. Wheaton, author of *Savoring the Past: The French Kitchen and Table from 1300 to 1789*, who cited the above observation from Lister, states that sweet potato "has never become important in the French kitchen."[149] And, given that La Varenne was said to have learned to cook in the kitchen of Queen Marie de' Medici (1575–1642), a member of the powerful House of Medici in Tuscany, the absence of the sweet potato in his *Cookery* could also imply that the root was not a common ingredient in Italian cuisine either.

By comparison, from the sixteenth century onward, sweet potatoes were welcomed in the British Isles with a "sweet" attitude. The union of Prince Arthur and Catherine of Aragon (1485–1536), the daughter of King Ferdinand and Queen Isabell of Spain, and of Catherine and Henry VIII (1491–1547) after Arthur's demise, could have had a significant impact. Henry, in particular, was credited with helping "to popularise the sweet potato." He was said to be "very fond of cubes of sweet potatoes that had been candied in sugar," which, as shown below, is exactly how it was made in Spain as a sweetmeat. After his brother's passing, Henry also planted salad gardens for Catherine, which most likely contained sweet potatoes since they could be eaten uncooked like chestnuts.[150]

It is no surprise, then, that when Thomas Dawson compiled his *The Good Huswife's Jewell* a few decades later, he included an early recipe for making a sweet potato tart. And he was not the only one who described the root as a nutritious food. Thomas Muffet, a naturalist who wrote *Health's Improvement* in the late sixteenth century, said this about the sweet potato's nutritional value: "Pottato-roots are now so common and known amongst us, that even the husbandman buys them to please his wife. They nourish mightily, being either sodden, baked or roasted. The newest and heaviest be of best worth, engendering much flesh, blood and seed, but withall encreasing wind and lust. Clusius thinks them to be Indian Skirrets, and verily in taste and operation they resemble them not a little."[151]

Tobias Venner, a physician by training who shared Thomas Muffet's interest in educating the public about health improvement, praised the sweet potato as well. Venner recommended the root as a healthy food in his 1620 book *Via recta ad vitam longam* (The right to long life):

> Potato roots are of a temperate quality, and of a strong nourishing parts: the nutriment which they yeeld, is, though somewhat windy, very substantiall, good and restorative, surpassing the nourishment of all other roots or fruits. They were diversly dressed and prepared, according to everymans taste and liking: Some use to eat them, being roasted in the embers, sopped in wine, which way is specially good: but in what manner soever they be dressed, they are very pleasant to

the taste, and do wonderfully comfort, nourish and strengthen the body, and they are very wholesome and good for every age and constitution, especially for them that be past their consistent age.[152]

Muffet and Venner, both health writers, confirmed the nutritional value of the sweet potato. However, their descriptions of how the root was prepared were hazy, though sufficient to suggest that what they described were sweet potatoes rather than white potatoes. Moreover, the warm reception sweet potato received in Britain led to it being used as an ingredient in more dishes than just a simple tart by Thomas Dawson. *The English Huswife*, written in 1615 by Gervase Markham (ca. 1568–1637), provided detailed descriptions of how the sweet potato from Spain was accepted as a food by the English:

> To make an excellent *Olepotrige*, which is the onely principall dish of boild meate which is esteemed in all *Spaine,* you shall take a very large vessell, pot or kettell, and filling it with water, you shall set it on the fire, and first put in good thicke gobbets of well fed Beefe, and being ready to boile, skumme your pot; when the Beefe is halfe boiled, you shall put in Potato roots, Turneps, and Skirrets: also like gobbets of the best Mutton, and the best Porke; after they haue boyled a while, you shall put in the like gobbets of Venison red, and Fallow, if you haue them; then the like gobbets, of Veale, Kidde, and Lamb; a little space after these, the foreparts of a fat Pigge, and a crambd Pullet; then put in Spinage, Endiue, Succory, Marigold leaues & flowers, Lettice, Violet leaues, Strawberry leaues, Buglosse and Scallions, all whole and vnchoot; then when they haue boiled a while, put in a Partridge and a Chicken chopt in peeces, with Quailes, Rails, Blackbirds, Larkes, Sparrowes and other small birds, all being well and tenderly boiled, season vp the broth with good store of Sugar, Cloues, Mace, Cinamon, Ginger and Nutmegge mixt together in a good quantity of Veriuice and salt, and so stirre vp the pot well from the bottome, then dish it vp vpon great Chargers, or long Spanish dishes made in the fashion of our English woodden trayes, with good store of sippets in the bottome; then couer the meate all ouer with Prunes, Raisins, Currants, and blaunch't Almonds, boiled in a thing by themselues; then couer the fruite and the whole boiled hearbes, and the hearbes with slices of Orenges and

Lemmons, and lay the roots round about the sides of the dish, and strew good store of Sugar ouer all, and so serue it foorth.¹⁵³

Based on the context, it is clear that the *olepotrige* described by Markham was the popular Spanish dish *olla podrida*. As previously stated, Robert May's *The Accomplisht Cook*, published in the same century, also suggested using sweet potato to make what he called *olio podrida*. In contrast, while *olla podrida* was included in earlier Spanish cookbooks by Diego Granado and Domingo Hernández de Maceras, as well as a Portuguese cookbook by Domingos Rodrigues, we can only speculate that it might have used sweet potatoes as one of the ingredients.

In addition to the *olepotrige* recipe, Markham's *The English Huswife* included another that called for sweet potato in a pastry with meat and vegetables. It was written as follows:

> Next to these already rehearsed, our *English Hous-wife* must be skilfull in Pasterie, and know how and in what manner to bake all sorts of meate, and what Paste is fit for euerie meate, and how to handle and compound such Pastes: As for example, red Deere Venison, wilde Boare, Gammons of Bacon, Swannes, Elkes, Porpas, and such like standing dishes, which must bee kept long, would be bak't in a moist, thicke, tough, course, and long lasting crust, and therefore of all other your *Rie* paste is best for that purpose: your Turkie, Capon, Pheasant, Partridge, Veale, Peacocks, Lambe, and all sorts of water-fowle which are to come to the table more then once (yet not many dayes) would be bak't in a good white crust, somewhat thick; therefore your *Wheate* is fit for them: your Chickens, Calues-feet, Oliues, Potatoes, Quinces, Fallow deere and such like, which are most commonly eaten hot, would be in the finest, shortest & thinnest crust, therefore your fine wheat flower which is a litte baked in the ouen before it be kneaded is the best for that purpose.¹⁵⁴

Sweet potato was present in even more English meat dishes a few decades later, when Robert May offered his recipe book. How do we account for potatoes' warm reception in the British Isles and the root's widespread culinary appeal to the English in the sixteenth and seventeenth centuries? Using John Gerard's contemporary *Herball* as an example, David Gentilcore, an expert on early modern European gastronomy, believes Gerard's categorization of potatoes provided a clue, as Gerard noted in his pioneering book the existence of potatoes from Spain and Peru, in addition to those from Virginia. As a result, Gerard may have mistakenly classified the sweet potato and white potato as belonging to the same botanical family. Gentilcore suggests another possibility: Because the root crop is associated with the English colony of Virginia, in North America, English authors may have considered it "less foreign" and "less dangerous for English constitutions."[155] Gentilcore's words reflect a widely held belief among food scholars that the English seemed to embrace sweet and white potatoes more readily. That is, while Gerard made two mistakes in his *Herball*—treating the two root plants as the same and confusing their origins—the English may have accepted them more willingly than other Europeans because of this dual error. Ken Albala writes in his book *Eating Right in the Renaissance* that the English seized on the root plants from the Americas because they "firmly associated potato with Virginia, rather than with 'foreign' colonies, and so could consider them their own."[156] Likewise, Alan Davidson points out that "It is possible that this double confusion [by John Gerard] may have delayed a proper assessment of the merits of the potato in Europe."[157] However, Gerard was not the only one who made such mistakes; as previously stated, his contemporaries John Parkinson and John Forster also identified multiple origins of potato in America.

Were there other botanical and/or culinary factors, besides the English's familiarity with their Virginia colony, that led them to incorporate New World roots into their cuisine? The answer may be in the affirmative, especially with regard to the sweet potato. Both Albala and Gentilcore observe that, in comparison to other Europeans, the English believed that their bodily makeup enabled them to deal with New World roots more easily; in Albala's words, the English agreed that their "constitutions were well suited to digesting rough and heavy foods, and there was no real reason to refuse them."[158]

Why? It seems that the juxtaposition of sweet potato and skirret in the aforementioned texts suggests the reason. Not only did Thomas Muffet cite Carolus Clusius, the Latin translator of Nicolás Monardes's *Historia medicinal de las cosas que se traen de nuestras Indias Occidentales*, in his *Health's Improvement*, but Gervase Markham's *The English Huswife* and Robert May's *The Accomplisht Cook* also compared sweet potatoes to skirrets in their recipes. Nonetheless, it was John Gerard who first made the association between sweet potato and skirrets in his *Herball*:

> This plant (which is called of some Skyrrets of Peru) is generally of us called Potatus or Potato's. It hath long rough flexible branches trailing upon the ground like unto those of Pompions, whereupon are set greene three cornered leaves very like those of the wilde Cucumber.
>
> The roots are many, thicke, and knobby, like unto the roots of Peonies, or rather of the white Asphodill, joined together at the top into one head, in maner of the Skyrret, which being divided into divers parts and planted, do make a great increase, especially if the greatest roots be cut into divers goblets, and planted in good and fertile ground.[159]

If sweet potato and skirret could be compared, it was because the two had similar propensities. They were both considered rough and heavy foods as root plants.[160] Take a look at how Gerard described skirret:

> Sisarum. Skirrets. The roots of the Skirret be moderately hot and moist; they be easily concocted; they nourish meanly, and yeeld a reasonable good iuice: but they are something windie, by reason whereof they also prouoke lust. They be eaten boiled, with vineger, salt, and a little oile, after the manner of a sallad, and oftentimes they be fried in oile and butter, and also dressed after other fashions, according to the skil of the cooke, and the taste of the eater.[161]

In other words, skirret, thought to have originated in China, was regarded as a nourishing food that was both windy and aphrodisiac—"provoking lust"—both of which were well-known qualities associated with the sweet potato. When describing a man meeting his lover, George Chapman, an English dramatist and poet contemporaneous with William Shakespeare, used sweet potato, skirret, and eryngo as "whetstones of venery."[162]

Furthermore, these root plants were prepared in the same way as foods. John Gerard suggested dressing the sweet potato in the following manner: "They are used to be eaten rosted in the ashes. Some when they be so rosted infuse and sop them in wine: and others to give them the greater grace in eating, do boile them with prunes and so eat them: likewise others dresse them (being first rosted) with oile, vineger, and salt, every man according to his owne taste and liking. Notwithstanding howsoever they be dressed, they comfort, nourish, and strengthen the body."[163]

Figure 2.3 The illustration of skirrets in John Gerard's *Herball* (1597), another popular root plant in the British Isles. The dietary familiarity with skirrets may have eased the British acceptance of sweet potatoes and later white potatoes.

That is, in addition to roasting, sweet potato and skirret could both be boiled with oil, vinegar, and salt. One could speculate that the similarities between the two root plants, one having been used in English kitchens for centuries and the other a newcomer, were another factor in the English's early acceptance of the sweet potato. Skirret had long been used as an ingredient in fritters in England, as evidenced by *The Forme of Cury*, a 1390 manuscript used by master cooks for King Richard II (1367–1400).[164] Peter Brears's comprehensive study of medieval English recipes includes one that used skirrets in a vegetable pottage. Skirrets, as a root vegetable, were "cooked like parsnips and turnips," he says, and "pottages made of various mixtures of green herbs were very popular."[165] Skirrets, parsnips, and sweet potatoes were frequently mentioned in Robert May's *The Accomplisht Cook* in the preparation of meat and sweet dishes.[166] Of course, prior to the sixteenth century, skirret was also consumed in continental Europe. However, the British Isles appear to be the region where the sweet potato arrived earlier than other parts of Europe. The absence of the New World root mentioned earlier in contemporary German and French cookbooks is evidence of this.

If the English's familiarity with skirret and other similar root crops, such as eryngo and parsnip, played a role in their acceptance of the sweet potato, it was not unusual. Food experts have long stated that whether a food resembled something familiar to Europeans at the time was an important factor in the reception of New World foods. Turkey is a case in point. When turkey first arrived in Europe from West Africa and the Americas, it reminded Europeans of pheasant, and they quickly incorporated it into their diet. Turkey had become a popular Christmas dish in England by 1573.[167] Many dieticians lauded the bird as "one of the best" and "healthiest meats." Interestingly, as with the sweet potato, most of them misidentified its origin. Beans from the Americas, which resembled the black-eyed pea or cowpea that Europeans had eaten for centuries, were also readily accepted as New World vegetables. Indeed, beans from the New World were not considered "foreign," nor were they classified in a new, distinct botanical category. In a word, concludes Ken Albala, "the key appears to be whether the new food was considered analogous to something already standard in the diet or could be substituted in a recipe with comparable results."[168]

All the same, while sweet potatoes were used in a variety of dishes in Tudor and early Stuart England, their sweetness made them ideal for use as a sweet delicacy. If Thomas Dawson only recommended a potato tart at the end of the sixteenth century, several decades later, John Forster's *England's Happiness Increased* suggested a variety of potato pastries. In his *The Whole Body of Cookery Dissected*, Forster's contemporary William Rabisha, a Catholic cook, provided a detailed recipe for a sweet potato pie:

> Boyl your Spanish Potatoes (not overmuch) cut them forth in slices as thick as your thumb, season them with Nutmeg, Cinamon, Ginger, and Sugar; your Coffin being ready, put them in, over the bottom; add to them the Marrow of about three Marrow-bones, seasoned as aforesaid, a handful of stoned Raisons of the Sun, some quartered Dates, Orangado, Cittern, with Ringo-roots sliced, put butter over it, and bake them: let their lear be a little Vinegar, Sack and Sugar, beaten up with the yolk of an Egg, and a little drawn butter; when your pie is enough, pour it in, shake it together, scrape on Sugar, garnish it, and serve it up.[169]

Intent to integrate traditional English gastronomy with outside influences, Rabisha suggested using a mix of local and exotic ingredients to make a traditional pie by following the medieval method, implying that in seventeenth-century England, as in the rest of Europe, sweet potato was viewed as a perfect material to be candied into sweets. In addition to being made into a tart (Thomas Dawson) or a pie (William Rabisha), the New World root was prepared in the same way that skirret, turnip, and parsnip pottages had been in medieval England—"when dished, they were sprinkled with a little sugar and cinnamon."[170] After first being boiled and then simmered, the sweet potatoes should be transferred to "a hot dish with sippers, pour on the remaining juice beaten into the remaining butter, stick with the candied peels, and finally add a sprinkling of caster sugar," according to a Tudor recipe.[171] These culinary practices, such as making a sweet potato compost/conserve or a sweet potato tart or pie, would spread to English colonies in North America. Over time, sweet potatoes cooked as a side dish or dessert would appear on the Thanksgiving dinner menus of many American families.

In short, after making its first appearance in the Old World via Spain and southern Europe, the sweet potato was embraced across the modern West and incorporated into a variety of confections. For example, *El libro de las familias: Novísimo manual práctico de cocina Española, Francesa y Americana*, *higiene y economía doméstica* (The book of families: New practical manual of Spanish, French, and American cuisine, hygiene, and home Economics) included a few such recipes. It was a useful and popular handbook for many families and went through several editions throughout the nineteenth century. Sweet potato could be fried (*boniato frito criollo*), roasted (*boniato asado*), or sauteed (*boniato salcochado sin agua*, which literally means "sweet potato sauteed without water"), according to the recipes in *El libro de las familias*. Sweet potato fritter (*buñuelo de boniato*) and dried sweet potato (*boniatillo seco, ó paota*) were also available. The most popular way to eat the sweet potato, according to the book, was to slice it into pieces and bake it with water, sugar, or honey, along with other spices, into *dulce de camote* or *batata de Málaga*, which are still popular among Spaniards today.[172]

Sweet potato in the American South

By identifying another New World plant—distinct from the sweet potato—known as the "potato of Virginia" and found in the West Indies and Spain, John Gerard's *Herball* arguably provides the earliest recorded reference to the introduction of the white potato to Europe. Yet, not only did Gerard misidentify the tuber's origin, but he also perplexed his readers by claiming that Carolus Clusius had reported on its discovery.[173] Of course, Clusius was one of the first European botanists to describe the sweet potato. However, according to John Reader, author of *Potato: A History of the Propitious Esculent*, his report on the white potato was not published until 1601, four years after Gerard's *Herball*. "How did he [Gerard] learn about Clusius' report?" Reader wonders. Reader doubts that the potato Gerard grew in his garden came from Virginia: "It is confusing that Gerard should claim his tubers came from Virginia, for the potato was unknown there until introduced by settlers after his time."[174] He believes that the potato encountered in North America by Sir Walter Raleigh's settlers—Raleigh himself never landed in Virginia—may have been the sweet potato. Likewise, Clarissa Wright suspects that the alleged story about Raleigh's dinner of potato leaves could only be those of the sweet potato, which are edible and even palatable to many Asians and Africans, because white potato leaves are poisonous.[175]

Before Europeans arrived, both cultivated and wild sweet potatoes had likely been grown in Virginia; in the early seventeenth century, the

Virginia colony also covered a wider territory than the present-day U.S. state of the same name, and thus included more regions in the American South.[176] Sweet potato, a semitropical plant that is frost tolerant, grows best at an average temperature of 24 degrees Celsius (75 degrees Fahrenheit). The British Isles are in the temperate climate zone, between latitudes 49° and 61° north. While the sweet potato could be planted as a summer crop with caution, it would not grow as well in Britain as it would in Spain or Virginia, where the temperatures are higher. As previously stated, after being introduced to Europe, the root thrived in southwestern Spain, particularly in Málaga and Seville, in Andalusia.

Like Andalusia, Virginia also has a warm-temperate climate as the two regions are both located between latitudes 36° and 38° north, which is certainly suitable for growing the sweet potato. Several European expeditions arrived in Virginia during the sixteenth century, including a group of Spanish Jesuits, followed by Sir Walter Raleigh's exploration of North America's Atlantic coast at the end of the century. Virginia, established in 1607 as the first English colony in North America, experienced rapid economic growth in its early years, owing to the development of tobacco plantations, for which the colonists imported slaves from Africa as laborers. Meanwhile, a mix of Old and New World food crops and vegetables were planted to meet the needs of both the plantation owners and their workers. "The sweet potato was grown by European settlers in Virginia as early as 1648, in Carolina by 1723," according to botanist Vincent Lebot.[177] Thomas Jefferson (1743–1826) was interested in cultivating sweet potatoes in his Monticello garden, though it is unclear if the root was also planted elsewhere in the area. His slaves, on the other hand, grew sweet potatoes in their gardens and sold them to the Jeffersons.[178] When Jefferson served as the U.S. minister to France in Paris at the end of the eighteenth century, he hired James Hemings—with whose sister, Sally, Jefferson had a long affair—as his chef. Jefferson grew sweet potatoes and other vegetables in the garden of the Paris hotel where he stayed.[179]

Mary Randolph's 1824 book *The Virginia Housewife* contained three sweet potato recipes. Randolph, a Virginia native and the sister of Thomas Mann Randolph, Thomas Jefferson's son-in-law, had a long and close relationship with the Jeffersons. Sweet potatoes, according to Mary Randolph, can be roasted on a tin sheet, stewed with boiled ham, chicken, and herbs with pepper and salt, or broiled in slices on a griddle and served with butter.[180] Her recipes resembled those offered by George Washington's wife, Martha Washington, a few decades earlier. The latter's recipes also suggested boiling, roasting, and baking sweet potatoes. More intriguingly, Washington provided a fourth option, "french fried sweet potatoes." She also mentioned that the same recipe could be used to fry white potatoes. In other words, Washington's *Cook Book*, which she was said to use throughout her life before giving it to her granddaughter Eleanor Park Custis in 1799 for the latter's marriage to Lawrence Lewis, could have made the first mention of french fried white and sweet potatoes in print, a half century earlier than Eliza Warren's *Cookery for Maids of All Work*, written in the 1850s. When comparing the two recipes, Washington simply stated that the potatoes should be boiled for half an hour before frying, whereas Warren provided a more specific instruction, very similar to how french fries are cooked today: "cut new potatoes in thin slices, put them in boiling fat, and a little salt; fry both sides of a light golden brown colour; drain dry from fat, and serve hot."[181]

Regardless of when Martha Washington wrote the cookbook that she later gave to her granddaughter as a wedding gift, it is evident that by the late eighteenth century, the English and early Americans on both sides of the Atlantic were employing a clear and correct terminology in reference to sweet and white potatoes—"correct" in the sense that these are terms used to distinguish the two plants in today's English. Both Martha Washington and Mary Randolph included recipes for cooking white potatoes in their cookbooks, in addition to those for the sweet potato, mentioned above. The fact that white potatoes were now planted in Virginia during the age, or "during Jefferson's time," undoubtedly contributed to the development of this sense of distinction.[182] After leaving the White House, George Washington, for example, grew white potatoes in his garden at Mount Vernon. In 1799, the same year Martha Washington allegedly gave her cookbook to her granddaughter, Joshua Brookes, an English visitor to the mansion, noted that potatoes were on the menu.[183]

Examining English cookbooks from the seventeenth and eighteenth centuries reveals that there was a change in terminology in the anglophone world when referring to the two potatoes. When Robert May wrote *The Accomplisht Cook* in the mid-seventeenth century, he called sweet potato "potato" and white potato "Virginia potato." Around 1727, Richard Bradley, a self-taught botanist who later became Cambridge University's first professor of botany, published *The Country Housewife and Lady's Director*, which included an intriguing recipe for "biscuits of potatoes." It said to grind potatoes with sugar and butter and combine them to make a cake. The potatoes used in the recipe could still be sweet potatoes rather than white potatoes. Bradley also included a recipe for "candied eringo-roots" in the same book, which followed a similar procedure.[184] The similarity of the two recipes demonstrated, in a sense, that the English practice of eating roots like skirret, eryngo, and parsnip as sweets had influenced them to embrace the sweet potato.

White potatoes were making significant inroads into English gastronomy on both sides of the Atlantic Ocean in the same decade that Bradley selectively compiled recipes from his time, including being used as a main ingredient in sweets. This progress was reflected in the nomenclatural shift in which it became known as potato, or white potato, whereas sweet potato received the qualifier "Spanish" and later "sweet." However, it appears that the white potato was introduced into English cooking because it could be used in place of the sweet potato. Eliza Smith, one of the most popular female cookbook authors of the eighteenth century, wrote *The Complete Housewife*, which first appeared in print in 1727 and went through several editions, including an American edition in 1742. As such, Smith is regarded as the first cookbook author in colonial America. *The Complete Housewife* presented several potato recipes, each of which Smith clearly distinguished. Smith suggested using sweet potatoes, which she referred to as "Spanish potatoes," like many of her contemporaries, in two sweet meat dishes: sweet lamb pie and sweet chicken pie, in which sweet potatoes were a main ingredient and baked together with spices, sugar, and butter. Meanwhile, Smith suggested making a potato pie and a potato cheesecake with the white potato. However, because the potato cheesecake recipe called for sugar, a sweet potato cheesecake could be made instead, as many people do today.[185]

In *The Complete Housewife*, Eliza Smith, unlike John Forster a century earlier, distinguished between sweet and white potatoes. While her recipes suggested that the two could be used interchangeably in a sweet confection, they also attested to the white potato's growing appeal as a food during her time. The fact that the aforementioned potato pie recipe called for white potatoes was an example, because sweet potatoes would have been used to make the confection in Tudor and early Stuart England.[186] Moreover, Smith stated that the potato pie was made for Lent, a major religious observance during which fasting is required, as per Christian tradition. In this dish, potato was used as a substitute for meat, which was avoided by many Christians as part of their Lenten sacrifice. Immediately following the recipe, Smith added another, "Artificial Potatoes for Lent: A Side-Dish, Second Course." As its name suggest, the dish did not use potato; rather, one takes "a pound of butter, put it into a stone mortar, half a pound of Naples biskets grated, and half a pound of Jordan almonds beat small after they are blanched, eight yolks of eggs, four whites, a little sack and orange-flower water." Then one should add sugar to "your taste; pound all together till you don't know what it is" to make a paste. After the paste becomes stiff, it should be cut into pieces the size of chestnuts and pan-fried in boiling lard and then, after draining the fat, served with some sack, melted butter, and refined sugar.[187] This "artificial potato dish" recipe could be seen as proof of the tuber's rising popularity over the course of the century.

Except for the Lent potato pie (not the artificial one described above), all of Eliza Smith's potato recipes were sweet dishes cooked and served with sugar. As a result, both sweet and white potatoes could be used in these recipes. However, by the time Hannah Glasse published *The Art of Cookery Made Plain and Easy*, the trend of replacing sweet potatoes with white potatoes was well underway. *The Art of Cookery Made Plain and Easy*, published in 1747, was a best seller of the time; like Eliza Smith's *The Complete Housewife*, it was reprinted numerous times well into the nineteenth century. Hannah Glasse provided a half dozen potato recipes, including one that used "Spanish potato" for a sweet lamb or veal pie, very similar to the one in Smith's book. Then the rest used potato, or "white potato," as Glasse referred to it in one of them. Furthermore, while some of the recipes used the tuber to make sweet dishes, such as potato cake and potato pudding, others used it to cook with other ingredients, such as "Beans Ragoo'd with Potatoes," "An Onion Pye," and "A Cheshire Pork Pye for Sea." The latter two recommended that one of the pie's fillings be a layer of potatoes. Salt was suggested as an alternative to sugar in the dishes. There were also recipes for broiling, mashing, and frying potatoes.[188] One could argue that by the mid-eighteenth century, white potato was no longer regarded as a substitute for sweet potato in Britain and North America; rather, it was cooked for its own flavor or to absorb the flavors of other foods in the dish.

The transition from sweet to white potato appeared to be complete in English and American cuisines by the second half of the seventeenth century. This shift reflected the overall decline in Europeans' early enthusiasm for sugary foods, as discussed in the previous section. Indeed, the sweet meat recipe cited above by Eliza Smith and Hannah Glasse may have been the last one to teach the English to use sweet potato as its main ingredient. *The Experienced English Housekeeper* by Elizabeth Raffald, published in 1769, mentioned potatoes several times, including a recipe that taught people how to "scollop potatoes." It was written as follows: "Boil your Potatoes, then beat them fine in a Bowl with good Cream, a Lump of Butter and Salt, put them into scollop'd Shells, make them smooth on the Top, score them with a Knife, lay thin Slices of Butter on the Top of them, put them in a Dutch Oven to brown before the Fire.—Three Shells is enough for a dish."[189] The white potato, rather than the sweet potato, was most likely used in this recipe. In fact, Raffald's book omitted the sweet potato entirely, a telling testimony to the aforementioned transition. Raffald began her career as a maid at the age of fifteen and later worked as a housekeeper for most of her life, providing "a correct list of everything in season in every month of the year." Except for March and April, potato appeared in the "roots" category almost all year. However, after the spring interregnum, potatoes reappeared as "early potatoes" in May and continued all the way through February.[190] The prominence of the potato on the list suggested that many English households had mastered the technique of storing the tuber through the winter. A similar list, "the product of the kitchen and fruit of the garden," composed by Hannah Glasse in her *The Art of Cookery Made Plain and Easy* a few decades earlier mentioned potato only in the months of August, October, and November.[191] It appears that the English did not learn how to store potatoes for the winter until the second half of the eighteenth century, as evidenced by naturalist Gilbert White's *The Natural History and Antiquities of Selborne*, published in 1789.[192]

In 1788, a tavern cook named Richard Briggs compiled *The English Art of Cookery*, a comprehensive recipe book with clear imprints of the three cookbooks mentioned above. *The English Art of Cookery* offered a number of potato recipes, the majority of which were identical or similar to those provided by Eliza Smith and Hannah Glasse. There were recipes for broiling, baking, frying, and mashing potatoes on their own, as well as making potato cake, pie, and pudding, which came in three varieties with slightly different cooking methods. Others included potato as one of the ingredients cooked with other foods, such as "French beans ragoued with potatoes," "A Cheshire Pork Pye for Sea," and "An onion pie." Briggs also mentioned a sweet meat dish—sweet lamb or veal pie—and suggested using

"Spanish potatoes" for it. Briggs provided an intriguing recipe for "Potatoes in imitation of a collar of veal or mutton" to use as a meat substitute. If all of this reminds us of Smith's and Glasse's cookbooks, *The English Art of Cookery* also provided a monthly list of food ingredients. Potato was listed as a primary root and vegetable available almost all year around, except during the spring season, in Elizabeth Raffald's *The Experienced English Housekeeper*.[193]

There was also something new in Richard Briggs's *The English Art of Cookery*, which attested to the emerging trend of using potato not only in sweets but also in a variety of savory dishes. It suggested using potato in a variety of fish dishes, including "to boil Cod Sounds," "to boil Barrel or Salt Cod," "to boil Salt Ling," "to boil Scotch or Salt Haddocks," and "British or pickled Herrings boiled."[194] Then, in Amelia Simmons's *American Cookery*, the first cookbook published in the United States, in 1796, potato was listed as the first vegetable, with Simmons writing that potatoes "take rank for universal use, profit and easy acquirement." More specifically, she described its varieties: "The smooth skin, known by the name of How's Potato, is the most mealy and richest flavor'd; the yellow rusticoat next best; the red, and red rusticoat are tolerable; and the yellow Spanish have their value—those cultivated from imported seed on sandy or dry loamy lands, are best for table use." The "yellow Spanish" variety seemed to be the sweet potato. Like her predecessors, Simmons included recipes for "potato pudding" and "potato cake," which, as one supposes, could use both potatoes. Yet her following comments are perhaps more notable: "A roast Potato is brought on with roast Beef, a Steak, a Chop, or Fricassee; good boiled with a boiled dish; make an excellent stuffing for a turkey, water or wild fowl; make a good pie, and a good starch for many uses."[195] That is, during Simmons's time, potatoes became a common ingredient in American cooking.

And the trend continued in both Europe and North America well into the nineteenth century. Maria Eliza Rundell's *A New System of Domestic Cookery*, appearing in 1806, and Eliza Acton's *Modern Cookery for Private Families*, in 1845, are two examples. Though both authors were English, their works were published on both sides of the Atlantic. The former suggested that potatoes be boiled, broiled, roasted, fried, and mashed, in addition to making potato pudding, cake, cheesecake, and fritters, while the latter demonstrated the potato's universal use—it was cooked in almost all dishes, including stews, soups, fish, salads, and, of course, meat.[196] In other words, the white potato now established itself as a primary vegetable in ordinary family kitchens, a position it has held ever since in the modern Western world and beyond.

Sweet potatoes, by comparison, continued to be a popular ingredient in sweet dishes. This is reflected in R. H. Price's popular pamphlet *Sweet Potato Culture for Profit*, which was published in 1896 and provided a thorough discussion of how to cultivate the root plant. Price, a horticulturist who taught in Texas, clearly intended his writing to promote the root plant among southern American farmers. His suggestions for cooking sweet potatoes included fried sweet potatoes, sweet potato pie, and pudding.[197] Over time, the root would become a staple—as either a side dish or a dessert—on the Thanksgiving dinner table for most American families. This achievement, in my opinion, reflected the culinary tradition's influence in the American South, combining significant contributions from both African slaves and Amerindians. As many experts testify, the development of southern cuisine in the United States combined three equally important culinary traditions: European, Amerindian, and African. In this context, three, as Jessica B. Harris puts it, "is a magic number." In the antebellum South, she writes enthusiastically, "The early foodways and their subsequent evolutions are the result of an intricate braiding of three cultures: Native American, European, and African. The result, as we all know, is savory and varied indeed. It has resulted in such dishes as succulent Virginia hams (occasionally stuffed with collard greens), corn pone and sweet-potato pone, and even New Orleans' turducken!"[198]

It is not surprising that Jessica Harris, a well-known culinary historian and major promoter of African American cuisine, mentions sweet potato in her remarks on southern cuisine. Indeed, the root plays an important role in culinary tradition, particularly in the creation of the "soul food" legacy among African American communities in the American South. Traditional soul foods, according to experts, include the following: "Chicken and fish rolled in meal or batter and deep fried, greens and cowpeas boiled with pork and served with pot liquor, okra cooked to a low gravy, sweet potatoes baked to a golden brown, and cornbread in many varieties form the basis of a quasi-ethnic cuisine whose roots are, arguably, as African as they are American."[199] That is, despite being native to the Americas and grown by Amerindians in North America, sweet potato was not initially accepted as necessary for holiday celebrations by all residents of the continent. It was an accomplishment that the root later became a standard on the Thanksgiving dinner menu, as it served as a good example of the creolization of southern cuisine shaped by European settlers, Amerindians, and African diasporas in the antebellum South, and its subsequent acceptance by the entire nation.

Thanksgiving, sweet potato, and African American foodways

To explain how the sweet potato became a staple Thanksgiving food, a brief history of the national holiday is probably in order. While the exact origins of Thanksgiving are unknown, it began about a decade after Virginia was established as an English colony. In 1620, 102 Pilgrims on the ship *Mayflower* arrived in Cape Cod, Massachusetts, after sailing from Virginia; of these, 55 survived until the end of 1621. Their survival was aided by local Amerindians known as the Wampanoag, particularly Tisquantum (ca. 1585–1622), also known as Squanto, who had visited England. Squanto, a member of the Patuxet tribe, had lived with the Wampanoag during the Pilgrims' arrival and taught the English how to plant corn, pumpkin, and squash. On December 13, 1621, the Pilgrims invited Squanto and other Amerindians to a three-day feast to thank nature for their success in the New World. According to the few existing records, the latter provided them with "dear, fish, beans, squash, corn soup, pumpkin, corn bread, berries, wild turkeys, ducks, geese, and swans."[200] As one can see, many of the items, with the notable exception of the sweet potato, later became standard foods for the Thanksgiving celebration in the United States.

Neither the holiday nor the apparent unity between English settlers and Amerindians was established until 1621. However, the concept survived. When another group of English settlers arrived safely in America in 1630, all plantations held a Thanksgiving celebration. At Valley Forge in 1777, George Washington ordered the army to hold a Thanksgiving celebration for the newly formed United States of America. In 1789, during his first presidency, Washington issued the first Thanksgiving Proclamation for the nation, hoping to establish a day for "Public Thanksgiving and Prayer." With Abraham Lincoln's proclamation during the Civil War, the celebration of Thanksgiving gradually became a national tradition. Yet it was not until 1870, under President Ulysses S. Grant (1822–1885), that Thanksgiving became a federal holiday, first in Washington, DC, and later throughout the country.[201]

President Lincoln's proclamation clearly advocated for national unity during the war, a sentiment shared by Sarah Josepha Hale, the primary campaigner for making Thanksgiving a national holiday. In the 1830s, Hale had begun a lobbying effort aimed at uniting the entire country.[202] Hale was a well-known public figure at the time, having written several books on cooking and household management that advocated for the integration of northern and southern culinary traditions. This effort, one might argue, contributed to the sweet potato's inclusion as a Thanksgiving food. Sweet potato was best grown in warm-temperate regions like Virginia and the Carolinas in the American South, for the reasons stated above. Hale included recipes for "potato pudding" and "sweet potato pudding" in one of her cookbooks, *The Good Housekeeper*. Both called for the root to be sweetened, the former with "a tablespoon of brown sugar" and the latter with "one cup of sugar."[203] The sweet potato pudding was apparently intended to be a confection. The inclusion of the sweet potato pudding recipe reflected Hale's keen interest in southern cuisine, as contemporary texts, such as Eliza Acton's *Modern Cookery for Private Families*, featured a plethora of potato recipes but none for the sweet potato. In comparison, Hale's *The New Household Recipe-Book* included recipes for cooking gumbo, jambalaya, and johnnycakes, all of which were exemplary southern dishes showing Amerindian and African influences.[204]

Furthermore, in a special issue of *Godey's Lady's Book* edited by Sarah Hale as part of her campaign to establish the Thanksgiving holiday, she wrote an editorial recommending new dishes to serve alongside turkey, including "Indian Pudding with Frumenty sauce" and "ham soaked in cider three weeks, stuffed with sweet potatoes, and baked in maple syrup."[205] Needless to say, these were well-known foods from the American South, particularly Virginia.

Since the middle of the eighteenth century, sweet potato has also been grown in New England, where Thanksgiving was first celebrated. Yet it was only during the nineteenth century that it gradually became a national delicacy.[206] In 1839, for example, the *Housekeeper's Annual & Ladies Register*, an American magazine aimed at middle-class white women, included a delightful recipe for sweet potato pudding:

Oh, bring me from far in a Southern clime,
The sweetest potatoes that ever grew;
Such apples of earth as the olden time
In its visions and prophecy envied the new.
And wash them with lady-like lily hands,
Till they look as pure as the saffron light
That falls in the summer on fairy lands,
From the moon in the depth of a cloudless night.
And let them be next of their skins beguiled,
But tenderly strip off the earthy vest,
As if you were flaying a sleeping child,
And were cautious of breaking its gentle rest;
And let them be pulverized next by the skill
Of the same white hands and the grater's power;
And a heaping up table spoon five times fill
With the precious result of their golden flour;
Of boiling hot milk add a full quart cup,

And next with five eggs, in a separate bowl,
Beat five table spoonfuls of sugar up,
And stir them well in with the foaming whole.
And one table of spoonful of *eau de rose*
Of salt a tea spoonful, and after these
Of butter an egg-sized morsel; and close
With a flavor of nutmeg, as much as you please.
Then bake it—'t is putting—I pause at the name
To reflect on the puddings of days that are past
And prospects of more, which aspiring to fame
And failing, I've lost to go hungry at last.[207]

Despite Sarah Hale's recommendation and the promotion of the root as "apples of earth" in women's magazines, the sweet potato's establishment as a staple food for Thanksgiving had more to do with emancipated African Americans and their migration from the South to the North and beyond in the aftermath of the Civil War. "Despite its modern popularity as a Thanksgiving standard," writes historian James W. Baker, "this southern favorite [sweet potato] was not included in standard Thanksgiving bills of fare until later in the nineteenth century."[208] This was far from a coincidence.

The acceptance of the sweet potato as a prominent food in the American South served as one of the best examples of the three-in-one creolization of African American cooking. Enslaved Africans working on plantations, in particular, embraced the New World plant as a food in their daily lives. Africans brought yams and other African foods to North America, including rice (*Oryza glaberrima*), okra, and black-eyed beans/peas (cowpea), but they also quickly adopted sweet potatoes as a substitute for yams, such as roasting possums with sweet potatoes instead of stewing cutting grass rat with yams.[209] Some of the Africans had most likely seen the sweet potato before because Portuguese colonists introduced it to West Africa beginning in the late fifteenth century. Sweet potatoes were roasted in ashes when Africans cooked their one-pot meal for subsistence as an important supplement to rice, which was a staple food for them and was planted in plantations alongside tobacco. "African American cooks continue," writes cultural historian Frederick Douglass Opie, "to grow and cook with yams and sweet potatoes. They used these staples like bread, just as their descendants had done in West Africa."[210] Despite their botanical differences, yams and sweet potatoes are both root plants. Their similarity, as in the previous case with European acceptance of turkey and beans, likely explained why Africans in the American South readily accepted both into their cooking. As Jessica B. Harris, the aforementioned expert on African food diaspora, points out, this resulted in "the eternal confusion" of yam and sweet potato among many Americans today. "In African American parlance and from there into Southern usage, they retained the name of the African tuber that they replaced—yam," she writes.[211] Meanwhile, Africans spread yams throughout the New World. As a result, according to food scholars Judith Carney and Richard N. Rosomuff, "the African yam remains one of the most important crops that slaves pioneered in the Americas."[212]

Needless to say, yam and sweet potato differ in appearance: the former is usually larger and less sugary, earning it the nickname "elephant's foot" in Africa, whereas the latter is juicier and softer, with a more reddish and smooth skin. However, their differences did not prevent them from being liked by everyone who helped shape the American South. As previously stated, sweet potato recipes were collected by Mary Randolph in *The Virginia Housewife*, and "Thomas Jefferson bought cucumbers, sweet potatoes, and squash from his slaves."[213] Indeed, one could argue that the sweet potato's sweetness made it more appealing to the people of the time, including Africans.

When discussing how African Americans favored sweet potatoes, it is natural to bring up the life and career of George Washington Carver (ca. 1864–1943). Carver was dubbed a "black Leonardo [da Vinci]," and his accomplishments were truly astounding and admirable given his humble upbringing. Carver was born a slave in Missouri, lost his father before his birth, and was kidnapped along with his mother and sister when he was only a week old. He was later rescued, but he grew up without a mother. Because slavery was abolished during his childhood, Carver was able to attend school in his early teens. His schooling, though, was far from smooth or systematic. After trying several other subjects at other schools and hoping to find "the key to unlock the golden door of freedom to our people," he eventually decided to study scientific agriculture at what was then Iowa State College. This proved to be the right decision. Despite the hostility he faced as the campus's first and only African American student, Carver excelled in botany and performed above average in other related courses. While living in an abandoned office from time to time during his undergraduate years, he managed to present a paper at the annual meeting of the Iowa Horticultural Society. Upon graduation, he was offered a job to earn money while enrolled in a master of science program.[214]

Indeed, Carver was one of the few Black Americans who received graduate education during his lifetime. As a result, after earning his degree, he was recruited to teach and later head the Agricultural Department at the Tuskegee Institute, a historically Black college in Alabama, by Booker T. Washington (1856–1915), a distinguished African American educator serving as president of the institute. Carver experimented with innovative ways to cultivate peanuts, sweet potatoes, and other food plants at Tuskegee, where he later taught for nearly half a century. His initial goal was to investigate how these plants could help preserve soil fertility. Soon after, he shifted his focus to developing new products from these crops. Throughout his career, Carver created over a hundred peanut projects, ranging from cream, soap, and lubricating oil to brittle, cake, and punch. Concerning his experiments with the sweet potato, he remarked in a 1921 interview that "The sweet potato products numbered 107 up to date, I have not finished working with them yet."[215] Its ingredients list included nonfood items such as alcohol, dyes, wood fillers, synthetic cotton and silk, and writing ink, as well as foods such as mints, powder, chocolate, candy, coffee, sugar, and yeast. Carver's success in producing sweet potato flour seemed to be the most notable of the latter. He experimented with making sweet potato flour during World War I, when the country was suffering from a wheat shortage. He also published *How to Make Sweet Potato Flour, Starch, Sugar Bread, and Mock Cocoanut* in 1918. His experiments demonstrated that combining sweet potato flour with wheat flour significantly increased the supply of the latter. While attracting the attention of the Department of Agriculture, Carver's success also earned him a national reputation as "one of the world's foremost scientists for all time." Following his death, U.S. President Franklin D. Roosevelt (1882–1945) praised him as follows: "The versatility of his genius and his achievements in diverse branches of the arts and sciences were truly amazing. All mankind is the beneficiary of his discoveries in the field of agricultural chemistry."[216]

The fact that the world's most innovative sweet potato scientist was African American is no coincidence, as the root plant played an important role in the lives of many African Americans. Indeed, as a native American plant brought to North America primarily by European settlers and embraced by enslaved Africans due to its resemblance to yam, sweet potato epitomized the previously discussed creolization of African American food culture. Michael W. Twitty writes about the mixed influences on Edna Lewis, an iconic figure in shaping the culinary tradition of the American South:

> In the New World, both Creole-born Africans and Europeans living in the Americas would come to greater cultivation and appreciation of their sweet tooth. The traditional African palette favored more hot, bitter, spicy, and oily dishes rather than sweet ones. In the Americas, however, Africans would not only cut cane but embrace the confections of the societies that enslaved them. The soothing nature of sucrose was one of the few comforts knowns to bondspeople. Great-grandmother was a stranger to sweet, but great-granddaughter would be jumbles, biscuits, cakes, sweet bread, jams, and other delights.[217]

Sweet potato is prominently featured in Edna Lewis's 1976 autobiography *The Taste of Country Cooking*, where Lewis describes the food she grew up with in Freetown, Virginia, in an African American family. The sweet potatoes she ate were cooked with sugar and butter, such as sweet potato casserole and sweet potato pie, except for pan-fried sweet potato, which she recommended eating as a side dish with "ham in heavy cream sauce" and "covered fried eggs" for an early summer breakfast.[218] Sweet potatoes are mostly cooked as desserts in her 2003 recipe book *The Gift of Southern Cooking*, which she coauthored with Scott Peacock. Lewis even teaches her readers about "the sweetness in sweet potatoes," or how to choose different types of sweet potatoes for their varying sweetness in order to achieve the desired result.[219]

It should come as no surprise that the way African Americans in the American South prepared sweet potato remind us of how the root was dressed and consumed in Europe. Since the region's early settlement days, the aforementioned culinary creolization has been in motion. Edna Lewis's demonstration of how to cook sweet potatoes reflected the influence of English gastronomy. For in the English culinary tradition, pie and pudding making has a long tradition, so much so that some experts on the Thanksgiving tradition argue that pumpkin and mincemeat pies should take precedence over roasted turkey as a standard food for the celebration.[220] For the Puritans, pumpkin had an "iconic status as the symbol of New England frugality and self-sufficiency," according to James W. Baker. Early nineteenth-century recipe books by Sarah Josepha Hale included roasted and boiled turkeys. She, on the other hand, did not associate them with Thanksgiving. In comparison, Hale thought "mince pies" were "indispensable" for Thanksgiving and Christmas, as was the "family mince pie," which was made with "one pound of brown sugar" and other spices.[221] Nonetheless, the English's emphasis on pie and pudding making influenced African Americans in the American South. "From the British, African Americans acquired a taste for and the ability to make pies and puddings, which they made with both the African yam and the American sweet potato," according to Frederick Douglass Opie. "They also took pie making to another level with the baking of fruit cobblers from cast-off and foraged fruit and scraps of dough leftover from pie making done in the big-house kitchen."[222]

While African Americans absorbed European culinary influences, Opie points out, the planter class also "took great delight in the dishes of their slaves." Among the foods were "roasted yams," although they may have been sweet potatoes, as the two were often confused in the American South.[223] As slaves, who had quickly become the majority of the region's population, prepared foods, more and more African food traditions entered the culinary practices that came to shape southern cuisine. That is, the influence was reciprocal. Examples of this were found in Mary Randolph's

1824 *The Virginia Housewife*. Her cookbook included, perhaps unbeknownst to her, ingredients from Africa, such as field pies (black-eyed peas) and okra, which she spelled inconsistently as either "ochra" or "ocra." Randolph not only recommended okra soup but also used it in her dish "Gumbs—A West India Dish." Other African and Amerindian influences were also discernible, such as roasting and barbecuing pig/shoat, making pepper pot, boiling rice, and cooking rice journey or johnnycake (which Randolph thought was as good as cassava bread), all of which were well adopted and developed by African Americans and marked their soul food tradition.[224] Eugene D. Genovese, a twentieth-century historian of the American South and slavery, observed that slave masters "imbibed much of their slaves' culture and sensibility while imparting to their slaves much of their own. . . . Slavery, especially in the plantation setting and in its paternalistic aspect, made white and black southerners one while making them two." The negotiation of various foods and their ever-changing preparations was crucial in that process of blending. Before becoming a well-known landscape architect, Frederick Law Olmsted (1822–1903) was a journalist who documented in his journals his trips to the South, where enslaved Africans prepared "hot corn bread, sweet potatoes roasted in ashes, and fried eggs" at plantations. Olmsted tasted the meals on a Virginia plantation and chose sweet potato as his staple. "There was no other bread, and but one vegetable served—sweet potato, roasted in ashes, and this, I thought, was the best sweet potato, also, I ever had eaten," he said.[225] Indeed, by the nineteenth century, "the majority of Southerners ate like enslaved Africans and free African Americans, enjoying all parts of the hog, corn bread, greens, sweet potato pie, candied yams, and black-eyed peas and rice," concludes Opie.[226]

Thanks to the emancipated African slaves who relocated nationwide after the Civil War, the southern foodway, in which sweet potato played a central role, made great strides in reshaping American gastronomy. "The Great Migration of the first part of the 20th century," explains Jessica B. Harris, "moved many African Americans from the South to homes elsewhere around the country. They brought with them the tastes of Africa that they had made southern. In both the South and the North, African American women found work as housekeepers and cooks, spreading the tastes of African American food still more widely in our national culture."[227] Harris includes a recipe for the aforementioned "possums with sweet potatoes" in her book *High on the Hog*. She also describes how African Americans in Harlem, New York City, sold street foods they remembered from their enslavement days in the South, such as fried chicken and pig trotters, whereas "sweet potatoes, mistakenly called yams, ... were another reminiscence of things Southern."[228] Then, in his famous *Invisible Man*, Ralph Ellison (1913–1994), the African American writer whom Harris frequently quotes in recounting the migration of southern foods, provided a vivid depiction of African Americans' nostalgia for the root after leaving the South for the North.

Such nostalgia is also addressed in Frederick Douglass Opie's *Hog and Hominy*, which depicts the evolution of soul food in the United States. "Collards and cabbage seasoned with salt pork, fried chicken, and sweet potato pie carried with them similar memories of childhoods in the South," he writes of southern migrants. And the memory was usually passed down through the family for generations. Adam Clayton Powell Jr. (1908–1972), the first African American from New York to become a member of the United States House of Representatives, recalled fondly the variety of southern foods his parents cooked for him when he was a child. Among those of a similar background, sweet potato pie would be a must for holidays as well as in daily life. For example, in the household of Fred and Lucy Opie, the author's relatives who settled in New York City, Lucy baked "biscuits; hot cross buns; cake; mincemeat, rhubarb, sweet potato, and cherry pies; and peach cobbler" on a regular basis. Sweet potato pie was also remembered as a regular Christmas food for Ruth and Roy Miller, another African American family featured in Opie's book, whose parents had moved to New York from the South in the 1920s.[229]

Over time, sweet potato evolved into more than just a southern specialty. D. M. Nesbit of *Farmer's Bulletin* observed in 1902 that improved technology and better transportation made it available to people in the North after the month of November. He also mentioned that sweet potato, also known as "Jersey sweets," was produced in New Jersey at the time.[230] As a result, the root was chosen as a street food by city food vendors, who frequently mixed food offerings from different traditions in order to cater to and cultivate their customers' palates. "In urban areas in and around New York," according to Opie, "Europeans, Caribbeans, and southerners interacted in tenement houses, on the job, and on the streets, where they purchased new types of inexpensive good-tasting foods. Italian vendors sold pepper-onion-sausage sandwiches and Italian ices. Yiddish-speaking Jews sold arbis, knishes, and sweet potatoes."[231] Sweet potatoes were found on the menus of the catering businesses pioneered by African Americans in big cities, facilitating the progress by which sweet potatoes became accepted among all residents. The Dutrieuilles family, for example, established Albert E. Dutrieuilles Catering in the late 1800s and prospered in the catering business in Philadelphia. Sweet potato was on their menu as a side dish or dessert, and it was frequently cooked and served to their customers.[232]

By the early twentieth century, most American families would have included sweet potatoes on their Thanksgiving dinner tables. Indeed, according to Elizabeth Pleck, who studied the holiday's perception among newcomers to the country, the "entire American menu" for the holiday's celebration should include "chestnut stuffing and sweet potatoes" in addition to turkey, pumpkin pie, cranberry sauce, and other traditional fare.[233] When Kathryn and Anthony Blue published their *Thanksgiving Dinner* near the end of the twentieth century, they included three recipes for the root: candied yam (sweet potato), praline sweet potato, and mashed sweet potato.[234] Needless to say, whether as a side dish or a dessert, the sweet potato is cooked with enough butter and sugar to caramelize. Sweet potato with marshmallows is popular among many American families to enhance its sweetness, but in recent years, there appears to be a new trend to leave the latter out because it may be too sweet even for a candied dish, especially among those who are health-conscious.[235] All the same, a variety of candied sweet potatoes are regarded as a standard dish among the majority of families in the United States and Canada today.[236]

Sweet potato in modern Latin America

How important is sweet potato in Mexico, another major North American country? The answer is a little more complicated. On the one hand, Mexico was home to both the Maya and Aztec civilizations, where sweet potato consumption has had a much longer history than both the United States and Canada. On the other hand, if one assumes that sweet potato is a daily component of the Mexican diet, and even that Mexicans today necessarily consume more sweet potato than their northern neighbors, such an assumption may be somewhat false, because the root clearly pales in comparison to corn, another native food plant in Mexico that, in the form of tortilla bread, is a daily staple for the majority of Mexicans. Sweet potato, in comparison, is more of a snack, side dish, and dessert for many Mexicans. As a result, the root, prepared in various ways, is commonly sold at food stands on streets across the country, alongside other ground provisions such as onions, cabbage, zucchini, carrots, and white potatoes. One reason Mexicans do not eat sweet potatoes on a regular basis is that the root is usually made into a concoction, such as the popular street food *camotes de Santa Clara*. According to legend, the sweet, which is essentially a candy made of the root, was invented in the sixteenth century by nuns in Santa Clara, Puebla. As simple as it is, the candy has remained popular among Mexicans and their Mesoamerican neighbors for the past six centuries. However, most people in the region regard candy as a treat and do not consume it daily.[237]

Where the sweet potato originated and was first consumed as food, in Central and South America, as well as in the Caribbean, it was received differently than in North America. The root is also eaten as a holiday food in some areas, such as the Caribbean. *Habichuelas con dulce*, for example, is a famous traditional drink/soup in the Dominican Republic, or Hispaniola, where Christopher Columbus landed on his first voyage to the Americas. It is a sweet bean liquid dessert that is usually served with milk cookies or, if you want to be truly authentic, with *casabe*, which is a flatbread made of yuca (cassava) flour. To make it, the locals usually boil red beans with cinnamon, nutmeg, coconut milk, raison, sugar, and salt for about thirty minutes, or until everything is well blended and consistent. Then cooked sweet potato chunks and other spices for flavoring are added. Sweet potato pudding (*pain patate*) is almost essential for several religious festivals in Haiti, where sweet potato has been a staple for centuries. On November 2, the Roman Catholic All Souls' Day, or Fête Gede (Festival of the Dead), a typical menu consists of greens, yams, macaroni and cheese, corn bread, red beans, rice, cabbage, baked chicken, fried red snapper, and, last but not least, sweet potato pudding. Most families include the pudding as a featured dessert when celebrating Christmas. Then, in Argentina, a major South American country, some regions celebrate Midsummer, known as the Night of Saint John, by building large bonfires in both villages and towns. Sweet potatoes are cooked in ashes and eaten warm as snack foods, which is a common practice among many peoples, past and present.[238]

However, the sweet potato's popularity extends beyond its use as a holiday food in the Caribbean, Central, and South America. This is not surprising given that the root is native to these places. Again, throughout the Caribbean, where Columbus first encountered the root plant, and most of

Central America and beyond, sweet potatoes remain a staple food.[239] The root is particularly popular in island countries and regions such as Haiti, Cuba, Barbados, Dominica, Dominican Republic, Jamaica, Grenada, Trinidad and Tobago, and Puerto Rico. Aside from rice and beans, tubers, and roots (sweet potato included) are commonly referred to as *viandas* and are an important part of the local cuisine. They are eaten daily for lunch and dinner, whether boiled, fried in slices, or mashed. *Pain patate*, the aforementioned popular sweet potato pudding, is also prepared for both special occasions and everyday consumption. The same can be said for the sweet potato's widespread popularity in countries such as Belize, Guyana, Suriname, and French Guiana, which are in Central and/or South America but are part of the Caribbean Community. In both Guyana and French Guiana, it appears that locals prefer starchy foods like manioc, sweet potato, and taro over rice, even though rice is consumed on a daily basis. In Suriname, the daily dish is called *Her'heri*, and it combines sweet potato and taro with salt cod, also known as *bakkeljauw*. The following is the recipe:

> Put a large pot of water on the stove, and bring to a boil. Meanwhile, peel the plantains [cooking banana]. Add salt to the water, and put the plantains in the pot. Bring to a boil. Add the pieces of Chinese taro, pomtajer, frozen cassava, napi, and sweet potatoes. With a slotted spoon, remove the plantains after thirty minutes. Boil the rest of the tubers about thirty more minutes, and until tender.
>
> Meanwhile, rinse the salt cod under cold running water. Cook it for ten minutes in two quarts of boiling water. With a slotted spoon, remove the fish from the pot, and set aside.
>
> Divide the fish into pieces. Use a slotted spoon to remove the tuber pieces from the pot. Serve the plantains and tubers with the fish and a little bit of its cooking liquid.[240]

The above is just one example of how sweet potato, along with other ground provisions, is used in many daily dishes in the Caribbean and Central America. For example, in Martinique and Guadeloupe, two of France's overseas territories in the Caribbean, a fish dish called *Blaff de Poisson Blancs* (fish blaff) is considered a typical meal for the residents. It is traditionally served with sweet potato, taro, or yam. The Indigenous Miskitu people perceive and divide foods into two categories in Nicaragua, the largest country in Central America: protein and starch, or *upan* and *tama*. The

Miskitu believe that daily food consumption should include both *upan* and *tama*; the former typically consists of turtle and fish meat, while the latter includes all starchy foods such as cassava, sweet potato, and cocoyam, as well as rice, maize, and plantains. Locals in Guatemala, the most populous country in Central America, which once hosted both the Maya and Aztec civilizations, still refer to sweet potatoes as *camotes*, as do Mexicans. Instead of making sweet potato pudding for dessert, Guatemalans, like their Mexican neighbors, eat *camotes* soaked in honey or syrup as an after-dinner treat, which likely reflects centuries-old Mayan and Aztecan dietary practices.[241]

Sweet potato is a native plant in South America, particularly in the Andean countries of Peru, Colombia, Bolivia, and Ecuador, where it has been cultivated and consumed for millennia. As a result, there are thousands of cultivars of both the sweet potato and the white potato, which is also native to the continent, readily available at markets for people to choose from. In fact, potatoes are so common in Peru that one may argue that Peruvians do not differentiate between white and sweet potatoes, as people tend to do in the rest of the world. This is not to say they do not know the difference, but it is less important in everyday life because, as Jennifer Woolfe noted back in 1992 in *Sweet Potato: An Untapped Food Resource*, genetic variations of the plant have been so rich and abundant. Not only were there, she writes, "diverse names given to sweet potato types in the *quechua* language: the sweet and moist types being known as *apichu* and the drier types as *kumara*. Sweet potatoes of different skin colours were also known by individual names." In other words, given the variety of potatoes grown and consumed in Peru, distinguishing white and sweet potatoes solely by skin or flesh color is not as useful as it is elsewhere. Perhaps because of Peru's richness in potato varieties, sweet potato consumption there has not been as robust over the past century as one might expect: The root plant, though grown throughout the country, was regarded as merely one of the food crops people consumed. Lima, the capital city, was a major sweet potato consumer, with people buying them not only for themselves but also for their dogs. Despite this, the shipment of sweet potatoes to the city was the third-largest in terms of volume, indicating that the residents had other food options. Prior to the 1990s, when Woolfe conducted her research, only 10 percent of sweet potato production went into processing to make starch and snack chips in the country. Peruvian sweet potato production also fell by more than 26 percent between 1971 and 1987.[242]

Interestingly, it was also during the last quarter of the twentieth century that another new trend began to emerge. Concerns about climate change and overreliance on imported grains such as wheat prompted countries like Peru, and South America in general, to promote native food plants.[243] The incentive was partly reflected in the establishment of the International Potato Center (Centro Internacional de la Papa, or CIP) in Lima by Peruvian government decree in 1971. The CIP evolved from two potato research centers in Peru previously funded by the Rockefeller Foundation and USAID. Since its inception in 1971, the CIP has worked in close collaboration with the Consultative Group for International Agricultural Research, which it joined in 1972, by cosponsoring international symposia and launching joint projects to promote both white and sweet potatoes as valuable food plants around the world. And the work of the CIP is focused on both potatoes concurrently.[244]

Peruvian chefs have also made efforts in recent decades to combine the use of native and imported food ingredients in the hopes of developing the country's unique cuisine, or *cocina criolla*. "Even if all you had were cheap ingredients, like rice or potatoes with some gravy or bits of pork; we [Peruvians] could depend on our good *sazón* [seasoning]." This is how a Peruvian journalist describes the ideal combination of Spanish/European recipes with native Latin American food ingredients and cooking styles.[245] As one might expect, both potatoes are widely used in the preparation of people's daily meals. Ceviche, for example, is arguably Peruvian cuisine's most popular dish, frequently served as the first course in restaurants as well as by street food vendors. It even has its own public holiday! Ceviche is a raw fish dish that typically features striped bass or grouper as the main ingredient. The marinade consists primarily of lime or other citrus juice, pepper, red onions, and salt. Ceviche is traditionally served with a slice of boiled sweet potato, a few kernels of corn, or one to two plantains as garnish.[246] And sweet potatoes, and lime, can also be cooked with other types of fish, such as sole. In terms of cooking method, boiling is the most common way to prepare both sweet and white potatoes across South America. After boiling, the potatoes can be served whole or mashed in almost any dish, but they are especially good as the base for "refreshing salads and spicy cold sauces."[247] In sum, sweet and white potatoes, as well as other types of potatoes with varying skin and flesh colors, are essential foods for Lima residents. "Potatoes are," Michel Ian Collins finds, "one of the principal components of the diet in Lima. Together with rice, bread and noodles they compromise the main staples of the Lima diet."[248]

Despite being a native plant with a long history of being consumed as food in South America, the sweet potato's popularity appears to vary from region to region and country to country. The availability of a large number of sweet potato cultivars on the continent also means that consumers can choose a specific type of sweet potato that they prefer. In general, sweet potatoes grown in South America are classified as either sweet or non-sweet. For example, while Peruvians prefer the former, many Colombians may prefer the latter. Sweet potato, which was introduced by the Portuguese centuries ago, is also very popular in Brazil. Aside from their preference for purple-skinned varieties, most Brazilians consume sweet potatoes in the same way that Europeans and North Americans do. That is, while the root is popular, it is not a staple among them. However, research has revealed that sweet potato consumption patterns in South America are influenced by socioeconomic conditions and corresponding family income levels. In the Brazilian Amazon, for example, roots such as sweet potato and manioc are common crops grown by local villagers who eat them on a daily basis.[249] Another instance can be found in Peru. *Amarillo verdadero* (yellow type) and *Morado legitimo* (purple type) are the two most popular sweet potato varieties in the country, with over two thousand varieties available. The former is buttery, soft, and starchy, whereas the latter contains less water and is thus more resistant to damage. According to Jennifer Woolfe's research, *Morado legitimo* (purple type) is between 80 and 100 percent more expensive than *Amarillo verdadero* (yellow type), making the latter much more popular, particularly among low-income families. However, as in Brazil, middle- and upper-income families prefer the purple variety of sweet potato to garnish their ceviche. While more expensive, the purple type is more appealing because it is usually round in shape, whereas the yellow is irregular in size and more perishable, both of which affect the latter's price negatively. In general, high-income families regard sweet potato as a supplementary food, whereas low-income families in Central and South America regard it as a staple. For example, in Lima, there are a variety of street food vendors from whom workers frequently purchase sweet potato with bread for breakfast. Throughout the day, city residents and tourists can sample the delectable *picarones*, a sweet potato and pumpkin doughnut that is extremely popular in Peru, Ecuador, and neighboring countries. As if the snack was not already sweet enough, many people eat it with *chancaca*, a sugarcane syrup. Of course, the doughnut can be dipped in honey or other syrups to increase its sweetness.[250] In comparison, "more than 50% of local rural and small town dwellers eat sweet potato daily or almost daily" in the Cañete Valley, one of the Central Coast sweet potato production areas that supplies the majority of Lima's sweet potato consumption. A similar situation can be found in Peru's northern highlands, in such places as Cajamarca City and its surrounding countryside, where low-income families eat boiled sweet potatoes with rice as their daily staples, whereas middle-income families consume the root as dessert, as a puree or pie filling. And high-income families feed the root, particularly the yellow variety, to their dogs, whereas low-income families only use the vines for this purpose.[251]

Since the turn of the sixteenth century, when European explorers brought the sweet potato back from the Americas, the plant has been a sweet gift linking the Old and New Worlds. That is, even though in parts of the Americas, such as the Caribbean, where it has been consumed more or less on a daily basis by people, particularly those from the lower social strata, the root has been traditionally processed as a sweet in most culinary practices in Europe and the Americas from then until now. Although most of the root's varieties are sugary, making them an ideal candidate for a confection, there are also non-sweet cultivars, which are more widely available in South America, where the plant originated, than elsewhere, as I hope to have demonstrated in this chapter. To some extent, the distinctions between *batata* and *aji* discussed earlier serve as examples, as the former is juicy and sweet, whereas the latter is dry and starchy. Because of their strong interest in sweet and even saccharine foods by today's standards, which had been fostered by their prior contacts with the Middle East since the late medieval period, Europeans, including Christopher Columbus, were drawn to the sweet potato in the early modern era primarily because of its sugar content. The *batata* variety—"potato" in English, as the root was known before the qualifier "sweet" was added to distinguish it from the white potato—had become most favored in the plant's transfer from the Americas to Europe, from the Iberian Peninsula, where it first landed, to the British Isles and continuing across the European continent and farther on to North America. Many professional cooks and ordinary families experimented with sweet potato desserts, including puddings, tarts, cakes, casseroles, purees, and pies, to capitalize on its sweetness. While interest in sweets has waned across Europe and North America, related culinary practices and dietary preferences have generally persisted to the present day. As a result, despite its importance as a food plant, sweet potato remains a supplementary food—a side dish, snack, and dessert—in the mainstays of Euro-American cuisine. But we will see that the root has met a very different fate when we turn to Asia and Africa.

CHAPTER III

Mundane or Miracle?

Feeding and Fueling China's Population

China has been the world's largest sweet potato producer since the early twentieth century, a status it has maintained to this day.[1] In 1961, the total sown area of the sweet potato in China reached 10.89 million hectares.[2] The same year, Wang Jiaqi, a historical geographer, claimed in *Wenwu* (Cultural relics), a major journal in the fields of history and archaeology, that the sweet potato was a native plant in China. Using historical sources, Wang concluded that *ganshu* 甘薯 (lit. sweet yam/taro), which appeared in various texts beginning in the third century, was the sweet potato. The texts on which he based his argument were included in the *Ganshu lu* 甘薯錄 (Records of the sweet potato), a book compiled in the eighteenth century during the high Qing. Soon after his article was published, prominent archaeologist Xia Nai disputed Wang's claim. Xia emphasized that, while *ganshu* was one of the names for the sweet potato, it actually referred to *shanyao* 山药 (*Dioscorea polystachya*), or Chinese yam, a native plant of East Asia. While some *shanyao* varieties are sweet, it is not sweet potato. In the same issue of *Wenwu*, another historian, Wu Deduo, expressed his support for Xia Nai, while Wang Jiaqi responded to both of them.[3]

However, the debate was not over, nor, as Manuel Perez Garcia recently argued, was the attempt to create a "national narrative" on the origin and spread of the root plant in China.[4] In 1983, after the country had recovered from the devastation of the Cultural Revolution (1966–1976), Zhou

Yuanhe, a historical geographer at Fudan University's famed Institute of Historical Geography, came forward to agree with Wang Jiaqi. Citing additional Chinese sources, Zhou stated that the *ganshu* mentioned in ancient texts could be the sweet potato, despite the contrary opinion expressed by many others that the sweet potato originated in the Americas and was introduced into China in the sixteenth century. In fact, by the time Zhou published his article and reignited the debate, China's sweet potato production had reached 111 million tonnes, while global production was 133,360,000 tonnes.[5] That is, China then contributed more than 80 percent of the world's sweet potatoes. Currently, despite a significant decline, China remains the world's largest producer of the root plant, with 46.6 million tons produced in 2022.[6]

Early records of the sweet potato's introduction

Academic research in China advanced significantly between 1961 and 1983, or from Wang Jiaqi's article to Zhou Yuanhe's. Chinese scholars became far more up to date on scholarly developments from around the world. Sweet potato, like other food plants from the Americas, had also received a lot more attention from academics both inside and outside of China. Alfred Crosby's *The Columbian Exchange*, which appeared in 1972, was, without a doubt, a seminal example. Crosby referred to the work of Ping-ti Ho, an acclaimed Asian scholar in the United States who in the 1950s had studied the introduction of American plants into China and their impact on population growth.[7] Ho's studies were apparently unknown to Wang Jiaqi, but they were cited by Zhou Yuanhe because, while originally published in English, they had been translated into Chinese only a few years before Zhou wrote his article.[8] In addition, Quan Hansheng, a renowned economic historian based in Hong Kong, investigated the role of American plants such as sweet potato and maize in changing Chinese agriculture from the sixteenth century onward.[9]

While there were still disagreements in China about the origin of the sweet potato in the 1980s, few disputed its significant impact on the Chinese economy and population, which this chapter will discuss in detail. Zhou Yuanhe's support for Wang Jiaqi was unusual at the time because, among others, Ping-ti Ho and Quan Hansheng had tended to the thesis that the sweet potato is a native plant of the Americas and was brought to China in the late sixteenth century. Ho cited sinologist L. C. Goodrich's earlier contention that the sweet potato had entered China in 1594, which he elaborated in his 1955 article "The Introduction of American Food Plants into China," though he pointed out that the root had probably arrived in China via Yunnan a few decades earlier, based on other sources. In other words, according to Ho's research, there was more than one route for the sweet potato to enter the Asian mainland; in addition to the one through Yunnan from Burma (Myanmar), he wrote, there were at least two additional routes for the plant to reach Fujian on the southeastern coast before continuing northward and northwestward on the continent.[10]

Except for Zhou Yuanhe, most Chinese scholars from the 1980s to the present agreed with Ho that sweet potato was imported to China during the late Ming period (1368–1644) and that there were multiple routes for the transfer. Chen Shuping, an economist, writes at the beginning of his 1980 article on the spread of maize and sweet potato in China that "both maize and sweet potato were originally domesticated in Central and South America," and that "these two important food plants spread from the Americas to the rest of the world after Columbus' voyages from 1492 and, via different routes, also reached our country [China]." The same year, Liang Jiamian and Qi Jingwen, two agriculturalists, published their research on the sweet potato's introduction to China. While they agreed that the plant was from the Americas, they suggested that, in addition to the route through Fujian, the root also arrived from Vietnam via two routes through the province of Guangdong. In 1988, Cao Shuji, then a doctoral student in history, proposed his thesis on the introduction of the sweet potato to China. Cao disagreed with Ping-ti Ho and others that sweet potato could have spread into China via land routes (Burma to Yunnan); his argument was that the root could only have arrived from the sea.[11] In recent years, the subject has resulted in book-length studies, master's and doctoral theses, and it appears that Cao's general hypothesis has garnered the most support among the younger generation of scholars—many believe that sweet potato entered Fujian and Guangdong via sea routes, rather than the land route suggested by Ping-ti Ho.[12]

But the question remains: Which Chinese province, Guangdong or Fujian, was the first to cultivate the American plant? The fundamental problem is that almost all of the above theories about the introduction of the sweet potato into China, including the land route from Burma to

Yunnan, are based on readings of local gazetteers, whose compilation was an important branch of Chinese historiography. These gazetteers frequently provide encyclopedic coverage of the area, depicting both notable figures and significant events. Obviously, the information contained in such local histories is valuable. However, because those works were typically created by a group of people, the content recorded in them is inconsistent. In the case of the sweet potato, the plant was referenced by several names in those gazetteers. Ping-ti Ho, Liang Jiamian, and Qi Jingwen proposed a land route from Burma to Yunnan, basing their argument on the use of *Yunnan tongzhi* (General history of Yunnan). Later scholars argue that, while the name *ganshu* was used in *Yunnan tongzhi*, it actually referred to Chinese yam rather than sweet potato.[13] After all, the term *ganshu* had appeared in historical sources for many centuries—in fact, it was how Wang Jiaqi in 1961 made his case for the plant being a Chinese native, as *ganshu* was found in several texts dating back to the third century.

However, a new name, *fanshu* 番薯 (lit. foreign yam/taro), was recorded in two other local gazetteers in reference to a root encountered by Chinese travelers in Vietnam. The prefix *fan*, which was traditionally used to mean "foreign" or "barbarian," suggested that the plant's origin was outside China. More specifically, the gazetteers recorded two rather dramatic stories about how the sweet potato was smuggled into Guangdong, proving that it was a foreign crop. The hero of the first story, which appeared in *Dongguan Xianzhi* (Gazetteer of Dongguan County), was a scholar-official named Chen Yi, whose father and grandfather were both government officials. Chen took advantage of an opportunity to visit An'nan (Vietnam), where he was served sweet potatoes. The gazetteer wrote that he found them so delicious that he paid someone there to bring back a sample. Chen planted the root in his garden after returning from Vietnam in 1582, and it grew quickly. He named it *fanshu* to distinguish it from the native roots because it originated outside of China.[14]

The other story, which appeared in the *Dianbai Xianzhi* (Gazetteer of Dianbai County), was more theatrical. It began by stating that sweet potato export was strictly prohibited in Vietnam, and that anyone who brought it to China would face the death penalty. The hero in the story was a doctor named Lin Huailan (or Lin Huaizhi, as some others have believed), who traveled to Vietnam and cured the illness of the king's princess. When given a cooked sweet potato, he requested a raw one and hid it in his sleeve. Lin was stopped at the border as he attempted to return to China. He begged the border officer to release him, but the latter responded that despite being impressed by Lin's skill as a doctor, he could not disobey his king. The officer eventually relented, letting Lin go, and then threw himself into a river to make amends with the king. Lin promoted the sweet potato in Guangdong after bringing it back.[15] While Lin Huailan's story was captivating, it did not provide a date for his departure from Vietnam. Yang Baolin, a modern agriculturalist, therefore, proclaims that Chen Yi, instead of Lin Huailian, was the first to introduce sweet potato and that the plant first grew in China in Dongguan, Guangdong.[16]

Both episodes, referring to the sweet potato as *fanshu*, were supposedly recorded before 1594, which L. C. Goodrich, who was aware of the aforementioned sources on its transfer to Guangdong, thought was the first year the root was introduced to China. Ping-ti Ho, as previously stated, believed that the sweet potato could have entered the country before 1594. However, the year remained significant in terms of the root's introduction to Fujian. And, for reasons discussed below, the record about the transfer to Fujian was more detailed and widely circulated than the previous two. It was presented in *Jinshu chuanxilu* (On the cultivation of the sweet potato), an early text on the sweet potato's propensities and farming method compiled by Chen Shiyuan in memory of his great grandfather Chen Zhenlong's contribution to bringing the root to Fujian. According to Chen Shiyuan, Chen Zhenlong was a merchant who spent a significant amount of time in Luzon, in the Philippines, and saw and tasted the sweet potato in a place where it had already become a local staple. "When he [Chen Zhenlong] asked," Chen Shiyuan wrote, "the residents all said that sweet potatoes have six benefits and eight advantages, on a par with the five grains. It is the country's treasure, on which people rely for a living." Chen thought it was a good plant to introduce to Fujian for its fast ripening and high yield, as well as its palatable flavor, because the soil in the province is too poor to grow other food plants well. Chen Zhenlong, like Chen Yi and Lin Huailan, smuggled the root back to China despite the Spanish government's prohibition. This supposedly happened in 1593. The following year, Fujian was struck by a famine. Chen Zhenlong presented the sweet potato to the province's governor, Jin Xuezeng (fl. 1568–1597), who ordered that it be planted throughout Fujian to help alleviate the famine. The survivors, grateful for Jin's policy, named it *Jin gong shu* (Mr. Jin's spud) or simply *jinshu* after Jin. Because *jin* also means gold in Chinese, the name *jinshu* reflected the people's liking for the plant as a famine food.[17]

In subsequent centuries, shrines were built in both Guangdong and Fujian to commemorate the sweet potato's introduction to China. The one for Lin Huailan was called Lin Gong Miao (Master Lin's Shrine), and it was located in Dianbai, Guangdong. It had been destroyed but was rebuilt in recent decades. The shrine for Chen Zhenlong and Jin Xuezeng was first established in the early nineteenth century and was called Xian Shu Ci (Shrine for Sweet Potato Ancestors). It is still in Fuzhou, Fujian.[18] Both are now popular tourist attractions. However, it remains unclear whether sweet potato entered China first through Guangdong or Fujian. Of course, it is possible that the root was introduced to the country by multiple people and via multiple routes, as some scholars have suggested.[19] In terms of its transfer to Guangdong, the aforementioned sources appear to indicate that Vietnam received the plant from the Philippines several decades before China, which, however, has lacked convincing evidence.[20] Furthermore, even if Guangdong had seen the sweet potato before 1594, it was not widely known because many local sources still claimed that the root originated in Luzon, not Vietnam.[21] Similarly, Ping-ti Ho's conclusion about the land route for the sweet potato's transfer from Burma/Myanmar to Yunnan lacked persuasive evidence. Of course, it was possible that the Portuguese introduced sweet potato to South and Southeast Asia in the early sixteenth century. However, according to research, sweet potato was not seen in India until 1616, and in the centuries since it has not been received as well as it has been in China after that.[22] Thus, Chen Shiyuan's *Jinshu chuanxilu* appears to be a more credible source, attesting that the American plant was transferred to China via a sea route from the Philippines in the 1590s. Indeed, the Philippines, as a Spanish colony and the destination of the so-called Manila galleons from Mesoamerica, was most likely the source for the sweet potato's spread in Asia. By the turn of the seventeenth century, the number of Chinese residents in the Philippines was estimated to be over twenty thousand, making it the largest Chinese diaspora community in Maritime Southeast Asia.[23] One can easily imagine that if not Chen Zhenlong, other Chinese could have brought the sweet potato back to China.

A few words about *Jinshu chuanxilu* are in order. It was a collection of essays and poems about the spread of the sweet potato in China from its introduction in the late sixteenth century to the eighteenth century, when the book was edited by Chen Shiyuan, about whom little is known. While the content was somewhat repetitive and the organization was a bit perplexing, it was a valuable primary source collection about the plant's early acceptance in China. At first glance, it appeared that the editor's main goal was to request official recognition for his family's contribution to the root's cultivation across the country. Notwithstanding these efforts to satisfy personal and familial vanity, the book offers a thorough examination of the sweet potato. Similar books had been produced in China and East Asia, as we will see below, but such focused study was uncommon anywhere in the world during the period.

The compilation of the *Jinshu chuanxilu* was a family affair that began with Chen Shiyuan and was continued and completed by his two sons, Chen Yun and Chen Shu; Chen Yun also wrote a short treatise in the book discussing the benefits of the sweet potato. *Jinshu chuanxilu* was divided into two volumes, the first of which contained essays and the second of which contained poems. The first volume stated that during the 1590s, Fujian suffered a famine, prompting Chen Zhenlong, who had sojourned in Luzon, to look for the sweet potato because he knew the plant was very popular among Filipinos as a useful food. The root was then tested in his garden by Chen Jinglun, his son, and in 1594, Chen presented it to Governor Jin Xuezeng, who ordered its cultivation throughout the province to help alleviate the famine.

The preceding account constituted the first part of *Jinshu chuanxilu*. Its second section described how the Chen family's subsequent generations helped spread sweet potato farming beyond Fujian. The principal editor, Chen Shiyuan, appeared to have traveled to Central and North China in the mid-eighteenth century, where he witnessed famine. He appealed to local officials, such as Li Wei, the governor of Shandong, describing the benefits of the sweet potato as an easy-to-grow plant and how it could become a valuable food on par with all grain crops. These descriptions, written not only by Chen Shiyuan but also by others within and outside the family, provided valuable information about the root's spread from Fujian on the southeastern coast to the Yangzi River Delta and then to the provinces of Shandong, Henan, and Hebei in Central and North China during the late seventeenth and eighteenth centuries.

The third section of *Jinshu chuanxilu* is equally valuable; it provided quite detailed information about sweet potato cultivation, consumption, and preservation. It taught readers that the plant was best grown from spring to fall, that the root and vines could be eaten raw or cooked, and that it could not survive the winter except by keeping the plant's old vines inside the house as a method of propagation for the following year. Other methods of sweet potato propagation were also known to the contributors. What they liked best about the plant was its edaphic flexibility, particularly its ability to grow in sandy loam and on hillsides; its relative immunity to locusts and other insects as a root plant; its high yield, including the fact that both its tuberous roots and vines could be eaten as needed throughout the growing season; and its low fertilizer demand, though they did mention that some fertilization with animal manures could improve the root's growth. All of these advantages were described in both prose and verse in the book's two volumes.[24]

Jinshu chuanxilu was thus a thorough and knowledgeable account of the sweet potato from both a botanical and historical standpoint. Following its compilation, it was thought to have been reprinted twice in eighteenth-century China, during which time the American plant experienced significant growth in both the South, where it first arrived, and the North, where it was accepted as an equally valuable crop on par with other traditional food crops like millet and wheat. Indeed, Chinese sources show that during the reigns of Emperor Qianlong (1735–1796) and Jiaqing (1796–1820) of the Qing dynasty (1644–1912), sweet potato gradually became regarded as valuable as *gu* 穀, the term used to refer to a grain crop in the Chinese language.[25]

Despite the Chen family's multigenerational effort and initial success in the eighteenth century, *Jinshu chuanxilu* appeared to have faded into obscurity in China by the following century. Only one copy of the book remained in the country when Wu Deduo, the aforementioned historian, discovered it in the mid-twentieth century. It's no surprise that the copy was kept in the Provincial Library in Fujian, where the Chen family lived. According to Wu, who discovered it, the reason it was nearly lost was that while Chinese literati and the government usually made efforts to preserve knowledge, they tended to overlook the importance of agricultural books. The stately *Siku quanshu* (Complete library in four sections) bibliographic project, launched during Emperor Qianlong's reign, aimed to catalog all printed books found in the country. And yet there were only ten agricultural books in its otherwise impressively comprehensive collection, lamented Wu.[26]

However, *Jinshu chuanxilu*, published in the eighteenth century, was not the first Chinese book on the sweet potato, nor was it the only agricultural book that was nearly lost. Another notable example was Xu Guangqi's *Ganshu shu* (Explanations of the sweet potato). A high-ranking official in the Ming dynasty who was also a Christian, Xu was well-known for his broad interest in different branches of science, ranging from agronomy and astronomy to mathematics. It was believed that in the early sixteenth century, when the sweet potato spread from South China to Southeast China, where he was born and raised, Xu carefully observed and discussed its botanical characteristics as a food plant. *Ganshu shu* was the result of his research and was published in 1608. It became the first Chinese text on the American plant, written more than two centuries before the compilation of *Jinshu chuanxilu*. Unfortunately, despite Xu Guangqi's prominence—which ensured the preservation of many of his other writings—*Ganshu shu* was nearly lost within a decade of its printing, likely because, as a book focused on an agricultural plant, it was considered insignificant in representing Xu's broader intellectual legacy. When *Xu Guangqi ji* (Xu Guangqi's collected essays) was edited in 1962, it only included Xu's preface and two surviving lines of *Ganshu shu*; the rest of the text was missing.[27] In addition to his essays collected in *Xu Guangqi ji*, Xu's interest in botany and agriculture also led him to author the multivolume *Nongzheng quanshu* (Complete treatise on agriculture), in which he discussed the sweet potato in the section titled "melons and gourds."[28] Needless to say, his discussion, a concise description of the root, its botanical proclivities, and cultivating methods, should have drawn on his previous *Ganshu shu* writing. Its existence piqued people's interest in locating Xu's original text.

Such interest was pursued not only within China, but also beyond its borders. Shinoda Osamu, a Japanese expert on Asian food plants, published an article on a sweet potato text written by Seo Yugu, the nineteenth-century Korean scholar, in 1944. Seo, like Xu Guangqi, was a high-ranking official in his country's government who was also passionate about agricultural development. Seo wrote *Jongjeobo* (Manuals for sweet potato cultivation) in 1824, which was based primarily on three texts, two by his Korean predecessors and the other, lo and behold, Xu Guangqi's *Ganshu shu*—all three were sweet potato studies. According to Shinoda, Xu's *Ganshu shu* did, in fact, account for one-third of the content in Seo's *Jongjeobo*. In other words, Seo Yugu's *Jongjeobo* helped preserve a version of *Ganshu shu*, albeit incompletely.[29]

When the two versions of *Ganshu shu*, one found in Xu's own *Nongzheng quanshu* and the other in Seo's *Jongjeobo*, were compared, the latter contained more details and was also better organized. However, the two accounts were comparable in terms of essential content. Before we get into

地瓜兒苗

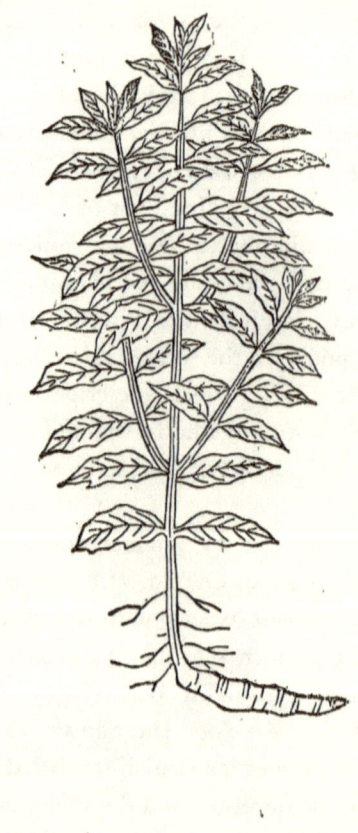

地瓜兒苗。生田野中。苗高二尺餘。莖方四稜。葉似薄荷葉微長大。又似澤蘭葉。拈莖而生。根名地瓜。形類甘露兒更長。味甘。救飢。掘根洗淨。煠熟。油鹽調食。生醃食亦可。

Figure 3.1 The illustration of the sweet potato, referred to as 地瓜 (*digua*, a regional name for the plant in China), in Xu Guangqi's 徐光啓 *Nongzheng quanshu* 農政全書 (Complete treatise on agriculture) (Zhonghua shuju, 1956). Xu, an early seventeenth-century scholar-official, was among the first to promote sweet potato cultivation in China. The Chinese caption highlights the plant's versatility as both root and vine are edible, either raw or cooked.

the details, it's worth noting some interesting differences between the two. Although the section on the sweet potato in *Nongzheng quanshu* was not the original *Ganshu shu*, it was revised in the first half of the sixteenth century by Xu Guangqi himself. Seo Yugu's compilation of his *Jongjeobo*, on the other hand, was thought to have occurred around two hundred years later, in 1834. Xu included two earlier mentions of the term *ganshu* in pre-sixteenth-century Chinese texts at the start of the sweet potato section. Then he quickly added that there were two varieties of the plant: one native and one foreign, known as *fanshu*, which, according to Xu, was brought back secretly from overseas against the foreign government's ban.[30] That is, Xu implied to his readers that the plant he was about to discuss was a foreign variety, rather than the Chinese yam and its relatives. However, as mentioned at the beginning of this chapter, some Chinese people believed that sweet potato was a native plant. Around 1776, Xu Guangqi's *Ganshu shu* inspired Lu Yao, a Qing official in Shandong, to write a brief pamphlet called *Ganshu lu* (Records of the sweet potato). Despite consulting Chen Shiyuan's *Jinshu chuanxilu* and other texts, Lu maintained that sweet potato, or *ganshu*, had been seen in China before, rather than being brought back from overseas in the sixteenth century. In fact, Wang Jiaqi relied heavily on Lu Yao's text when making the same argument in 1961.[31]

Seo Yugu cited earlier Chinese texts in which the word *ganshu* appeared when compiling his *Jongjeobo*. However, he quickly cited Xu Guangqi's statement about sweet potatoes being imported into China. He also informed his readers that Ryukyu and Japan introduced sweet potato, or *fanshu*, to Korea. Seo's *Jongjeobo* consulted a total of seventeen earlier texts on these subjects, with the goal of presenting knowledge of the sweet potato's propagating, sprouting, transplanting, storage, and advantages as it was known by Asian farmers in the nineteenth century. Xu Guangqi's *Ganshu shu* was also frequently cited by Seo in beginning his discussion. While brief, Xu's description in the *Nongzheng quanshu* did cover these topics two centuries before. Reading both Xu's and Seo's texts, then, provides us with useful information about the level and progression of knowledge Asian agriculturalists gained in growing the American plant after its introduction to Asia.

Jongjeobo's contents, which is divided into fifteen sections, can be summarized with reference to four categories. First, it describes the sweet potato's appearance and dissemination, as well as its introduction to the Korean Peninsula. According to Seo Yugu, while sweet potato is similar to some

Asian taros and yams, it comes in a variety of shapes and sizes. It has purple skin and a sweet and fragrant white and yellow interior. Sweet potato was grown in Ryukyu and southwestern Japan, including Satsuma and Nagasaki, before it was introduced to Koran in 1766.

Second, it discusses the various methods of sweet potato cultivation. Sweet potato seeds can be used to reproduce sexually, but this is uncommon. *Jongjeobo*'s propagation methods were asexual, involving the use of its roots and vines rather than its seeds. The book recommends selecting a healthy root, washing, and drying it, and covering it with dried rice stems before planting it in the soil to allow it to sprout. When using its vines, it recommends cutting an old vine and growing it in a jar or vat before moving it to the field. To ensure the efficacy of these methods, the book also discusses the need for root and vine preservation in light of climatic differences in southern and northern regions, as East Asia is located in both tropical and temperate zones.

Third, in the book's main section, Xu Guangqi, Seo Yugu, and other Asian authors describe soil preparation, farming season, transplanting methods and density, and sweet potato harvesting. They observe that sweet potato grows well in sandy loam and on hillsides, and that the soil is better fertilized with animal manure prior to cultivation. However, because the root is sweet, it attracts insects, so any insects and their eggs in both the fertilizer and the soil should be killed first. After the vine has grown to about three feet in length, cut about one and a half feet off the top and plant it in the field with a density that corresponds to the season of transplanting—lower density in early spring and higher density in later, warmer seasons. Cutting vines is also required to help the plant develop properly during the growing season. The text then explains that because sweet potatoes cannot survive the winter in the ground, fall is the time to harvest them all, no later than the winter solstice, even in southern regions.

Fourth and final, *Jongjeobo* discusses the various uses of the sweet potato, including its use as a famine food, animal fodder, as well as in starch production and the brewing of alcoholic beverages. Xu Guangqi had also noted almost all of these two hundred years before. Indeed, Xu had listed "thirteen advantageous utilities" of the sweet potato, which Seo Yugu quoted verbatim in *Jongjeobo*. Xu also praised the root as a good substitute for grain food because it is easy to reproduce and care for, is drought tolerant, has a pleasant taste, a high yield, and is impervious to locusts.[32]

To sum up, Xu Guangqi and Seo Yugu were two Asian experts on the sweet potato; even though they wrote their works centuries ago, their knowledge of the plant was undoubtedly comparable to that of modern scholars. For example, in terms of sweet potato reproduction, John C. Bouwkamp, a horticulture professor at the University of Maryland, observes that "sweet potatoes are vegetatively propagated," which "gives the crop an advantage in several ways." These benefits were extensively discussed in Xu's and Seo's works. Despite their use of different units of measurements, Jennifer Woolfe's instructions in *Sweet Potato: An Untapped Food Resource* are almost identical in discussing the transplanting method, such as the length of the vine and sprouts to be transplanted.[33]

Cultivation and development from south to north

Asian farmers accumulated sufficient knowledge about the American plant after its transplantation to the continent, as seen all the way from Xu Guangqi's writing of *Ganshu shu* in the early sixteenth century up to Seo Yugu's *Jongjeobo* in the mid-eighteenth century. Needless to say, such knowledge attested to the sweet potato's widespread popularity as a valuable food resource throughout East Asia. Its introduction remains a popular topic that is widely covered in the Chinese media to this day. During the first half of 2021, for example, a news report with an intriguing title swept the country: "China gained over 300 million more people as a result of a single vine smuggled into the country!" The story was about Chen Zhenlong's bringing the sweet potato back from the Philippine island of Luzon in 1594. According to some reports, Chen was compared to Yuan Longping, the renowned "father of hybrid rice" who significantly increased the crop's per unit yield beginning in the 1970s. Indeed, the media frenzy surrounding Chen Zhenlong was not coincidental; Yuan's death in 2021 reminded people of his predecessor, who had played a comparable role in significantly improving the country's food supply.[34]

However, there are distinct variations to the various modern accounts how Chen Zhenlong actually completed the act. Chen Shiyuan only vaguely mentioned in *Jinshu chuanxilu* that his great-grandfather brought the plant back from Luzon and experimented with its cultivation in his own field before presenting it to Governor Jin Xuezeng. In his *Ganshu shu*, Xu Guangqi described Chen Zhenlong braiding a vine onto a water bucket rope and smuggled it out of the Philippines, despite the Spanish ban on the plant's export. However, modern storytellers claim that he braided the vine on a mooring rope or cable and covered it with dirt before sailing back home.[35] Regardless, while scholars believe that there were multiple routes for the sweet potato to enter China, it appears that popular memory has rewarded Chen Zhenlong and his family for their introduction.

One of the main reasons for the Chen family receiving credit, perhaps, relates to the significant impact the sweet potato's introduction had on the Fujian economy and people's livelihoods from the sixteenth century onward. Indeed, in agricultural production, Fujian paled in comparison to Guangdong, its close neighbor on the southeastern coast whose Pearl River Delta is known for its fertile alluvium for growing grain crops, particularly rice, the dominant food plant in China since the Ming period.[36] Known for its jagged topography, Fujian is said to be "eight parts mountain, one part water, and one part farmland," which means that rice, though a popular food crop, cannot grow everywhere. Furthermore, because Fujian's climate is subtropical, it is not an ideal location for wheat cultivation. In other words, while Guangdong was another early destination for China's reception of the sweet potato—some of its regions have since become one of the country's main producers—the plant's impact was not as transformative as it was in Fujian.

In Fujian, on the other hand, the introduction of the sweet potato had a significant and consequential impact that reshaped the province's economic life from the sixteenth century forward. Many studies have been conducted over the centuries on the role that the American plant, which was indeed treated as a staple food across much of the province, played in significantly improving the people's food supply. There were several reasons why the American root was more popular in Fujian than elsewhere in South China. The first reason is because of its natural surroundings. Fujian, as a mountainous coastal province, has limited space for growing traditional food crops. A twelfth-century text described the topography of the province as follows: "The land in Fujian's 'seven prefectures' is small and not very fertile. Rivers and springs are shallow and inaccessible. The people work extremely hard, but their means of subsistence are inferior to those found elsewhere. Rich people created terraces on the hills for cultivation, which rise level after level like the steps of a staircase."[37]

Sweet potato has become an ideal dryland plant for the region due to its edaphic tolerance, as it grows well on hillsides and in sandy loam. Second, the increased demand for food in Fujian caused its people to quickly adopt this high-yielding plant, first as a famine food and later as a reliable staple, particularly in the southern parts of the province. Food demand in Fujian has been steadily increasing since the tenth century, owing to the influx of people from the North fleeing the devastation caused by the demise of the Tang dynasty (618–907). In a nutshell, prior to the introduction of the American plant, Fujian was experiencing increasing population pressure.[38]

Several efforts were made to relieve the pressure, including land reclamation from both the sea and the mountains. However, it appeared that little could be done about the former after the Ming period, as most beaches had already been converted to rice paddies. People then chose different paths, such as trade and commerce. "While grain crops could ripen," commented a local gazetteer from Haicheng County in Zhangzhou, one of the first regions in sweet potato cultivation, "their harvest was poor; it was a hard life to tender the rice paddies. Some brave souls turned to the sea for a living. The rich provided the money, while the poor provided the labor; they traveled long distances to trade Chinese goods for foreign ones in order to make a profit."[39] Indeed, Fujian has been China's window to the rest of the world since the Song period. Quanzhou in particular, known as Zaiton or Zayton outside of China, rose to prominence as an intercontinental trade hub. It is a port city in southern Fujian near Zhangzhou that flourished during the Song and Yuan dynasties (960–1368), boasting a large number of foreign merchants. It was believed that Marco Polo had visited the city, as had Ibn Battuta, an Arab traveler; both were apparently impressed by Quanzhou's charm and prosperity as China's largest port city at the time. It is no surprise that, as mentioned in the introduction, Christopher Columbus spent his first and second voyages in the Americas searching for the city.

Another attempt was made to improve life by diversifying crop cultivation on the province's limited farmland. Fujian's peasants had developed techniques for cultivating early ripening rice, such as the Champa variety, that could be harvested two or even three seasons in a year since the Song period. They also tested wheat varieties that were better suited to the region's climate and could be harvested twice or three times a year. To improve their living conditions, Fujian peasants farmed cash crops such as tea, cotton, true indigo, tobacco, sugarcane, and tropical fruits (longan, lychee, oranges, pomelo, and so on) beginning in the Ming period. Some of these plants can grow on hillsides, making them appropriate for Fujian's terrain.⁴⁰ However, because of these activities, the province became increasingly reliant on grain imports to sustain its food supply for the people. When a natural disaster struck, food supply became a serious issue. Fujian is geographically and historically prone to both drought and flood. Because it is mountainous, the province's water resources are plentiful, but they frequently flow directly to the sea, making them difficult to harness for agricultural use. Hurricanes and heavy rains frequently cause flooding in the summer, whereas in other seasons, its coastal regions suffer from drought.⁴¹ Sweet potato, a drought-resistant crop, was thus the first American dryland plant to be adopted in southern Fujian.

Government policy during both the Ming and Qing dynasties exacerbated Fujian's economic hardship at times. From the fourteenth to the seventeenth centuries, robust commercial trade with countries outside China aggravated the so-called *wokou* problem, which challenged Chinese border control. The term *wokou* literally meant "Japanese pirates" or "dwarf pirates," the latter being a derogatory Chinese name for Japanese. Though pirate activities frequently originated in the Sea of Japan, pirates also included Chinese and Koreans. To deal with the problem, both dynasties imposed "coast clearance" policies at times, forcing people to evacuate coastal regions and move inward to mountainous areas, causing even more population pressure and fierce demand for food. Many Fujianese defied the policy and chose to emigrate to Taiwan and Maritime Southeast Asia in order to survive. Some of them may have also been involved in piracy during the period. Overall, when the sweet potato was introduced to Fujian, the province was facing a number of economic and demographic challenges. These challenges, as well as the people's tradition of engaging in foreign trade and their openness to receiving exotic plants and goods, were important factors in this coastal province's warm reception of the American plant.

Chen Shiyuan's *Jinshu chuanxilu* and Xu Guangqi's *Ganshu shu* both contained information about the early cultivation of sweet potatoes in Fujian. Furthermore, about a half century after the introduction of the sweet potato in the province, Zhou Lianggong, an ex-governor of Fujian, wrote *Min xiaoji* (A brief account of Fujian), in which he observed the following:

> Sweet potato, a plant brought back from Luzon, could reproduce in infertile sandy loam.... Sweet potato can be interplanted with the five grain crops because it grows on hillsides and field margins and

does not compete for farmland. Its sprouts grow quickly when fertilized, and its roots spread rapidly when it rains. Even during a drought, the roots can mature. People in Quanzhou purchase them for one mace per *jin* [about half a kilogram]. Then two *jin* of sweet potato will suffice to fill one's stomach. As a result, people of all ages, including beggars, consume sweet potatoes. It satisfies hunger and will not harm you if consumed in excess. The leftovers can then be fed to dogs and chickens.[42]

"Introduced in the late Ming, sweet potato was from overseas," said another seventeenth-century text. "The people of Quanzhou were the first to learn to grow it. Rich people used to treasure the root and use it to treat their guests with just a few strips in a box. However, as of now [1650], it is grown in Xinghua, Quanzhou, and Zhangzhou. Because it is widely available, its price has dropped significantly, and ordinary people now eat it three times a day as a staple."[43] In 1612, the Quanzhou gazetteer noted that, "imported from overseas, sweet potato propagates in vines and its roots are abundant, one *mu* can reap several *shi* [picul = 50 kilogram]. Sweet potato is a better variety than native roots because it is easier to grow and yields more. Poor people rely on it for food." These early seventeenth-century texts, along with Chen Shiyuan's account of *Jinshu chuanxilu*, documented the sweet potato's early diffusion in southern Fujian. Incidentally, Chen Zhenlong, the reputed transferer who figured primarily in *Jinshu chuanxilu*, was from Fuqing (or nearby Changle), which is just to the north of Xinghua (today's Putian), and Quanzhou and Zhangzhou are just to the south. Sweet potato received a warm welcome in the southern regions of Fujian such as Fuzhou, Xinghua, and Quanzhou because, according to a Song text, these areas "have poor soils and, even in the best harvest years, produce enough rice to last only half a year."[44]

The sweet potato spread throughout the province thanks to Jin Xuezeng, who served in the Fujian provincial government from the late sixteenth century. According to *Jinshu chuanxilu*, after witnessing the Chen family's success with sweet potato transplantation, Jin Xuezeng wrote a manual and distributed it throughout the province, teaching farmers how to cultivate the root in order to alleviate the famine. Though brief, the sweet potato knowledge contained in the manual covered essential aspects comparable in scope to that discussed by Xu Guangqi and Seo Yugu in their works. Topics covered in Jin's pamphlet included the following: the sweet potato's size, color, and taste; its higher yield (several thousand *jin* per *mu*—Chinese traditional unit of farmland measurement, equaling one-fifteenth of a hectare) than the five grain crops; its multiple ways of consumption, raw and cooked, milled and sliced, as well as fermenting it to make alcohol; its flexible adaption in different soils; its method of cultivation—sprouting in the early spring and transplanting after the sprouts grow to about two feet; plowing the field before planting the sweet potato sprouts and maintaining certain density between them for a better growth; harvesting the roots through the growing season, no later than the winter; and, finally, how to store the roots inside and covering them with grass or hay.[45] In recognition of Jin Xuezeng's efforts to promote the sweet potato, several prefectural gazetteers from this period and afterward went so far as to (incorrectly) credit him with introducing the crop to China from abroad. However, these local records also demonstrated that sweet potatoes had been grown all over Fujian by the end of the eighteenth century. Jin earned the distinction of being responsible for the first sweet potato cultivation in China by penning and distributing the aforementioned manual.[46]

And, because of its widespread propagation of the American root, Fujian became the most well-known source of sweet potato dispersal across the country.[47] What about Guangdong, another South Chinese province where sweet potato was allegedly introduced? Interestingly, despite the dramatic stories provided by two prefectural gazetteers about the sweet potato transfer mentioned in the previous section, few additional sources describing the root's early dissemination in Guangdong have been found. Of course, it is possible that sweet potato was grown concurrently in both Guangdong and Fujian—some have argued forcefully that Dongguan was the first place in China where the American root was grown. If the sweet potato was introduced to Guangdong before Fujian, it was likely cultivated in private gardens rather than on a larger agricultural scale.[48] However, sources in Guangdong began to record sweet potato as a food resource toward the end of the seventeenth century. For example, Wu Zhenfang, a Qing official from Zhejiang, described his trip to Guangdong in *Lingnan zaji* (Miscellaneous notes on Guangdong) during the period. Wu claimed that sweet potato was grown all over Guangdong and was widely available in Zhejiang, his home province. He also noticed that when rice prices spiked in Guangdong in 1699, people relied on it for food. In 1687, the gazetteer of Huaxian, near Guangzhou, also noted that sweet potato was easy to grow and that the poor ate it with rice to combat hunger.⁴⁹ While demonstrating the development of sweet potato in Guangdong, both sources also suggest that, unlike in Fujian, where famines were more common, rice was the main food crop on which people relied in Guangdong; it was when rice prices increased that there was greater need for sweet potato as an expedient supplement. This explains why sweet potato cultivation was not as common in Guangdong as it was in Fujian during the seventeenth century. Indeed, if sweet potato were grown in Dongguan and Huaxian, it would not pose a serious threat to rice's dominance as a food crop, as both are located in the Pearl River Delta, a major rice-producing area for both the province and China as a whole. A thirteenth-century text praised the delta's successful rice-farming history, saying, "It has much fertile land. It has the best 'Long-waist' and 'Jade Kernel' rice in the South. It also provides a significant portion of supplies to neighboring prefectures."⁵⁰ Sweet potato cultivation was common in Chaozhou, which bordered southern Fujian, as it was in Zhangzhou and Quanzhou. Locals regarded its roots as a preferred supplementary food in hard times, prizing it over other native root plants. Sweet potato vines, on the other hand, were traditionally valued as hog feed.⁵¹

Sweet potato had already spread to Zhejiang, or Jiangnan, by the end of the seventeenth century, according to Wu Zhenfang's *Lingnan zaiji*. This is not surprising given that the root was introduced to the region in the early years of the century by Xu Guangqi, a native of Shanghai, north of Zhejiang. If the sweet potato spread northward from South China, when did the rest of the country get the American plant? Wan Guoding provided a time line for sweet potato cultivation in China in his 1961 book *Wugu shihua* (Stories of the five grains). According to his research, which was based primarily on plant records in gazetteers, sweet potato had dispersed throughout the country by 1768, except for the northwestern provinces of Shanxi and Gansu.⁵² Others, such as Chen Shuping, later revised this chronology; Chen believed that sweet potato was grown in Taiwan in 1712, a few years earlier than Wan's date of 1717. The sweet potato arrived in Taiwan in 1696, according to the Taiwan gazetteer. Taiwan's sweet potato cultivation probably began shortly after Fujian and Guangdong. The island is just across the Taiwan Strait from Fujian, and many Fujianese had chosen it as their destination for emigration. It is equally possibly that Taiwan might have received the sweet potato around the same time from nearby Brunei, in Maritime Southeast Asia.⁵³ Taiwan was a key region for sweet potato production for the majority of its modern history—the root played a significant role in shaping the island's economic and sociocultural life, as will be discussed in chapter 5.

Returning to the northern transfer route for sweet potato dispersal in mainland China, it should be noted that the diffusion did not always take a predictable path. According to the time lines developed by Wan Guoding and Chen Shuping, due to individual interventions, certain provinces may have developed the sweet potato earlier than their southern neighbors on the supposed south–north diffusion path. In other words, if Fujian saw wider sweet potato cultivation than Guangdong, Jin Xuezeng's role was critical. Similarly, if Jiangnan as a region cultivated the sweet potato before Guangxi and Hunan, which are immediate neighbors of Guangdong and Fujian, it was largely due to Xu Guangqi's promotion. And Jin Xuezeng and Xu Guangqi were not the only people who contributed to the spread of the American plant in China. Shandong's sweet potato cultivation was another good example. It began about a decade before Anhui and Guizhou, even though the latter were closer to South China.[54] As alluded to earlier, Chen Shiyuan, the chief editor of *Jinshu chuanxilu*, was instrumental in introducing the sweet potato to Shandong. According to his own recollection, Chen had lived in Shandong, then known as Jiaozhou, for over a decade and had witnessed the famines that occurred between 1746 and 1748. The resulting misery prompted Chen to recall and emulate what his ancestors had done to alleviate the famine in Fujian. In 1749, he and his friends raised funds to assist with the transportation of sweet potato roots and vines from the South to Shandong. Chen also successfully taught the people how to cultivate the plant in Shandong's sandy loam, which he knew could support sweet potato growth. Then, in 1766, Chen Shiyuan, following in the footsteps of Xu Guangqi, enumerated sweet potato's "Eight Benefits" as a food plant, further promoting its cultivation in Shandong.[55]

Chen Shiyuan received official support for his effort to transplant sweet potato to Shandong, as had his ancestors. That is, Jin Xuezeng was not alone on the government side in encouraging sweet potato cultivation. Li Wei, a provincial official, was another. Born in Hebei, Li rose through the ranks in the Qing government, and at the time Chen Shiyuan was attempting to promote the sweet potato, he was governor of Shandong. Convinced that sweet potato was a valuable crop, Li wrote *Zhongzhi hongshu faze shiertiao* (Twelve methods in growing the sweet potato) and distributed it throughout the province to encourage cultivation around 1749. Li's methods were, for the most part, comparable to those offered by Jin Xuezeng in his manual, which was written for and distributed in Fujian one and a half centuries before. They taught readers how to choose and nurture storage roots, how to transplant them using sprouts, when to grow and harvest them, and so on. Meanwhile, Li Wei emphasized the sweet potato's advantage as a famine relief crop because the dryland plant, which grows from early spring to late fall, is drought and flood resistant. Specifically, he stated that "as it can be planted on hillsides, sweet potato can grow with some water in a drought. If a flood occurs, sweet potatoes can be planted as late as July after the water has receded, when no other crops can be grown." Moreover, Li stated that "sweet potato vines can be a good animal feed, and dried sweet potato could feed hogs to help them grow fat in the winter months." Li Wei's predecessor, Jin Xuezeng, had not discussed these two aspects of the sweet potato in depth.[56] Last but not least, Li Wei's efforts influenced Lu Yao, who served as governor of Shandong in the mid-eighteenth century. Lu compiled the *Ganshu lu*, mentioned at the beginning of this chapter, which was a brief synthesis (3,300 characters) of earlier texts on the American plant that he used to further promote sweet potato cultivation in the province and country.

Chen Hongmou was the third Chinese official who played a significant role in promoting sweet potato cultivation. Shaanxi, a province in northwestern China where Chen was governor from 1743 to 1746, began developing sweet potato in 1745, almost simultaneously with Hebei and Shandong and earlier than the province's southern neighbors like Shanxi and Anhui. Chen's preference for sweet potatoes appears to be the result of three factors. The first is that Chen should have known the plant's value as a food crop because he was born and raised in a peasant family in Jiangxi that grew it in 1736, if not earlier. Chen was known for his attention to agricultural development and its importance to people's livelihoods since 1733, when he began holding provincial posts across the country.[57]

Second, in 1737 Emperor Qianlong, who had trusted and promoted Chen, ordered the compilation of *Shoushi tongkao* (Agricultural handbook), and when the book was distributed among government officials in 1742, the emperor wrote his own preface to it, emphasizing agricultural development as the dynasty's foundation. The compilation of *Shoushi tongkao*, like Emperor Qianlong's other book projects, aimed to absorb the knowledge accumulated in previous agricultural books while also surpassing all of its predecessors. *Shoushi tongkao*, which was about ten times the size of *Nongsang jiyao* (Essential knowledge of agriculture and sericulture), compiled during the Yuan dynasty, and contained about one million characters, was the largest book of its kind in Chinese history. Chen Hongmou, the emperor's loyal administrator, read the book immediately after its publication and learned how to grow the sweet potato from it.[58]

The third factor that prompted Chen Hongmou to introduce the American plant to Shaanxi was a severe drought that hit Xi'an, the provincial capital and its suburbs, in July 1744, only four months into his governorship. In fact, the drought had occurred a year before and had affected the entire North China region, resulting in widespread wheat failure.[59] Chen learned that sweet potato could grow on hillsides and in sandy loam, and that it was relatively resistant to natural disasters, locusts, and other insect hazards after reading *Shoushi tongkao*, which drew on earlier texts such as Xu Guangqi's *Ganshu shu*. In his directive to local subordinates, he explained that Shaanxi, as a northern province located at high altitude, was unsuitable for rice paddy farming. Traditionally, the people of Shaanxi grew what was known as *zaliang* (various grains) for food. Sweet potato, which was already grown in many provinces at the time, should be a much better alternative to all other *zaliang*, according to Chen, because Shaanxi's loess-covered land is comparable to Fujian's sandy soil, where sweet potato should grow rather well and produce a much higher yield. "While it began in the southern provinces, sweet potato is now also cultivated north of the Yellow River," he said. "It can produce several *shi* per *mu* and a family of a few can live comfortably for a year with just one *mu*'s produce." In other words, he wrote, "sweet potato is as good as rice and wheat as a food crop.... Many Chinese people refer to sweet potatoes as *shuliang* (root food)."[60]

Shaanxi thus became one of the first northern provinces to grow the sweet potato because of Chen Hongmou's initiative. Chen himself left the province in November 1946 for his home province of Jiangxi, shortly after introducing the American plant. However, before departing, he urged his subordinates, many of whom were from Fujian, to carry on the project of recommending the plant as a new and viable alternative to more traditional food crops. In other words, despite his brief stay in Shaanxi, Chen Hongmou was instrumental in establishing sweet potato cultivation in northwestern China, even though the region's climate is better suited to growing maize and white potatoes. Later, white potato became a more cultivated crop in the Northwest than the rest of China. Furthermore, maize was a common food plant in the area. In other words, maize and white potato surpassed sweet potato in popularity among Shaanxi peasants over time.⁶¹ Nonetheless, long after his departure, Shaanxi farmers remembered Chen fondly for his keen interest in the sweet potato. Sweet potato was dubbed *Chen gong shu* (Mr. Chen's spud) in local parlance, like the homage paid to Jin Xuezeng by Fujianese in the late sixteenth century.⁶²

Each of the individuals profiled above aided in the promotion of sweet potato cultivation in their respective provinces as provincial governors. Aside from their contributions, the trend of sweet potato diffusion in China extended beyond provincial borders. Indeed, as Chinese researchers have observed, the pace and breadth of dissemination varied markedly by region, both within and outside of a given province.⁶³ Many factors influenced the differences. The first was the region's varying levels of food demand. Though sweet potato was introduced to both southern Fujian and southern Guangdong around the same time, it became more readily accepted as a main food in the former than the latter due to higher food demand in southern Fujian than in the lush Pearl River Delta of southern Guangdong. Sweet potato easily outperformed other food crops as a high-yielding food plant when food demand was high due to either population density or frequent famines. That is, despite being introduced to North China later than South China, sweet potato was generally accepted as a staple in the former. In comparison, sweet potato has not been accepted as a daily food source along the Yangzi River regions long after its arrival.

The second factor was topography. While the sweet potato spread across both plains and hills, it was more welcomed as a food crop in the latter. Indeed, sweet potato was frequently transported by sea and, within China, by canal during the sixteenth and eighteenth centuries. The root was probably introduced from South China to Jiangnan by Xu Guangqi via a sea route along the coast, as was Chen Shiyuan's transplantation in Shandong. Meanwhile, the Grand Canal helped transport sweet potatoes from the South to the North.⁶⁴ However, as previously stated, sweet potato was frequently regarded as a snack rather than a staple in such plains as Jiangnan. On the other hand, sweet potato was quickly adopted as a major food crop once introduced to hilly regions in Fujian, western Henan and Sichuan, and southern Shaanxi, because it grew well on sandy loam and clay soil, which were not ideal for cultivating other crops. In other words, sweet potato, as a dryland plant, served as both an "insurance crop" for famine relief and a daily food source for people living in mountainous areas. Ridge farming became the "most popular method used in China" of the various planting techniques used in developing the sweet potato.⁶⁵

The third factor was the climate. Sweet potato is a root plant native of subtropical Central and South America that sprouts in the spring and withers in the winter. Following its arrival, Asian writers ranging from Xu Guangqi to Lu Yao and Seo Yugu informed their readers about the sweet potato's farming season: It should be planted after the last frost in the spring and harvested in the late fall. As the root plant spread south to north, more emphasis was placed on methods for preserving the storage roots. A quarter of Li Wei's *Zhongshi hongshu faze shiertiao*, for example, was devoted to the topic.[66] Obviously, he intended to introduce the southern crop to the North. If sweet potato root does not survive the winter, it will not be a useful crop in the region. Thanks perhaps to his attention, it was in Shandong that the peasants developed successful solutions to the problem. According to one source, this occurred in the early eighteenth century, paving the way for the sweet potato's diffusion in North and Northwest China.[67] Nonetheless, the sweet potato's high-producing areas remained in subtropical Fujian and Guangdong, as well as temperate Sichuan, Henan, Shandong, Anhui, eastern Zhejiang, and northern Jiangsu. For example, despite Governor Chen Hongmou's promotional efforts, discussed above, Laura Murray found in her 1985 doctoral dissertation that sweet potato paled beside maize and white potatoes in its appeal to farmers in Shaanxi's Wei River Valley from the early nineteenth century.[68] According to a more recent study, there are five major sweet potato farming regions in China: the northern spring region, the Yellow-Huai River spring–summer region, the Yangzi River Valley summer region, the southern summer–autumn region, and the southern autumn–winter region. These regions have all become sweet potato–producing areas because the root plant can grow well if "the air temperature is stable at 15°C," though the length of the sweet potato growing season varies from 120 to 185 days due to temperature differences. Regardless, sweet potato is grown throughout most of China as a dryland plant that thrives in both subtropical and temperate climates.[69]

Sweet potato and population explosion

Ping-ti Ho, the Chinese American historian, made three important observations in his influential 1959 book *Studies on the Population of*

China, 1368–1953. The first was that prior to the tenth century, China's population fluctuated at around fifty million, even during the Han and Tang dynasties, which were widely regarded as China's two golden eras. The introduction of early ripening rice, or the Champa variety, which could mature in sixty days, from mainland Southeast Asia increased the population to an unprecedented level. More specifically, around 1103, as Ho and many others believe, China's population surpassed 100 million for the first time. This was what Ho considered "the first agricultural revolution" in Chinese history, one that has also been recognized by such experts as Eugene N. Anderson, who, in his *The Food of China*, acknowledges that "rice was a miracle crop of Sung [Song]."[70] In subsequent centuries, more effort was made in both North and South China to increase the cultivation of fast-ripening rice and to disseminate more dryland crops such as wheat, millet, and particularly sorghum.

Second, Ho argued that, while rice culture in China proper did not reach "its saturation point" until the mid-nineteenth century, there had been a "second agricultural revolution" beginning in the early sixteenth century, marked by the spread of New World dryland food plants such as maize, sweet potato, and peanuts, followed by white potato. Peanuts, according to his observations, were the first of these plants to be cultivated in China, grow in sandy loam near Shanghai, among other places. Maize was first grown in the Southwest before spreading to North and Northwest China. Needless to say, there was also the sweet potato, which Ho believed was first cultivated in South and Southwest China from the late sixteenth century. Ho wrote the following about the introduction and impact of this "second agricultural revolution":

> Early-ripening rice aided the conquest of relatively well-watered hills. American food plants have enabled the Chinese, historically a plain and valley folk, to use dry hills and mountains and sandy loams too light for rice and other native cereal crops. There is evidence that the dry hills and mountains of the Yangtze [Yangzi] region and north China were still largely virgin about 1700. Since then they have gradually been turned into maize and sweet potato farms. In fact, during the last two centuries, when rice culture was gradually approaching its limits and encountering the law of diminishing returns, the various dryland food crops introduced from America have contributed

the most to the increase in national food production and have made possible a continual growth of population.[71]

According to Ping-ti Ho's analysis, the spread of rice culture has nearly depleted "well-watered hills" and farmland. From the sixteenth century onward, Chinese farmers were able to continue land utilization and increase food supply due to the introduction of American dryland plants. The story of the sweet potato was one of the most impressive of the American food plants. Sweet potato, which has remained a supplementary food in Europe and North America, as discussed in chapter 2, became the most favored American food plant in China due to its high yield proclivity and the promotion of Chinese officials who regarded the crop as "the primary famine plant" (*jiuhuang diyiyi*—lit. the top priority in staving off famine).[72] Ho observed that from the early twentieth century, "next to rice and wheat, the sweet potato is now the most important source of food for the Chinese."[73]

Third, thanks to the spread of American dryland crops, China's land utilization improved significantly, propelling its population growth even further. In Ho's words, the cultivation of these dryland plants enabled Chinese farmers to "use dry hills, lofty mountains, and sandy loams. The opening-up of millions of acres of highlands in modern centuries must have helped to redress the old agricultural balance so badly upset by the dissemination of early-ripening rice." But "by the last quarter of the eighteenth century," Ho warned, "there was every indication that the Chinese economy, at its prevailing technological level, could no longer gainfully sustain an ever-increasing population without overstraining itself."[74] Leading scholars in the China studies field have generally agreed with Ho's observation. Susan Mann and Philip Kuhn, for example, acknowledged at the outset of their chapter in *The Cambridge History of China* in 1978 that "demographic pressure" was a root cause of the Qing dynasty's subsequent series of rebellions in the nineteenth century. Susan Naquin and Evelyn Rawski agreed nine years later in their book *Chinese Society in the Eighteenth Century*, quoting Ping-ti Ho's above statement, and claiming that what Ho argued was agreed upon by "most historians." Then, in 2002, for the new edition of *The Cambridge History of China*, William T. Rowe stated that "the single dominant factor of early and mid-Ch'ing [Qing] social history is population growth." Rowe also believed that Ho's population estimate was among the "most widely accepted."[75]

So, what are the widely accepted estimates of China's population growth during the Ming and Qing periods, or before and after the introduction of American food plants into China around the sixteenth century? Since Ho published his study, if not before, this has been the subject of scholarly debates among demographers and historians in China and abroad. The debates were attributed to the fact that, while governments attempted to collect census data during both the Ming and Qing dynasties, their methods were inconsistent. Sometimes the data was based on the number of *hu* (households), while other times it was based on the number of *ding* (male adults). Furthermore, because of entrenched sociopolitical discrimination against women, the composition of a household frequently excluded female members. As a result, most historical data on Chinese population figures during the Ming and Qing periods were at best approximations.

Nonetheless, we can point to two possible exceptions. One is that when the Ming dynasty was established in 1368, Zhu Yuanzhang (1328–1398), or the Hongwu Emperor, directed the compilation of the Yellow Records in order to ensure government revenue from the collection of land taxes. The Yellow Records, a nationwide registration of both population and taxation records under the Hongwu Emperor, provided a relatively reliable number of households in the country, and thus a good base for later scholars to estimate the population figure at the beginning of the Ming.[76] The other was the population figure used by the aforementioned Qing historians to discuss "population growth" and "demographic pressure" at the end of the Emperor Qianlong's reign. For example, in 1712, during the reign of Emperor Kangxi (1661–1722), the emperor decreed that the *ding* tax would henceforth be frozen at the 1711 level, in order to proclaim the dynasty's flourishing economy. According to Jonathan Spence, a prominent historian of the Qing, this decision eliminated the "reluctance on the part of the local officials to report any population increases" and encouraged them to "report actual population figures."[77] Then, in the latter half of Emperor Qianlong's reign, or the second half of the eighteenth century, serious efforts were made by imperial fiat to conduct census surveys, not for tax collection, but for social stability. While *hu* was still used as the basic unit for data collection, the records now included information such as the age and gender of each household member. As a result, Ge Jianxiong, a contemporary Chinese historical geographer, believes that population data from 1776 to 1850 was mostly reliable.[78]

According to these two relatively reliable census data, what, then, was China's population during the early Ming, or the late fourteenth century, and how much did it grow to at the turn of the eighteenth and nineteenth centuries? The answer to the former was around 70 million and the latter over 300 million (discussions of these two figures will be given below). In other words, during these four centuries, China saw its population grow more than fourfold, turning the country into the most populous nation on earth. And the growth continued. By the mid-nineteenth century, China's population grew still further to around 430 million before it was checked by a series of crises, caused by both external challenges from Western powers and internal peasant rebellions. Needless to say, the growth rate certainly qualified as an "explosion." John K. Fairbank, the doyen of the China studies field in the United States during the mid- to late twentieth century, remarked that during Qing's rule, China's population increased "from 141 million in 1741 to 432 million in 1851," which to him was "amazing" because the country then remained "an ancient and thickly agricultural state."[79]

How significant a role did the cultivation of sweet potatoes, and American food crops in general, play in this impressive growth? Before answering this question, it seems that two caveats regarding the late imperial Chinese population surge must be addressed. One is that during the four centuries under consideration, population growth was not linear or steadily upward; several major events occurred in between that either contributed to or undermined growth. The other is that, despite being introduced in the early sixteenth century, American plants, the sweet potato in particular, gradually expanded their participation in China's demographic expansion over the four centuries.

More specifically, the fall of the Ming and rise of the Qing may have had the greatest impact on the linear path of population growth. There have been debates about the impact, as there have been about population growth throughout the Ming dynasty. As previously stated, a few decades after the establishment of the Ming dynasty, China's population stood at around 70 million—Ping-ti Ho estimated it at over 65 million, while most Chinese demographers believe it was between 60 and 70 million. Ming historian Martin Heijdra's contribution to *The Cambridge History of China* provided the highest estimate of 85 million. Regardless of the differences, all scholars seem to agree that during the Ming period, China's population grew steadily and peaked around 1600, or around the time when American plants such as sweet potatoes and peanuts entered China. However, they differ greatly in terms of the peak number. According to Ping-ti Ho, the Ming population doubled and reached 150 million by 1600, whereas Chinese demographers believe it was closer to 200 million and Martin Heijdra believes it should be closer to 230 million.[80] Despite differing estimates of the Ming's population growth, owing primarily to the continued advancement of the rice culture that began with the Song, most scholars agree that when discussing the population explosion in late imperial China, it should be noted that it had already begun in the Ming. "All evidence points to the fact," writes Heijdra, "that much of the late imperial era's population 'explosion,' which some Ch'ing [Qing] economic and social historians use as a general solution to explain a wide variety of social and economic phenomena, was also a Ming, not exclusively a Ch'ing, phenomenon."[81]

If the Ming population was already 200 million in 1600 and grew to over 300 million two centuries later, at the end of the eighteenth century, then the late imperial population explosion would be more of a Ming than a Qing story. However, it should be noted that growth did not occur in a straight line—there were significant ups and downs in between. China's population began to decline several decades after reaching its peak. Natural disasters, widespread disease, and government incompetence sparked a peasant rebellion in Shaanxi in 1628, led by Li Zicheng, a former imperial courier. Within a decade, the rebellion had spread like wildfire across the country, with Li's rebel army sacking several major cities, including Fengyang in Anhui, the hometown of Ming founder Zhu Yuanzhang. Worse, Li was not the only one. Other rebellions erupted; together, these rebel forces defeated the Ming army, bringing the dynasty to an end in 1644. Seizing the opportunity, the Manchus from the Northeast marched south, aided by Wu Sangui, a former Ming general. They established the Qing dynasty in 1644 after defeating Li Zicheng, while continuing their military campaigns throughout the country. The extensive warfare that occurred during the Ming–Qing dynastic transition resulted in significant population loss. According to contemporary records, the Manchu army massacred 800,000 people in Yangzhou, located in the lower reaches of the Yangzi River, in a ten-day period while seizing the city. The battle for Jiangyin, a neighboring city, also resulted in 100,000 casualties.[82]

What was the exact population loss in mid-seventeenth-century China during this apocalyptic epochal change? Again, different estimates have been provided by demographers. Ge Jianxiong, for example, calculates a loss of up to 80 million people, or a 40 percent drop from the peak of the Ming population of 200 million to around 120 million at the start of the Qing in the late seventeenth century. Others agree with Ge that there was a substantial population decline in the Ming–Qing transition, though their estimates of the early Qing population vary greatly; it has been estimated to be as low as 80 million and as high as 150 million.[83] In other words, if the Qing population at the end of the eighteenth century surpassed 300 million, as experts both inside and outside China seem to agree, then, in the words of William Rowe, "the population considerably more than doubled again (other scholars have it tripling) in the eighteenth century, reaching 313 million in 1794."[84] Under the Qing, the Chinese population certainly exploded, whether it doubled or tripled.

And the explosion had a lot to do with the expansion of American food plants, particularly sweet potatoes, in the Chinese agricultural system, enriching and significantly increasing the country's food supply. In other words, if population growth in imperial China began in the Ming, it was primarily due to improved rice supply thanks to the cultivation of early ripening varieties, as the introduction of American food plants occurred only in the last fifty years of the Ming dynasty. However, the continued population growth of Qing China throughout the twentieth century, which was as impressive if not more so than the previous wave of growth, was most likely due to the presence of American food crops.

Sweet potato, maize, chile pepper, and tomato are unquestionably the most important American food plants in Chinese gastronomy today—so much so that many Chinese take them for granted, as if they have always been a part of their diet. The debate among Chinese scholars on whether sweet potato was a native plant, mentioned at the beginning of this chapter, is but one example. In recent decades, several popular TV dramas have shown corn and chile pepper in episodes ostensibly depicting historical scenes from pre-sixteenth-century China.[85] Tomato has long been a favorite vegetable among Chinese consumers. Tomato with scrambled eggs, for example, is a popular dish that can be served with either rice or noodles and is almost a daily food throughout the country. When China hosted the Olympic Games in 2008, the red and yellow dress worn by its athletes at the opening ceremony was compared to the iconic dish by many in the Chinese media.[86] Spicy food dishes from Sichuan and Hunan have gained enormous popularity in the recent development of Chinese cuisine; to cook them, chile pepper is used liberally as an indispensable spice. Indeed, for people both inside and outside China, there has been an ingrained impression of an intrinsic bond between Sichuan food dishes and chile pepper, despite the fact that prior to the latter's arrival in the sixteenth century, the spices used in Sichuan cuisine were Sichuan pepper and black pepper, neither of which is as pungent as the chile.[87]

The sweet potato's status in China has changed several times, indicating its growing importance as a popular food. Despite their widespread popularity, tomatoes and chile peppers have always been classified as vegetables. Sweet potato was initially classified as a vegetable because of its botanical resemblance to *shanyao*, or Chinese yam. One of the earliest records of sweet potato cultivation, for example, was in the Quanzhou Prefecture gazetteer of 1612. The sweet potato was mentioned under the heading "vegetables" at the time. In 1696, when the gazetteer of Taiwan Prefecture described the root, it was also placed under the same rubric: "Sweet potato's [*fanshu*] skin is either red or white. [The root] grows in vines with a lot of roots."[88] A gazetteer in Hua County, Guangdong, considered sweet potato a vegetable about a decade earlier, even calling it "sweet potato," or *tianshu* 甜薯, and comparing its dark red color to that of pig liver. Indeed, sweet potato acquired over a dozen names in Chinese after entering the country. And it appears that these names influenced the root's shifting categorization as well. For example, *digua* 地瓜 (lit. earth melon) is a common Chinese term for the sweet potato, especially in Central China. Perhaps because of the name, some gazetteers, such as those of Shangcai County in 1690 and Dongming County in 1756, both in Henan Province, considered the root to be a fruit or melon. While naming it *fanshu*, the gazetteer of Fujian's Shouning County in 1686 also classified the sweet potato as a fruit.[89]

The above nomenclatural variation suggests that when sweet potato was first cultivated in China, local historians initially found it difficult to incorporate it into the existing foodway and Chinese gastronomy. However, an intriguing and discernible trend seemed to emerge over time in their attempt at taxonomic designation of the American plant. Because of its high yield, which was widely acknowledged by all sources, some gazetteer compilers began classifying the sweet potato as a grain food, comparable, if not superior, to rice and wheat, because it played the same or a better role in feeding the people, particularly the poor. Beginning in the 1740s, several local gazetteers in both Fujian and Guangdong began not only to recognize that sweet potato was as valuable as other grain crops, but also to classify it as a grain (*gu*) rather than a vegetable or fruit. For example, the Dong'an County gazetteer of Guangdong in 1740 stated that "sweet potato is a type of grain among many varieties." The Yongfu County gazetteer of Fujian explained in 1748 that "taro, *shanyao*, adlay, and sweet potato are the four that can be eaten as food, so they are in the grain category." In the decades that followed, more and more gazetteers, particularly in Fujian, adopted the trend of classifying the sweet potato as a grain. The Anxi County gazetteer of 1757 correctly stated that "sweet potato, known either as *fanshu* or *digua*, was a foreign crop of many varieties," classifying it as a grain food.[90]

Two interesting examples should be mentioned in describing the reclassification trend. One was that, almost at the same time as in Fujian, the Xianyang gazetteer in Shaanxi followed suit by classifying the root plant as a grain. Its compiler did not forget to credit Chen Hongmou with the claim that, "introduced by Governor Chen, sweet potato, which is coarser than yam but crisper than taro, began its cultivation here." Because Xianyang is in the cold North, growing the root there is difficult. "Our people have yet to master its farming techniques as well. However, the plant's advantage is obvious, and there are numerous varieties to choose from. As such, this is a profitable addition to the region's traditional produce." The other was a notable description found in the 1783 gazetteer of Fu'an County in Fujian. Aside from classifying the sweet potato as a grain, it also gave it a new name, *fanshu mi* 番薯米, which translates as "sweet potato rice," and recommended that "one peel off its skin, slice it into cubes, and cook it with rice." While many people still consider sweet potato to be a vegetable, and others describe it as one of the local forms of produce, as the nineteenth century progressed, more and more gazetteers reclassified it as a grain crop. This change occurred throughout the country, not just in Fujian and Guangdong.[91]

Needless to say, the reclassification of the sweet potato as a grain food in local historical sources signaled the integration of the American plant into the Chinese agricultural system. This integration had to do with the fact that, as in Oceania, Chinese farmers were accustomed to growing root crops such as taro and *shanyao*. However, sweet potato readily differentiated itself from these traditional root crops because of its high yield, edaphic flexibility, soft texture, and sweet taste. In contrast, taro and *shanyao*, except in a few regions where they were grown in relatively large quantities, were/

are not considered grain foods by the general population. Indeed, of the three American food crops introduced to China, the transformation of the sweet potato from a fruit or vegetable to a grain food within a century of its introduction was truly miraculous, something rarely seen in the rest of the world. According to Ping-ti Ho's research, peanuts were the first to gain acceptance among Chinese farmers when it came to the cultivation of American food plants. Peanuts first appeared in Jiangnan in the early sixteenth century and gradually spread to other parts of South China. However, progress was slow: Until the end of the eighteenth century, peanuts remained a southern crop, rarely planted in the North, even though its value in producing cooking oil had become widely recognized. Maize has always been regarded as a grain crop throughout its history, and its acceptance in China is no exception. However, in comparison to sweet potato, maize production in China was relatively slow. Both plants were introduced to southeastern coastal regions, but "maize remained relatively neglected; the people preferred rice and sweet potatoes, and, more importantly, maize competed with native cereal plants for good land." Maize did not begin to spread to the upper reaches of the Yangzi River and the Han River, a tributary of the Yangzi, until the late eighteenth century, following the migration of people from coastal provinces to inland areas. Needless to say, maize eventually became a major food crop throughout North and Northwest China, successfully displacing millet, sorghum, and even wheat. However, it competed with both the sweet potato and the white potato; after its late development, the latter appealed to Chinese farmers because it could grow in poor soils at high altitudes and in cold weather regions that were unsuitable even for farming maize.[92] Nonetheless, in most of China, white potatoes have long been considered a vegetable rather than a food staple. As mentioned in the introduction, it was not until the beginning of the twenty-first century that the government began to promote it as a "staple food" (*zhuliang*).

In comparison to all of the above plants, sweet potato has been a miracle food for the Chinese, from Xu Guangqi to modern researchers, because it possesses some unique qualities not found in other crops. To summarize what these scholars have written, the benefits of the sweet potato are various. First, of course, is its extraordinarily high yield, which is unrivaled by most other food plants. According to one estimate, the sweet potato could be turned into 250 *jin* of food, equivalent to 500 *jin* of rice and 417 *jin* of millet, with a yield of over 1,000 *jin* per mu. Growing sweet potato, as was well-known among Chinese farmers, could easily provide a half a year's food supply for a family.[93] Second is its extended farming season, which can take place from spring to fall, unlike other cereal crops, which are typically planted in one or two seasons. Third, it can be grown almost anywhere or in areas that were not previously considered farmland, as well as interplanted with other crops in fields without competing for good land. Fourth is its usefulness as food—both its swollen roots and green vines are edible—which is unusual even among other high-yield root plants like the white potato. Fifth, it is a fast-ripening food plant; according to one Chinese observer, "sweet potato can be taken as food almost as soon as it propagates and, in addition to eating it roasted, boiled, and steamed, it can also be eaten raw," making it ideal as a famine-relief crop.[94] And the sixth advantage, which was known earlier but became more popular in the twentieth century, is the root's ability to be processed into starch, which can be eaten alone (e.g., glass noodles) or used to thicken a soupy dish, as well as fermented to make either an alcoholic drink or industrial alcohol.

In sum, despite notable ups and downs, China's population followed a previously unseen trend of continuous growth from the tenth to the eighteenth centuries.[95] By the end of the period, the country had easily become the most populous nation in the world, with around 300 million people. Ping-ti Ho's seminal studies, with which we started this section, drew our attention to what he called the period's two "agricultural revolutions," which he saw as causes of this exponential demographic upsurge. If the cultivation of early ripening rice propelled the Chinese population to surpass 100 million in the early eleventh century, the dispersal of American food plants from the sixteenth century helped push it significantly higher. Our preceding discussion explains that, in terms of the impact of New World crops on Chinese agriculture, the country's rise to become the world's most populous nation by the late eighteenth century was largely attributed to sweet potato farming, because, despite being introduced earlier, peanuts remained a minor food crop and, despite its widespread appeal, maize cultivation was more of a nineteenth-century story than an eighteenth-century one in China. Geographically, maize also grew mainly in the western part of the country. In comparison, although originally a southern crop, sweet potato became widely cultivated in southern, central, and northern parts of the country, where the majority of the population lived. It played a notable role in sustaining the China's continual population growth into the twentieth century.

The burden of success? China's Malthusian discussion

When Lord George Macartney and his entourage embarked on their historic trip to China in 1792, it appeared that they were also well aware that the country they were about to visit had the world's largest population. It is hard to say if this served as one of the main factors for launching the trip, but it was clear that the British envoy was curious enough to inquire about the matter because it was related to the United Kingdom's interest in trading with China. According to his own recollection, before his trip, Macartney had learned about China's population estimate from his friends. Upon his arrival and at his request, a high-ranking Chinese official named "Chow-ta-zhin/Chou-ta-gin," or Mr. Qiao, offered him a figure of over 333 million. Macartney expressed some skepticism, but also thought it likely, because he saw with his own eyes that the number of people living in China was greater than he had anticipated.[96]

Aside from Macartney, more members of his entourage have left us with memoirs about China. And they, too, recorded the population figure given by the Chinese official, either in surprise or disbelief. Nonetheless, they appeared to prefer to accept the estimated figure rather than challenge it. In fact, they were drawn to explain how China could support such a large population and whether the country could do so in the future. Macartney's right-hand man, George Staunton, who served as secretary for the trip, was more interested in explaining how China could do it in the past, while Macartney's comptroller, John Barrow, shared his suspicions on whether it could continue it into the future. Staunton appeared to believe that there were reasons for China's enormous population, despite wars, infanticides, epidemics, and famines, and noted differences in marriage and family system, among other things. "In general," he wrote, "there seems to be no other bounds to Chinese populousness [sic], than those which the necessity of subsistence may put to it." He then hastened to add that "These bounds are certainly more enlarged than in other countries," for "No arable land lies fallow. The soil, under a hot and fertilizing sun, yields annually, in most instances, double crops, in consequence of adapting the culture to the soil; and supplying its defects by mixture with other earths, by manure, by irrigation, by careful and judicious industry of every kind."[97] In other words, Staunton believed that the only potentially harmful factor limiting China's population growth was food supply. However, thanks to the tenacity of

Chinese farmers, the land's produce appeared to be sufficient to feed the country's massive population.

John Barrow agreed with Staunton but was less optimistic about the future. Barrow, like Macartney and Staunton, chose to believe Mr. Qiao, whom he described as "a plain, unaffected, and honest man." From there, he offered explanations similar to Staunton's: China was able to provide enough food for its large population primarily because its farmers utilized every inch of land and farmed it efficiently. Indeed, he explained, in China "the population is not yet arrived at a level with the means which the country affords of subsistence." British counterparts, on the other hand, were unable to do so due to "idleness or bad management." Meanwhile, Barrow believed that there was more land to be developed in China, that Chinese peasants could improve their farming tools, and that they could grow high-yield crops such as the white potato rather than rice, the latter being "the poorest of all grain" and prone to failure in bad seasons.[98] In a nutshell, Barrow saw a potential challenge for China to support its overburdened population lest it improve its food supply system further.

While George Staunton and John Barrow were more or less admitting that the country they were visiting appeared to have the resources to support its population, they, along with George Macartney, also noted and recorded visible signs of poverty in China in their writings. Macartney wrote that "A common weaver, joiner or other tradesman earns little more than a bare subsistence. . . . It does not appear that there is always sufficient employment for the people, whose multitude is so great as to exceed the means of subsistence by labour. . . . It is affirmed that in every year vast numbers perish of hunger and cold."[99] And when the British, who seemed to be oversupplied by the generosity of the Qing court and officials, threw animals "of disorder" [sick or spoiled animals] from the provisions overboard, "the Chinese eagerly picked them up, washed them clean, and laid them in salt," according to John Barrow and Aeneas Anderson, another attendant on the British Embassy who also provided a detailed account of the trip.[100] Indeed, there were far too many cheap laborers eager to find work, a sign of overpopulation. "In China it is not merely that the labour of men is cheaper; but it does not seem to occur to spare it," Staunton observed. Barrow agreed, citing a specific incident in which the embassy required porters to load their luggage onto the boats before entering Beijing, which required approximately three thousand men. But it was quite simple to double the number, "as there seemed to be ten times the number of idle spectators as of persons employed," in his words.[101]

By advising Chinese farmers to grow white potatoes instead of rice, it appears John Barrow was unaware that the American crop had already arrived in China prior to his party's arrival, except that widespread cultivation, as previously mentioned, would not begin for decades after the British Embassy left China. Yet his general observation of the tension between food supply and population growth was correct. Barrow also noted in his journal that, while rice was the most important cereal crop in South China, millet, wheat, buckwheat, and sorghum were its northern counterparts. George Staunton and Aeneas Anderson seemed more observant, in that they also noticed that corn was another food crop planted in the North; Anderson even recalled that before their arrival in Tongzhou, south of Beijing, "In the evening we took a walk along the banks of the river. The corn was now almost ripe; agriculture appeared in its best form; and copious plenty seemed to countenance and support the immense population we every where [sic] observed."[102] He appears to have attributed China's success in feeding its enormous population to corn production.

What about sweet potatoes, another American food plant that is popular among Chinese farmers in both South and North China? Samuel Holmes, who provided yet another account of the embassy, mentioned it a couple of times. As a guardsman for Lord Macartney, Holmes, described as "a worthy, sensible but unlearned man," kept a journal of the voyage. Perhaps because of his lower social status relative to George Staunton, John Barrow, and Aeneas Anderson, Holmes not only saw how food plants were grown in the fields like the others (which may have caused them to miss the cultivation of the sweet potato as an underground plant), but he also tasted the foods, including the sweet potato, in both Vietnam and China, most likely alongside the locals. During his time in Vietnam, he wrote that "we could get nothing but a few ducks, a little fish, some sweet potatoes, some sugar" for about three weeks. He admitted that food was provided "in great profusion" after arriving in China, despite his initial discomfort with the cuisine's method of cooking, which involved "all their meat and vegetables being hashed up in such medley confusion." He also enjoyed the local flavors: "Beef, mutton, and pork were excellent, as were their vegetables, such as sweet potatoes, cabbages, pumpkins, onions, and a great variety of others, common to both Europe and Asia." Holmes shared the others' shock at the size of the Chinese population, especially when he was in the South, noting that China's "population exceeds even belief; it is impossible for any one [sic] to conceive it, the whole country is absolutely covered with people, and every river is full of floating houses," he wrote.[103]

Samuel Holmes's journal was published in 1798, just one year after those of George Staunton and Aeneas Anderson, even though the author was in a lower position in the embassy. Lord Macartney's embassy was undoubtedly a cause célèbre in Britain and Europe at the time, despite its failure to open China to trade, which most likely explains why the works of Holmes and other members of Macartney's retinue were published in such a timely manner. (The only exception might be John Barrow's journal, which was not printed until 1804.) It is interesting to note that Thomas Malthus published his influential *An Essay on the Principle of Population* contemporaneously in 1798. Unlike his compatriots, Malthus had never been to China, but he was interested in its population growth and used it as a crucial example to support his main argument that population growth always outstrips food production, barring periodic "positive checks" such as war and famine.[104]

As a demographer by training, Malthus was clearly aware that by his time China had become the most populous nation on earth. He also stated that China was not only "the most fertile" but also "the richest country in the world, without any other." This was due to the fact that "almost all the land is in tillage; and that a great part of it bears two crops every year; and further, that the people live very frugally, we may infer with certainty that the population must be immense." In other words, if what Malthus said was based on inference, then it was quite consistent with what his countrymen who had accompanied Macartney saw with their own eyes. He then immediately moved on to his second inference, which was the lower classes' ability to survive in a nation with a high population density like China. As Malthus explained, "In some countries population appears to have been forced, that is the people have been habituated by degrees to live almost upon the smallest possible quantity of food. There must have been periods in such countries when population increased permanently, without an increase in the means of subsistence. China seems to answer to this description."[105] Again, his observations seemed to be in general agreement with those of the various members of the Macartney Embassy, including its namesake. For none of them had seen any significant challenges to the country's progress, except John Barrow, who, perhaps worried that China's potential food shortages would prevent it from feeding its swelling population, recommended high-yielding crops such as the white potato to Chinese farmers.

If Barrow was wary, Malthus was utterly pessimistic because he foresaw the potential danger should China fail to manage its population pressure. "If the accounts we have of it are to be trusted," Malthus wrote, and "the lower classes of people are in the habit of living almost upon the smallest possible quantity of food," then, he continued emphatically, "a nation in this state must necessarily be subject to famines. Where a country is so populous in proportion to the means of subsistence that the average produce of it is but barely sufficient to support the lives of the inhabitants, any deficiency from the badness of seasons must be fatal."[106] Malthus did not specify what "accounts" he used or consulted in developing his thesis. Nevertheless, his attitude toward China differed from that of his contemporaries. In particular, he turned Barrow's worry into a doomsday prediction: "positive checks" (such as famine, disease, and warfare) would inevitably occur in China, causing mortality to rise significantly to reduce its population size. To prevent this disastrous outcome, Malthus also suggested that such "preventive measures" as abortion, birth control, late marriage, and celibacy be implemented to address the inevitable imbalance between population growth and food production.[107]

Regardless of the validity of Malthus's prediction, there were people in late eighteenth-century China who agreed with him. In fact, some educated Chinese people had voiced concerns exactly like those of the English demographer after observing the country's exponential population growth. Malthus's senior by twenty years, Hong Liangji, was one such person. Hong published his first essay on population in 1793, five years before Malthus published his own.[108] Coincidentally, it was also the year the British Embassy in China was launched. In other words, while the British tourists recorded their observations of China's expanding population, the Chinese actually felt its effects. In Hong Liangji's case, it was the experience itself that had caused him to consider the problem. In his *Shengji pian* (On livelihood), one of his essays on the effects of high inflation brought on by population growth, Hong compared and contrasted his father and great-grandfather's lifestyles. He claimed that during the time of his great-grandfather, one person could provide enough food and clothing for a family of ten, whereas during the time of his father, there were ten times as many people who needed to be fed and clothed. The prices of cotton, silk, millet, and rice consequently increased significantly on the market, while "literati, peasants, artisans, and merchants all reduced the value for their worth." What was even scarier, in Hong's opinion, was that as the overall population grew, so did the number of "loafers and vagabonds." "These people would not simply wait to die, but would [cause trouble] when such disasters as flood, drought, and epidemic hit," the author wrote.[109]

As a result, Hong Liangji believed that the greatest threat to one's livelihood was social instability caused by population growth. According to his calculations, China's population could grow more than tenfold in a half century—roughly from the time of one's great grandfather to the time of one's father. This was, of course, an estimate. What earned him the moniker "China's Malthus" was his reasoning that population growth was an unstoppable trend in human history. Hong expanded on his analysis of demographic impact in his oft-quoted essay *Zhiping pian* (On the rule of peace). He began the essay by stating that everyone desires and appreciates peace in their lives. "And a peaceful age that lasts more than a century is long. However, in terms of population growth, it expands approximately five times in thirty years, ten times in sixty years, and more than twenty times in one hundred years." Of course, he admitted that there had been some vacant land and extra fields in the past that could accommodate increased housing and farming to match the greater population. However, Hong, like his English counterpart, maintained that the rate of population growth far outstrips the rate of land reclamation. His reasoning was as follows:

> Speaking of households, there are now twenty times as many as there were a century ago.... Some may argue that there would be enough wild land to cultivate and still leave room for housing. However, they can only be doubled, tripled, or increased five times, whereas the population can be ten to twenty times larger at the same time. As a result, housing and crop fields are in short supply, while the population is always in excess. Given that some households become monopolists, it is no surprise that so many people have suffered from cold and hunger, and even died here and there.[110]

Is there any way to halt the seemingly unstoppable trend of population growth? Anticipating his English counterpart, Hong believed that there were two methods for keeping the population under control. The first was the "method of nature" (*tiandi you fa*), and the second was the "method of government" (*junxiang you fa*). More specifically, the "method of nature" referred to natural disasters such as droughts, floods, and diseases, which he believed typically affected between 10 and 20 percent of the population. And "method of government" referred to a variety of government policies aimed at alleviating problems caused by population growth. These policies included increasing farmland cultivation, providing social relief, lowering land and other taxes, imposing sumptuary laws, and reducing waste. Hong did not believe that all of these measures, as the "method of nature," would work in the end, because everyone wants peace, and peace causes population growth. "People grow naturally in the world," he concluded, "but the world can only support a certain number of people."[111]

If Hong Liangji held similar views to Thomas Malthus, he was far from alone among Chinese scholars. One of Hong's forefathers was Xu Guangqi, a Chinese Christian scholar who highlighted the importance of the sweet potato. Aside from his praise for the American plant, Xu estimated the rate of population growth in his writings. Like Hong Liangji, he believed that a population doubles every thirty years, a "natural tendency" that must be checked by "major warfare." He then questioned the national economy's capacity to sustain such natural population growth, ultimately concluding that this would be difficult—if not impossible—since a growing population does not necessarily result in increased grain production.[112]

Following Xu Guangqi, Song Yingxing, a noted seventeenth-century scientist, expressed similar concerns about population growth. Song witnessed China's population growth during the late Ming period as the author of *Tiangong kaiwu* [Explorations of the works of nature], an impressive encyclopedia covering a wide range of technical issues, including those related to agriculture. Of course, New World crops had already been introduced to China during his time, but they had not yet made a visible impact because sweet potato and corn were at that point only grown in South China. Song thus attributed the Ming population increase to widespread rice cultivation. According to his estimate, rice constituted up to 70 percent of total grain food consumed by the Chinese at the time. Nonetheless, Song seemed concerned in one of his poems about the people's insatiable desire to produce offspring.[113]

Xu Guangqi and Song Yingxing were each born in areas that experienced rapid population growth during their lifetimes. As previously stated, Hong Liangji's personal experience played a role in drawing his attention to the issue of population. Indeed, Hong witnessed an 80 percent increase in the Chinese population from the time of his birth in 1746 to the publication of his essay in 1793. And the growth rate in Jiangsu, where he was raised, was higher than the national average. "As a Jiangsu native," a Chinese study noted, "it was quite natural for Hong Liangji to be extremely concerned about the imbalance between population explosion and the province's reduced ratio of farmland per person."[114]

Despite Hong Liangji's "extreme concern," China's population continued to grow, from over 300 million at the beginning of the nineteenth century to around 430 million by the time Wang Shiduo, the nineteenth-century scholar, expressed his concerns. Throughout that century, New World crops, particularly sweet potatoes, were grown across much of the country, which undoubtedly contributed to this increase. Wang, one of Hong Liangji's most ardent supporters, grew up in Jiangsu and witnessed the Taiping Rebellion firsthand, a widespread peasant uprising that swept across a quarter of the country in the middle of the century. Taiping rebels controlled nearly half of China's population at the peak of their power, nearly toppling the reigning Qing dynasty. Wang's population discussions were thus his musings on the causes of that historic peasant revolt. Indeed, despite being born in Anhui, Wang spent his childhood in Nanjing, which had long served as Jiangsu's capital but was seized by the rebels to serve as the capital of their decade-plus-long Heavenly Kingdom (1851–1864).

According to some Chinese scholars, Wang Shiduo's strong emphasis on population control has made him another, and more radical, "Malthusian scholar" of nineteenth-century China.[115] In two ways Wang appeared closer to the English demographer than his predecessors. One was that, while he had obviously suffered from the Taiping Rebellion's devastation, Wang believed that the disaster, like other disasters such as epidemics and famines, actually reflected heaven's will in slowing the rate of population growth. If Wang Shiduo's line of thought was similar to Malthus's "positive check," then Wang also advocated for "preventive checks," which was to control population growth through late marriage, contraception, and infanticide, particularly of female infants. Wang, in particular, advocated for the government to impose restrictions on men marrying under the age of twenty-five and women marrying under the age of twenty. Later, he raised the minimum age for marriage to thirty for men and twenty-five for women, demonstrating his commitment to halting population growth. As for his motivation for female infanticide, Wang emphasized that it should be a top priority, because female infanticide could effectively limit human reproduction as the fewer women there were, the fewer people would be born, and women, he unfairly charged, were the culprit in causing unwanted population growth. To that end, he went so far as to argue that the government should teach people how to kill female babies after birth—usually by submerging them in bathwater—and reward those who do so. Furthermore, he proposed that the government double the taxation of any family with more than three sons or more than two daughters.[116]

If Wang Shiduo appeared to have taken a more radical stance than his predecessors to tackle the consequences of late imperial China's population increase, his radicalism arose from his living in a rather different time and experiencing the country's bloodiest civil war. Did the outbreak of the Taiping Rebellion in the mid-nineteenth century, roughly fifty years after China's population surpassed 300 million, validate the warnings of Thomas Malthus and his Chinese counterparts about the dreadful but unavoidable consequences of uncontrollable population growth? To put it another way, did the Taiping Rebellion act as a "positive check" on the country's growing population? Scholarly opinions have varied. But many have pointed to an increase in "population pressure" during the period.[117] Ping-ti Ho is most outspoken about the correlation between the population increase and the outbreak of the Taiping Rebellion. In his classic *Studies on the Population of China*, he writes that "The Taiping Rebellion is deservedly called the greatest civil war in world history. . . . In sheer brutality and destruction it has few peers in the annals of history." While he does not provide his own estimate of total population loss during the period, Ho states that by 1953, the year he completed his study of the Chinese population, the combined population of the four provinces most affected by the war was still approximately 14 percent lower than in 1850. And the Taiping war was just one of the events that he believed had deterred the Chinese population's rapid growth; other devastating military conflicts had of course taken place in the country, and no doubt would again. Ho's use of the term "deterrents" in naming the chapter suggests that he was sympathetic to Malthus's theory. In concluding the chapter, he mentions Malthus as well as his contemporary William Godwin, the British social thinker, and concludes that, of all the "catastrophes" such as famines and diseases, the impact of "wars and civil disorder" was much greater during the Qing period.[118]

In mentioning Thomas Malthus's theory, Ping-ti Ho also did not forget the latter's Chinese counterparts.[119] For like Malthus, Chinese demographers from Xu Guangqi to Wang Shiduo were similarly concerned with how to deal with food shortages in the face of the seemingly inevitable trend of population growth. Wang Shiduo's above-mentioned proposal to limit family size, for example, was comparable to Malthus's idea of implementing "preventive measures." Interestingly, while his recommendation of female infanticide was appalling, Wang's suggestion of maintaining a family with no more than three sons and two daughters actually reflected reality throughout Chinese history, according to contemporary studies. Jiang Tao, the late twentieth-century demographer, asserts in his survey of imperial China's demographic structure that since ancient times, most Chinese have desired more children, particularly male offspring. On the other hand, "despite the desire to have more children, it was by no means the more the merrier," he emphasizes. Instead, "it was restricted, as was the average family size." Jiang discovers, based on a rich corpus of literature, that most Chinese consider a family to have between seven and eight members—three sons and two or three daughters in addition to the parents. However, he quickly adds that the average size of a Chinese family "was by no means of eight people." Rather, it was only five, which he claimed was also a number commonly used by governments in calculating taxation from early to late imperial China.[120]

In a 1998 article, historians William Lavely and R. Bin Wong also argue that China was not an example of "demographic profligacy," nor was population pressure the only cause of widespread peasant uprisings in the mid-nineteenth century. They present evidence that low marital fertility and female infanticide aided in the reduction of family size in China.[121] Their argument explains why, despite the entrenched practice of early marriage, the average Chinese family size has remained small. Expanding on an earlier case study, another historian, James Z. Lee (and his collaborators), provides a detailed analysis of how infanticide served as the Chinese (preventive) check on population growth, refuting Malthus's thesis that war and famine were the only options for China to control its swelling population, as witnessed by the English demographer and his fellow countrymen in the Macartney Embassy at the end of the eighteenth century. Lee observes that "In some Chinese populations, families regularly practice infanticide to regulate the number and sex of their children." In addition to infanticide, he and his colleagues discovered that marital fertility in Chinese families was lower than expected. His demographic study backs up Lavely and Wong's argument: Instead of a counterexample of Europe, which allegedly practiced "demographic restraint," as suggested by Malthus and his followers, China had not allowed its population to grow out of control. Instead, to quote James Lee once more, "In contrast to the Malthusian paradigm, human agency in China was not restricted to nuptiality." That is, while the Chinese tended to marry early and maintain an almost 100 percent nuptiality rate, especially among women, they developed other methods, known as the "Chinese check," to limit their family size and slow the country's overall population growth.[122]

The preceding discussion demonstrates that, while seemingly radical, even outrageous, Wang Shiduo's recommendation of female infanticide as a means of preventing overpopulation did not come out of thin air. In fact, research has shown that the ideas and practices surrounding infanticide have a long history in China and became more prevalent in the Ming and Qing periods as the population grew. Historian Ann Waltner observes at the outset of a related article that "infanticide, particularly female infanticide, was part of a population strategy in traditional China." That is, families in China manipulated the mortality rate to regulate family size based on economic condition and gender preference under the influence of Confucian teaching. Female infants were often victims of this practice, despite official efforts to curb it. Although offenders were theoretically subject to government punishment, and local officials regularly issued admonitions and regulations against it, the offense was frequently carried out by midwives.[123] In their 1997 case study of a rural community in Liaoning, in northeastern China, between 1774 and 1873, demographers James Lee and Cameron Campbell calculate that female infanticide could account for up to 20 percent of all female babies who died in the area.[124]

Asian scholars such as Chang Jianhua, Lin Liyue, Ogawa Yoshiyuki, and Yi Ruolan agree with the findings presented above. Their own research shows that, while female infanticide was a national phenomenon, it was more prevalent in the South and occurred more frequently in the Qing period than in the previous Ming period. Chang Jianhua and Lin Liyue, both experts on Ming history, acknowledge that, relatively speaking, infanticide in the Ming period received less attention than during the subsequent Qing. Chang and Lin both claim that female infanticide was common in Fujian, Jiangxi, Zhejiang, and Jiangnan, citing local gazetteers, magistrates' manuals, and morality pamphlets. These were apparently areas with the highest population density in the country. Chang discovered that some families in these regions killed not only female but also male infants in order to reduce family size due to economic hardship.¹²⁵ Ogawa Yoshiyuki notes in his case study of the phenomenon in Jiangxi during the Qing period that female infanticide was also irrespective of a family's social class, because due to rising dowry prices, both the rich and poor thought girls were financial liabilities to a family. The high cost of marriage expenses was, ironically, attributed to a shortage of marriage-age women, a cause also identified by Ann Waltner.¹²⁶ In other words, while there were many reasons for the heinous practice of female infanticide, demographic factors were among the most important in explaining both its occurrence and impact. Indeed, there appeared to be a vicious circle: The more prevalent female infanticide became, the more expensive dowry and marriage grew, owing to the obvious lack of women in a given location. Yi Ruolan's study of the incidence in Taiwan adds to the evidence. Taiwanese society, which was founded primarily by mainland immigrants during the Qing dynasty, did not see many cases of infanticide at first due to a lack of women in the early waves of immigration. By the nineteenth century, however, female infanticide had become as prevalent on the island as it was on the mainland, as population growth gradually corrected the sex imbalance among residents.¹²⁷

Sweet potato's impact on Chinese agriculture and society

Though cruel and inhumane to women, the long and widespread practice of female infanticide provided strong evidence that China was not a "demographically profligate" country; the "preventive checks" on overpopulation coined by Thomas Malthus to extol European population behavior also existed in China. Meanwhile, as previously stated, widespread peasant rebellions in the mid-nineteenth century caused catastrophic losses of human life in the country.¹²⁸ Nonetheless, the Chinese population recovered and continued to grow, reaching approximately 500 million in the early twentieth century and more than doubling by the turn of the twenty-first, despite the country suffering from natural disasters such as floods, locusts, and droughts, as well as wars and revolutions. Evidently, Malthus's theory is insufficient to explain China's demographic development. Ester Boserup, the Danish economist, argued that population is not only an independent variable, it also drives agricultural development and food production as needed. Her argument, to paraphrase a well-known proverb, is that "necessity is the mother of invention." That is, as the population grows, inventions and improvements are sought and usually achieved, allowing people to catch up with the rising demography; or, as she puts it in her classic study, *The Conditions of Agricultural Growth*, "population growth is here regarded as the independent variable which in its turn is a major factor determining agricultural development."[129]

The introduction and cultivation of New World crops such as sweet potato and maize beginning in the sixteenth century, as it were, represented Chinese farmers' invention and improvement in "agricultural development," per Boserup. The expansion of their cultivation not only supported the country's continued population increase to a new, higher level in more recent centuries, it also helped sustain the demographic explosion from the eighteenth century. Indeed, as Laura Murray found while completing her 1985 dissertation, farmers in the Wei River Valley of Northwest China turned to sweet potatoes to cope with the increased population in the mid-1800s; her study thus modified Ping-ti Ho's thesis that the introduction of New World crops led to the eighteenth-century population explosion. Rather, the appeal of the sweet potato as a valuable crop followed a northward trajectory that supported population growth as it spread across the country. In recent years, publications by Wang Siming and Li Xinsheng of Nanjing Agricultural University have further argued that while the introduction and cultivation of New World crops such as maize and sweet potatoes aided China's agricultural expansion, their impact was more pronounced in the nineteenth century than in the seventeenth and eighteenth centuries, as suggested by Ho. In particular, they emphasize that, contrary to common perception, maize was not widely cultivated in China until the nineteenth century or even later, a finding consistent with Murray's research.[130] These studies contribute to our understanding that sweet potato, and New World crops in general, followed an upward course in terms of their importance in Chinese agriculture and gastronomy between the seventeenth and twentieth centuries.

The growing importance of the sweet potato as a food for the Chinese in modern times is evidenced by the steady expansion of its cultivation in the country. Although its production has declined since the late twentieth century, the root crop is grown in almost the entire country, consisting of five major production zones based on the variations in its growing season and ripening period.[131] The first is the "southern autumn–winter region," in China's extreme South, which includes the Pearl River Delta, Hainan Island, and the southern Fujian cities of Zhangzhou, Quanzhou, and Xinghua. The plant can grow all year in this region, which has a humid tropical monsoon climate and a growing season of 185 days per year. Autumn sweet potatoes are planted in July and August and harvested in November and December, whereas winter sweet potatoes are planted in November and harvested in April and May of the following year. This was also one of the first areas where the plant was grown, and thanks to the promotion of Governor Jin Xuezeng, it spread to neighboring regions in the same climate zone. Chen Hong, a member of the local intelligentsia, noted that the sweet potato was grown everywhere in Xinghua, Quanzhou, and Zhangzhou as early as 1650.[132] At the turn of the seventeenth century, the sweet potato overtook rice as the second most important food crop in neighboring Chaozhou, southern Guangdong. And the trend continued—research shows that in the mid-twentieth century, the importance of wheat as a food crop, which had less than half the per unit yield of sweet potato, was falling further behind, occupying only one-tenth of the cultivated areas of the American plant in Chaozhou. And its regional production was only one-tenth that of sweet potatoes.[133]

Sweet potato is planted in May in the "southern summer–autumn region," which stretches from Yunnan and Guangxi to northern Guangdong and Fujian, just above the "southern autumn–winter region" in South China, and harvested between August and October. Because of the region's semitropical climate, the root can then be planted between July and August and harvested between late November and early December. Consider sweet potato cultivation in Guangxi: Scholars generally agree that the plant arrived in the province from Fujian in the eighteenth century, and by the first half of the twentieth century, it had become the most cultivated dry-land plant and the third most important food plant in the province. According to local sources, sweet potato was seen everywhere in certain counties, and some villages even enacted ordinances specifically targeting anyone who stole its roots. Sweet potato was cultivated in sandy loam on hillsides due to its edaphic tolerance, as seen in other places. Scholars observed that this advantage made it even more appealing than maize—"wherever maize is grown, one can also plant the sweet potato." Sweet potato also yields twice as much as maize per unit. More importantly, sweet potato's appeal in Guangxi and elsewhere in the zone stemmed from the fact that it was an ideal candidate for growing in the fall and winter seasons, after rice, the most commonly grown food plant, had been harvested. According to local historians, the root plant was popular in rice-growing areas because "it could grow in homestead alongside vegetables while yielding as many as several thousand jin, which was enough to supply food for a family of eight over several months."[134] In other words, sweet potato became a daily food even in areas where rice was the primary crop.

The Yangzi River Valley spring–summer region is China's largest sweet potato farming region in modern times. Because of the area's subtropical climate, the sweet potato growing season can last up to 155 days. It can be planted after the last frost between April and June and harvested between late October and mid-November. Hunan, which borders Fujian and Guangdong, may have been the first province to grow sweet potato in a region where rice is the dominant food crop. Its mountainous terrain also encouraged Hunanese farmers to grow the root plant, which became increasingly valued as a significant addition to the food supply beginning in the late eighteenth century. During the early twentieth century, Hunan's sweet potato land acreage more than doubled that of Hubei, its neighbor. Hunan most likely influenced sweet potato cultivation in Hubei. However, some of its cultivars may have originated in Jiangnan, where sweet potato was grown as early as the sixteenth century due to Xu Guangqi's promotion.[135] In comparison, Sichuan's case may be more eye-catching; despite being a major rice-producing province, it has seen a marked increase in sweet potato's popularity. Introduced in the late eighteenth century by a Guangdong official, the root was dubbed *hongshao* 紅苕 by the locals instead of *fanshu* and was widely accepted as a tasty food throughout the province. While roast sweet potato became a popular street food in cities, the root became a daily staple for the people living in the mountains surrounding the Sichuan Basin. "Instant Starch Noodles" (*fangbian fensi* 方便粉丝), which became a registered brand in Sichuan in 2009, is a fast food made from sweet potato starch, a technique developed long ago by the region's residents. It is sold throughout the country, along with other sweet potato foods from the province.[136]

Sweet potato cultivation in the Yellow-Huai River Valley spring–summer region is equally impressive. In modern-day China, the region accounts for roughly 40 percent of sweet potato production, with Shandong and Henan contributing the most. Sweet potatoes are typically planted in the spring and summer and harvested in October. Take, for example, Shandong, where the early cultivation of the sweet potato, as mentioned above, was introduced by the Chen Shiyuan, the compiler of *Jinshu chuanxilu*, in the mid-eighteenth century. However, it was not until Shandong farmers discovered ways to store the root over the winter that the American plant gained popularity. Sweet potato had been cultivated throughout the province since the early twentieth century. And the expansion continued throughout World War II. In Shandong, a major battleground, both Japanese invaders and Chinese resistors turned to the high-yielding sweet potato to deal with food shortages during the war. Sweet potato farming reached new heights after the war and the regime change in 1949. In the 1960s, for example, while the root plant was the second most cultivated food plant after wheat, its total production was equal to the sum of wheat, maize, and millet.[137] Sweet potato's popularity as a daily food is reflected in the invention and availability of a variety of sweet potato foods in Shandong and its neighboring regions, as seen in Sichuan of the Yangzi River Valley spring–summer region. Perhaps the most well-known dessert is "silk-pulling candied sweet potatoes" (*basi fanshu/digua* 拔絲番薯/地瓜).

The northern spring region in temperate and cold-temperate climate stretches from North to Northeast China. It has approximately 170 frost-free days, allowing farmers to grow the sweet potato in the spring and harvest between September and October, owing to its growing season of 130–140 days. However, for most of their reign, the Qing rulers prohibited immigration to Manchuria, which was a major reason why sweet potato cultivation in the region did not begin until the early twentieth century. After the Qing dynasty fell, immigration accelerated; farmers from Shandong and elsewhere brought sweet potato to the region. However, due to its late development, sweet potato farming is not as developed as it is in the other areas.[138] The cold climate on the west side of the region also makes the white potato, rather than the sweet potato, a more suitable food plant.

What role has sweet potato farming played in the modern development of Chinese agriculture? There are a number of factors to take into consideration when answering this question. The first is its role, along with maize and peanuts, in increasing the country's arable land area. *Agricultural Development in China, 1368–1968*, by Dwight H. Perkins, published in 1969, remains useful in providing a general picture. Extending somewhat Ester Boserup's theory, Perkins maintained that due to the period's population growth, which he described as following an unprecedented upward path, Chinese agriculture expanded significantly to accommodate the demographic explosion. Perkins calculated that over the course of six centuries, Chinese farmers were able to double the country's food production. However, increasing the size of arable land to grow more food grains was only one side of the story, because, as Perkins observed, Chinese farmers had largely exhausted China's arable land by the nineteenth century. The other side of his research involved the efforts of Chinese farmers/agronomists to improve farming technologies, introduce new crops, and increase the level of land productivity. And the two appeared to be inextricably linked. That is, due to the scarcity of arable land, Chinese peasants were forced to select high-yielding food plants while attempting to increase the acreage of cultivated land. Perkins discovered that during the twentieth century, the country saw a significant increase in maize and sweet potato cultivation, a moderate increase in wheat and rice cultivation, and a decline in barley and sorghum cultivation. Indeed, he noted that the majority of the 400-million-*mu* increase during the period was in the form of maize and sweet potato farming.[139]

Recent works by agriculturalists Shi Zhihong, Wang Siming, and Li Xinsheng expand on the fact that widespread cultivation of sweet potato and maize as a means of improving agricultural production began in the eighteenth century, when China's population reached an unprecedented level, and continued into the twentieth century. This development coincided with an increase in the country's cultivated acreage. In his *Agricultural Development in Qing China*, Shi Zhihong has reviewed a number of earlier studies and concluded that "the area of cultivated land in China in 1850 had increased by about one-third in comparison with the area in the early 18th century. In traditional China, this really was a great achievement." Furthermore, the expansion continued into the following century. Shi estimated in 1952 that the cultivated land area had increased from 1,300 million *mu* in 1850 to around 1,600 million *mu* in 1952—a 20 percent increase.[140] Wang Siming's research, echoing the findings of Dwight Perkins, states that the increase was primarily due to the expansion of sweet potato and maize farming. He calculates that during the first half of the twentieth century, maize cultivation increased by 150 percent and sweet potato cultivation increased by a whopping 528 percent! As a result, the sweet potato's importance in the country's total food supply increased from 11.2 percent between 1924 and 1929 to 16.2 percent between 1938 and 1947. Needless to say, this result was achieved because sweet potato and maize are both high-yielding food plants, easily outperforming rice and wheat in per unit yield. Sweet potato, with an average of 1,000 *jin* per *mu* (dried sweet potato is about 250 *jin* per *mu*), is also significantly higher than maize, which has an average of 180 *jin* per *mu*. The cultivation of the two America crops added about 17 *jin* per *mu* to Qing China's average grain yield.[141]

Second, it goes without saying that the increase in sweet potato cultivation in China from the seventeenth to the twentieth centuries was impressive. However, that fact only revealed one side of the sweet potato's importance as a food plant in the country—while the increase was significant, sweet potato farming occupied a very small percentage of China's total area of food production. While scholars differ in their calculations, the general estimate has been that from the eighteenth century to the twentieth, sweet potato cultivation accounted for no more than 2 percent of the country's cultivated acreage (maize accounted for about 5 percent).[142] In other words, rice and wheat remained the two most important food crops grown by Chinese peasants. On the other hand, the fact that sweet potato farming was limited in the country did not diminish the importance of the root plant; rather, it highlighted its exceptional value as a food resource in Chinese agriculture. The main reason is that sweet potato is not only a high-yielding food plant that can be grown in almost any soil, including those unsuitable for grain crops like rice and wheat—it can also grow alongside other food grains via intercropping without competing for farmland. Indeed, while intercropping and/or multicropping had previously been practiced in China, Shi Zhihong believes the technique "never played a [major] role until the 18th century," or after the promotion of sweet potato and maize.[143] Li Xinsheng and Wang Siming tend to give a low estimate of the total area of maize and sweet potato farming in the country, but they also readily acknowledge that "American plants fully used peripheral land that had been either unsuitable or less suitable for growing other grain crops. As a result, their cultivation increased the size of arable land, which was significant, because without maize and sweet potato, the peripheral land would not have been used, particularly in non-plain regions." That "Sweet potato could be planted in any leftover land" was also confirmed by historical records.[144] In a word, the significance of the sweet potato's contribution to food supply is grossly underestimated if only the small percentage of its farming area in China's total area of food production is considered.

Third, the appeal of sweet potato cultivation to Chinese farmers stemmed from the fact that it was both a high-yielding and an easy-to-grow food plant. American historian Sucheta Mazumdar, who refers to sweet potatoes as "miracle foods," explains:

The sweet potato need not be transplanted like rice, and although maximum size is obtained by fertilizing the seedlings a little, it can be grown without fertilizers. Planting a new crop is relatively cheap, for one tuber can provide a dozen or so shoots for new plants. Twentieth-century studies in Taiwan show that whereas one crop of rice requires 85.92 male labor days, 18.75 female labor days, and 15.79 animal labor days per hectare (2.47 acres), the sweet potato requires only 9.47 male labor days, 2.47 female labor days, and 2.05 animal labor days per hectare.[145]

Put simply, while rice farming requires a lot of labor, sweet potato farming does not. This advantage of the sweet potato as a food plant brought another benefit: Because food supply is now guaranteed with this "insurance crop," Chinese farmers were able to grow other cash crops to improve their living conditions, as Philip Huang observed in his study of the peasant economy in North China. As a result, sweet potato farming aided agricultural commercialization. It appeared that the two were mutually beneficial—in the twentieth century, agricultural commercialization accelerated in the country, coinciding with a significant increase in sweet potato cultivation.[146]

Finally, after its arrival and dispersal in China, sweet potato farming had a significant impact on inducing environmental and demographic changes. As previously stated, sweet potato was initially welcomed as a valuable food plant in Fujian due to its ability to grow on hillsides. As such, it helped alleviate the food shortage in the province, which is known for its poor soil and mountainous terrain, with only 10 percent of land suitable for rice farming remaining. One poem extols the virtues of the plant like this: "Since the arrival of the sweet potato in Fujian in the sixteenth century, it is difficult to find hungry people anywhere in the realm."[147] Of course, this was an exaggeration. Nonetheless, because hills and mountains cover two-thirds of China's landscape, it appeared that Fujian's initial success with sweet potato cultivation was replicated across much of the country in the subsequent centuries. Zhejiang, for example, is an important part of Jiangnan known for its rice production. However, because its western portion is covered by hills, the only way to reclaim land was to open up the mountains. "To increase the local supply of grain, some immigrants, or 'shed people' (*pengmin*), cleared mountain land to plant sweet potato and corn," writes Bozhong Li.[148] They did so because sweet potato requires little fertilizer and grows in almost any soil. According to local gazetteers, these "shed people" or "vagrants" (*liumin*) mostly migrated from South China to the mountainous regions, set up tents or sheds, and planted sweet potato and maize on the slopes. They tended to move elsewhere after several harvests when yields declined, causing complaints among local residents, because, as Li describes, "opening mountain land caused water loss and soil erosion and was therefore often prohibited by the local people and government."[149]

Local gazetteers and officials criticized the "wandering households" and "shed people" from Fujian and Jiangxi, where sweet potato was first cultivated, for changing the landscape. These people were part of a vanguard who helped spread the root plant from south to north, armed with knowledge of sweet potato farming. The complaints thus traced their migration route, which also helped map how the American plant dispersed over time. Local gazetteers in northern Fujian, for example, described "shed people" from Quanzhou and Zhangzhou, in the southern parts of the province, opening the mountains to plant the sweet potato all over the slopes in the early nineteenth century. Then, in the late nineteenth century, the "shed people," now identified as coming from Fujian, Jiangxi, and Hunan, appeared in Zhejiang and Sichuan, and sweet potatoes were grown in the Yangzi River regions. Knowing that sweet potatoes can grow without fertilizer, the "shed people" frequently practiced slash-and-burn techniques in the mountains in order to cultivate the roots. After the soil was loosened by root spread and soil maintenance was neglected, the cultivated area became more vulnerable to landslides and floods during rainstorms, which were recorded as causing damage to farmland beneath the mountains.[150] According to Chinese records, the cultivation of maize, which can also grow without fertilizer in the first few crops, caused more environmental damage than sweet potato cultivation. Southern Shaanxi was a prime example. Because of the region's large-scale maize farming in the second half of the nineteenth century, "the timber base had been destroyed and the soil had been exhausted," write William Lavely and R. Bin Wong.[151]

The migrant waves of "shed people" and "vagrants" caused concern among local residents and resulted in government prohibitions, partly because their activities disrupted the existing social order. That is, they were chastised not only for their sweet potato farming, but also for their appearance, which raised security concerns among officials. In their reports to the central government, these officials described how the migrants opened up mountains without permission and clashed with local farmers when the latter tried to confront them about their behavior.[152] "Many localities," according to historians Susan Naquin and Evelyn Rawski, "saw intense conflict between new settlers and original inhabitants. . . . To the light hand of local authority was added the explosive demographics of a population with a large proportion of rootless and footloose young men, organized into peer groups and free from familial restraint."[153]

Did the sweet potato's proclivity as an easy-to-grow and high-yielding crop cause the aforementioned demographic shifts? According to Naquin and Rawski, if mountains were opened up for planting tea, indigo, and timber, these plants required long-term cultivation and stable settlement, whereas sweet potato is a perennial plant but is often planted annually or even seasonally.[154] While derogatory, terms like "shed people" and "vagrants" indicated that these immigrants were temporary hill dwellers, living in flimsy huts and shanties. However, the conflict between the "shed people" and the locals was not constant, nor did it take place everywhere. According to Chinese demographers, the concerns of the "shed people" mostly occurred during the early stages of the migration trend that swept most of the country; many "shed people" chose more permanent settlement, even if some of them were constantly referred to as Hakkas (lit. guest people) by the native residents.[155]

More importantly, these demographic activities demonstrated the importance of the sweet potato in sustaining China's population growth. The fact that the "shed people" and "wandering households" originated in Fujian and Jiangxi, where sweet potato was first grown, provided proof. According to one estimate, the population in the hilly regions of Fujian, Jiangxi, and Zhejiang doubled to over ten million from the seventeenth to the eighteenth centuries.[156] This was, as shown in the gazetteer of Yushan County in Jiangxi, mainly because farmers planted not only tea, indigo and bamboo, but also sweet potato and corn. Furthermore, sweet potato replaced beans in the region's intercropping system, being planted in the summer after rice harvest and before wheat planting in the early winter.[157] Sichuan is another example. The province's rapid spread of sweet potato cultivation coincided with its impressive demographic expansion in the nineteenth century. According to one study, Sichuan's population increased from 1.36 million in 1752 to 21.43 million in 1812.[158] In other words, even if they began as "shed people" and "vagrants," many of the migrants chose to stay and contributed to the growth in the local population. While their initial slash-and-burn practices raised environmental concerns, the adventurous land reclamation efforts of these migrants invariably increased the country's cultivated acreage. Indeed, the massive movement of people from the South and Southeast, as well as their relocation in the North and Northwest, planting sweet potato and maize, effectively opened up the hills and mountains along the Yangzi River's middle and upper reaches, while continuing to occupy the Han River regions. And the *Völkerwanderung*, or mass migration, picked up more veracity as the Qing dynasty weakened its rule. Taiwan (as well as Maritime Southeast Asia) and Manchuria experienced large waves of new immigrants between the early nineteenth and early twentieth centuries.[159] Taiwan eventually became a major sweet potato producer throughout the twentieth century, thanks to the arrival of farmers from southern Fujian and Guangdong.

Sweet potato's changing fortunes in the twentieth century

You should have seen our land here then! All tiny plots. Hilltops like a bald man's head. The rest was vegetables and sweet potatoes. We attacked the hills and terraced them, carried up tea plants growing, orchid trees and cotton, and now even those white quartz sandhills are producing wealth! When the masses take charge, change comes.[160]

When Rewi Alley, a political activist and writer from New Zealand, was touring the eastern part of the country in the 1960s and 1970s, he was told the above by a Chinese peasant in Hunan. Alley had joined the Chinese Communist Party in his thirties and lived in the country until his death. His travel journals, simply titled *Travels in China, 1966–1971*, documented the changing lives of ordinary Chinese people under Communist rule. Without a doubt, the establishment of the People's Republic of China (PRC) in 1949 altered many aspects of Chinese life. However, the preceding conversation suggested that there was one notable continuity: the Chinese fascination with the sweet potato as a valuable food persisted. During his travels through the country, Alley frequently encountered instances where the American plant was grown, consumed, and processed by the Chinese. He was served "a big bowl of steaming hot sweet potatoes, which people like to eat in winter sitting around a fire, and which they at times further roast in hot embers," when he called on a family in southern Hunan, for example. In neighboring Hubei Province, Alley visited a distillery in Hengshui, which he described as "a convivial place." The distillery produced sweet potato liquors that were stronger than well-known national brands like Maotai and Fen. "For the white spirits," Alley wrote, "a sweet potato is fermented and distilled. The residue goes for pig feed. The season for sweet potatoes is the five months between October and March, when the full 419 workers are employed."[161]

Sweet potato farming and production were at their peak during the time when Alley crisscrossed the country.[162] However, the trend had begun earlier. According to John Lossing Buck's detailed survey of land use in 1920s and 1930s China, sweet potato farming made significant progress beginning in the early twentieth century, at the expense of more traditional food plants such as sorghum and even wheat. His study divides the country into two regions based on different areas of food crop cultivation: the "wheat region" and the "rice region," with subregions such as "spring wheat," "winter wheat-millet," and "winter wheat-sorghum (*gaoliang*)" in the former and "Yangzi-valley rice-wheat," "rice-tea," "Sichuan rice," "double-cropping rice," and "Southwestern rice" in the latter. Between 1929 and 1933, sweet potato was the number three food crop in the entire country, following rice and wheat, in terms of the percentage of family food consumption. In fact, in the "wheat region," it was just slightly lower than wheat (59 percent compared to 64 percent), whereas in the "rice region," sweet potato outpaced wheat by a wide margin in both the "Sichuan rice" and "double-cropping rice" subregions. Obviously, the "double-cropping rice" region was in South China, encompassing Fujian and Guangdong. According to a recent study that revised Buck's survey, Guangdong farmers then customarily practiced intercropping of rice, sweet potato, and sugarcane.[163]

Aside from these statistical data, Martin Yang's anthropological investigation in Taitou, Shandong, provided detailed information about how the American root plant was used in people's daily lives during the first half of the twentieth century. Shandong, located in North China, was classified as a "wheat region" by John Buck, where wheat, millet, and sorghum were commonly grown. However, Yang discovered that Taitou, a village in northern Shandong, grew sweet potato as one of its most cultivated crops during this period: "From June to October, sweet potato, peanuts, and soybeans occupy almost 50 and 60 percent of the crop land. Next in importance is millet, to which 30 percent is given, leaving only 10 percent for other crops and vegetables." Outsiders referred to Taitou residents as "sweet potato eaters" because, with the exception of a small percentage of the population who could afford wheat-flour foods, the rest consumed the root on a daily basis, with some even including it in all meals.[164]

China was embroiled in wars and revolutions during the first half of the twentieth century, when John Lossing Buck and Martin Yang conducted their research. After the 1911 Revolution deposed the Qing dynasty, the resulting Republic of China failed to bring the country the necessary peace and stability. In order to combat rampant warlordism in the 1920s, Nationalists and Communists collaborated to form a united front. Chiang Kai-shek (1887–1975) unified the country in 1928, thanks to the success of the "Northern Expedition"—Buck's survey of land use in China reflected Chiang's government's initiative to restore socioeconomic order. However, challenges remained. Japan's desire to conquer China led to World War II, which was followed by the Chinese Civil War, a bloody conflict between Chiang's Nationalist forces and Communist forces led by Mao Zedong (1893–1976). The Civil War came to an end with Mao's victory and the establishment of the People's Republic in 1949.

Sweet potato fared extremely well during those turbulent years of war and revolution. If Martin Yang's intensive research revealed that the root became a daily food in Taitou, Chen Dongsheng's research reveals that a similar situation was found throughout the province of Shandong, and possibly throughout most of North China. From the early to the mid-twentieth century, the sown area of sweet potato farming often exceeded that of all other food crops in Shandong's many counties, according to Chen. And, as previously mentioned, not only did Japanese occupying authorities in Shandong promote it, but Chinese resistance forces led by the Communists also turned to the root and increased its cultivation in their controlled areas during World War II.[165]

Indeed, in looking at the Chinese Communist movement's history, sweet potato seems to have continually occupied a prominent position as a valuable food source. In the aforementioned text, Rewi Alley recalled Mao Zedong's speech in Jiangxi after the establishment of the first Communist base in Jinggangshan, in which Mao encouraged his followers that even though the Red Army was small at the time, the force would grow if the soldiers could demonstrate good discipline, such as by not taking "a piece of sweet potato from the people."[166] Mao, of course, referred to sweet potato in this context as a figure of speech. Yet after the root was first cultivated in Fujian and Guangdong, Jiangxi was undoubtedly one of the major sweet potato–producing provinces.

Mao became more serious about the root crop as a useful food for the people after establishing the People's Republic. During the republic's first decade, Mao led a series of political and social campaigns aimed at eliminating dissent in society in order to strengthen Communist rule. He shifted his focus to the economy near the end of the 1950s, spearheading the Great Leap Forward. While the primary goal of the Great Leap Forward was to dramatically increase iron and steel production in order to industrialize the country and catch up with the United Kingdom and the United States, Mao also hoped to significantly increase food production in order for people to "open their belly and eat" and taste the success of socialism. To this end, writes Hanchao Lu, "The single most important crop used to boast about the astronomical increases in agricultural output during the Great Leap Forward was sweet potatoes." When the government declared that the country had officially entered the phase of socialism in 1958, Mao invited a peasant and a local cadre from Henan, another major sweet potato producing province, to his Beijing residence for dinner. The peasant mentioned the root because of its high yield when the three discussed how to increase food production. It elicited a positive response from Mao: "The sweet potato is a treasure. It's delicious. It's my favorite food. It would be nice to include sweet potatoes as part of everyone's food ration."[167]

The story of Mao having dinner with the two was widely publicized, and it was interpreted as Mao's official endorsement of the sweet potato by the entire country. Under various auspices, Chinese scientists quickly compiled handbooks on sweet potato farming, processing, and utilization. Three of them, with titles like *How to Cultivate the Sweet Potato*, *Storage and Utilization of the Sweet Potato*, and *The Experience in Multifarious Utilization of the Sweet Potato*, appeared in 1959 or the following year after Mao openly stated how sweet potato appealed to him as food. All three were put together quickly and collectively for quick distribution. The first was conducted by a local college in Fujian Province's Putian County, the second by an ad hoc committee in the Department of Agriculture, and the third by the Food Section of the Department of Light Industry.[168] A year earlier, a Research Institute of the Sweet Potato had been established in the Chinese Academy of Agricultural Science. Because the academy had only been founded in 1957, this was one of its very first institutes. After hosting a series of presentations training agricultural technicians around the country, the Research Institute of the Sweet Potato compiled the lecture notes into a pamphlet on sweet potato farming in 1960.[169]

As a result, the late 1950s heralded a "great leap forward" in sweet potato cultivation in China. How effective was it? Two statistical studies could provide some answers. In 1966, John Lossing Buck collaborated with others to publish *Food and Agriculture in Communist China*, a sequel to his *Land Utilization in China*. It retained the categories of "rich region" and "wheat region," along with all their subregions. Nonetheless, when assessing food production and consumption in the PRC, Buck and his coauthors included "potatoes (grain equivalent)" alongside "rice," "wheat," and "miscellaneous"; the latter, likely *zaliang* 雜糧 in Chinese, should include maize, sorghum, and all other food plants. The "potatoes" category is more intriguing, as it is most likely a translation of the Chinese term *shulei* 薯類, which refers to root and tuber crops. However, his inclusion of potatoes under "grain equivalent" implies that sweet potato should dominate the category, because, as previously discussed, white potato had not been cultivated nationwide at the time, whereas sweet potatoes had already been considered a food crop equivalent to rice and wheat by many local gazetteers as early as the eighteenth century. Beginning in 1955, the Chinese government also implemented a state monopoly on the purchase and marketing of all food grains (*tonggou tongxiao* 統購統銷). Sweet potato was officially included in the system as a grain equivalent to rice and wheat.[170] As Isabel and David Crook found during their fieldwork in the Yangyi Commune of Hebei at the time, four *jin* of sweet potatoes were counted as the equivalent of one *jin* of grain.[171]

There appeared to be another reason for Buck to include the "potatoes" category in the sequel to his earlier *Land Utilization in China*: his statistics show that entering the PRC period, "potatoes" had the greatest gain in terms of "crop hectare in food grains," or sown area, while the increases of "rice," "wheat," and "miscellaneous" were insignificant. Indeed, the gain experienced by "potatoes" between 1949 and 1958 doubled or even tripled that between 1929 and 1937, as Buck had shown in his previous book. The year 1958 saw the greatest increase, with 16.3 million people compared to 10.5 million the previous year. Also in 1958, the production of "potatoes" in food grains surpassed that of "wheat" for the first time, trailing only "rice" and "miscellaneous."[172] Buck's statistics ended in 1958, which is unfortunate because, given Mao Zedong's personal recommendation of the plant, one would expect sweet potato production to increase significantly in subsequent years.

Surprisingly, however, such a predicted rise did not occur. According to another statistical study on root and tuber crop production in the PRC, which covered the period 1949 to 1984, root and tuber crops, primarily sweet potato, white potato, and cassava, set records in sown area, production, and yield in 1958. But it experienced a significant decline in all three areas over the next decade. The study also reveals that cassava was a distant third among the three root and tuber crops, grown only in Guangdong and Guangxi. Thus, the decline was primarily caused by the sweet potato, because white potato was still not widely cultivated nationwide in the 1960s. There are two likely explanations for the sharp decrease in sweet potato production following the Great Leap Forward. One was that, because of Mao's enthusiastic endorsement of sweet potato consumption, local officials likely exaggerated its farming figures in 1958, which was common in most areas of data collection during the Great Leap Forward. In other words, the astronomical increase in sweet potato production reported to the central government in 1958 was not as large as previously assumed.[173] The other was due to the Great Leap Forward's catastrophic failure, which, much to Mao's chagrin, resulted in widespread famine rather than rapid industrialization for the country. The precise number of casualties from the Great Famine is still debated among historians, but one thing is certain: it disproportionately affected the rural population. Without a doubt, the Great Leap Forward's devastation of China's countryside reduced food production. In terms of sweet potato production, it did not recover until well into the 1970s.[174]

However, the ups and downs of the sweet potato in contemporary China not only persisted but also underwent more significant twists. Despite a reduction in production following the Great Leap Forward, the root became increasingly important for the Chinese to survive the famine. After the famine ended in 1962, it remained a highly valued food crop on which many people relied until the end of the 1970s, or until China began its Reform and Opening-Up period under Deng Xiaoping. Following Mao's promotion of it in 1958, sweet potato rose to prominence among the country's food crops. As previously discussed, the numbers reported to the government regarding the dramatic increase in sweet potato production in 1958 were most likely exaggerated. People in the countryside, however, recall that, by government decree, local officials made an extra effort to forcefully encourage its cultivation over other food crops. Historian Frank Dikötter, author of the 2010 book *Mao's Great Famine*, writes that "the proportion of sweet potatoes grew during the years of famine, as cadres responded to pressure to increase the yield by switching to the tuber, which was easy to cultivate. More often than not farmers were left with potatoes only." Dikötter's findings are corroborated by an official report filed contemporaneously by a commune in Taizhou, Jiangsu: "since last year, the local cadres have forced peasants to use 60 percent of the agricultural fields to grow sweet potatoes. There was no room for discussion. . . . To make the fields available for growing sweet potatoes, they even ordered peasants to plow under the already sprouting sorghum crop." Local officials in Suiyang, Guizhou, went even further, declaring war on the dead and converting a graveyard into a field for growing the root. A slogan was sounded in championing the campaign: "The ancient hero did not lead the way; the real hero is born today. Thousands of years of tradition and superstition must be wiped out completely; turn the dead people's graveyard into a mountain full of potatoes."[175]

Sweet potatoes are a high-yielding crop that Mao and the government promoted. However, they did not always grow well everywhere they were planted. This was partly due to the crop being introduced quickly and forcefully and partly because those who were ordered to plant it sometimes lacked adequate knowledge about it. According to the same report from Taizhou, in order to strictly follow the order from above, the local brigade leader instructed the peasants to transplant sweet potato sprouts prematurely in one hot day. As a result of being exposed to harsh sunlight, all of the transplanted sweet potatoes died a few days later.[176] Local farmers in Yangyi, Hebei, where the Crooks stayed in 1958 and 1962, also struggled to learn how to sow the root crop because it was new to the area—it had only been introduced to them in 1953.[177]

Nonetheless, as the famine worsened, sweet potato quickly became the only food available to many people, supporting them in fighting hunger and surviving the hard times. While the sweet potato was new to the region, people in Yangyi, for example, ate it "day in and day out," especially in late spring and early summer when other grain food became scarce.[178] Zhou Xun of the University of Essex asked those who had lived through those days to recall their experiences while compiling an oral history of the period. A college graduate who was sent down to Liangshan, Sichuan, recalled that sweet potatoes were the food served in the canteens in his home village for an entire year: "sweet potato for breakfast, sweet potato for lunch and sweet potato for dinner. When there were no sweet potatoes left, the villagers resorted to food substitutes such as sweet potato leaves and wild herbs." People ate not only sweet potato roots but also their leaves because they were the only substitute for vegetables at the time. Worse still, for some, a bowl of the leaves was the only food they could eat for the day. Another recalled that as a child growing up in Liangshan, he was frequently awakened by hunger, and his parents had nothing for him to eat except "a few sweet potato leaves to chew on." If sweet potato leaves were so prized, the roots became even more valuable; they became a target of theft and prized items some exchanged for rice, and which corrupt officials grabbed and selectively distributed in a display of favoritism.[179]

Sichuan is China's major sweet potato farming region, and John Lossing Buck observed that the purple-brown forest soil was ideal for growing the root plant.[180] However, the situation described by Zhou Xun's interlocutors did not occur only in Sichuan, but throughout China. A Jiangxi party official discovered that peasants in his province "cooked sweet potato leaves to supplement their daily meals." Furthermore, a common practice during the famine was known as *chiqing* 吃青, which refers to peasants scavenging sweet potato leaves and other greens in addition to unharvested wheat and corn crops. A peasant in Fujian recalled that "during the great famine, we secretly ate raw food grain crops standing in the crop fields. We also ate raw sweet potatoes while they were still small and in the fertilization period." According to Ellen Oxfeld, who carried out anthropological fieldwork in Meixian, sweet potato was also the main food on which the people relied during those years in neighboring Guangdong. "All we ate was sweet potatoes!" When Oxfeld asked the villagers about their lives in the 1950s and 1960s, she frequently heard this response.[181]

Sweet potato was thus an important food substitute for cereal crops at the time, as American agricultural scientists recognized.[182] Though the famine had the greatest impact in rural areas, it also had an impact on urban life. A study by historian Gao Hua unveils that in the city of Shanghai, grain shops had to distribute "sweet potato strips and maize noodles to serve as fixed rations." Centering on the state initiative to make "food substitutes" (*daishipin* 代食品) in the period to allay people's hunger, Gao's study found that the attempt was also extended from dried sweet potato roots, which were cut into strips, to sweet potato seedlings, stems, and vines.

Chinese scientists determined that sweet potato stems, which had previously been used primarily as hog feed, and corn husks were suitable for human consumption after conducting research. Actually, the use of sweet potato as a substitute for other food materials began with the start of the Great Leap Forward. For example, Li Shuangshuang, a peasant woman who was turned into a heroine by the government for ratcheting up the Great Leap Forward, allegedly discovered how to make delicious noodles using sweet potato flour instead of wheat flour. Her recipe was dubbed "Great Leap Noodles," and it was widely publicized and promoted across the country. In Yangyi, where sweet potato had become a staple of the people's diet, the head cook of the commune's canteen, Wang Guei-ling, made an effort to learn different sweet potato recipes so that he could cook seven different sweet potato dishes (noodles, dumplings, cakes, and so on) in a week.[183]

Ironically, despite Mao Zedong's praise of the sweet potato as a delicacy, the widespread starvation caused by his Great Leap Forward turned the root back into a famine food, just as it had been at the turn of the seventeenth century, when it was first cultivated in the country. As a result, wrote anthropologist Eugene Anderson in his *Food of China* in the 1980s, "Sweet potatoes have never become popular in China; they are regarded as the worst of all foods everywhere they grow."[184] This statement certainly requires qualification, because if sweet potato was despised by many Chinese, it was precisely because, as the preceding discussion reveals, it had been exceptionally popular as a must-eat food throughout the country during China's "most devastating catastrophe," as historian Frank Dikötter calls it in the subtitle of his book on the famine. One example was that sweet potatoes became a "mainstay" in Yangyi in those days, despite the fact that the root had only been known to the locals a few years before. Another was that the root "figured prominently" in constructing the fabric of the Meixian people's memory of the period, as anthropologist Ellen Oxfeld reported. In fact, similar stories were common in other regions where the root was not grown as a primary food crop or was not commonly consumed daily. During the famine, sweet potato cooked with rice porridge became a staple food for Beijing residents.[185] Yang Jisheng's *Tombstone*, a monumental account of the colossal food crisis resulting from the Great Leap Forward, provided overwhelming evidence about the changing fate of the sweet potato in the period: how it was first hailed as a symbol of socialism's success—a miraculous yield of 500,000 kilos of sweet potatoes per *mu* was proudly reported in 1958—only to quickly descend to the status of a despised substitute for people's survival.[186]

In other words, the sweet potato suffered from success. It was a lifesaver for the Chinese during times of food scarcity, but it was also remembered as a symbol of poverty. After the famine was declared over in 1962, many people in the countryside continued to live in poverty. "Through the 1950s, 1960s, even into the 1970s," a peasant woman in Meixian told Ellen Oxfeld, "we had sweet potatoes for one or two meals per day, then congee [*zhou*, rice porridge] was for other meals. We never had *fan* [dried cooked rice]." This occurred not only in Guangdong, but also in other parts of the country. According to a local historical study, sweet potato made up between 10 and 40 percent of the food ration in Anhui Province in 1967.[187] Indeed, it was not until the 1980s, when the state monopoly on purchasing and marketing all food grains was gradually phased out, that life in China's countryside began to improve noticeably. Farmers benefited from the change in government policy because they were no longer required to hand over the majority of their food produce in a year. Instead, the state assigned them a quota, based on the previous year's target, then allowed them to keep the remainder if more was produced. Chinese peasants, who had been released from the commune at the same time, became motivated to grow more and different food grains; sweet potato and maize were no longer preferred simply because of their high yield.

As a result, the land devoted to sweet potato decreased significantly in the post-Mao years. The total sown area of both root and tuber crops was 15,382 (thousand hectares) in 1958 and 9,363 (thousand hectares) in 1982, when the Reform and Opening-Up period began, a drop of nearly 65 percent in the country. Consequently, sweet potato, which was previously the third most cultivated food plant in the country, became the fourth after rice, wheat, and maize. Guangdong and Guangxi in South China, where sweet potato was first grown, saw a more than 50 percent decline during the same period. And the pattern continued.[188] Oxfeld discovered that in Meixian, Guangdong, sweet potato production fell from 47,000 tons (42,638 tonnes) in 1955 to 26,100 tons (23,678 tonnes) in 1987, a whopping 80 percent drop recorded in the county gazetteer.[189] China's significant decline in sweet potato production had a global impact. Sweet potato, which was once the sixth most cultivated crop in the world in the 1970s, is now ranked seventh, with cassava taking its place.[190]

Indeed, during her fieldwork in the early twenty-first century, Ellen Oxfeld found that people in Meixian could recall the period when they had only sweet potatoes as a staple food because those days were long gone; they were just a memory that older generations would draw on to teach the young about the hard times they had lived through.[191] In China today, sweet potato is no longer regarded as a "coarse food" that people ate to fill their stomachs. About half of the root's harvest in the country today is used for pig feed. Another significant portion (roughly one-sixth) is converted into starch.[192] Thus, in most of the country, sweet potato has ceased to be the food that people relied on as a daily staple in the 1950s and 1960s. Its roots, usually roasted or steamed, are now more commonly consumed as a snack between meals, which was once a common sight in city streets in winter prior to the Great Leap Forward. Shiu-ying Hu, a botanist at Harvard University, describes sweet potato consumption in China as follows:

> In urban areas the sweet potato is used as a treat. In northern China it is sold by itinerant vendors who carry a charcoal-burning, earthenware urn to bake the potatoes. In southern China sweet potato cubes are boiled with brown sugar for refreshments. For the Chinese New Year, deep fried sweet potato chips are prepared by paring the potatoes, and cutting them diagonally to 3 mm slices, which are then steamed, dried, cooked in vegetable oil.

Hu wrote her book *Food Plants of China* in the 1980s. She was well aware that in previous decades and/or centuries, sweet potato had been a survival food for many Chinese. She grew up in North China and ate sweet potato as a child, which was a minor crop in the region compared to crops like wheat and sorghum.[193] In other words, Hu has witnessed the sweet potato's transformation in modern China. Of course, her descriptions of how the root is eaten are illustrative. They are, however, far from exhaustive. Geographer Frederick Simoons acknowledges in his book *Food in China* that sweet potato once had a bad reputation. However, this did not stop it from becoming a "common snack" in the country. Simoons observes that roasted and steamed sweet potatoes are "quite satisfying" on a cold day in North China. In South China, in places like Guangdong, he continues, "sweet potato balls" is a popular dessert that is made by first simmering cubed and mashed sweet potato roots in sugar and then shaping them into balls. Before being served, the balls are deep fried until golden brown. In Sichuan, where sweet potato is popular, the root has been used in dishes such as "steamed beef with sweet potato." Last but not least, "fragrant duck with braised sweet potatoes" is a well-known dish that was once served at the imperial court.[194]

Although not a staple, sweet potatoes remain popular in China today. Chain stores specializing in sweet potato products have emerged, often boasting a long list of sweet potato foods and drinks. The Shulifang 薯立方 (Sweet Potato Cubes) fast-food chain, which originated in Shandong, is one example, serving a variety of products made from dried sweet potatoes. Fujian, one of the first places in China to see the American root, has its Kuaile fanshu 快樂蕃薯 (Happy Sweet Potato) chain, which caters to the young with a wide range of foods and drinks. Aside from these national food chains, there are countless small food stores throughout the country that serve and develop more sweet potato–based products on a daily basis. All of this demonstrates the sweet potato's persistent appeal. In fact, the sweet potato's nutritional value is increasingly being recognized in the country. Once considered a "famine food," it is now regarded as a "fitness food" by many.[195]

CHAPTER IV

Hunger Food? Daily Meals?

Sweet Potato in Japan and Korea

Sweet potato is known by several names in the Japanese archipelago. *Karaimo*, which literally means "foreign yam/taro" or "Chinese yam/taro," is one of them. The root word *imo* is a rubric term in Japanese, referring to root crops such as yam, taro, or potato, whereas *kara*, written in Chinese character 唐 and also pronounced *tō* in Japanese, refers to the Tang dynasty (618–907) in China, which had a significant influence in medieval Japan. Since then, *kara* or *tō* has been interchangeably used in Japanese to denote foreign objects, with the word acting as a prefix, as in the case of the sweet potato. On the Korean Peninsula, a similar linguistic practice exists. For example, the main component of the famous *japchae* dish *dangmyeon* has the prefix *dang*, which is the Korean pronunciation of the character 唐. *Dangmyeon* is a cellophane or glass noodle made from sweet potato starch. As mentioned in the previous chapter, the cellophane noodle, also known as *fensi*, is widely thought to have originated in China. It is well-liked by many people in East and Southeast Asia.

However, despite its name, sweet potato did not reach Korea or Japan directly from China, nor did the root crop achieve the same level of popularity as it did in China. According to Seo Yugu's *Jongjeobo*, an early nineteenth-century Korean text on sweet potato farming discussed extensively in the previous chapter, sweet potato arrived in Korea from Japan between 1763 and 1765 through the hands of a few Korean officials, particularly Jo Eom and Gang Pilri. Jo was a Korean diplomat who discovered the root while visiting Tsushima, an island located roughly halfway between the Korean Peninsula and Japanese island of Kyushu. He brought it back to Busan because of its pleasant taste and high yield, and Gang successfully cultivated it. Gang also authored *Gamjeobo* (Sweet potato manual), one of the earliest texts on the root plant in Korea. The text was partially excerpted in Seo Yugu's work, but otherwise is no longer available.[1]

Tsushima, near Nagasaki, is in the region where sweet potato was first adopted as a food crop in Japan. Indeed, southwestern Japan, and particularly Ryukyu/Okinawa, were entry points for the American root plant to spread throughout the Japanese archipelago. When Basil Hall Chamberlain, a prominent Japanologist who spent most of his career teaching at Tokyo University, visited Ryukyu in 1895, he noticed that the root, like rice, was already harvested twice a year. "This invaluable plant, now the staple food of the people, was only introduced here from southern China in the year 1605, for which reason it is called 'Kara-mmu,' that is, the 'Chinese potato,'" he said of the sweet potato's origins and significance in Ryukyu. Captain Basil Hall, the British naval officer who arguably left the first European account of the Ryukyu Islands, was Chamberlain's grandfather. Prior to the sixteenth century, Ryukyu was not a part of Japan; it sent tributes to both China and Satsuma, a nearby Japanese domain in Kyushu that threatened its existence at times. Aside from *karaimo*, *satsumaimo* is the most-used term for sweet potato in Japanese. Chamberlain elucidated the reason: "From Luchu [Ryukyu] it was carried north to the Japanese province Satsuma, where it is accordingly known as the 'Luchuan potato'; and thence it spread to Central and Eastern Japan, where the people call it '*Satsuma-imo*,' from a mistaken impression of its being indigenous to that province."[2] Nonetheless, the name *Ryukyuimo* has been used in Japan since Ryukyu saw its cultivation. That sweet potato took a circular route to reach Japan was agreed by anthropological studies in the early twentieth century, and was in turn echoed by many, including the popular writer Mock Joya. Joya stated simply in his best-selling book *Japan and Things Japanese*, first published in 1958 and reprinted several times since, that sweet potato "came to be known by many different names—*Satsuma-imo*, *Ryukyu-imo*, and *Kara-imo*, bespeaking of its introduction from China first to Ryukyu, and then to Satsuma in Kyushu."[3]

Several stories have circulated about how the sweet potato made its way from South China to Ryukyu. One story goes that in 1594, the same year that Chen Zhenlong supposedly brought sweet potato from Luzon to

Fujian, China, a low-ranking official named Chōshin Sunagawashinya from Miyako Island in Ryukyu, on a tour, was washed ashore in Fujian and brought back the root in 1597. Sweet potato quickly became a staple on the island because of his promotion. As discussed in the previous chapter, there was a chance that sweet potato arrived in China before 1594. However, in terms of the root's arrival in Fujian, most scholars credit Chen Zhenlong with bringing the root from the Philippines in 1593 and Governor Jin Xuezeng's endorsement of it as a famine relief the following year. It would be too soon for Chōshin Sunagawashinya to do for his fellow islanders what Chen Zhenlong did in 1597. However, his story is still told today, and his statue is enshrined on Miyako Island.[4]

According to another story, in 1604, Nokuni Sōkan, in charge of Ryukyu's tributary fleet to China, arrived in Fujian and discovered that sweet potato was a popular food among the local Chinese. In fact, Nokuni Sōkan was not his real name; Nokuni was the name of the village he was from, and Sōkan was the title of his position. Little, therefore, is known about his real life. Nonetheless, Nokuni is remembered fondly by both Okinawans and Japanese for his initial interest in the American root. It is said that because Ryukyu was frequently hit by hurricanes, Nokuni thought that a root crop like the sweet potato could be a valuable subsidiary to ensure the people's food supply if grain crops were destroyed by storms. After successfully growing the root in his and neighboring villages, he passed on the knowledge to Gima Shinjō (1557–1644), a local nobleman and self-made agriculturalist. Sweet potato became a major food crop across Ryukyu as a result of the latter's enthusiastic promotion. Islanders later built shrines to honor both for their roles in introducing the new root crop. Nokuni Sōkan was known as the *imo ufusu*, or sweet potato king, while Gima Shinjō was regarded as one of Ryukyu's "five great men."[5]

The second of these two stories is believed to be more credible; it has been frequently told in accounts of the sweet potato's early history in Japan.[6] Yet the three Ryukyu men, or Okinawans, are far from the only people who receive recognition for spreading the sweet potato in Japan today. As the root hopped from island to island in the Japanese archipelago, a variety of people played their roles in assisting its journey. Like in China, where its cultivation was supported by the imperial court, particularly Emperor Qianlong, Ryukyu kings promoted the crop by giving the roots as gifts to other islanders. However, they sometimes did it for different reasons. The Satsuma daimyo (feudal lord) launched an invasion of Ryukyu in 1609. The

Ryukyu king Shō Nei (1564–1620) petitioned for peace and lavished banquets on the invaders. It was said that sweet potato was one of the foods served at his court, which made a favorable impression on the invaders. That could be the first time the root crop was transmitted from Ryukyu to Satsuma, from which it spread to other islands such as Tsushima. Kagoshima, Satsuma's former capital and now a prefecture, is Japan's leading producer of sweet potatoes today. A stone stele stands in a local Shinto shrine on Tanega Island in Kagoshima Prefecture. Its inscription reads, "Japan's Birthplace of Sweet Potato Cultivation." This could be considered a historically accurate statement because Ryukyu/Okinawa, which had grown the root earlier, was not ruled by Japan until 1879.

Sweet potato in Tokugawa Japan

As described in chapter 2, the English were the next Europeans to encounter the sweet potato after the Spaniards and Portuguese, and they quickly developed a love for the root and began preparing it with other ingredients in a variety of dishes. It is interesting to note that two Englishmen stood out in the history of sweet potato farming in Japan because of their unique contributions. The first was named William Adams, who is widely regarded as the first European to "discover" Japan. Adams, who was born in Kent in 1564, lost his father when he was a child. He worked as an apprentice in a shipyard in Limehouse, East London, learning not only shipbuilding but also navigation. With this knowledge, Adams joined the Royal Navy and fought in the war against Spain. In fact, when the English fought the legendary Spanish Armada in 1588, Adams had been in charge of a resupply ship commanded by Sir Francie Drake, the naval officer who was mistakenly credited with introducing the white potato to Europe. Adams worked for an English trading company before joining the Dutch East India Company after the battle, in which the English triumphed and replaced Spain as a new sea power.

While serving the Dutch, William Adams arrived in East Asia in the spring of 1600, a pivotal year in Japanese history that saw the island nation's unification after a two-century period of constant warfare known as the Warring States period (*Sengoku jidai*). The unification process was credited to three powerful military figures. Oda Nobunaga was the first (1534–1582). Oda, a powerful daimyo and ruthless warrior, laid the groundwork by defeating several major warlords in Honshu before being betrayed by one of his retainers, forcing him to commit ritual suicide in Kyoto. The second was Toyotomi Hideyoshi (1537–1598), an equally if not more brutal warrior who continued to conquer most of Japan after avenging Oda Nobunaga's death. Toyotomi launched invasions against Korea, only to face crushing defeats when the Koreans allied with Ming China to strengthen their defenses. Tokugawa Ieyasu (1543–1616), the third unifier, known for his patience and resolve, took over after Toyotomi died. Despite being a trusted ally of Toyotomi Hideyoshi, Tokugawa decided to defy his son, Toyotomi Hideyori, who had been installed by his father on his deathbed, and declare himself the new ruler of Japan.

William Adams's arrival in Japan, which resulted from a near-death experience while sailing for nineteen months, was fortunate. Tokugawa waited about a half year before deciding to fight Toyotomi Hideyori and his supporters in the famous Battle of Sekigahara in the fall. Though Adams and his crew were initially accused of piracy and imprisoned, he was eventually brought to Tokugawa's castle to see him. The two met several times, and Adams clearly won the trust of Japan's future shogun, or regent who ruled Japan for the imperial house. The nineteen bronze cannons aboard Adams's ship were also unloaded and used in the battle to aid Tokugawa's army in defeating Toyotomi Hideyori. Adams became Tokugawa Ieyasu's trusted confidant during the period when he was consolidating his power and becoming the shogun—so much so that he was one of the first Europeans to be presented with samurai status and a fief worth 250 *koku* of rice at the time. As a result, William Adams, the former English sailor, was now Miura Anjin, a Japanese name bestowed upon him by his new master. Tokugawa Ieyasu retained Adams's services, which included the construction of Japan's first shipyard, the establishment of trading factories by the Netherlands and England, and the exploration of trading relationships in Southeast Asia. As one of the most influential foreigners during the Tokugawa period, he has left a cherished legacy in Japan. A district in central Tokyo is now named after him. The train station in his former fiefdom, Hemi, in modern Yokosuka, is also named after him. And on Sakigata Hill in Hirado, where he helped build the trading factory, there is the William Adams Memorial Park. More recently, the 2024 U.S. television series *Shōgun*, based on the 1975 novel by James Clavell, featured William Adams as the character Blackthorne.

A lesser-known aspect of William Adams's contributions to Japanese history is his instrumental role in moving the sweet potato from Ryukyu to Honshu. Adams was initially denied permission to return to his homeland after earning the trust of Tokugawa Ieyasu. He eventually returned to sailing, working for the British East India Company under Richard Cocks. Captain Adams, commanding the ship *Sea Adventure*, landed in Naha, Ryukyu's capital, in 1615 to repair some leaks during one of his many exploratory voyages aimed at cultivating new trading routes with Thailand and the Philippines. He saw the sweet potato being farmed extensively in Ryukyu and bought some on his way back to Hirado, near Nagasaki, where the English trading factory was located.[7] Except for a few letters to his family and friends in England, Adams did not leave us with many of his own writings. However, the English merchant in charge of the trading factory, Richard Cocks, kept a detailed diary in which Adams appears frequently. Cocks wrote on June 2, 1615, that "Capt. Adams and Ed. Sayer wrote me 2 letters from Goto on 30th of May; and Mr. Adams sent me a bag of potatoes," implying that Adams and his *Sea Adventure* had arrived from Ryukyu. Cocks went on to say that the potatoes were from "Liquea," or Ryukyu—"I tooke a garden this day and planted with pottatos brought from the Liquea, a thing not yet planted in Japan." Cocks appeared to be quite enthusiastic about growing the sweet potato. On June 22, he wrote that he had paid the potato garden a year's rent. In later years, he continued and expanded his farming. In 1618, he wrote in his diary that he "set 500 small sweet potato roots" in his garden, which had been brought to him from Ryukyu.[8]

While William Adams brought the sweet potato back from Ryukyu, Richard Cocks was the first to grow the root in Japan. According to Miyamoto Tsuneichi, arguably the first Japanese historian to write about sweet potato farming, if Cocks planted five hundred sweet potato roots in his private garden, those would be sweet potato sprouts—when they were faced down in the soil, new roots would grow quickly and be ready to eat by the end of the Obon festival in August. According to his observations, this method of propagation is only used in Hirado, while sweet potato vines are more commonly used for propagation elsewhere.[9] Regardless, Cocks was the first to cultivate the American root in Japan. In terms of the introduction of sweet potato into the country, he thus played a role similar to that of Gima Shinjō, who had promoted it in Ryukyu a decade or so earlier; and William Adams's role was comparable to Nokuni Sōkan's. While

Cocks kept much information about his fellow Englishman in his diary, little is known about him, and he is not as well-known in Japan as his friend. We do know that he led the Hirado Trading Company, a new branch of the British East India Company, for a decade, from 1613 to 1623. Cocks met the Japanese shogun through William Adams and offered his assistance in exploring the possibility of trading with Southeast Asia, specifically the Philippines. His trading factory thus competed with the Dutch East India Company. The main reason for his company's bankruptcy was the rivalry between England and the Netherlands in East Asia. Another reason was the Tokugawa government decided to implement the "lockup"—*sakoku*—policy, first implemented in 1623, which prohibited Japanese from traveling abroad and foreigners from residing in the country except in Nagasaki. Cocks sailed back to England in 1624, frustrated and disheartened, as evidenced by his diary, only to die en route and be buried at sea somewhere in the Indian Ocean.

However, sweet potato cultivation continued, albeit at a slower pace than Richard Cocks had hoped. The root was mentioned in passing in a series of texts throughout the seventeenth century, implying that it gradually became known to the Japanese. However, there were some exceptions. One was Miyazaki Yasusada, an acclaimed agriculturalist in early Tokugawa Japan who compiled *Nōgyō zensho* (Complete work on agriculture). Miyazaki was born in a samurai family in 1623, the same year that Cocks's trading factory in Hirado was forced to close. Miyazaki left his hometown of Hiroshima in his mid-twenties to work as a forestry official in Fukuoka, where he met Kaibara Ekken, an upcoming Confucian scholar, and his elder brother Kaibara Rakuken. The Kaibara brothers were also known for their knowledge of botany. Miyazaki eventually left his post and retired to his estate to pursue his growing interest in agriculture, devoting his time to improving agricultural technology. He traveled across western Japan, learning from experienced farmers and inspecting the soil conditions of mountains, forests, dryland, and paddy fields. He also initiated land reclamation and afforestation projects. He wrote *Nōgyō zensho* in 1695, which was Japan's first work on agricultural studies. Its original Chinese title was nearly a verbatim translation of Xu Guangqi's *Nongzheng quanshu*. Miyazaki apparently hoped to emulate Xu Guangqi's success in writing his work. The Kaibara brothers also assisted in its final publication in 1697, with Kaibara Rakuken contributing the eleventh and final chapter and Kaibara Ekken providing a preface.

Principally, Miyazaki Yasusada's *Nōgyō zensho* is a Japanese-language revision of Xu Guangqi's *Nongzheng quanshu*. His description of the sweet potato is a telling example, which is compared in detail with Xu Guangqi's by Barry R. Duell, a sweet potato expert at Tokyo International University. Miyazaki named the sweet potato *bansho*, or *fanshu* in Chinese, after its Chinese predecessor, as well as *Ryukyuimo* (Ryukyu taro) and *akaimo* in Japanese (lit. red taro). He also stated that sweet potato looked like taro (round or tubular) but tasted better. He told Japanese readers that it grew well near Nagasaki and that people ate it raw as well as cooked in a variety of ways, including with rice. As such, sweet potato could be considered a "main food" (*shushoku*); in China, it was already a common food upon which people relied in their daily lives. Miyazaki, as an agriculturalist, also described the sweet potato's proclivities. For example, he correctly observed that the leaves of sweet potato were similar to those of *asagao*, a plant in the same morning-glory family that Japan already possessed. Miyazaki then made the following recommendation:

> "*Bansho*" has a similar shape to "*yamanoimo*," but it has light purple skin. It has a thin skin and white meat. It is sweeter than "*yamanoimo*," and it looks and tastes good. It can be eaten as a sweet or used in various types of cooking. When grown in large quantities, it is useful for feeding people when other food sources are scarce, such as during a famine. It is especially beneficial for extending one's life when consumed over a long period of time. The sweet potato is lavishly praised in *Nongzheng quanshu*. Sweet potatoes grown in Satsuma and Nagasaki are known as "*Ryukyu-imo*" or "*akaimo*," and they are abundant there. However, they are not yet widely grown throughout Japan. They are, however, particularly suited to growing in warm climates, and if grown in soft fertile soil using standard growing methods, they should do well.[10]

Miyazaki Yasusada died the same year that his *Nōgyō zensho* was published. On the surface, his recommendation of sweet potato as a food crop was largely a reiteration of what Xu Guangqi had stated in his work. However, as a hands-on agriculturalist, Miyazaki's work could have also reflected his own firsthand knowledge of the root's cultivation in Kyushu. Indeed, while the army from Satsuma that conquered Ryukyu in the early seventeenth century may have brought back the sweet potato, it was not until the end of the century, or around 1697, that the American plant began to take root in southwestern Japan. Sweet potato was transformed from a rare and treasured food to a readily prepared insurance crop that could help allay hunger in famines during the century, as Yamada Shōzi (Syoji) analyzes while studying its dissemination in southeastern Japan.[11]

Aside from Miyazaki and the Kaibara brothers, three other people should be mentioned for their contributions to the transformation of the sweet potato in Japan. Tanegashima Hisamoto, a high-ranking samurai appointed by Satsuma's daimyo to administer Tanega Island, was the first. In 1698, he was said to have received a basket of sweet potatoes from Ryukyu's King Shō Tei (1646–1709). The latter informed him that they were good food during a famine. Tanegashima directed his samurai Nishimura Tokinori to lead the farmers in an experiment with cultivation of the root crop. It proved to be a huge success. Within a few years, the plant had spread throughout the island and into other parts of the Satsuma domain. And, as Miyazaki described in his *Nōgyō zensho*, the farmers in the region experimented with a variety of consumption methods, using sweet potatoes to make vinegar, soy sauce, sugar, *shōchū* (a type of liquor), and sweets, in addition to milling them into starch for other uses.[12] As mentioned before, there is a stone stele on Tanega Island today inscribed "Japan's Birthplace of Sweet Potato Cultivation." It stands on the site where Nishimura and the villagers first grew the plant.

Maeda Riemon, an ordinary farmer about which we know little, was the second. He was born in the 1670s and died young in 1707. He was known only as Riemon during his lifetime because, as a commoner, he had no surname—"Maeda" was given posthumously by his descendants during the Meiji period. According to local legend, Riemon saw sweet potatoes growing in the Ryukyu Islands, where they thrived on land with poor conditions, in 1705 while traveling back and forth between the Satsuma domain and the Ryukyu Islands on business. He was inspired to bring sweet potatoes to Satsuma by growing them in pots and bringing them home. Riemon distributed seed potatoes and seedlings to his farming neighbors after successfully growing sweet potatoes. Though he was killed in a shipwreck on his way to Ryukyu not long after, his introduction of the sweet potato was well remembered because the new food crop helped the locals survive the famine caused by multiple rice crop failures in the early eighteenth century. Maeda Riemon earned the nickname "sweet potato master" (*kanshoō*) among the locals after his death. In present-day Kagoshima, he is worshipped in the Tokumitsu Shrine, nicknamed the "Sweet Potato Shrine" (Karaimo jinja). There is also a memorial hall dedicated to him, called Tōimoden or Karaimoden. In addition, "Riemon" is the brand name of a popular sweet potato *shōchū* produced by a local brewery.[13]

And the third was Aoki Konyō, known in Japan as "Mr. Sweet Potato" (*Kansho sensei*). In comparison to Nokuni Skan, the "sweet potato king," and Maeda Riemon, the "sweet potato master," Aoki Konyō as "Mr. Sweet Potato" is well-known throughout Japan. That is because, as the name *satsumaimo* suggested, sweet potato was primarily cultivated in the Satsuma domain of Kyushu during the early eighteenth century, following its cultivation in Ryukyu. As previously stated, sweet potato was initially regarded as something exotic in both regions, as suggested by its name *karaimo*. The fact that Ryukyu kings sent the root as a gift and presented it at a court banquet are examples of its exalted status. Miyazaki Yasusada recommended the crop to his readers by paraphrasing Xu Guangqi's *Nongzheng quanshu*, allowing them to recognize sweet potato's usefulness as a subsidiary food plant. Then Tanegashima Hisamoto and Maeda Riemon both demonstrated to the people that sweet potato could be an insurance food crop in times of emergency.

Aoki Konyō came to play his remarkable role because, during his lifetime, Japan experienced one of the four worst famines in the nearly three-century-long history of Tokugawa rule. Aoki excelled in his studies as a child, despite being born and raised as a commoner in central Edo, today's Tokyo. His interest in Confucian learning, which was popular during the Tokugawa period, led him to become a disciple of Itō Tōgai (1670–1736), son of Itō Jinsai (1627–1705), both of whom were eminent Confucian scholars of the time. Aoki was also a serious student of Dutch learning, science, and medicine. Aoki received attention from the Tokugawa government during a time of crisis due to the breadth of his knowledge. The famine, known as the "Great Kyōhō Famine" (*Kyōhō no daikikin*), began in Kyushu but spread throughout the country. It was caused by a series of poor harvests that began in 1716 and culminated in 1732 because of cold rain that hit the region relentlessly for two months, destroying wheat and barley harvests, which lost roughly half of their harvests. Planthopper and locust infestations were also caused by the bad weather. The former was known as *unka* in Japanese, and it was an insect that was particularly damaging to rice paddies, causing widespread failure. Due to widespread crop destruction—worsened by locust infestations and severe weather—it was estimated that, in 1732, western Japan produced only 10 percent, and the nation as a whole just 27 percent, of the average annual rice yield. According to contemporary sources, two and a half million people, or one-tenth of the population, suffered from hunger, and one million died from starvation. Meanwhile, hundreds of thousands of people flooded into major cities, hoping for government assistance to survive.

As the famine started in Kyushu, where sweet potato was grown, people quickly realized the value of the root plant as an emergency food crop. In addition to Aoki Konyō, whose role will be discussed further below, Asami Kichijūrō and Ido Masaakira were credited with spreading the root from Kyushu to Shikoku and western Honshu. The former was born into a wealthy family but went on to become a monk. During his pilgrimage to Satsuma in 1711, he came across sweet potatoes and brought them back to Ōmishima, Shikoku. When the famine struck, the latter was an official in Chūgoku, western Honshu, who tried to bring sweet potato from Satsuma to the region. The two were also remembered fondly by the locals for saving lives. The famine was said to have claimed up to one million lives in the areas surrounding Ōmishima. Remarkably, no one died on the island itself—thanks to the cultivation of sweet potatoes. Ido Masaakira, too, saved many people in his district by promoting sweet potato as a relief food.

If Asami Kichijūrō and Ido Masaakira assisted in getting the sweet potato out of Satsuma, Aoki Konyō was the one who spread it to the Kantō region of Honshu, which was the political and economic center of Tokugawa Japan. Japan was ruled by Tokugawa Yoshimune shogun (1684–1751), the eighth-generation scion of the Tokugawa family, at the time of the famine. The shogun was well-known for his open-mindedness and the implementation of reform policies aimed at improving government finances. When Tokugawa Yoshimune learned that Ōmishima had escaped death due to Asami Kichijūrō's efforts, he charged Aoki Konyō, a learned scholar with a commoner's background, with replicating Asami's success and experimenting with sweet potato cultivation. The Koishikawa Herbal Garden (Koishikawa yakuen), now managed by Tokyo University, was chosen as the location for Aoki's experiments. Aoki, who was serious about the project, was said to walk six kilometers from his house to the site every day. The cultivation was a huge success right away. According to Aoki's report, they produced 565 sweet potatoes from 180 seeds planted. The sweet potatoes were then transplanted elsewhere, yielding an impressive harvest.

The resulting sweet potatoes were distributed by the government to the Izu Islands outside of Tokyo for further cultivation.[14]

Aoki Konyō was chosen by the shogun to serve as the "sweet potato commissioner" (Satsumaimo goyōgakari) in his administration for yet another reason. Aoki's fellow student under Itō Tōgai, Matsuoka Jo'an, created a small pamphlet titled *Bansho roku* (Sweet potato book) over a decade earlier. As Matsuoka noted at the beginning of his book, he focused on sweet potatoes because they were grown in Ryukyu. Matsuoka quoted extensively from Miyazaki Yasusada's account of the plant when putting together his work, repeating the latter's assertion that sweet potatoes were "a necessary famine crop that could benefit the people in addition to the five grains." He also provided several accounts from local gazetteers in South China about sweet potato farming methods.[15] Aoki was quite impressed by the *Bansho roku*. Meanwhile, he felt compelled to create his own text because Matsuoka left out some other accounts on the root plant. His *Bansho kō* (Sweet potato study), written in Chinese, included, for example, Xu Guangqi's description in his *Nongzheng quanshu*, which was missing in Matsuoka's account. Aoki's training in Dutch learning, which promoted modern science in the country, also influenced him to pay more attention to the country's farming methods—his writing of this section was both detailed in description and conveyed in accessible language. Ōoka Tadasuke (1677–1752), the magistrate of Edo at the time, obtained a manuscript copy of Aoki's *Bansho kō* before it was printed in 1735. Impressed by Aoki's knowledge, Ōoka passed it on to Tokugawa Yoshimune. Equally impressed, the shogun decided to create an ad hoc position for Aoki to cultivate and promote sweet potato throughout Japan.[16]

Aoki Konyō, like Xu Guangqi, enumerated thirteen benefits of the sweet potato as a food plant in his *Bansho kō*, refuting the unfounded rumor, which had spread among some Japanese, that the root was poisonous. His action received Tokugawa Yoshimune's endorsement.[17] Aoki used the Buddhist concept of "benefit or grace" (*kudoku*) to describe the plant's proclivities, which usually refers to the goodness in a person on which his/her karma is accrued. The first is the sweet potato's high yield—a single *mu* can produce hundreds of *shi*. The second characteristic is that its flesh is white and quite tasty. The third benefit is that sweet potato, like *rokoko*, a Japanese yam, is medicinal. The fourth advantage is that the stem can be cut and used to propagate many more *shi* the following year. The fifth advantage is that once the stem reaches the ground, it transforms into the root, which is not easily damaged by bad weather. Sixth, if famine occurs, sweet potato can be a good food substitute. The seventh advantage is that it can be consumed as a confection (*kashi*). The eighth feature is that it can produce alcoholic beverages. The ninth benefit is that it can be milled into flour and used to make pancakes. The tenth benefit is that sweet potatoes can be eaten raw or cooked. The eleventh benefit is that because sweet potato grows well in any type of soil, it can be planted in the same field year after year. The twelfth advantage is that sweet potatoes can be harvested as late as October if planted in the spring and summer. And the thirteenth is that sweet potato, a root crop, is resistant to the damage caused by *unka*, or planthoppers, and other insects. When his thirteen "benefits" are compared to Xu Guangqi's thirteen "advantages" of the sweet potato, it is clear that the two are nearly identical. Aoki largely agrees with Xu, save for his praise of sweet potato's edaphic tolerance in the eleventh paragraph. In the thirteenth, he also substitutes *unka* for locust, possibly because the latter was more familiar to Japanese farmers.[18]

Famines, plants, and population

Aoki Konyō not only compiled the *Bansho kō*, which provides a detailed description of sweet potato farming, but also successfully experimented with planting the crop himself. What role did his promotion of the sweet potato play in relieving Japan's suffering during the Great Kyōhō Famine? It is difficult to say for certain. However, a known fact was that in 1735, the year Aoki's *Bansho kō* was printed, the Tokugawa shogun allowed farmers to diversify crop farming by growing alternative food plants, demonstrating the influence of Aoki's study of the American root plant. After completing his mission as the "sweet potato commissioner," Aoki resigned in 1739 and was encouraged to pursue his scholarly interests. Oka Tadasuke asked him to oversee the culling and cataloging of old texts for the Tokugawa family. Later, he traveled to Nagasaki to further his studies in Dutch learning and the Dutch language. In his later years, he was appointed as an official librarian (*shomotsu bugyō*), a prestigious position reserved for someone known for his erudition. However, it appeared that the best recognition came from ordinary people. Aoki's name was already associated with the introduction of sweet potato from Satsuma to Honshu during his lifetime. Following his death, the location where he had experimented with sweet potato cultivation was turned into a Shinto shrine in his honor, where he was enshrined as the "sweet potato *kami* [deity]." Aoki Konyō is still referred to as "Mr. Sweet Potato" in Japan.

Aoki Konyō earned the nationwide moniker "Mr. Sweet Potato" in part because famines were common in premodern Japan. According to Mikiso Hane, an expert on Tokugawa and modern Japan, "there were at least fifteen serious famines during the Tokugawa era."[19] After Aoki became the official librarian in the mid-eighteenth century, the Horeki famine occurred, which was caused by rice failure and killed between 50,000 and 60,000 people, mostly in northern Honshu. A decade or so after Aoki's death, Japan was hit by another severe famine known as the Great Tenmei Famine. It lasted six years, from 1782 to 1788, and was the largest and deadliest of the Tokugawa period's four great famines. The main cause was cold weather, which caused massive rice and wheat failure and a sharp decline in food supply. Worse, Mt. Iwaki and Mt. Asama erupted one after the other, disrupting weather patterns and contributing to the crop disaster. It was estimated that at least 200,000 people died of starvation during the Great Tenmei Famine (although other estimates range as high as 2 million), with northern Honshu bearing the brunt of the toll.

In other words, according to some research, crop failure caused by inclement weather was a major cause of famines. Official reports on crop failure were numerous from the eighteenth to the mid-nineteenth centuries, or before Japan's modern period, such as those that appeared in 1755–1756, 1762, 1767–1768, 1776, the 1780s, 1792, 1813, 1821, 1828, 1835–1836, and 1850.[20] During the Great Tenmei Famine of the 1780s, an itinerant scholar saw "mounts of bleached bones" in a village in northern Honshu. A peasant informed him of what had occurred:

> These are the bones of the people who starved to death. During the winter and spring of the year before last, these people collapsed in the snow. Some of them were still breathing as they lay on the ground. Their bodies blocked the road for miles and miles, and passersby had to tread around them carefully. At dusk and at night, one had to be careful not to step on corpses and snap bones or step into rotting guts. You probably cannot imagine the terrible stench that filled the air. In order to keep from starving to death we used to catch the horses roaming about, tie ropes around their necks, bind them to

posts, cut into their flesh with swords or knives, cook the bloody meat with some grass and eat it. We also used to catch chickens and dogs running around in the open and eat them. When we ran out of animals, we stabbed and killed our children, our brothers, or other people who were on death's door with some disease, and ate their flesh. We would bite into the flesh from their breasts to stay alive. The eyes of people who have eaten human flesh glisten in an eerie manner, like the eyes of wolves. The harvest of this year is likely to be poor again so we will no doubt experience another winter of famine.[21]

Without a doubt, it was because of the severity of the famine, characterized by such terrifying cannibalism, that Aoki Konyō and all those who helped spread the sweet potato were sincerely appreciated throughout Japan. As specified by Aoki and practiced by Asian farmers, the American root plant became a much-desired food crop because, once sown in the spring and/or summer, it (including both its roots and vines) could be eaten through the fall. After the fall harvest, dried sweet potato strips could be easily stored as food for the people during the winter when other food crops failed due to bad weather. Cold weather frequently struck Japan in the eighteenth century, which climatologists refer to as a "Little Ice Age," damaging the rice and wheat crops that had historically been the nation's primary sources of food. When this happened, sweet potato started to play an important function. Ishige Naomichi, a prolific writer on Japanese food culture, noted that "the sweet potato was the most influential of the new crops," and that beginning in the early modern era, the Japanese diet began to show signs of European influence, or *nanban ryōri* (lit. southern barbarian cuisine). This influence was primarily caused by the fact that "in locales where rice production was low and habitation was sparse, there was perceptible population growth following the introduction of the sweet potato."[22] The advantages of growing sweet potatoes were more obvious when other food crops failed. Economic historian Susan Hanley in her contribution to *The Cambridge History of Japan* notes that the Great Kyōhō Famine, which was primarily brought on by a lack of rice, occurred during this time: "the death rate was low in both Satsuma and Nagasaki because people were not relying entirely on grain; now they had sweet potatoes to fall back on. Sweet potatoes could be grown upland, in contrast with rice, and they produced more calories per acre than did almost any other crop.

Sweet potatoes may well have been an important factor in maintaining a dense population in Japan in the eighteenth and nineteenth centuries and also in explaining why the population in western Japan grew faster than did that of the rest of the country."[23]

To put it another way, in order to better and more fully assess Aoki Konyō's role, and the impact of sweet potato farming in general, in the course of Japanese history, we must revisit the agrarian life and demographic situation in eighteenth- and nineteenth-century Japan. Until the 1970s, scholarship generally portrayed Tokugawa Japan as a backward society, one doomed to be replaced with the arrival of modernity. However, a more balanced viewpoint has gradually emerged among both Japanese and Western academics since then. Scholars discovered that during the first hundred years of Tokugawa rule, the Japanese population grew at an impressive rate, based on surveys conducted by both the Tokugawa government and reginal daimyo lords. It is widely assumed that after Japan was unified and placed under Tokugawa rule in the seventeenth century, its population increased from around twelve million in 1600 to over thirty million in the early eighteenth century.[24] In sharp contrast to China, where the population suffered a significant loss during the Ming–Qing dynastic transition, the demographic expansion in Japan was attributed to the peace brought about by the unification. It was also because Japanese peasants worked hard during the seventeenth century to reclaim land and improve agricultural technology. Many now point out that, contrary to popular belief, Tokugawa Japan was not an economic backwater: "the amount of arable land is estimated to have doubled during the Tokugawa period, and much of the increase occurred during the seventeenth century."[25] That is, the demographic advance was based on the period's economic progress.

In the eighteenth century, Japan and China once again diverged in significant ways. While the Chinese population skyrocketed, Japan's population remained stagnant. At the start of the nineteenth century, it was still around 30 million, and this number remained stable until the Meiji Restoration in 1868. Hayami Akira, a distinguished historical demographer, estimates that the figures were likely even lower based on a careful examination of the data available from the Tokugawa period. Japan's population was 26.1 million in 1721 and 26.9 million in 1846.[26] In other words, there was no discernible population growth in Japan during the one and a half centuries preceding Japan's modernization from the mid-nineteenth century. The authors of a detailed study of Japanese economic history describe the phenomenon as follows:

> The extensive economic growth at the start of the early modern era, with its explosive increase in population and arable land and its expansion of the total volume of production, began to approach its limit around the turn of the eighteenth century. From the 1730s until the beginning of the nineteenth century there was almost no increase in the aggregate national population, and the expansion of cultivated land slowed markedly.[27]

What happened? As previously stated, there were over a dozen famines in Tokugawa Japan, the two deadliest of which, the Great Kyōhō and Tenmei Famines, occurred in the eighteenth century. Did Japan then fall into a Malthusian trap, in which the population growth in the previous century had surpassed the food supply? It appeared so, except that, as mentioned above, weather played a significant role in causing both the Kyōhō and Tenmei famines, which will be discussed further below. Hayami Akira and his colleagues made the most significant contribution by challenging the Malthusian thesis in explaining the stagnation of Japanese population in the eighteenth and early nineteenth centuries. Hayami Akira's meticulous research revealed that, while famine was a factor that slowed Japan's population growth in the eighteenth century, there were significant regional differences. Southwestern Japan, which included Kyushu, Shikoku, and western Honshu, actually experienced steady population growth during the period; the reason Japan's overall population did not advance was due to a significant decline in the rest of Honshu, which included major cities such as Edo and Osaka. Indeed, between 1721 and 1804, the population of southwestern Japan increased by 15–20 percent on average. More notably, even when disasters struck, the population in southern Kyushu (for example, Satsuma and Nagasaki) grew at an annual rate of 11.9 percent between 1721 and 1846. In sum, per the caution by Japanese economic historians: "we must keep in mind that the label 'stagnation' only applies to the national population figures. When we take a closer look at each region, considerable differences become readily apparent."[28]

How do we explain the distinctions? We can probably explain them in three ways, based on the valuable research of Hayami Akira and others. The first reason was due to the nature and level of economic development in those regions. Hayami observed that population growth in southwestern Japan occurred not only in the eighteenth century, but also in the first half of the nineteenth century. In contrast, areas around Edo, Kyoto, and Osaka, the nation's heartland, did not see significant population growth until the late nineteenth century. He maintained that the development of village industry, with a focus on handicraft production, aided southwestern Japan and the Inland Sea area. Another positive factor in the Satsuma domain was the daimyo's decision to monopolize the sugar market. Second, in comparison to the economic growth in southwestern Japan, central Honshu, which was under the Tokugawa government's direct control, faced increasing economic and financial challenges. These difficulties included a persistent government deficit and an increase in the samurai population in the cities of the Kinai region, particularly Edo, which burdened the Tokugawa economy as a whole. The death rate, which disproportionately affected the urban poor rather than the samurai class, was also higher in cities than in rural areas such as Kyushu. Third, crop failure caused by cold weather, such as during the Great Tenmei Famine, was more severe in the North than in the South. Cold weather did not cause as much damage in southwestern Japan, which is located next to the Sea of Japan, as it did in the Northeast, which is located on the Pacific Ocean.[29]

Of course, the cold weather had an impact on agronomy in southwestern Japan. The Great Kyōhō Famine was proof. However, as previously stated, there was another cause—an infestation of *unka*, the rice insect, which ransacked rice fields. The fact that *unka* could not harm the sweet potato, which grows underground as a root crop, contributed to the root's usefulness. Another indication of its advantage as a food resource was that southern Kyushu, where it had been cultivated more extensively than elsewhere, suffered fewer deaths during the famine. In other words, while Hayami Akira did not include sweet potato farming in his explanation of regional differences in Japan's population fluctuation at the time, the cultivation of the root plant was certainly a factor to consider, as Susan Hanley's observation above demonstrated. Yamada Syoji, a historian who researched the sweet potato cultivation in Ryukyu and Satsuma, observed that during the eighteenth century, while Japan as a whole was experiencing population decline, the Satsuma domain was an exception—its successful sweet potato cultivation enabled it to export surplus population to other regions. He explains that because the root had become a staple food for the farmers in the region, it had a significant impact on the domain's population growth.[30]

Arizono Shōichirō, another historical geographer, conducted a focused study on how sweet potato farming in southwestern Japan, including Satsuma, had a positive impact on population growth. His main thesis is that in eighteenth-century Japan there was a significant correlation between sweet potato cultivation and consumption and regional differences in population increase or decrease. Ryukyu, Satsuma, and Ōmura, a domain near Nagasaki, were the areas he investigated. According to his research, the sweet potato became a major food crop in the region for three reasons. The first is its high yield, which he claims has improved over time since sweet potato became a major crop in the region. The yield buoyed southwestern Japan's population growth because, in 1883, the indexes of population supported by per unit of arable land in Ryukyu and Ōmura were five and two times higher, respectively, than the national average (excluding that of Hokkaido). In other words, sweet potato significantly increased the food supply, thereby sustaining local population growth. The second benefit is that sweet potato, a tropical plant, is both drought-tolerant and typhoon-resistant, which is especially important in southwestern Japan due to the frequency of storms. It mitigates drought damage because, as an underground crop, sweet potato vines cover the ground surface, protecting the soil moisture needed for root growth. The third advantage is that its vines are easy to propagate and provide a good yield even with extensive farming. However, with a little care, it produces an even better harvest.

According to Arizono, sweet potatoes also have three drawbacks. The first is that it has a heavier unit weight than rice; according to his calculations, a person would need to consume about five times as much sweet potato to absorb the same amount of protein from eating rice.[31] The second is that it contains a lot of water per unit, which makes it challenging for farmers to store and move the roots. The third is its sweetness, which can cloy people if they rely on the root as their daily staple. That is why sweet potatoes are usually consumed with other foods in areas where they are a staple. Japanese farmers frequently eat sweet potatoes with rice and other grain foods, then down them with miso paste and other soybean-based non-grain foods in order to account for both their benefits and drawbacks. They also include salted sardines in their meals because there are many fishermen living in southwestern Japan. The saltiness of these non-grain foods helps counteract the sweetness of the sweet potato.

Which of the three places saw the most positive impact from sweet potato farming on population growth? According to Arizono's research, the best example was Ryukyu, where the root first arrived in the region. In the early seventeenth century, Ryukyu farmers began to cultivate sweet potato. The root plant could grow all year round on the islands because the average temperature in the winter is 16 degrees Celsius, which is one degree higher than the 15 degrees Celsius required to sow the plant. Sweet potatoes were frequently grown in rotation with other crops by Ryukyu farmers, either as a double crop or as part of an intercropping regimen, such as five crops in two years. If the two-year, three-crop system is used, the yield could be even higher. Sweet potatoes constituted roughly 90 percent of the food intake among the islands' inhabitants beginning in the seventeenth century, providing enough energy for them due to their high yield, which was six times that of rice. Ryukyu's population grew rapidly in tandem with the expansion of sweet potato farming. Sai On (1682–1762), a capable chief minister to the Ryukyu court, was aware of the kingdom's remarkable population growth. In one of his works from the mid-eighteenth century, he stated that the population had increased from between 70,000 and 80,000 to 200,000. Sai On did not specifically attribute the increase to sweet potato farming. However, Arizono observes that Sai's remark was made precisely after the introduction of the sweet potato to the islands. That is, thanks to the cultivation of the sweet potato, Ryukyu's population increased by 300 percent between the early seventeenth and mid-eighteenth centuries.

What about Ōmura and Satsuma? Sweet potato was not cultivated as a food crop in either domain until the early to mid-eighteenth century. Because the temperature in both domains was slightly colder than in Ryukyu, farmers usually planted the root only once a year in the summer, resulting in lower production than in Ryukyu. Nonetheless, the root was an important food plant in both domains' agrimonies. It accounted for about 15–20 percent of cultivated acreage in Satsuma and was consumed as a staple, whereas in Ōmura, where irrigated rice paddies were scarce, locals grew to rely heavily on the sweet potato to supplement their food intake. As a result, according to Arizono, the root was a significant factor in helping to grow the regional population. Satsuma's population grew at a rate of 167 percent between the early eighteenth and late nineteenth centuries, whereas Ōmura's population grew at a rate of 180 percent between 1721 and 1856.[32] These were impressive increases, as the rest of Japan's population growth during that time did not advance noticeably.

More importantly, Arizono Shōichirō's three case studies show that, beginning in the eighteenth century, sweet potato, which was initially accepted as a backup food crop in hard times, gradually became a common, if not a staple, food in Japan. And it appears that sweet potato will play an even larger role in the Japanese food system in the future. For example, in twentieth-century wartime Japan, most Japanese relied on boiled sweet potatoes for survival, according to popular writer Mock Joya, who lived through the period.[33] The next section will discuss how sweet potatoes have fared in modern Japanese society.

How did the root fare on the Korean Peninsula if sweet potato was able to successfully increase its acceptance in Japan from the early modern to the modern era? As pointed out at the beginning of this chapter, rather than traveling through China in the middle of the eighteenth century, the root plant reached Korea via the Japanese island of Tsushima. Tsushima is located halfway between Korea and Kyushu, roughly fifty miles from either location, meaning that sweet potato entered Korea around the same time that it started spreading in Kyushu. Because of its location, Tsushima has historically served as a bridge connecting the two nations. However, near the end of Japan's unification process, Toyotomi Hideyoshi launched unsuccessful invasions against Korea. Then, in the early seventeenth century, after Japan was unified under Tokugawa Ieyasu, the "lockup" policy was implemented, which prohibited exchanges between the two countries as well as Japan's overall foreign trade (limited trades between Japan and China and the Netherlands were permitted). Needless to say, these two events harmed Tsushima's economy and its people's livelihoods.

Suyama Don'ō, a Confucian scholar, played a similar role in Tsushima sweet potato cultivation as Aoki Konyō did in Honshu. Suyama, whose father was an herbalist, received his early training in Confucian teaching from a distinguished scholar, as did Aoki. His teacher was Kinoshita Jun'an, a court tutor hired by the Tokugawa family who advocated for the practical application of knowledge. Suyama returned to Tsushima after the latter's death and, having been hired by the local lord, was eager to put his knowledge to use. Tokugawa's "lockup" policy effectively halted trade between Korea and Japan. However, approximately eight thousand merchants remained in Tsushima and smuggled goods between the two countries. Their presence exacerbated the residents' food shortage, and Suyama worked to reduce the presence of those outside merchants in order to solve the problem. Meanwhile, he launched a campaign to eradicate wild boars from the island, claiming that the boars harmed food crops. His actions earned him a national reputation, in part because the reigning shogun had issued a decree calling for the pitying of all living beings.

Suyama turned to sweet potato in the 1710s after reading Miyazaki Yasusada's *Nōgyō zensho*, promoting the plant's cultivation in Tsushima as another way to alleviate food shortages. There was a possibility that sweet potato arrived in Tsushima directly from Ryukyu. However, many people believe that the root plant arrived on the island through the hands of Harada Saburōemon, a peasant who studied with Suyama. Suyama encouraged Harada to bring sweet potato from Satsuma and cultivate it in his garden. Because Tsushima Island is very mountainous and has few irrigated fields, the root plant was well received as an alternative food crop—so much so that the islanders gratefully referred to it as "filial yam/taro" (*kōkōimo*), because even the samurai class had to go to the hills from time to time to look for sweet potatoes to feed the elders and fulfill their filial duties to their families. Suyama, on the other hand, traced the name back to China in his writing, claiming that it derived from a poor Chinese peasant son who used it to feed his sick and starving father.

Suyama told this story in his *Kansho seshi* (Discourse on the sweet potato), one of the earliest works on the root plant in Japan, written in 1724, before Aoki Konyō's *Bansho kō*.[34] Suyama stated at the beginning of his book that sweet potato was already widely grown in both Satsuma and Nagasaki, and that sweet potato was known as *Ryukyuimo* or *akaimo*—the latter describing its red color. He also stated that the root was first planted in Ryukyu before spreading to Satsuma and then Nagasaki. Suyama provided quite detailed instructions on sweet potato farming and storage, drawing on Harada's and other farmers' experience with the plant. He suggested planting sweet potato in the early spring, right after spring wheat. The soil should be moist, and if there is no rain for several days, the ground should be watered. He also advocated for sweet potato and wheat intercropping—for example, by planting one crop of the former with two crops of the latter. To store sweet potatoes, he recommended washing them, cutting them into three to four pieces, and drying them thoroughly in sun-dried bales.

Suyama Don'ō's *Kansho seshi* stood out among other texts on the root plant because it included instructions for making sweet potato starch, a local specialty for which Tsushima is well-known throughout Japan. A preserved food made of starch and dietary fiber from fermented sweet potatoes, it was known as *sen* or sweet potato *sen* by the locals in Tsushima before spreading throughout the country. Many people have seen sweet potato processed into starch. Jennifer Woolfe writes in *Sweet Potato: An Untapped Food Resource* that "Starch is quantitatively the most important component of sweet potato root dry matter. The extraction of this starch on a home, village and commercial scale, and its use in various food products is widely practised in some countries, for example China."[35] It is not known if Tsushima farmers learned how to make *sen* from their Chinese counterparts, as turning sweet potato into starch is a well-established practice in villages throughout South China. Regardless, Suyama's description, which must have come from his own observations on the island, could be the first record of the technique. While brief, it consists of three steps for extracting starch from sweet potatoes. The first step is to select large or medium-sized sweet potatoes, peel them, and grate them in a mortar. The second step is to wash the grated sweet potatoes in a kapok bag and drain the water in a vat, stirring and changing the water three times to cause sedimentation. The third step is to drain the water from the vat. The sediments that settle at the bottom, when dried, become starch, which can be mixed with rice flour to make foods like *dango*, a Japanese dumpling, among other things.[36]

Suyama's method for making sweet potato starch appears to be an abbreviated version, as the process is slightly more complicated. But it was enough to show that sweet potato had become widely accepted as a staple food in Tsushima. This could be the impression Jo Eom, a Korean diplomat, had when he visited the island on his way to Edo in 1763. His mission was to improve relations between the two countries. On the island, Jo not only tasted the sweet potato, noting that it was comparable to chestnut but sweeter, but also described sweet potatoes and Chinese yams as root plants. Though it was his first time seeing the plant, Jo had heard about it before. Yi Gwangryeo, a learned Confucian scholar Jo knew, had asked him to find the root and bring it back to Korea. Yi believed that cultivating sweet potatoes would be beneficial for Korea after reading Xu Guangqi's section in the *Nongzheng quanshu*. "I heard there is an interesting thing going on in China right now," Yi wrote around the mid-eighteenth century. "Its name is 'sweet potato,' and it was introduced to China about a hundred years ago. Sweet potato has spread to the inner regions from Fujian and the Yangzi River area. The exceptional value of the sweet potato is that it eliminates the need to be concerned about drought or flood, good or bad harvests. Many of the benefits are described in [Xu Guangqi's] *Nongzheng quanshu*. The benefits of bringing this plant to Korea would be incalculable. It just so happened that a good friend's son was chosen to accompany a legation to Japan. So I asked him to bring the seeds back."[37] Jo Eom, apparently, was his friend's son. When Jo first saw the sweet potato on Tsushima in the fall of 1763, he asked someone to send the seeds back to Yi while he continued his journey onto Honshu to meet the shogun. Jo was concerned that Yi would miss the growing season the following spring. Jo returned to Tsushima the following year and bought more sweet potatoes to take home with him.[38]

But Yi Gwangryeo's attempts to grow sweet potatoes in his garden were not very successful. At this point, Gang Pilli, the third person to play a significant role in the introduction of the sweet potato to the Korean Peninsula, entered the scene. In the Busan neighborhood of Dongnae, where Jo Eom returned with the sweet potato he had acquired from Tsushima, Gang served as the local magistrate. Gang planted the seeds after receiving them, and they flourished in his garden. After that, he produced *Gamjeobo*, an early Korean study of the root plant used as a reference source by Seo Yu-go in the early nineteenth century when compiling *Jongjeobo*. *Gamjeobo* has since been lost, though, as mentioned at the opening of this chapter.

This is the widely accepted account of how sweet potato arrived in Korea: The American root plant was introduced to the country via Tsushima, Japan, thanks to the efforts of Korean emissaries such as Jo Eom. In the Korean language, sweet potato is known by several names. One of them is *goguma*, which should have derived from the local term *kōkōimo*, which means "filial yam/taro" in Tsushima, where Jo Eom discovered the plant. *Goguma* was Jo's transliteration of the Japanese term in Korean. Meanwhile, sweet potato is also known as *beonseo*, *gamjeo*, and *jigwa* in Korea, which correspond to the Chinese names for the root plant, *fanshu*, *ganshu*, and *digua*.

Since Korea shares a land border with China, could sweet potato have reached the peninsula overland from somewhere in North China, such as Shandong, where *digua* is the plant's known local name? Ogura Shinpei, a renowned early twentieth-century Japanese philologist specializing in the Korean language, proposed that Koreans, because their name for sweet potato followed Chinese pronunciation, might have seen the plant prior to Jo Eom's introduction in 1763. No Seonghwan, a modern Korean scholar who studies the sweet potato's migration from Japan to Korea, concurs with his hypothesis. While Korean emissaries and officials such as Jo Eom and Gang Pilri are credited with introducing sweet potato to the peninsula, No believes sweet potato may have arrived in Korea from China before the mid-eighteenth century. Felix Siegmund, a contemporary German scholar who has written about the origins of the sweet potato in Korea, agrees. Using Korean historical sources, Siegmund notes several incidents indicating that sweet potato may have arrived in Korea, and specifically Jeju Island, around the 1720s. Jeju Island, like Tsushima Island, is located in the Korean Strait between Korea and Japan. It was said that after a shipwreck, Jeju fishermen traveled to Kyushu, where they discovered that sweet potato had become a staple for the locals. Another shipwreck, this time involving a Chinese fishing boat, landed a group of Chinese fishermen from southern Fujian in Jeju, where they could have brought the roots with them or, at the very least, informed the Koreans about the plant. To this day, according to No Seonghwan, *gamjeo*, or *ganshu* in Chinese, is the preferred name for the sweet potato in Jeju and nearby regions, whereas *goguma* is more commonly used in the rest of Korea.[39]

If *gamjeo* predated *goguma* in Korea, this implies that, in addition to Jo Eom's introduction from Japan, Koreans learned about the root plant from Chinese sources. Yi Gwangryeo, for example, used *gamjeo* in his remarks on the sweet potato mentioned above. Bak Jega (1750–1805), a well-known scholar known for his support of *Silhak* (Practical Learning), mentioned *gamjeo* in one of his key texts. Yi Gwangryeo and Bak Jega, both fluent in Chinese, enthusiastically recommended sweet potato as a valuable alternative to grain crops. "The sweet potato is extraordinary," wrote Bak, "among the measures against famine. It would be fitting to order the officials in charge of military settlements to . . . plant sweet potatoes. . . . Furthermore, the people should be brought to plant sweet potatoes themselves. Then it will not be necessary to worry this year. One simply lets the sweet potatoes sprout and watches out for dampness and frost."[40]

There appeared to be numerous reasons for Korean scholars such as Bak Jega and Yi Gwangryeo to encourage their compatriots to plant sweet potato. Despite having attended different schools, they were both interested in putting their academic knowledge to use in the real world. Bak, as previously stated, was a major proponent of *Silhak*/Practical Learning, a popular intellectual trend that spanned the seventeenth and early nineteenth centuries. Yi, a follower of the Wang Yangming school in China, emphasized the importance of seeking unity between knowledge and action, a key tenet of Wang Yangming's teaching passed down to him through his teacher Jeong Jedu. And they were not the only ones. A similar recommendation for sweet potato farming was recorded in the *Joseon Wangjo Sillok* (Veritable records of the Joseon dynasty), an annual chronicle of historical events compiled by court historians from 1392 to 1865. Seo Yeongbo, a provincial governor from a distinguished family, recommended to Emperor Jeongjo (r. 1776–1800) in 1794 that sweet potato farming, whose method was taught by Xu Guangqi in his *Nongzheng quanshu*, had taken root in coastal regions of southern Korea for some time and had achieved a similar success as it had in South China and southwestern Japan. However, due to the ignorance and incompetence of local officials, the plant, which he believed would be a good famine food crop, did not thrive in Korea as it should have. Seo pleaded for a court order so that farmers could be incentivized to grow the root plant without being harassed by officials, particularly in Jeju, which has a climate similar to Tsushima.[41]

But it appeared that this was the only case mentioned in the chronicle. This suggests that by the end of the eighteenth century, despite the recommendations of learned scholars such as Bak Jega and Yi Gwangryeo, as well as the efforts of Jo Eom and Gang Pilri to introduce and cultivate it, sweet potato remained a rare, exotic, and somewhat luxurious food for many Koreans. In fact, rather than referring to the sweet potato, as it did in historical texts from the eighteenth century, *gamjeo* is the name for the white potato in Korea today. White potato, unlike sweet potato, was introduced later to the Korean Peninsula and received a better reception, particularly in the North, where it is now a common food.[42]

Why was the sweet potato not as well received in Korea as it was in China and Japan? There are several plausible explanations. The first was that after Seo Yeongbo passionately made his plea to the court, explaining the benefits of the sweet potato as a food crop, there was no reaction recorded in the chronicle about whether Emperor Jeongjo, who was otherwise known for his inclinations toward reform, heeded and reacted to his imploration. On the other hand, sweet potato received special attention from the rulers of both China and Japan. Qing China's Emperor Qianlong recommended the root plant in the *Shoushi tongkao*, which was distributed to all officials throughout the country, whereas Japan's reigning shogun, Tokugawa Yoshimune, was persuaded of sweet potato's usefulness as an alternative food crop by Aoki Konyō. The second was related to the previously mentioned improved reception of the white potato in Korea.

Impressed by the success of sweet potato farming in both China and Japan, Korean scholars, the majority of whom were well-versed in Chinese, hoped to replicate it in their own country. However, the climate on the Korean Peninsula appears to be too cold for the sweet potato to thrive; it is certainly not a plant suitable for the North, which has more snow-covered high mountains. The early nineteenth-century acceptance of the white potato, which is a temperate crop rather than a semitropical crop like its sweet counterpart, is evidence.

According to Felix Siegmund, the third reason could be that sweet potato, as a root crop, has not been broadly accepted in a grain culture like that of Korea.[43] However, the same could be said of China and Japan, where rice and wheat, particularly the former, are widely preferred. Nonetheless, sweet potato initially and primarily appealed to the Chinese and Japanese because its value as a grain substitute was recognized in hard times. Food crises occurred in premodern Korea as well. When one searches the *Joseon Wangjo Sillok*, one finds frequent mentions of famines on the peninsula, indicating that the chroniclers took the matter seriously when compiling the dynasty's five-century history. Indeed, records of famines in Korea go back almost as long as the written chronicles. Famines occurred on a regular basis during the eighteenth century, prompting those Korean scholar-officials to search for viable famine foods and recommend the sweet potato. Interestingly, no famines were recorded in the *Joseon Wangjo Sillok* from 1798 to 1837. In the early nineteenth century, the well-respected scholar-official Jeong Yakyong's influential *Mongmin simseo* (Admonitions on governing the people) gave detailed instructions on how local officials should prepare for famine relief. However, Jeong did not recommend the sweet potato as an alternative food plant; instead, he proposed that in times of famine, the production of wine be prohibited in order to save grain.[44] The indifference shown by Jeongm, Emperor Jeongjo's trusted confidant, to the efficacy of the sweet potato as a famine plant may have contributed to the emperor's inaction after hearing Seo Youngbo's recommendation. A few years later, in 1800, Emperor Jeongjo died unexpectedly, resulting in Jeong Yakyong's exile. Seo Yugu compiled *Jongjeobo* during this period, synthesizing previous Asian scholarship on the sweet potato. While Jeong and Seo were both prominent exponents of *Silhak*/Practical Learning in their respective eras, they were also scholarly rivals. There was reason to believe that by assembling his *Jongjeobo* in 1834, Seo hoped to change the government's stance on the root plant. However, there is no evidence that he had succeeded. According to the *Joseon Wangjo Sillok*, the absence of major famines in early nineteenth-century Korea may have played a role. To summarize, the American plant did not become more prevalent in Korean food culture until the early twentieth century.

Sweet potato vs. rice as food

As explained in chapter 3, when Chinese farmers of the eighteenth century embraced sweet potatoes, local gazetteers started to regard the root as equivalent to grain crops. And the trend has continued into modern times, with the sweet potato becoming a popular food in both North and South China. In comparison, the root plant made no progress on the Korean Peninsula. In this Korea also deviated from Japan, where the root became increasingly popular as a valuable alternative to grain food across the Japanese archipelago in the modern period. Tokugawa Japan experienced yet another major famine in the first half of the nineteenth century, from 1833 to 1839, known as the Great Tenpō Famine. The causes of the famine were similar to previous ones: Cold weather combined with flooding resulted in widespread rice crop failure, particularly in central and northern Honshu. Then, in the midst of the famine, in 1834, a 7.6-magnitude earthquake struck northeastern Honshu. Tokugawa Nariaki, a conscientious and concerned daimyo of Mito, near Edo, described his time as one of "barbarians, thieves, and famine." His description accurately portrayed the challenges that the Tokugawa government faced during the nineteenth century: in addition to the "barbarians," or the presence of Western powers, socioeconomic challenges were mounting. Nariaki's attendants were said to have carefully remapped his travel route to and from Edo in order for him not to see the bodies of famine victims on the road.[45]

Grain culture ruled supreme in Japan, as it did in Korea. Many Japanese people regarded rice as the most important of all grains. Indeed, during the Tokugawa era, rice began to function as a currency; not only did peasants pay landlords' rents and government taxes in rice, but samurai also received daimyo payments in the cereal. The annual rice production of a domain was traditionally used as a barometer of its wealth. Even though rice cultivation had a long history in Japan, the grain was sensitive to climate change. The climate in northeastern Honshu, which shares the same latitude as North Korea and China's Manchuria, only "grudgingly" accommodated rice cultivation. Rice failure would occur whenever there was cold weather, because "rice growing, needing an average temperature of twenty degrees centigrade during the crucial months of July and August, had always been particularly hazardous." To some extent, ensuring the rice harvest in Japan remains a challenge today. "Rice production in Japan is under the control of the government," states a recent study, in order "to produce approximately 10 million tons [9.07 million tonnes] annually which just meet the national demand. This production plan of rice, however, is occasionally disturbed by climatic variations. . . . Except for typhoons which occasionally attack some district of Japan, variations in temperature and solar radiation during rice growing seasons are the major factors that bring yearly variations of Japanese rice production."[46]

When the weather turned colder than usual in 1833, the region only produced 35 percent of its normal crop, resulting in the Great Tenpō Famine. It got worse in 1836, when the region's output was only 28 percent of the average. In the Mito domain, located north of Edo in central Japan, 75 percent of the rice crop and 50 percent of the wheat and barley crop were lost in the same year. Grain food prices, particularly rice, skyrocketed, leaving many people hungry. In a nutshell, the Great Tenpō Famine was just as bad as the previous Great Tenmei Famine in the late eighteenth century. If, as previously described, cannibalism occurred in some places, horrifying observers, the Tenpō famine reduced people in affected areas to eating "leaves and weeds, or even straw raincoats." The government also advised people to "bury corpses found on the roadside as quickly as possible, without waiting for official permission."[47]

In response to the crisis, the Tokugawa government implemented some reform measures, albeit with limited success. As the famine spread to southwestern Japan, domains such as Satsuma and Chōshū instituted reforms. Historians generally agree that regional reforms, particularly in Satsuma, were more successful than the central government's attempt. Notably, regional reforms included promoting new crops to avert famine and, as previously stated, monopolizing the sugar market in Satsuma.[48] In other words, sweet potato farming has once again aided the agronomy of southwestern Japan. Yamada Syoji writes in his history of the sweet potato's spread in Japan that during the eighteenth century, the root plant was planted here and there in many regions. Sweet potato was a reserve food plant for the majority of Japanese people in case of famine. However, in Satsuma, sweet potato had already evolved from a "backup" crop to a staple food, thanks in part to the daimyo's promotion and in part because, as a tropical plant, sweet potato thrived in the region due to its suitable climate and soil conditions.[49]

Chūman Katsumi, who has written on the cultivation of the sweet potato in Japan over the period, agrees that in both the Great Kyōhō and Tenpō Famines, southwestern Japan managed to avoid widespread starvation because the people, particularly farmers, in the region had adopted sweet potato as the "main food" (*shushoku*). According to Chūman, the root crop became appealing because it could produce a harvest even when all other food crops died due to drought or storms. By the early nineteenth century, the region's people had developed a number of ways to consume the root. They discovered that it could be eaten raw and that its juice could quench thirst. On the other hand, eating an excessive amount of (raw) sweet potato in a day can cause stomach discomfort, even diarrhea. Thus, the most common way to consume sweet potatoes was to steam or boil the root, skin removed. They also combined sweet potato with other grains, such as in pancakes and dumplings (*dango*) made with sweet potato and wheat or millet flour. Making sweet potato starch, like the Tsushima islanders, was popular, putting every bit of the roots to use. They also made sweet potato liquor, or *shōchū*, by fermenting the starch. The locals on the islands south of Satsuma and north of Ryukyu had developed even more sweet potato consumption methods, such as using its lees, or sediment extracted from the starch, to make miso paste and various types of sweet potato–based sweets or candies. Sweet potato was the mainstay of peasant families' three daily meals on the island, even during holidays and festivals.[50]

What is more notable is that, due to the severity of the Great Tenpō Famine in the early nineteenth century, sweet potato was cultivated in more regions of Japan. More specifically, the root spread from Kyushu to Shikoku before eventually taking root in Kantō, on central Honshu. In terms of sweet potato farming on Honshu, Japan's main island, three great famines during the eighteenth and nineteenth centuries each played a role in educating people about the crop's value. As previously discussed, Aoki Konyō made the effort to introduce the root plant to the region in order to save people from hunger during the Great Kyōhō Famine. His success alleviated some of the previous concern that sweet potatoes were poisonous. However, in some places, such as coastal areas, the rumor seemed to have persisted, preventing its widespread cultivation for a time. Then came the Great Tenmei Famine, which invariably led to increased cultivation of the plant. According to Chūman Katsumi, "sweet potato was widely grown in every village in both coastal and hilly areas" in Kantō, by the 1830s and 1840s, when the Great Tenpō Famine occurred.[51] At the end of the nineteenth century, when the Japanese anthropologist Tsuboi Shōgorō visited the Izu Islands adjacent to Kantō, where the root plant had been cultivated since the early eighteenth century, it had become a staple food for the islanders.[52]

In other words, while rice was historically a preferred staple food for most Japanese, frequent famines caused by crop failure due to bad weather had also weakened its dominance in the country's food system. Indeed, scholars such as Emiko Ohnuki-Tierney and Charlotte von Verschuer have argued that the notion that rice was always a staple consumed daily by most Japanese during the Tokugawa and modern periods is essentially a myth. According to Verschuer, rice's primacy in Japanese food culture was an "invented tradition" that did not exist prior to the Meiji Restoration in 1868. To commemorate Emperor Meiji's accession to the throne, the government issued the following proclamation that year:

> In accordance with the will of the great goddess Amaterasu Ōmikami, rice, in our Empire, is a food that comes from a fine green plant. It was rice that the goddess had planted in her long and narrow heavenly rice fields; she gave it to the divine grandson when he descended to earth. We shall never forget this divine favor. Moreover, in order to ward off droughts and tempests, monarchs have continued, from generation to generation since Emperor Jinmu, to celebrate the offering of the new grain harvest to the heavenly and earthly gods, on the second day of the hare in the eleventh month. This has continued for three thousand years.[53]

Thus, at the same time that Meiji leaders led the nation into a new era, they also promoted rice as a kind of national grain linked to the imperial house and its divine ancestor. Ironically, Emperor Meiji himself enjoyed roasted sweet potatoes, according to his personal records. Indeed, according to Verschuer, rice never enjoyed such a high status in historical texts, making rice worship a nineteenth-century invention. Instead, it was frequently mentioned alongside other common grain foods constituting the "five grains" (*gokoku*), which included rice, barley, wheat, millet, and beans. Interestingly, the idea of the "five grains" was also present in ancient China, though the specifics of their makeup could vary.[54] However, when compared to China, Japan has a more elevated topography—mountains cover 70 percent of the country's land. Rice cultivation thus took up a small amount of land; in mountainous areas, swidden agriculture was a long-standing tradition that lasted well into modern times. Furthermore, while rice cultivation was extremely labor-intensive, Japanese farmers were also motivated over the centuries to plant other dryland crops to avoid being taxed by the government. In fact, rice was not only used as a currency, but its harvest was also taxed. Other food crops, on the other hand, were not subject to tax until the seventeenth century. Needless to say, all of this aided the introduction and spread of sweet potato as an alternative food plant in Japan, because, as Verschuer concludes from the study of one eighth-century case, "rice was not the main staple food of the Japanese population in the eighth century and probably never was, at any time during the medieval period."[55]

In her 1993 book *Rice as Self: Japanese Identities Through Time*, anthropologist Emiko Ohnuki-Tierney articulated her position on the status of rice in Japanese culture. While she acknowledges rice as a "dominant metaphor" for Japanese identity, she quickly adds that rice's cultural significance in Japan was not due to some notion that "rice was 'the food' to fill [the Japanese] stomach." Rather, drawing on work on postwar Japan by scholars such as Charlotte von Verschuer, Ohnuki-Tierney emphasizes that "rice has not always been quantitatively important to many Japanese." Instead, she discovers that there has always been a culture of "miscellaneous grains" (*zakkoku*), which includes wheat, buckwheat, barley, and a few millet varieties. This culture was gradually undermined by rice culture as the grain gained more cultivated acreage. However, from the Tokugawa period to the modern period, the two cultures coexisted in the main. Rice is the principal food for most Japanese ritual ceremonies and major holidays, such as the New Year's celebration, when it is common to see people making and eating rice cakes (*mochi*) for celebratory feasts. Meanwhile, in some areas, rice was actually prohibited from being consumed during the New Year. Instead, people were used to celebrating "taro/yam New Year" (*imo shgatsu*). In many areas, people are also accustomed to eating all "five grains" during important holidays.[56]

The long-established "miscellaneous grain culture," as discussed by Emiko Ohnuki-Tierney, and Charlotte von Verschuer's "five grains" concept each suggest that, while rice was/is a preferred grain food in Japan, it has never been the staple food among ordinary Japanese. Rather, most Japanese, particularly farmers, have grown accustomed to eating other grain, or non-rice, foods such as barley, millet, wheat, buckwheat, and, since the early seventeenth century, sweet and white potatoes.[57] An account left by a samurai living in Edo during the early nineteenth century bears witness to this tradition of mixed carbohydrate staples. The author, calling himself Buyō Inshi, described food consumption in both cities and the countryside during the period, among other things. While townspeople, including the "idlers" he despises, took rice for granted as a staple food, farmers were accustomed to eating rice and other grains together in their meals. In fact, he bemoaned, "Farmers, especially, are in a dire state. Wearing their rough padded cotton garb, they mingle with oxen and horses, handle excrement and manure, suffer cold and heat, lose their takings in taxes and expenses, lack sufficient food, and are unable to feed themselves even on miscellaneous grains. They truly suffer the worst kind of deprivations and toil throughout the year without respite."[58]

Needless to say, sweet potato was a prized food not only in hard times but also on a daily basis for these impoverished farmers. Meanwhile, sweet potato was prepared as a sweet to accompany tea ceremonies and formal banquets among Japanese aristocrats, thanks to the influence of *nanban ryōri*. According to Eric Rath, the practice of making sweet potato confections for a tea ceremony dates back to the first half of the seventeenth century. The daimyo of Satsuma hosted a multicourse banquet in the early eighteenth century at which he served a special after-meal dish that involved simmering the in soy sauce. It suggested that the Japanese had developed a wide range of ways to eat sweet potatoes. In the later years of the same century, a series of cookbooks known as "hundred tricks" became best sellers, and, lo and behold, *Imo hyakuchin* (One hundred tricks with sweet potato) was one of them.[59]

Sweet potato's popularity grew over the centuries, and it was eventually accepted as a common food throughout most of Japan. Robert Fortune, a renowned Scottish botanist who visited both China and Japan in the mid-nineteenth century and introduced a large number of plants from Asia to Europe, left behind a rather detailed research journal on the agricultural development of both countries during the nineteenth century. When Fortune arrived in Nagasaki, Kyushu, he noticed that sweet potato, buckwheat, and maize were grown in "the dry hill soil," while rice and taro were grown in the lowlands. As he traveled north, he noticed that sweet potato was a common summer crop planted in May in areas around Kanagawa, just south of Edo. From the standpoint of a botanist, he considered Japan's rainy season, in late spring and early summer, to be quite favorable to both rice and sweet potato. "The air becomes loaded with moisture and the rain comes down in torrents," he wrote. "Every hill stream is filled with water, and thus the means of irrigating the rice-fields are ready to the hands of husbandman." "Such excessive moisture," Fortune continued, "would have been fatal to the wheat and barley and rape, but it gives life and vigour to the paddies and sweet potatoes, and is necessary for their health and luxuriance." As such, he explained why sweet potato belonged in a group of dryland plants that Japanese farmers routinely grew in their fields.[60] Charlotte von Verschuer agrees that sweet potatoes and rice are comparable in terms of nutritional quality, adding that "the sweet potato *satsumaimo* was added to rice and the other cereals to sustain the demographic growth brought about by industrialization"—yet another justification for why the root was so well-liked in nineteenth-century Japan.[61]

Despite significant advances in sweet potato farming and consumption, there were notable regional differences in its weight in the respective food systems. Arizono Shōichirō has provided an inside look into the proportions of sweet potato consumption in relation to other foods in various regions after conducting a detailed survey of food consumption among ordinary people in premodern and modern Japan. His case studies look at food eaten by ordinary people in six prefectures in Ryukyu, Kyushu, central, and northeastern Honshu. Ryukyu and Kyushu are located in the Southwest, where sweet potato farming began. According to Basil Hall Chamberlain, an English anthropologist at Tokyo University, in his 1890 book *Things Japanese*, the root, "introduced as late as A.D. 1698, now forms the chief food of the common people."[62] Arizono concurs with Chamberlain's observation that, in the late nineteenth century, when it came to commoners' daily food, southwestern Japan belonged to the type of "sweet potato + grains" category, with the root as the main food, particularly in Ryukyu and Satsuma. In comparison, he claims that central Honshu belongs to the "rice + wheat + miscellaneous grains" category, while northeastern Honshu belongs to the "rice" and/or "rice + wheat" category. More specifically, fifty-three of the prefectures he studied were "rice" and/or "rice + wheat," sixty-one were "rice + miscellaneous grains," and twelve were "sweet potato + grains." When the latter two were added together, there were seventy-three prefectures where the common people did not eat rice alone, but rather rice in combination with a good proportion of "other grains" and sweet potato. According to Arizono, the average Japanese consumer consumed 52 percent rice, 28 percent wheat, barley, and rye, 6 percent millet, 9 percent miscellaneous grains, 3 percent sweet potato, 1 percent barnyard millet or Japanese millet (*Echinochloa crus-galli*), and 1 percent other foods. His findings are consistent with Shunsaku Nishikawa's study of "grain consumption" in both Tokugawa and Meiji Japan. Nishikawa states that rice's proportion as a staple in the Japanese diet in 1880 and 1886 was 53 percent and 51 percent, respectively, barley and wheat were 27 percent in both years, millet was between 13 and 14 percent, and sweet potato and other foods were 6 and 9 percent, respectively. These figures support the arguments of Emiko Ohnuki-Tierney and Charlotte von Verschuer that rice was not a staple food for the majority of people in modern Japan. In short, concludes Susan B. Hanley, "During both the Tokugawa and the Meiji period, the staple consumed varied by region. In the westernmost parts of Japan, people ate a higher proportion of *mugi* (wheat and barley) and sweet potatoes, while people in the mountainous areas ate more millet and *hie* (deccan grass). Only on the plains lying along the Japan Sea coast—Akita, Yamagata, Niigata, Toyama, Ishikawa, etc.—was rice the staple."[63]

As Japan's foodways are regionally diverse, so too are those of the Korean Peninsula. Rice, like rice in Japan, is said to have held a dominant position in the Korean diet in the past and present. However, the peninsula's hilly and mountainous terrain and climate are both unfavorable to rice cultivation. Indeed, if Japanese peasants in northern Honshu frequently faced harsh weather conditions for rice farming, Korean rice farmers presumably faced a similar challenge, given the peninsula's geographical proximity to Honshu. In other words, rice cultivation in Korea was historically concentrated in the South, whereas barley, wheat, and millet were more common in the North. As discussed previously, this geographical condition influenced sweet potato cultivation after the root appeared in eighteenth-century Korea. Sweet potato took root in South Korea as a subtropical plant, whereas white potato (called *gamjeo* or *bukgamjeo* in Korean, literally meaning "sweet potato" and "northern sweet potato") was quickly adopted as an alternative food plant in the North.[64] Sweet potato, however, seemed to have thrived in areas such as Jeolla Province, prompting Seo Yugu and other Chinese-influenced Korean scholars to recommend its cultivation as a useful food plant for the country. Indeed, historical records show that Jeolla and Gangjin, a county in southern Jeolla, were well-known for sweet potato farming in Korea as early as the early nineteenth century, or before Seo Yugu compiled his *Jongjeobo*. Because of its high yield, local farmers even turned over some wet land to grow the root in order to profit. However, due to geographical and climatic constraints, the root's impact as a food plant was more regional than national. Unlike in China or Japan, sweet potato was not a staple for many Koreans, and its cultivation did not contribute to the peninsula's population growth. Rather, the root plant was planted as a cash crop by its farmers and was traditionally regarded as a luxurious food by many across Korea for the majority of the nineteenth century.[65]

Sweet potato as the taste of war

In Japan, however, the situation was different. By the end of the nineteenth century, though not a daily staple for every Japanese, sweet potato had figured prominently among their preferred food choices. Sweet potato acreage in Japan reached 150,000 hectares in 1878, or a decade into the Meiji period, which was widely associated with the beginning of Japan's modernization, with a per unit yield of 560 kilograms. The total national output was 828,600 tons (751,693 tonnes). During the Meiji period in Japan, the root rapidly increased its importance in the country's agronomy as both food and feed; the annexation of Okinawa in 1879 may also have aided its advancement. The advancement was due to an obvious reason: After a period of stagnation or even decline throughout the majority of the eighteenth century, the country's population began to rise again in the second half of the nineteenth century. However, rice, though prized by the Japanese, accounted for only 63 percent of total crop value because it grew on approximately 53 percent of all arable land in the country. Other food crops, including sweet potato, accounted for 23 percent of total crop production, while industrial crops accounted for 12 percent. To keep up with the population growth, which increased by 45 percent between 1885 and 1920, efforts were made to improve rice unit yield. However, the outcome was less than stellar. According to one estimate, the national average rice yield in the 1870s was 1.6 *koku* per *tan* (0.245 acre). It had grown to only

1.9 *koku* by 1920, and the annual agricultural growth rate was only about 1 percent nationwide. One bright spot was that other food crops grew significantly more during the period.⁶⁶

Since the spread of sweet potato farming in Japan at the turn of the twentieth century, there has been increased interest in improving its cultivation method. In 1912, a farmer named Akazawa Nihei of Kawagoe, Saitama Prefecture, published *Gansho saibai ho* (Sweet potato cultivation method), which taught people how to grow sweet potatoes. His method was said to have evolved over the course of a half century of farming experience with the crop. As a result, Akazawa earned the nickname "Mr. Sweet Potato of Kawagoe" among the locals, who compared his contribution to Aoki Konyō's legendary promotion of the root plant in the eighteenth century. Other agriculture books published at the time also included sections on sweet potato farming. Meanwhile, foreign agricultural handbooks were translated into Japanese, expanding Japanese farmers' knowledge of the American plant.⁶⁷

Sweet potato received more attention in Meiji Japan because, as the country struggled to industrialize, the government put pressure on the agricultural sector to accumulate the necessary resources. The Meiji government introduced a new land tax shortly after "returning" power to the emperor in order to ensure and increase revenue. Although the rate was set at 3 percent of the assessed value of land, "the average landowning farmer had to market 30 percent of his produce in the 1870s to meet his tax obligations."⁶⁸ Worse, as previously stated, rice farming in Japan was frequently subject to weather fluctuations, particularly in the Northeast, where temperatures often dropped below normal. While the country avoided large-scale catastropes like the Great Kyōhō, Tenmei, and Tenpō Famines of the Tokugawa period, it still suffered from famines in the North for about twenty-four years during the Meiji period. When rice failed, peasants revolted, events that came to be known as "rice riots." For example, in the period 1882–1884 in Saitama, where Akazawa Nihei gained experience in sweet potato farming, raw silk prices dropped by 50 percent, followed by a severe crop failure. The Chichibu Uprising (named after a county in Saitama) followed. Although it was contained relatively quickly, the fallout from weather-induced crop failure was widespread and severe, resulting in nationwide starvation. In a word, the Japanese turned to sweet potato as an alternative source of food after improvements to rice production failed to make further progress. Mikiso Hane remarks that growing rice "was a labor-intensive undertaking not suited for large-scale mechanized farming. As a result, traditional means of production remained basically unchanged."[69]

The Meiji government used an aggressive foreign policy to appease and divert the anger of peasants during the modernization process. After taking over power, the government abolished feudal domains and forced the samurai class to give up their privileged sociopolitical status, allowing peasants to join the army and take up civil service positions for the first time. Because they bore most of the tax burden, Japanese peasants initially resented the government's introduction of national conscription. However, observes Mikiso Hane, "reluctance to serve in the army began to wane with the Sino-Japanese War of 1894–1895. The patriotic fervor and militarism that were fanned by the war, and the hero's welcome accorded to victorious soldiers, aroused enthusiasm for military service even among the peasantry."[70] Japan defeated China in the war, and was again victorious in the Russo-Japanese War of 1904–1905. In a nutshell, peasants turned soldiers fought alongside samurai turned officers to help transform Japan into a world power.

Aside from militaristic zeal, winning wars required rapid development of heavy industry and machinery. That is, victories came at a cost. It was noted that the national defense budget increased tenfold during the Russo-Japanese War compared to the previous Sino-Japanese War, which contributed greatly to the fast growth of the country's iron and steel industries. Indeed, "governments after the early Meiji era systematically used all the powers at their disposal to promote the growth of modern industry in what they saw as the national interest, even, if necessary, at the expense of the traditional sector," writes E. Sydney Crawcour in his contribution to *The Cambridge History of Japan*.[71] In other words, Japan's modernization came at the expense of the agricultural sector, which resulted in a lower living standard for most Japanese beginning in the early twentieth century, if not earlier. The Japanese diet at the time was high in carbohydrates, but most peasants could not afford rice, so they ate rice mixed with barley, sweet potatoes, millet, and barnyard grass. Meat was rarely eaten in the countryside, and in cities, it was considered a luxury by many.[72] Except in northeastern Honshu, where it was not widely grown, sweet potato was consumed as a main food throughout Japan during the Meiji era, according to Chūman Katsumi. It was a daily necessity not only in Okinawa, but also in Kyushu and Shikoku. And, as it had been during the

Tokugawa period, the root was regarded as a "valuable" (*kichō na*) food among residents of the Satsunan Islands north of Okinawa. In short, sweet potato occupied a visible place in the main food for the people in the Kantō and Tōkai regions of central Honshu, whose diets consisted of rice, barley, sweet potato, millet, and buckwheat. While pure rice was only consumed during holidays and festivals, sweet potato, whether steamed or roasted, dried or fresh, was consumed all year, either as a meal or a snack.[73]

All of this contributed to an increase in sweet potato production in Japan. Sweet potato acreage increased to 250,000 hectares nationwide in the late nineteenth century, a more than 60 percent growth from just a few decades before, due to rising demand. When the country welcomed the new Emperor Tashō in 1912, nationwide sweet potato production reached a record of over 400,000 tonnes. During the subsequent period, production averaged around 350,000 tonnes. Because of the country's industrialization, it became an important source of starch production in addition to its traditional use as food and fodder.[74] The outbreak of World War I proved to be a boon to Japan's modernization as well. As Western powers became engulfed in prolonged trench warfare, Japan quickly increased its industrial production, filling the void and increasing its global market share. "Between 1914 and 1918," notes Sydney Crawcour, "Japan's real gross national product rose by 40 percent, an average annual rate of nearly 9 percent." However, he points out that though export industry and investment were robust, "gains in personal consumption and average standards of living were much more modest." This was somewhat deliberate, as the Japanese government had long practiced a policy that "kept most Japanese living standards lower than they might have been."[75] As a result, the working class remained impoverished well into the 1930s. For example, after going to work in Osaka, a textile worker from Kyushu said that "what pleased me most was the fact that I could eat all I wanted. At home I used to eat a pasty mixture of yam [sweet potato?] and chestnut, and, instead of white rice, we had brown rice mixed with barley. . . . At home we got fish no more than once a year, on New Year's."[76]

In other words, while Japan's economy expanded significantly during World War I, it did little to improve the lives of ordinary people. Indeed, as the war came to an end, the government's efforts to increase Japan's share of the global market, including agricultural products such as raw silk export, resulted in high food price inflation. Large-scale rice riots erupted across the country in 1918 because of a sharp increase in rice prices, causing widespread discontent among urban residents. The government was forced to import cheaper rice and other staple foods from Korea and Taiwan, which were colonial territories. While Japan emerged as a new world power on par with its Western counterparts in the interwar years, the war boom was quickly followed by a damaging bust. The price of raw silk, for example, fell by more than half after World War I. The production of sweet potato starch, which had boomed during the war due to export demand, also experienced a significant decline, resulting in widespread bankruptcies. A devastating earthquake struck the country in 1923, exacerbating the country's economic woes.[77]

Poverty was widespread in the Japanese countryside during the 1930s in the aftermath of the Great Depression, which appeared to be a contributing reason for its military aggression overseas, leading to the outbreak of World War II in Asia in 1937. The 1934 famine, for example, dealt a severe blow to the lives of farmers. Peasants could only eat rice mixed with daikon (white radish), including the leaves, in the past. When the famine struck, rice and other grains such as millet and barnyard grass were no longer available. According to historical accounts, sweet potato significantly increased its presence in people's diets as they sought to ensure their survival. One report stated that hungry villagers had to scrounge "in the hills for edible plants, roots, and tree bark." The other described that they cooked "unhulled rice mixed with pumpkins and [sweet?] potatoes." There was yet another complaining that "we in the farm village have been eating dried [sweet?] potatoes and herring dregs, which are used for fertilizer, day after day, and have been on the verge of starvation."[78] The widespread misery in Japan's countryside, on the other hand, did not deter its government's expansionist foreign policy. According to Lizzie Collingham, author of *The Taste of War*, "militarists and nationalist groups exploited the agrarian crisis to press for policies which pushed Japan towards war."[79]

Sweet potato was not only a famine food for Japan's rural population during the interwar period; it also helped the growing working population in cities. To be sure, there was an obvious disparity between the urban and rural in Japan's modernization process. "By 1929," writes Collingham, "city dwellers were eating at least 25 percent more rice than those living in the countryside."[80] As previously stated, this helped to explain why a textile worker from Kyushu was content with her life in Osaka. However, textile workers in cities did not fare much better. A weekly menu from a dining hall in a factory staffed by female workers revealed that while their three meals each day included rice, what went with the grain was mostly different types of pickles, tofu, beans, and daikon, with no meat or even fish. To improve the nutritional level of the working-class diet, a textile factory owner named Ōhara Magozaburō invented a new wheat flour bun to replace the carbohydrate for his workers. It was called *Rōken mantō* (lit. labor study bun), and it was inspired by and modified the steamed buns that workers in North China used to eat as a staple food. The skin of the bun was made from leavened wheat flour, and it was wrapped around various fillings to improve its taste and nutrition. Sweet potato paste, along with soybean and red bean paste, was a favorite filling of the *Rōken mantō*. The bun became extremely popular in the mid-1930s, consumed not only by factory workers but also by students and soldiers in Japan and its then-colony of Korea.[81]

If sweet potato was used to improve the diet of the urban population in pre-1937 Japan, it became a daily necessity for all Japanese after the war began and into the 1940s. Several government policies had a negative impact on the country's food supply. For example, Japan issued an order of general mobilization in 1938, which was followed by an order of national conscription the ensuing year, resulting in a labor shortage throughout the countryside. Furthermore, beginning in 1940, Japanese peasants were ordered to provide rice for the country at their own expense, which significantly worsened their living conditions. But city dwellers fared no better: Food had already been rationed across the country before Japan launched its surprise attack on Pearl Harbor in 1941. More specifically, in January 1941, the government strictly regulated rice sales. In April, major cities like Tokyo, Osaka, Nagoya, Kyoto, and Kobe required residents to use a special coupon to eat out. Then, in May, they could only eat meat once or twice a month, and beginning in July, people had to wait in long lines in order to buy vegetables. At the same time, sweet potatoes were added to the daily food portion distributed to residents due to a severe rice shortage.[82]

After Japan went to war with the United States in December 1941, sweet potato became a staple food for most Japanese—"sweet potato bread" (*imo pan*) was created to reduce rice consumption and prevent starvation. Apple orchards and even rice paddies were converted to sweet potato fields to meet the high demand. "We grew sweet potato in all the rice fields . . . as a winter crop," recalled a woman from Shinohata village, north of Tokyo. "You cut them in strips and dried them and that was all we had for snacks."

The reason was simple: The root plant contains "30 percent more calories than rice and double the number of calories found in wheat." And caloric intake was indeed what most Japanese desperately needed in wartime. "The standard daily Japanese caloric intake was 2,200 calories [on average], including 70 grams of protein. But the actual daily intake in 1941 was 2,105 calories and 64.7 grams of protein. By January 1944, rations provided a mere 1,405 calories."[83] In the 1940s, the government launched nationwide campaigns to increase sweet potato production in order to combat the deficit. Suzuki Teiichi, a lieutenant general in the Japanese army in charge of wartime economy, addressed the dire need for ensuring the food supply in his statement as president of the Planning Board: "concerning rice, I think it will be necessary to consider substitute food, such as soybeans, minor cereals, and sweet potatoes."[84] Indeed, the root was planted not only in rice fields, such as in the Kantō Plain village of Shinohata, but in virtually every space people could find, including residential areas, gardens, parks, sports fields, school grounds, factory lots, deserted places, dams, golf courses, race fields, and even roadsides.[85]

The war-induced starvation affected not only Japan but Korea as well. Japan took control of the Korean Peninsula after defeating China in the Sino-Japanese War of 1894–1895. It deposed the Yi family of the Joseon dynasty (1392–1897/1910), which had ruled Korea for five centuries, and made the country its colony in 1910. Sweet potato farming became more important in Korean agriculture and diet during Japan's rule of Korea, thanks to the Japanese government's promotion of the plant. The root, for example, was used as a relief food in the spring season after farmers harvested the barley planted the previous year and waited for the new one to ripen. Sweet potato became a staple food for many Koreans after World War II broke out in Asia. By the 1940s, "it was impossible to imagine the diet without it," writes Felix Siegmund. This was in stark contrast to the previous century, when the root was mostly regarded as a regional specialty in Korea.[86]

Sweet potato starch was also in high demand during the war. Thanks to local government incentives, over a hundred factories popped up in Kagoshima, which was traditionally known as an important sweet potato–producing area, after 1937. And, after the United States and other Allied countries imposed oil restrictions to hinder the Japanese war effort, these factories began producing industrial alcohol as well, because the technology was nearly identical. A similar phenomenon was observed across the country, not just in Kagoshima. After the war, sweet potato starch factories sprang up not only in southern and southwestern Japan, in such places as Nagasaki and Shikoku, but also in central Honshu. By 1944, the number of starch factories in Chiba Prefecture, which is located near Tokyo, had increased to sixty-five. Many of these factories were also converted to produce industrial alcohol, which could be used to replace petroleum, thus alleviating Japan's oil shortage.[87]

It is not surprising, therefore, that sweet potato production reached new heights in wartime Japan. Since the root had become well established in Japan as a valuable famine food, its production had steadily increased since the turn of the century. To combat the 1918 rice shortage, for example, sweet potato acreage increased to over 300,000 hectares nationwide, with a total production of more than 4 million tonnes during the 1910s. During World War II, production increased again and easily surpassed the previous record. In 1943, the country's acreage reached 325,400 hectares, thanks in part to the government's promotion campaign, producing 4,540,000 tonnes of sweet potatoes in total. Two years later, as Japan surrendered to the United States and China, sweet potato acreage increased to 400,200 hectares, a new record. This also marked the culmination of the country's three-century-long sweet potato farming tradition. "By the end of the war," comments food historian Ishige Naomichi, "rice distribution was so meagre that there was actually stronger traffic in sweet potatoes. As in the seventeenth century when its introduction led to population growth in western Japan, the sweet potato came to the rescue once more."[88] That sweet potato became a popular substitute for rice as a daily starch was also etched in people's minds. Takezawa Shoji, then a hungry young girl looking for frogs to catch for food, recalled hearing the Japanese emperor announce the country's surrender on the radio. "His voice sounded as weak as I felt," she explained. "I wondered if the emperor, like us, had not yet eaten his lunch." She was thinking about "pumpkins, potatoes, and sweet-potato-flour cakes" as she listened.[89]

The year 1945 marked a watershed moment in Japanese sweet potato farming, as the crop's sown area began a gradual and steady decline thereafter. By 1960, for example, the cultivated area had fallen to 329,800 hectares—representing a 21 percent decrease from the 1945 figure. However, total national production was weighted at 6,277,000 tonnes, exceeding the 3,897,00 tonnes of 1945 by more than 61 percent.[90] Why? The reason for the increase in total national sweet potato production was that as soon as Japan began its aggression against China, it started making efforts to develop new cultigens for mass production. Okinawa, which had a longer history of sweet potato farming than the rest of the Japanese Empire, was naturally chosen for the experiment. The new cultigen "Okinawa 100" was successfully hybridized in 1939, followed by "Agricultural Forest 1" (Nōrin 1) and "Agricultural Forest 2" (Nōrin 2) in 1940. (I will go into more detail about their cultivation and impact when I trace the history of the sweet potato in Okinawa in the next chapter.) These new cultivars were known for their quick ripening period and high per unit yield, which were critical for sustaining the survival of a starving population both during and after the war. Yamakawa Osamu, author of a recent Japanese history of the sweet potato, is convinced that "without Okinawa 100, a large number of Japanese would have died from hunger." According to her calculations, the sown area of Okinawa 100 reached 80,000 hectares, or 20 percent of Japan's total sweet potato acreage in 1946.[91]

In other words, despite a decline sown area, sweet potato farming advanced in terms of total production in postwar Japan. For example, in 1950, the country produced 6,290,000 tonnes of the root. Five years later, in 1955, it produced a record-breaking total of 7,180,000 tonnes. This was due to the fact that sweet potato was a daily staple for most Japanese in the decade following Japan's defeat. Even though rice, wheat, and other food materials were being exported from the United States, the Japanese still relied on sweet potatoes as their daily staple, according to Chūman Katsumi. It was not until 1955 that the Japanese began to lose their reliance on the sweet potato.[92]

Sweet potato in present-day Japan and Korea

Starting in the mid-1950s and especially the 1960s, sweet potato lost its standing as the "main food" (*shushoku*) among most Japanese, and the root has never regained its former status to this day. According to Vaclav Smil and Kazuhiko Kobayashi, this shift has coincided with the country's dietary transition over the last century. According to their research, "rice production was only about three times the amount of harvested barley" in 1900, when sweet potato farming entered a period of ascendant growth. However, a century later, in 2010, "it reached roughly 40:1." Pork consumption experienced the same rate of change. In 1900, the country produced only

30 grams of pork per capita per year, implying that most people never ate pork, whereas in the early twenty-first century, the amount increased to 20 kilograms per year, with the country also importing 45 kilograms of pork per capita. In contrast to the rapid growth of rice and pork as evidence of improved living conditions, sweet potato production has experienced a significant decline. The national production of the root plant, for example, was 2.8 million tonnes in 1900. It increased to 5.5 million tonnes in 1946 and 7.2 million tonnes in 1955. In stark contrast, the figure hovered around 1 million tonnes during the first decade of the twenty-first century.[93] These figures are consistent with those mentioned previously.

Jennifer Woolfe's *Sweet Potato: An Untapped Food Resource* provides more specific statistics to illustrate the decline: It was primarily due to Japanese farmers who grew the plant consuming significantly less of the root than they did in the first half of the twentieth century. For example, in the 1930s, approximately 70 percent of the roots harvested were consumed on the farms where they were harvested. However, sweet potato farmers consumed only about 10 percent of the crop several decades later. Woolfe states that between 1955 and 1984, "the percentage of total production consumed on producer farms declined from about 27% to only 6%." During the same period, sweet potato consumption fell by nearly 600 percent nationwide, from 27 kilograms to 4.3 kilograms per capita per year. Regardless, sweet potato is still widely available in grocery stores across the country, and the root is increasingly being consumed as a vegetable. Indeed, the percentage sold in markets as fresh vegetables and for food processing increased from 8 percent to 35 percent between 1955 and 1984.[94]

While the sweet potato has declined as a staple food, the root remains a popular snack (*kanshoku*) among many in modern Japan. Writer Mock Joya describes how eating sweet potato has become ingrained in people's daily lives in his well-received *Japan and Things Japanese*. Among the "things Japanese" Joya introduces to his readers is *irori*, or hearth, which he describes as a traditional heating system installed in Japanese homes of all classes. It is a six-by-ten-foot rectangular hole in the center of the family room. Bordered by stone and wood boards, it is used to burn wood to warm the room and boil water for tea. As a result, in addition to the kitchen, people can cook something on the *irori*. "On cold evenings," Joya writes, "children love to place sweet potatoes in the hot ash of the *irori*, and eat them when properly baked, or warm their *mochi* over the fire. They eat their meals around the fire, where they also receive callers. Sometimes they sleep in the futon spread around the *irori*, so the *irori* room becomes the living room, the dining room, and the drawing room for the whole family in the winter season."[95]

In fact, eating roasted sweet potatoes during the winter has long been a popular pastime in Japan. The author Yamada Syoji, mentioned earlier, provides the following sentimental and nostalgic account of the early development of sweet potato farming in Japan: "I have heard that some people don't eat sweet potatoes at all these days. Some claim that in the postwar period, they only ate sweet potatoes, which made them sick of them. On the contrary, I will never forget the flavor of sweet potatoes that my mother warmed over the fire in the morning. I am still eating them for breakfast. The only difference is that in high school, I used to eat several large sweet potatoes for breakfast. I am now already full after just one."[96]

Roasted sweet potatoes are a popular winter food among Japanese because, if they don't have them at home, as Yamada did when he was younger, they can easily buy them from street vendors. Itō Shōji's book on the sweet potato contains several stories about how the root appears in the lives of ordinary Japanese people. One of them describes Yokoo Terukazu, who sells roasted sweet potato, or *yaki-imo*, on the streets of Chiba, a Tokyo suburb. Yokoo, who was born into a peasant family in 1940, had worked in a department store in Tokyo after graduating from high school. After marrying in 1965, he turned to selling roast sweet potatoes in the hopes of providing a better life for him and his wife. For the first three years, he rented a hand cart from a merchant, burning mostly wood scraps found along the roadside. Many of his customers bought roasted sweet potatoes from him after drinking at pubs late at night. Yokoo switched to a smaller truck and expanded his routes. He worked all year, not even taking time off for the holidays. "These were such great days of rapid economic development," Yokoo recalled. However, after four decades of running his business, he discovered in 2010 that few people wanted to take it over, including his two sons.[97]

Yokoo Terukazu's experience is not unique; it demonstrates the shifting fate of sweet potato as a food in Japanese households. However, *yaki-imo* is not going away; rather, as reported in a recent article titled "The Endurance of Japan's Simple Street Snack," it is still popular in Japan during the winter season. "'*Yaki-imo* . . . ' The forlorn cry of the roasted sweet potato vendor," the article begins, "echoed through the canyons of concrete and tiled buildings in a Tokyo suburb. The pre-recorded song, bookended with spoken claims of '*oishii, oishii*' (delicious, delicious), flowed from speakers on a stubby flatbed *kei* truck. This small vehicle, a ubiquitous part of working-class Japan, had been converted into a vessel for *idōhanbai* (literally, mobile sales)." It describes in detail the truck used by Yokoo Terukazu and others to sell *yaki-imo* to customers. The article acknowledges that these trucks, which have an oven and an awning and are often painted with colorful advertisements, are becoming increasingly rare in modern Japan. Yet people still enjoy sweet potatoes and frequently purchase them at convenience stores and supermarkets. Indeed, roasted sweet potato was a popular fast food for the common people in Japan from the early twentieth century until the 1970s, when American-style snack foods and fast-food restaurant chains eroded its popularity. To many Japanese, however, "*yaki-imo* is and will always be a heart-warming treat that holds many fond memories," as a British journalist puts it.[98]

Perhaps it is for this reason that, despite fewer vendors on the streets of Japanese cities, roasted sweet potato remains a popular form of sweet potato consumption, albeit through different venues. Barry Duell has investigated the role of sweet potato in the Japanese diet. Tokyo International University, where Professor Duell taught for many years, is located in Kawagoe, Saitama Prefecture, which was one of the first places in the Kantō region where the root plant was cultivated. Duell has been able to witness the ebb and flow of sweet potato consumption by the local population as well as the entire country from this location. He observes that when sweet potato was first introduced to the country, it provided a reliable source of food for Japanese farmers. Yet as early as the second half of the eighteenth century, roasted sweet potatoes became a favored winter snack for urbanites in Edo, "especially among merchants and common people. Laborers and students also liked them because they provided a cheap way for them to fill their bellies." Over time, Kawagoe established itself as a key supplier to Edo's sweet potato market.[99]

According to Duell, there are two types of *yaki-imo*. The first is *ishi yaki-imo*, or stone-roasted sweet potato, which refers to the hot pebbles used to heat the root inside the hand-pulled cart or light truck. The other is *tsubo yaki-imo*, or urn-roasted sweet potato—*tsubo yaki-imo* is frequently purchased at a shop where "sweet potatoes are connected with wire hooks to form a chain, then suspended into an urn heated with coke."

Duell acknowledges that both types of *yaki-imo* are becoming less common—in Kawagoe, for example, there is only one *tsubo yaki-imo* shop left. But they are by no means going away. Like the author of the online article cited above, Duell also notices that a modern form of *yaki-imo* appears in Tokyo, which is roasted by a propane-fueled stove in a decorative van that cruises around the city streets. In addition to substituting a van for the traditional cart, he writes, "modern touches include delivery service to customers who call in orders, as well as fashionably designed boxes for delivering the sweet potatoes in English newspapers." And the price is high, he adds.[100]

In Korea, roasted sweet potato is also a popular snack food. During the postwar years, as the country struggled to recover and modernize, South Korea experienced food shortages, particularly during the 1960s and typically in the spring season following the harvest of winter barley. Sweet potato became a popular carbohydrate that helped people avoid starvation. Sweet potato production in Korea began to decline in the 1970s, a decade or so later than in Japan, as food supplies improved. Sweet potato sown area in South Korea was over 111,229 hectares in 1971, but this has since decreased by nearly 90 percent to the current level of 14,000 hectares. However, it appears that rice has lost its status as the most desired grain food since then, with wheat flour and sweet potato foods increasingly gaining popularity among both older and younger generations.[101] Sweet potato barrel roasters can be found on many street corners in Korean cities from early autumn to winter, selling *gun gogkuma*, or roasted sweet potato, to pedestrians. The steaming sweet potatoes are wrapped in newspaper and consumed on the go. Two sweet potato varieties are frequently chosen by street vendors because they are popular with their customers. The chestnut sweet potato, or *bam goguma*, has a dry texture and yellow flesh. It tastes like chestnut when cooked. *Mul goguma*, or water sweet potato, is the other. It has a higher water content and a softer texture. While both can be roasted, the former is also excellent for steaming. As is the case elsewhere, it appears that the most common ways for Koreans to consume sweet potato roots are roasted or steamed. However, due to the country's cold winters, Koreans are especially fond of roasted sweet potatoes. *Gun goguma*'s popularity in recent years has resulted in the creation of *gun goguma* ice cream and smoothies for people to enjoy during the summer.

Despite their popularity, roasted sweet potatoes are not the only type of sweet potatoes available in Japan and Korea. Throughout the year, there are a variety of ways to consume the root plant at home and in restaurants. Sweet potato tempura is quite common in Japan; it can be found not only in home kitchens but also at street tempura stalls. The other is candied sweet potato, also known as *daigaku imo* (lit. university sweet potatoes), which is commonly found near college campuses. Barry Duell provides a detailed description of how sweet potato tempura is prepared at home and on the street: Slices of sweet potato are "dipped in a batter of flour, egg, and water, then deep-fat fried until golden brown. To eat, the tempura is dipped in a soy-sauce-based sauce to which grated daikon radish has been added, then eaten together with rice."[102] *Daigaku imo* is more of a street food. It is said to have originated in the 1930s and was popular among students at Tokyo University, Waseda University, and other institutions of higher learning. It is made with molasses and deep-fried sweet potatoes. It has always appealed to young and hungry college students as a sweetened starchy food, and still does today.[103]

Other types of sweet potato confection and snack are popular in Japan, often accompanying the tea-drinking tradition. *Chakin shibori*, which literally translates to "tea towel squeeze," is a popular dessert. It's made with cooked sweet potatoes and sugar. The mashed sweet potato is placed in the center of a moist dish cloth and squeezed to leave a decorative imprint on the ball, often in light green, resembling the color of green tea, thus pleasing the eyes of tea drinkers who consume it at home or in a tea house. The other popular confection is simply known as *suito poteto*, which is the Japanese transliteration of "sweet potato." However, it is a Japanese invention in the sense that it is a sweet bun in the shape of a sweet potato. Its typical recipe includes the following steps: (1) Combine the mashed sweet potatoes, sugar, milk, unsalted butter, flavorings (vanilla or cinnamon), and liquor to taste; (2) form the mashed sweet potatoes into an oval shape and place on a small aluminum foil plate or press out with a star-shaped squeeze bag; and (3) brush the surface with egg yolk so that, once baked, the bun assumes a golden-brow color.[104]

Sweet potato bread is a popular confection in Korea that resembles Japanese *suito poteto*. It has a purple or reddish exterior, similar to sweet potato. Instead of baking sweet potato mash into a bun, sweet potato bread in Korea uses it as a filling wrapped in a thin, light, and fluffy dough. *Goguma mattang*, a Korean version of candied sweet potato, is another popular sweet potato confection in Korea. It's made with deep-fried sweet potatoes that have been soaked in a thick caramelized sauce of oil and sugar. *Goguma mattang* is a favorite dessert and snack for many Koreans.[105]

In addition to being used directly in foods, the root has long been employed as a material and ingredient in the processing of other foods and beverages in both Japan and Korea. Sweet potato starch, as seen in China, is used to make *shōchū*, a type of liquor, in Japan. Over the past century, Japanese agriculturalists have been constantly searching for and developing new cultivars for the specific purpose of producing high-quality *shōchū*, which requires a sweet potato variety with a high starch content.[106] In recent years, there have also been brands of *soju*, a Korean distilled alcohol, made from sweet potato grown in South Korea. "The content of starch and other fermentable carbohydrates," notes Jennifer Woolfe, "in sweet potatoes also make them a suitable raw material for the production of alcoholic beverages, which takes place in such countries as Japan and Korea. About 5% of total production is used for making the distilled spirit *shochu* in Japan."[107] Sweet potato lattes, sweet potato yogurt, sweet potato pizza, sweet potato pancakes, sweet potato *yōkan* (a Japanese-style sweet that involves pressing mashed sweet potatoes in a container and then cutting them into cubes before serving), and sweet potato cheese sticks are examples of more creative ways to use sweet potato as an ingredient.

This chapter began with a mention of *dangmyeon*, a sweet potato noodle found in Korean *japchae* dishes. Sweet potato starch has long been used to make noodles in East and Southeast Asia, from which many different varieties have evolved. Because dangmyeon is the most important part of the dish, japchae—a royal delicacy in pre-twentieth-century Korea—reflected the high and exotic status once held by the root in Korean cuisine. *Japchae* still retains some of its "royal" flavor in modern Korea; it is a savory party dish prepared for weddings, birthdays, holidays, and other festive occasions. The way to make it is to combine *dangmyeon* (i.e., glass noodles, which when cooked are chewy, clear, and plump) with various vegetables and then seasoning the mixture with soy paste, sugar, and sesame oil. Furthermore, *dangmyeon* can be found in a variety of other Korean dishes.[108] In comparison, observes Yamakawa Osamu, a Japanese expert on processed sweet potato foods, cellophane or glass noodles have not played a significant role in the use of sweet potato starch in her country. However, in recent years, the root starch has been used in making Japanese udon noodles, a popular dish found throughout the island nation. Yamakawa recommends using 80 percent wheat and 20 percent sweet potato starch for noodles like udon, which are more elastic and chewier than wheat-only noodles.[109] All in all, while sweet potato production has declined in Japan and Korea over the last half century, numerous methods have been developed to produce a wide range of sweet potato foods that satisfy people's taste buds. Indeed, as will be discussed in the epilogue, these novel efforts have come to mark a new phase in the changing fate of the root plant today, not only in Japan and Korea, but also around the world.

CHAPTER V

Sweet Potato Islands, Sweet Potato Peoples

In a short essay titled "Sweet Potato," Lung Ying-tai, a well-known Taiwanese essayist and cultural critic, recalled meeting a Swiss woman at a market in Switzerland. Lung was educated in Taiwan and the United States, and she spent much of her life abroad. Her writings frequently reflected her experiences living in both the United States and Europe. A new mother at the time, Lung went to the market that day to buy some vegetables. She saw a bunch of sweet potatoes and decided to cook some for her ten-month-old son. As soon as she picked one up, the Swiss woman next to her inquired, "How did you cook sweet potatoes?" Before she could respond, the woman admitted, "I only knew how to boil them, but you Asians might have a better way to dress them." The Swiss woman assumed Lung was a sweet potato eater and knew a lot more about the root as a food.[1]

Perhaps to the woman's disappointment, Lung admitted that she, too, intended to boil the sweet potatoes after bringing them home that day. But the woman's assumption was not unfounded; sweet potato had long been an "Asian crop" by the late twentieth century. As mentioned in chapter 3, the root had become a staple food among Taiwanese as early as the eighteenth century. As such, Lung is far from the only Taiwanese writer to reflect on the sweet potato and describe its significance in her life and memory. Despite her embarrassment at not knowing more ways to prepare the sweet potato, Lung expressed excitement in her essay at the fact that her newborn son would be able to taste the root for the first time in his life. Other Taiwanese writers have recalled the notable role that sweet potato played in their lives, both fondly and sadly (for reasons that will be explained later). Indeed, Taiwanese authors have written numerous essays and novels with the word "sweet potato" in the title. Indeed, the sweet potato has frequently been used as a symbol for Taiwanese cultural identity—many islanders proudly refer to themselves as "sweet potato folks" (*Han-tsi-a-kiann in Hokkien*, which literally means "small sweet potatoes" or "children of the sweet potato"), including some who, like Lung Ying-tai, were not even born on the island. Lin Haiying, an accomplished twentieth-century Chinese writer, was one such example. Lin was born in Osaka, Japan, to two Taiwanese parents, and spent her childhood in Beijing after relocating there when she was young; it was not until 1948, on the eve of the Communist takeover in China, that she moved to Taiwan with her husband and spent the rest of her life there. Lin rose to literary prominence after publishing *Chengnan jiushi* (My memories of old Beijing) in 1960, a memoir of her childhood in the city. Without a doubt, her time in Beijing left an indelible mark on her literary endeavors. Nonetheless, Lin recalled in one of her essays how she and other Taiwanese in Beijing identified themselves as "sweet potato folks."[2]

Another example was Chang Kwang-chih. This former Harvard anthropology professor was born in Beijing in 1931, and his father was a Taiwanese student before becoming a writer and Peking University professor. Chang returned to Taiwan with his family when he was in his teens, and after graduating from Taiwan University, he earned his advanced degree and began his career at Harvard. Chang, like Lin Haiying, valued his cultural identity as a "sweet potato man" and titled his autobiography *Fanshuren de gushi* (Story of a sweet potato man).[3] In her book, Lin Haiying offers an explanation: "Since Taiwan is shaped like a sweet potato, it is customary to call [people] from the island 'sweet potato folks.'"[4] Interestingly, then, Taiwanese people identify as "sweet potato folks" not only because the root was a staple food for them, but also because the island they live on looks like a sweet potato! As amusing as this notion may be, this chapter argues that cultural identification with the sweet potato is not limited to Taiwan; similar cases can be found in Ryukyu/Okinawa, Papua New Guinea, and Hawaii. To some extent, given the variety of physical shapes the root is capable of assuming, the outlines of these countries/regions can indeed be compared to that of a sweet potato.

Figure 5.1 Map of Taiwan, which resembles the shape of a sweet potato. Many residents of Taiwan refer to themselves as "sweet potato folks," symbolizing the root's significance in the island's culture and history.

But a more important point to consider here is that, like Taiwan, sweet potato has long been a staple of these islanders' diet, and as such, it has become woven into the fabric of their agronomy, culture, history, and religion. In a nutshell, these four are case studies of the sweet potato's cultural memory in our world.

Sweet potato and Taiwan's bittersweet cultural identity

Let us start with Taiwan. The island is not only shaped like a sweet potato, but its subtropical climate and hilly terrain are also ideal for the root to thrive. Taiwanese call themselves "children of the sweet potato," according to Cai Chenghao and Yang Yunping, authors of *Taiwan fanshu wenhuazhi* (Cultural annals of the sweet potato in Taiwan), because from the turn of the twentieth century, when the island was colonized by Japan, until the 1970s, sweet potato was the number one food crop sown on the island. This is quite unusual in modern world history, they explain, because while sweet potato was widely grown in mainland China and the Philippines, it was one of several main food crops in their agronomies[5]

Can and Yang's comparison of sweet potato cultivation in Taiwan to China and the Philippines is relevant and reasonable because these are the two most likely sources of the island's sweet potato. There were various theories as to how the American root plant arrived on the island. Because Taiwan is on the outskirts of Oceania and its original inhabitants are from the same Austronesian group that colonized the vast region, some speculated that the root was brought by them from nearby Micronesia or Melanesia. The geographical location of the island also contributed to theories about the root plant being brought by Spanish or Portuguese sailors as they explored South and Southeast Asia. However, China and the Philippines remain the most likely places of origin for sweet potato on the island. Chinese fishermen discovered Taiwan very early on, and attempts to colonize the island by Chinese emigrants from South China, especially those from Fujian, began in the mid-sixteenth century. It was entirely possible that after sweet potato cultivation began in Fujian, Fujianese who crossed the Taiwan Strait and settled in Taiwan introduced the root to the island. Indeed, the root plant was first mentioned in Taiwan in Chen Di's *Dongfanji* (Records of eastern barbarians) in 1603. Chen visited Taiwan for twenty days while following a Chinese army chasing pirates out of the region. Despite its brevity, his *Dongfanji* was one of the first Chinese-language accounts of Taiwan's native population.

Another possibility is that it originally came from the Philippines or nearby Southeast Asian countries; perhaps the fact that a specific variety of sweet potato cultivated in Taiwan was named "Brunei sweet potato" is indicative—Brunei is about eight hundred miles south of the Philippines. If sweet potato could arrive in Taiwan from Brunei, it could also come from the Philippines, where sweet potato has been a major food plant since the sixteenth century. Furthermore, Taiwan is located between China and the Philippines; if sweet potato spread from the Philippines to South China, it may also have spread to Taiwan.

Though Chinese fishermen, farmers, and traders from Fujian were among the first nonindigenous people to settle on the island, Taiwan was not directly ruled by Chinese regimes until the late seventeenth century. Prior to Chinese rule, or Qing China's rule, the island was a place where European powers fought for control and colonization, albeit only partially. From the early seventeenth century, the Dutch East India Company established a trading port in Tainan, on the southern part of the island. The Spanish Empire occupied Keelung, in northern Taiwan, around the same time. In the decades that followed, the two powers clashed on the island, with the Dutch eventually triumphing over the Spaniards. During the same period, the island also saw an increase in the number of Chinese immigrants from Fujian. According to Dutch documents, sweet potato was one of the plants grown on the island by both the Chinese and Indigenous peoples. Both Chen Di's journal and Dutch records indicate that the introduction of the American root plant to Taiwan occurred concurrently with that of South China.[6]

A historical event prompted widespread sweet potato cultivation on the island. The seventeenth century was a turbulent time in East Asia, with not only the Dutch and Spaniards clashing in Taiwan, but also, as discussed in chapter 3, the dynastic transition on mainland China, which was much larger in scale and bloodier in process. The Ming dynasty, during which Chen Di lived, was overthrown by a peasant rebellion brought on by widespread famine; in its aftermath, Manchus invaded from the Northeast and defeated the rebels near Beijing, the Ming capital. The Manchus then continued their march across the rest of China after establishing the Qing dynasty in 1644. However, the remaining Ming forces resisted the conquest until the end of the century. These forces, known as the Southern Ming

(1644–1662), established bases in South and Southeast China, and one of them was led by Zheng Chenggong, also known as Koxinga (1624–1662), who aided Prince Tang (1602–1646) and later Prince Lu (1618–1662), both descendants of the former Ming court and now resistance regime leaders.

Despite Zheng's backing, the Southern Ming were unable to repel the Manchu army on the mainland. Zheng decided to retreat to Taiwan in 1661 and use it as a new base for his anti-Qing campaign. In doing so, he waged a war against the Dutch colonialists, eventually ending Dutch rule on the island. But, along the way, Zheng and his army ran into a serious food shortage. After arriving on the island, they had to collect food from the locals, of which sweet potatoes had already made up a sizable portion. As the army's war with the Dutch force dragged on, they needed to grow their own food; sweet potato was an obvious choice because it ripened much faster than other food plants like rice. It was said that Zheng Chenggong's army grew sweet potatoes all over the island in 1661 and 1662. Their efforts were rewarded in the end: After securing their food supply, they were able to force the Dutch to surrender, and ultimately to expel them from the island.

While sweet potato helped Zheng Chenggong defeat the Dutch, the story does not end there. Zheng died not long after achieving success. When he ordered his army to grow the root, he did not want them to compete with the locals for land; the soldiers thus grew sweet potato in areas not usually deemed suitable for farming. Zheng's son inherited power after his father died. Recognizing that he would be unable to return to the mainland with the army, the new ruler encouraged more food plant cultivation on the island. Meanwhile, following the fall of the Southern Ming to the Qing, many Chinese fled to Taiwan. Sweet potato, a drought-resistant crop that grows in almost any type of soil, in both flatland and hills, naturally became popular among newcomers to the island. In a word, thanks to the Ming–Qing dynastic transition, sweet potato cultivation in Taiwan not only continued but entered a new phase of development.

Taiwan gradually became a major sweet potato production area in Asia beginning in the late seventeenth century. Unlike in other parts of the world, where the root was regarded primarily as an insurance crop, farmers in Taiwan valued sweet potato not only as a critical source of food but also as a cash crop. These uprooted farmers were more eager to secure financial success than their counterparts elsewhere because most of them were new immigrants from the mainland. They saw sweet potatoes as a commodity they could sell back to the mainland for cash or in exchange for other goods because they grew well on the island. Some of these activities, however, were viewed as smuggling by the Qing government at the time, as Cai Chenghao and Yang Yunping document in their *Taiwan fanshu wenhuazhi*. Nonetheless, sweet potato exports from Taiwan to China, including Fujian, where the root was also widely grown, appeared to be common. Aside from sweet potato, another commodity that grew well in Taiwan was rice. According to Cai and Yang, during the eighteenth and nineteenth centuries the price of rice was often higher across the Taiwan Strait in Fujian due to the province's unfavorable hilly topography. Taiwanese farmers would thus profitably export rice while consuming only sweet potatoes. All of these factors contributed to the island's continued success in sweet potato cultivation. Meanwhile, in Taiwanese society, whether one eats rice or sweet potato has become an indication of social stratification.[7]

Taiwanese farmers also learned how to process sweet potato roots into strips for export (*fanshu qian*). The reason and method are straightforward, as already discussed in the previous chapter: Sweet potatoes have a high water content per unit, making them difficult to store, let alone transport over long distances. Sun-drying is a common method for reducing their water content. However, drying them whole takes much longer. At the time, the islanders began threading sweet potato roots into strips with a grater and drying them in the sun. The sweet potato strips would lose about one-third of their water content or weight in a couple of days, making them much easier to transport. Aside from selling them outside of Taiwan, the islanders cooked dried sweet potato strips with rice for their own consumption or fed the strips to hogs as forage. Indeed, as many memoirs and literary accounts show, eating rice or rice porridge cooked with sweet potato strips was such a common experience that it became etched into the memory of generations of Taiwanese growing up on the island.[8]

The Qing army was able to subdue the regime established on the island by Zheng Chenggong and his son near the end of the seventeenth century. Taiwan subsequently became a prefecture administered by the Fujian provincial government. While China faced Western powers in the mid-nineteenth century, the Qing government established Fujian-Taiwan Province in 1887. However, it was quickly demolished by Japan in the First Sino-Japanese War of 1894–1895. Taiwan was ceded to Japan and became one of its colonies.

Sweet potato cultivation advanced significantly during Japan's rule of the island, which lasted a half century until 1945, and the root also gained importance in Taiwan's food system. This was not surprising given that, as demonstrated in chapter 4, the root has played an important role in Japanese agronomy since the early nineteenth century. After seizing Taiwan, Japan promoted the cultivation of both sweet potato and sugarcane, because the two crops were well suited to the island's subtropical climate. In other words, sweet potato farming in Taiwan became an integral part of Japan's imperial ambition: As it sought to industrialize, Japan hoped to transform Taiwan into a granary that would aid in the goal of empire building.[9] To that end, it not only conducted agricultural research; it also relocated its farmers to Taiwan. Sweet potato was one of their favorite crops, along with rice and sugarcane, that they chose to grow on the land. The root dominated Taiwan's agricultural production beginning in the early nineteenth century, parallel to Japan's colonization of the island. Sweet potato production increased dramatically from 200 tonnes to 1,700 tonnes between 1900 and 1937, and sweet potato sown area increased from 39,000 to 139,000 hectares. In Taiwan, the average yield per hectare increased from 5,200 to 13,000 kilos. The trend would last more than a century, well into the post–World War II years.[10]

There were two main reasons why sweet potatoes thrived in Taiwan. First, the islanders discovered that it was a hardy dryland crop that could be planted on hillsides and ridges without competing with other food crops for land. However, when compared to indigo and tea trees, two other popular dryland plants, sweet potato performed better in terms of soil fertility preservation. In fact, some farmers chose to plant sweet potato after several years of growing indigo to replenish soil fertility depleted by indigo. Second, sweet potato was a good candidate for intercropping in addition to crop rotation. Indeed, sweet potato was not the only crop grown in the field, despite its popularity among Taiwanese farmers. It was frequently intercropped with other plants in Taiwan, as seen in other sweet potato–producing areas. Tea farmers, for example, were inclined to grow sweet potato alongside tea trees because it provided them with food in the mountains. Sweet potatoes were interplanted right before the harvest of rice in paddy fields where farmers grew rice as their main crop; at the time, rice paddies were no longer completely covered by water but remained moist enough for sweet potato transplantation. Sweet potato was frequently intercropped with either dryland rice or tobacco on higher ground. Tobacco, like indigo, depleted soil fertility, whereas sweet potato could help restore it.[11] All of this demonstrates why sweet potato became such a popular crop among Taiwanese farmers.

Prior to the 1980s, then, sweet potato production topped all other food plants in Taiwan—it possessed an indispensable position in the island's multiple cropping system. Thanks to the warm weather, sweet potato grew all year round with four cropping seasons: fall (August–September), spring (February–March), summer (June–July), and winter (September–October). "Because planting season is not a limiting factor," observed H. Wan, a researcher at the Taiwan Agricultural Research Institute, in 1982, "for growing sweet potato, its position in cropping systems can become quite complicated. Sweet potato can either be a major crop in a rotational system or a minor one in an interplanting pattern." More specifically, as a spring and summer crop, it grew on upland with little irrigation, where in the fall and winter seasons it was interplanted in paddy fields. In the latter, Wan identified five cropping systems in different regions: rice/rice/sweet potato; rice/sweet potato/rice; rice/sweet potato; sweet potato/rice/tobacco; and rice/sugarcane/sweet potato. While sweet potato seemed ubiquitous, it did not always function as the main crop in these systems. By contrast, on upland where either two-crop or three-crop systems were practiced, sweet potato was the major crop, whereas soybean, sugarcane, peanuts, and/or tobacco were rotational ones planted after its harvest or interplanted with it. Notwithstanding the varied yields of the root in the above cropping systems, sweet potato undoubtedly occupied a critical position in the island's agronomy.[12]

Taiwan was ruled by Chiang Kai-shek and his son Chiang Ching-kuo (1910–1988) from the end of World War II until the end of the 1980s, after he lost to the Communists on mainland China. During the immediate postwar years, particularly during the Korean War, the regime hoped to return to the mainland with the support of the United States and Western Europe. However, it appeared to the Chiangs that their hope would not be realized anytime soon. In the following decades, they shifted their focus—particularly Chiang Ching-kuo as his father aged—to Taiwan's industrialization. The policy change was successful. Taiwan became one of East Asia's "little dragons" (along with South Korea, Singapore, and Hong Kong) during the 1970s, known for its impressive economic progress.

Importantly, it was also during this decade that sweet potato farming and consumption in Taiwan reached a tipping point. Sweet potato production tripled between the end of World War II and 1970, going from 116,000 tonnes in 1945 to 344,000 tonnes in 1968. As a result, the last quarter century could be considered a golden age for sweet potato production on the island. However, as Taiwan developed more export-oriented industries and became more economically successful, an increasing number of Taiwanese no longer relied on sweet potato as a staple food. In 1977, the Asian Vegetable Research and Development Center conducted a survey of Taiwanese farmers regarding sweet potato cultivation. It began with the observation that "Between 1971 and 1975 the area planted to sweet potatoes in Taiwan declined dramatically. Yields per hectare showed a steady increase, averaging 214 kg gain per year but the hectarage declined so rapidly that the total production curve also showed a steep decline." Apparently, the precipitous drop in sweet potato production during this five-year period was what prompted the center to launch its study. It designed three types of questionnaires, which were distributed to sweet potato producers in six municipalities where sweet potato was grown in four different seasons. They covered aspects of the planting season, yield rank, soil type, fertilizer, cropping pattern, production costs and return, postharvest handling, sales products, and producers' and former producers' attitudes toward sweet potato farming.[13]

Meanwhile, the survey also shed light on some of the reasons for the decline of sweet potato production in Taiwan. The first was due to the increase of family income as a result of industrialization. Despite its ubiquitous presence in the island's food system, sweet potato was still viewed as a "low status food" that the islanders chose to dissociate with when they could afford other foods. The second was the government's guarantee of a minimum price for rice, which incentivized farmers to turn to rice farming at the expense of sweet potatoes. The third was that since hog raising became more concentrated in confinement operations, it reduced the need for hog feed, which sweet potato vines and roots used to supply. The fourth was the increase of irrigated fields, enabling farmers to cultivate more rice and sugarcane instead of sweet potato as before. And the fifth was that given the shortage of farm labor (caused by industrialization and urbanization), farmers chose to grow crops, like sugarcane, that could be harvested mechanically.[14]

In other words, as the island became a more and more industrialized society, the islanders were no longer so dependent on sweet potato as their staple food. Yet the change did not occur overnight. In fact, during the 1945–1970 period, when sweet potato production reached a new height on the island, a trend was already emerging in which sweet potato was increasingly used for hog feed rather than human consumption. Meanwhile, there was a marked increase in industrial use of the root during the same period. Thus viewed, we can see that a variety of factors contributed to the golden period of sweet potato production in postwar Taiwan. From 1970 onward, somewhat ironically, some of these factors, such as the industrialization of hog raising, which saw a turn to processed corn feed instead of sweet potatoes, also played their part in the sweet potato's decline. As early as 1965, for instance, human consumption of the sweet potato in Taiwan had already dropped to around 15 percent, compared to about 25 percent in the previous decade.[15]

Despite the obvious decline from its once central position in Taiwan's agronomy and gastronomy, sweet potato and its products are still commonplace on the island, playing an important role in people's lives. Taiwan is known for its vibrant nightlife, owing to its warm climate; islanders are accustomed to visiting "night markets" (*yeshi*) and sampling a variety of "small eats" (*xiaochi*) there. Night markets exist in every city on the island, and Taipei, Taiwan's capital, has several. Of the common "small eats" served at these night markets, oyster omelet (*O-a-tsian*) is probably the most well-known and popular, a true must-eat item among visitors to the island. The dish is a pan-fried omelet with small oysters as its filling. Sweet potato starch is an indispensable ingredient, mixed with the egg batter in order to give it the desired consistency. Savory or spicy sauces are often poured on the omelet before consuming it. Besides oyster omelet, meatballs and/or fish balls are also night market favorites in Taiwan; these also need sweet potato starch to give them the necessary consistency. The islanders also eat roasted sweet potato, like many others in Asia and beyond. In Taiwan, roasted sweet potatoes are more customarily called *kao digua* (*digua* literally means "earth melon") instead of *fanshu*. Buying roasted sweet potatoes from the peddlers who sold them on almost every city corner is a fond memory for many Taiwanese.

Indeed, for the baby boomer generation in Taiwan, eating sweet potato was an integral part of their childhoods. In the countryside, recalls Hui-tun Chuang, kids "loved to play in a harvest field and to bury sweet potatoes under a mound of preheated rocks. This became a pre-dinner snack and after-school entertainment. The sweet potato then could become a symbol of collective memory."[16] In towns and cities, roasted sweet potato was a common street food. When someone mentions it today, as Cai Chenghao and Yang Yunping write, many Taiwanese "wear a contagious smile on their face because it was such a pleasant childhood memory."[17] Xiao Xiao, an essayist and poet, was born in 1947. He recalls in *Fanshu de haizi* (Children of the sweet potato) that as a child, he liked sweet potato strips or chunks cooked with rice because the mixture had a sweet taste and a soft texture. However, the memory became somewhat bittersweet over time, as it reminded him of the poverty his family and others like them faced in the postwar years. He grew tired of eating sweet potato rice for every meal of the day. Worse, he discovered that the mixture contained little rice and increasingly more sweet potato. As a work-around, he begged his parents to pack lunch for him to take to school, even though it was only a few hundred meters away. The reason for this was that the packed lunch bento had to be filled with rice rather than sweet potato because the latter would spoil faster in the warm temperatures of subtropical Taiwan. What he didn't realize as a child was that in order for his parents to meet this demand, a greater portion—if not all—of their own meals would thus have to consist of sweet potato![18]

Xiao Xiao's experience was probably a universal one among his generation.[19] As previously stated, many Taiwanese identify as "sweet potato folks." This identification stems not only from the fact that Taiwan, their homeland, is shaped somewhat like a sweet potato, but also from a bittersweet and complex feeling about the formation of their cultural identity, in which the sweet potato played an almost indispensable role. Qing China annexed Taiwan near the end of the seventeenth century. However, it was not fully integrated or accepted by the Qing empire—whose map, incidentally, looked like a taro—over the next two centuries. When the Qing was defeated by Japan in 1895, it also surrendered the island as loot. This separation appears to have left a deep impression on the minds of many Taiwanese. Following Japan's defeat in 1945, Chinese troops led by Chiang Kai-shek seized control of Taiwan. The islanders began to refer to them as "taro folks," in contrast to themselves, "sweet potato folks." As mentioned above, sweet potato farming and consumption advanced significantly during Japan's rule of Taiwan. According to George H. Kerr, an American diplomat during this period who spent most of his life in East

Asia, sweet potato became "the standard diet" on the island after the Chinese takeover. Having literally grown up on the root, school pupils liked to refer to themselves as "sweet potato children" (*huan-tsi-a*) in order to differentiate themselves from those "taro folks" coming from mainland China.[20] In other words, it was the combination of the shape of their island and the importance of sweet potato in their life experience on the island that made many Taiwanese identify themselves as "children of the sweet potato" or "sweet potato folks." Of course, because sweet potato played a similarly important role in the lives of many Fujianese across the Taiwan Strait, it was also customary for them to refer to themselves as "sweet potato folks." But the epithet's use among Taiwanese registered a distinct sentiment about their island's uncertain peripheral status vis-à-vis the mainland. It reflected the islanders' at once sweet and sour memory of their island's modern history.[21]

Sweet potato and Ryukyu/Okinawa: An intertwined story

If sweet potato farming in Taiwan reflected the island's entangled relationship with both China and Japan, a similar case could be made for Okinawa, or Ryukyu, in modern history. Flanking the sea-lanes connecting Taiwan and China, the archipelago's fate was inextricably linked, both ways fortunate and unfortunate, with the histories of Taiwan and Japan. And, beginning in the early seventeenth century, sweet potato played an important role in the development of this triangular relationship. As discussed in chapter 4, sweet potato was most likely introduced to Ryukyu from Fujian, China, in 1604 by Nokuni Sōkan and promoted by Gima Shinjō as a food crop in the archipelago. The root plant then spread to Kyushu's Satsuma and Nagasaki, and then to the rest of Japan. However, the transfer was far from peaceful, and it marked the start of Ryukyu's modern history. In fact, the introduction of the American plant to the archipelago occurred during Japan's transition from the Warring States period to its unification under the Tokugawa government. Before the seventeenth century, Ryukyu had been influenced by both China and Japan. Politically, however, it maintained an independent entity. In the words of George Kerr in his *Okinawa: The History of an Island People*, one of the earliest modern historical accounts of the archipelago, Ryukyu was "an independent kingdom in the Eastern Seas." Toward the end of the thirteenth century, when the Mongols prepared their invasion of Japan, they demanded that the Ryukyu king make a contribution. But the demand was rejected. A century or so later, Mongol rule of the Asian mainland came to an end with the rise of the Ming dynasty, with which Ryukyu kings entered a tributary relationship. In 1372, the king, calling himself King Chūzan, sent his younger brother to Nanjing, then capital of the Ming, and accepted China's suzerainty over his kingdom. A Chinese official accompanied the brother on his return trip back to Ryukyu, delivering a seal of investiture to the king. In the following five centuries, thanks to its strategic location, Ryukyu was a hub of the wider East Asian trade network that grew up under the guise of the Chinese tribute system. In the early fifteenth century, for example, in Shuri Castle, where the royal court was located, a bell was cast and hung in the audience hall with the following inscription: "Ships are means of communication with all nations; the country is full of rare products and precious treasures." Indeed, during those centuries, argues George Kerr, Ryukyu was comparable to a European city-state like Venice, Genoa, and Lisbon, enjoying considerable benefits from lucrative trades linking various ports in East and Southeast Asia.[22]

This, however, did not last. In 1609, a few years after the Ryukyuans discovered and sowed the American root plant, the Shimazu clan from the Satsuma domain in southern Kyushu invaded the archipelago. Over the next two centuries, the Satsuma lord wielded power and monopolized Ryukyu's trade with other feudal domains in Japan. While culturally and economically connected to Ming China, the Ryukyu kings also had to maintain a good relationship and peaceful relations with their Japanese neighbor to the north. But the unification of Japan under the Tokugawa shogun changed everything. Shortly after, the Satsuma lord sent a request to Ryukyu on behalf of the shogun, demanding that the king submit to the new ruler in Japan. The king's refusal to grant the request became the pretext for the invasion, which had been sanctioned by the Tokugawa court in 1606. The invaders won quickly because, after centuries of peace, the Ryukyuans were no match for the Japanese warriors who had just survived the bloody, two-century-long Warring States period.

In most Japanese accounts of the sweet potato's role in ending the war in Ryukyu, the root, as a newly discovered and much-treasured food in

Ryukyu, was offered as a friendly gift to the Satsuma warriors by Ryukyu King Shō Nei. Yamada Syoji's *Satsuma imo* (Sweet potato), for example, quotes from another seventeenth-century text on Satsuma history, describing the occasion as follows:

> King Shō Nei invited the Satsuma generals to the royal castle one day, offering them a farewell party with a banquet, at which sweet potato cooked in a stew was presented to them. The generals were delighted with the delicacy. Upon their departure, they expressed the desire for bringing back local products including the sweet potato. The King wrapped a bunch of raw sweet potatoes as his present to them. This was how the root was first introduced to our domain.[23]

What actually happened in those days is unknown, but it was not peaceful or pleasant, at least not for the Ryukyuan king and his people. In fact, while the invasion was over quickly, it had long-term ramifications in Ryukyu. For example, King Shō Nei had been captured by the invaders and imprisoned for three years in Kagoshima, Satsuma's capital. His release back to Shuri Castle, where the banquet was held, was contingent on Ryukyu paying Satsuma an annual tribute equal to one-eighth of its revenue. And the king and his councilors took an oath, admitting not only that the invasion was justified, demonstrating the invaders' benevolence to the island, for which they were eternally grateful, but also that Ryukyu was to accept any obligations imposed by Satsuma from then on. When a councilor refused to take the oath, he was beheaded in full view of everyone. In a nutshell, when Ryukyu King Shō Nei gave sweet potatoes as a parting gift to the Satsuma generals, he was also subjected to severe humiliation by his visitors.

After the farewell party, the Ryukyu kingdom remained in existence for the subsequent two and a half centuries. But "Okinawa would never again know independence and prosperity," laments George Kerr. Mamoru Akamine, author of a recent history of the Ryukyu kingdom, concurs. After the conquest, he explains, the kingdom "took its place within Japan" and its feudal system. Since the archipelago was put under the overlordship of Satsuma, its royal family became "a retainer of a retainer to Japan's 'virtual king,' the Tokugawa shogun." The Ryukyu king was "stripped the title *kokuō* (lit. king of the land), and given instead of the title *kokushi* (provincial governor)."[24]

In other words, while Japan did not formally annex Ryukyu until 1879, the kingdom managed to retain only a nominal status as a "foreign state" in the eyes of the Tokugawa government after Satsuma's invasion of 1609. Notably, this political development coincided with the development of the sweet potato, followed by that of sugarcane, in the archipelago. To some extent, indeed, it also paved the way for both crops to become quickly adopted by the general population of Ryukyu. As mentioned above, after it subdued the Ryukyu Kingdom, Satsuma demanded that the latter pay an annual tribute. At first, the tribute included Ryukyuan products, such as banana fiber cloth, fine ramie cloth, coarse linen, ramie grass, cotton, hemp palm rope, black braided rope, woven mats, and oxhides. Since the quantities for each item were set so high, the Ryukyuans were unable to fulfill these obligations. Beginning in 1613, they were permitted to pay in silver and rice instead,[25] since rice had long been treated as a currency in Tokugawa Japan.

However, growing rice in Ryukyu was difficult. The archipelago, like Taiwan, is located in a subtropical zone, with warm days all year and an annual average temperature of around 23 degrees Celsius. Rice should grow well in this weather pattern, except that Ryukyu is subject to weather vagaries—typhoons every year and drought every two to three years, resulting in crop failures on a regular basis. More than half of the land is also covered by red and yellow soil that erodes easily in heavy rains. Consequently, it was challenging for Ryukyu farmers to maintain irrigated rice paddies; most rice grown on the islands was of dryland varieties, which also suited the islands' soil type. Meanwhile, like the natives of Taiwan and neighboring islands, the Ryukyuans were familiar with root crops as food substitutes—they were accustomed to growing and consuming taro and sago palms during lean times, although they were not only low-yielding crops but also coarse in texture. Each of these factors contributed to the quick and unhindered reception of the American root plant on the archipelago from the early seventeenth century. Ishige Naomichi, the renowned Japanese food scholar, commented that sweet potato was "the notable exception" to the agricultural handicap in Ryukyu because it "can be grown in large quantities throughout the year in the subtropical climate." The introduction of the sweet potato to the islands, therefore, was a "momentous event."[26] More interestingly, Ryukyuan farmers became so enamored of sweet potato cultivation that the plant became something of a barometer of the local climate:

According to their experience, "when the vines of sweet potatoes stand up instead of crawling on the soil, that is another sign of typhoons."[27]

Sweet potato was once considered a treasured exotic food, as evidenced by King Shō Nei's forced hospitality toward his capturers. However, it quickly became a daily staple for the majority of Ryukyuans. From the seventeenth to the twentieth centuries, Ryukyu farmers ate rice only on major holidays and sweet potatoes the rest of the year. More specifically, save for a few wealthy noble families who could afford rice for all three meals of the day, most others ate sweet potato for breakfast and lunch, and rice cooked with other miscellaneous grains for supper. And for many poor families who could not afford rice, sweet potato was their "main food" (*shushoku*) in a meal while their supplementary foods (*fukushoku*) included miso soup, fish, meat (usually pork, lamb, or chicken, with beef and horse meat served only on rare occasions). Then in rural areas, sweet potato was consumed three or four times a day, accompanied by miso soup, which was cooked with sweet potato leaves, scallions, and some seasonal vegetables. In other words, meat was largely absent in the peasant diet.[28]

In comparison with Taiwan, where sweet potato was a "supplemental staple food," cooked usually with rice as part of the islanders' daily meals,[29] the root was the staple food among most Okinawans after its introduction. Naturally, sweet potato production also topped that of all other food crops in the region. Gima Shinjō, the Ryukyuan official who appeared in the previous chapter for his promotion of the root plant, was given credit for improving the propagation method, leading to the ubiquity of sweet potato farming in Ryukyu. Sweet potato could be propagated using its roots, as exemplified by Richard Cocks in his garden in Hirado, also mentioned in the last chapter. But from very early on after the root was introduced into China, as instructed by Xu Guangqi, Chinese farmers had used sweet potato vine cuttings instead. Xu had recommended that sweet potato be transplanted to fields after the vines grew to about three inches. Whether or not Gima knew about Xu's instructions, he recommended that farmers transplant the sweet potato vine cuttings once they reached about one foot in length. This proved to be a successful method. After Ryukyu farmers transplanted the sweet potato vines in this way, writes Yamada Syoji, the plant grew exceedingly well in the loam soil of the island. And thanks to this initial success, more varieties of the root plant were brought in from South China. The earlier ones brought back by Nokuni Sōkan had had either a red or white skin and white flesh, known as *hansu* or *hansu-umu* (following the Chinese pronunciation of *fanshu*); the red-skinned variety was also called *akaimo*, describing its color. Then toward the end of the seventeenth century, another variety came from China; this one had red skin but orange flesh and was called either *tōimo* or *karaimo* (lit. Chinese taro/yam), even though the earlier varieties had also been transferred from China.[30]

Sweet potato was/is consumed in a variety of ways in Okinawa, as it is elsewhere. Aside from daily consumption, which will be discussed further below, the roots are cut into strips and dried or steamed and then cut and dried. It can be processed into starch or flour, and the sediment from the process is used to make MSG, an enhancer for producing the umami flavor in food. Okinawans knew how to make distilled liquor from rice and/or millet, a technique passed down from China in the fifteenth century. After sweet potato gained popularity, they were one of the first who used (or mixed) it instead to brew sweet potato liquor, and this technology later spread to other parts of Japan. The day-to-day consumption of the sweet potato in Okinawa, however, is mundane. The roots are usually cooked once daily and consumed over the course of the day—served hot at one meal and cold at the remaining two. After peeling off the skin, the sweet potato is boiled whole with some salt and at times also tree onion. It can also be cooked with less water and mixed with wheat flour or sago flour, in a preparation that looks more like a gruel. In fact, sweet potato porridge is another popular form. Unlike in Taiwan, though, the root is often mixed with red bean, mung bean, and/or hyacinth bean instead of rice. Lastly, there is sweet potato pancake, which is made by using good quality midsized sweet potatoes, carefully crushed in a *suribachi*, a Japanese mortar and pestle, and steamed in banana leaf wrap.[31]

The coincidence of Ryukyu's submission to Japan following the Satsuma invasion and its adoption of sweet potato as the main food crop suggests that sweet potato consumption was not always a pleasant experience for many Ryukyuans in the seventeenth and eighteenth centuries. Rice was regarded as a superior grain food in its culture, as it was in other parts of East Asia. Scholars have pointed out that though the Ryukyu Kingdom was subdued by Satsuma and incorporated partially into the feudal system of Japan, it still worked painstakingly to finesse a good relationship with China, including by repeatedly sending tribute missions to the Chinese ruler and students to learn and emulate its culture. Indeed, after their submission to Satsuma, it appeared as though there developed a sort of cultural and political schizophrenia among the Ryukyuans, at least among the members of the *yukatchu*, the formally recognized upper class, with respect to their identity. While they were forced to pledge their political allegiance to Satsuma and henceforth to the Tokugawa government, Ryukyuans persisted in maintaining their cultural affinity with China well into the late eighteenth century, or before Japan's formal annexation of the archipelago in 1879. "By the late eighteenth century," observes Gregory Smits, "Chinese cultural forms had spread from the narrow confines of Kumemura to the large *yukatchu* population in Shuri. Strange as it may sound at first glance, one result of Japanese *political* domination of early-modern Ryukyu was an increase in the *cultural* influence of China." In fact, he goes on to stress that during this time, Ryukyu was "living a lie" to the extent that it was trying to hide its affiliation with Japan from China.[32]

Surprisingly, living this lie also included the sweet potato. Ryukyu continued to send tribute and receive Chinese officials after the Ming dynasty fell and China was ruled by the Qing. As previously stated, because the kingdom was required to pay annual tribute to Satsuma in both silver and rice, the islanders were forced to make sweet potato their staple food. However, the locals appeared to want to keep this fact hidden. When the Qing Ambassador Wang Ji visited Ryukyu in 1683 and asked if sweet potato had become a local food, the Ryukyuans blushed in embarrassment and replied that only the very poor would eat the root. Wang, for his part, was well aware that this was not so, having lived in Ryukyu for several months and made many island friends during his stay.[33]

Another anecdote underscores the essential role sweet potatoes played in the diet of the Ryukyuans. By the late eighteenth century, the root plant had become even more important as a source of food for the islanders. The Chinese ambassador to Ryukyu at the time, Li Dingyuan, recorded in his journal that after he finished his tenure and was ready to return to China, a member of his staff purchased a lovely dog in Ryukyu. However, once at home, the dog refused to eat anything. After several trials, they discovered that the dog would consume only sweet potatoes. Li sighed that, "in Ryukyu, even the pets there are accustomed to the diet!"[34]

Despite its reputation as poor people's food, the root's widespread adoption as a staple in Ryukyu benefited the region's economy, contributing to population growth in both the seventeenth and eighteenth centuries.

According to the research of Japanese scholars such as Hayami Akira and Arizono Shōichirō, while Japan's population growth stagnated in the eighteenth century, Ryukyu and southwestern Japan saw continued demographic expansion due to sweet potato farming, as discussed in the previous chapter. Arizono estimates that the population of Ryukyu increased by 300 percent as a result of the cultivation of the root plant over the next two centuries. In his general history of Ryukyu, George Kerr observes that the demographic trend indeed continued from the eighteenth well into the nineteenth century. He estimates that the Ryukyu population in the first half of the eighteenth century stood somewhere around 150,000, certainly not exceeding 200,000. By 1879, when Ryukyu was annexed by Japan, it grew to around 310,000, increasing to 480,000 in 1903. In fact, from the second half of the eighteenth century, population pressure became a growing challenge to Ryukyu's economy.[35] This means that Ryukyuans—or "Okinawans" in Japanese, as the archipelago had become a part of Japan—had come to depend even more on sweet potato in their diet.

George Kerr includes an incident in his comprehensive account of Ryukyu's history about how sweet potato was provided to the entourage of the Macartney Embassy to China at the end of the eighteenth century. As stated in chapter 3, George Macartney led the first English mission to China in order to persuade the Chinese government to open its doors to trade with Britain. Though he ultimately failed, he and his colleagues gained firsthand knowledge of China's swelling population, which was fueled in part by sweet potato farming. Lord Macartney did not record if he saw sweet potato in China. But some of his staff did see and taste the tuberous root. George Staunton, Macartney's secretary, also wrote down his impression of two Ryukyuans whom he met while returning from Beijing to Canton as the two were on their way in the opposite direction. In his description, Staunton wrote that the two men, possibly students sent by the kingdom to study in China, were "genteel" and "well-looking, tho of a dark complexion, well-bred, conversable and communicative." He also noted that given Ryukyu's geographical position, it should "naturally belong to the Chinese or the Japanese."[36] A couple of years later, his countryman Captain William Robert Broughton, commanding HMS *Providence*, entered Okinawan waters after the ship foundered on a reef. Broughton was known for his role in assisting the Vancouver Expedition a few years earlier. In Ryukyu, he was equally impressed by the gentle manners and warm hospitality of the Ryukyuans. When his crewmen went ashore in search of food, Broughton joined them. He described the experience as follows:

> These good people were fully acquainted with our misfortune, and naturally conceived our greatest wants were articles of life which, such as they possessed, they parted with in a most friendly manner.... In the morning of the 23rd [of May] we received from our friends the remainder of their presents, which amounted in all to 50 bags of wheat, 20 of rice, and 3 of sweet potatoes; each bag containing 1 cwt [i.e., hundredweight]; also one bullock of 3 cwt; six large hogs, and plenty of poultry. Indeed, whatever we asked for they immediately sent us.[37]

The quality and quantity of provisions provided to the English clearly demonstrated the Ryukyuans' kindness and generosity; as previously stated, the island was far from prosperous when measured against contemporaneous standards of living. Perhaps the islanders' overwhelming friendliness (which moved the English captain and his crew) toward strangers was also a reflection of their interest and experience in international trade.

Despite the palpable interest in commerce (for example, after its introduction to the archipelago, sugarcane became a cash crop for the islanders, as opposed to sweet potato as the primary food source), Ryukyu remained an agricultural society.[38] At the turn of the twentieth century, Katō Sango, an anthropologist from Japan who spent many years on the island as a teacher, provided a detailed report on the festivals celebrated among the islanders. Month by month, the Ryukyuans observed holidays, most of them related to the food plants they cultivated. For example, in the first and second months of the year, when wheat was sown, there was a wheat festival. The islanders paid homage to wheat in addition to rice, millet, and taro with kowtows, prayers, and sacrifices. In the fifth and six months when rice was transplanted, they held a similar ceremony for its bumper harvest. At the end of his research notes, Katō wrote that though rice and wheat had been the main food crops, cultivated on the island from time immemorial, sweet potato was now the staple after it was introduced by Nokuni Sōkan and Gima Shinjō in 1605. Thenceforth, the root gradually became the most preferred food plant among Ryukyuan farmers because it was less labor-intensive and higher in yield than its counterparts. At the time of his writing, he concluded, sweet potato was the food that 400,000 plus Ryukyuans depended on—as enduring as those rice- and wheat-related festivals appeared to be, they were nothing but a vestige of a time long past.

Katō Sango wrote his chronicle in 1903. His nostalgic and somewhat melancholy remarks in summing up his observation attested to the primacy of the sweet potato in Ryukyu's food culture. Katō also described two types of holiday foods associated with Ryukyuans' celebrations of those grain-related festivals. As in many other cultures in Asia, making cakes as a holiday food also seemed to be a Ryukyuan tradition. Rice cake was particularly popular. Called *zongzi* in Chinese, it is usually made of glutinous rice stuffed with different fillings and wrapped together by bamboo or other plant leaves. For the Chinese, *zongzi* is the food for the Dragon Boat Festival, held on the fifth day of the fifth month of the lunar calendar; the festival, according to Katō, was also observed in Ryukyu. But *zongzi* is just one of the leaf-wrapped cakes Asians customarily make on holidays. Similar customs have been seen across East and Southeast Asia. In Vietnam, for example, leaf-wrapped rice cakes in various shapes are prepared for the Lunar New Year. Ryukyuans liked to make cakes too. Instead of rice cakes, *taanmu*, a taro cake, was a traditional food for the Fire God Holiday, held on the third day of the third month. Then, on the eighth day of the twelfth month, people made *muchi*, also known as *onimochi* (lit. ghost cake), a rice cake wrapped in shell ginger leaves, a common plant grown on the islands that gave the cake a distinct scent. According to local legend, it was called "ghost cake" because, several centuries before, a man from Shuri moved to Osato and lived in caves and became (or was possessed by) a ghost who ate people. His sister went to see him and was shocked by the cannibal behavior. She made the rice cake, hid it inside seven iron nails, and wrapped it with shell ginger leaves. The two sat on a cliff and the brother ate the cake and became frightened. In the end, he fell off the cliff and died in a pool of blood. Eating the cake (*mochi*), per the custom, enabled people to expel and exterminate the ghost (*oni*).[39] It is also noteworthy that after sweet potato became a staple food in Ryukyu, it has been used as a chief ingredient in making both cakes—the root gives the cakes not only their sweet taste (in addition to brown sugar, another popular local product on the islands) but also an attractive color—for example, purple if the powder of the famous purple-fleshed sweet potato variety is blended into the rice.

Yomitan village in the Okinawa archipelago is most well-known among international tourists for producing the purple sweet potato (*beni imo*).

Locals would most likely tell visitors that the variety originated in Taiwan over a thousand years ago.[40] However, this is folklore, as sweet potato does not have such a long history in Taiwan, and nor does *beni imo*, which, according to agriculturalist Yamakawa Osamu, was first cultivated in Okinawa in 1947.[41] In fact, despite its overwhelming popularity as a famous local variant in Okinawa today, the purple sweet potato as a hybrid pales beside another variety that was actually named after the archipelago. Known as "Okinawa 100," it was developed by Japanese agricultural scientists in 1939. As explained in previous chapters, following the outbreak of World War II in Asia, Japan became increasingly interested in cultivating the root plant as a means of increasing and ensuring its food supply. After occupying Shandong, China, for example, the Japanese army tried to expand sweet potato planting areas. The cultivation of Okinawa 100 had begun as early as 1931, the year Japan took over Manchuria in preparation for its large-scale invasion of China. The Japanese Department of Agriculture started a project to develop new sweet potato varieties in Okinawa. During the 1930s, through hybridization, the scientists fostered several new varieties. Okinawa 100 eventually trumped others for its quick ripening and large size—in Okinawa's subtropical climate, it could be planted three or four times in a year. Having received recognition from the department in 1939, it was recommended for Japanese farmers in Kyushu, Kantō, and rest of the country as the favored cultivar for sweet potato farming from then on.[42]

Yet the initial attempt to grow Okinawa 100 met with a serious setback, which might be interpreted as an ominous harbinger of the ultimate failure of Japan's war effort. Following the surprise attack on Pearl Harbor, Japan seized control of parts of the South Pacific, including the Dutch East Indies, New Guinea, the Solomon Islands, Manila, Kuala Lumpur, Rabaul, and Chuuk (Truk) Lagoon, where it also established naval bases. After the Battle of Midway in 1942, however, the Japanese advance was thwarted by the U.S. Navy and Allied forces. In the following years, it was gradually forced into a defensive position. On February 17–18, 1944, the U.S. Navy launched Operation Hailstone, inflicting heavy losses on the Japanese Navy and its base on Chuuk Lagoon. After the Battle of Saipan in June of the same year, Japanese forces in the area were essentially cut off from their food supplies. Having exhausted the breadfruits that native Kanakas relied on, the Japanese armed forces decided to grow sweet potato as the main food on the Chuuk Lagoon and its neighboring one hundred islands. They organized "sweet potato cultivation team," recruiting soldiers who hailed from Okinawa, and rode on "sweet potato boats" to cultivate Okinawa 100 on the islands, hoping to have two and a half harvests in a year. Contrary to their expectations, however, the cultivation failed miserably due to the invasion of cabbage moth (*imomoshi* in Japanese), which ate sweet potato vines at night and killed the plant. The expected harvest did not happen; many Japanese soldiers died of starvation as a result. Itō Shōji, the author who retold the story at the outset of his *Satsumaimo to Nihonjin* (Sweet potato and the Japanese) in 2010, laments the fact that in the "battlefield of food" (*shoku no senjō*), the Japanese army was "utterly defeated" (*kanbai*), for the U.S. Army's provisions then included bread, bacon, oatmeal, and cheese.[43]

Nonetheless, Okinawa 100 appeared to be an instant success throughout the Japanese archipelago. According to Chūman Katsumi, the cultivar was sown in every major sweet potato–producing region in 1940, one year after Japan's Department of Agriculture recommended it. And it occupied approximately 65 percent of the area devoted to growing sweet potato on Okinawa.[44] The success of Okinawa 100 at the time built on earlier research on sweet potato cultivation. Indeed, given the importance of the root plant in their agronomy and their interest in international trade, it had been a long tradition for Okinawan farmers to import and nurture new varieties from overseas. After the earliest ones brought back from South China, they continued to find better varieties from Europe and Australia. At the end of the nineteenth century, for instance, "seven happiness" (*shichifuku*) was cultivated in Okinawa. The cultivar was so named because it was edaphic tolerant, easy to grow, could be stored for extended periods, and was palatable; besides the above four qualities, its travel route also included three places—Italy, America, and Japan.

Unlike previous varieties, Okinawa 100 was developed through hybridization. "Seven Happiness" was its "mother," whereas "Chōshū" (Chaozhou in Chinese, named after the city in South China) was its "father." It was a technology that Japanese scientists gradually mastered beginning in the early twentieth century. Okinawa was one of the key regions where experimental fields were prepared and devoted to nurturing new and desirable cultivars due to its suitable climate for sweet potato cultivation. Okinawa is famous for its purple sweet potato (*beni imo*), for example. According to Yamakawa Osamu's research, the cultigen most likely originated in Kyushu and gradually spread to Kantō areas beginning in the early Meiji era. Its flesh was mostly white at the time, with a purple center. People appreciated it as a rare variety not only for its sweet taste but also for its distinct color. But it was in Okinawa where the cultivar evolved to become entirely purple, as seen in "Miyanō 36," which was first hybridized in 1947, with the technology becoming fully matured in 1995.[45] Apart from cultivating the purple variety, Okinawa also nurtured the red-fleshed *Kawagoe shu* (variety), which was first discovered in Kawagoe, in Saitama Prefecture, in 1919 and was subsequently sent to the field there for further cultivation. In other words, though Okinawa 100 might be the most popular cultivar of the time, it was not the only success story from the archipelago. In fact, at every sweet potato harvest, it was said, scientists looked continuously for good-quality sweet potatoes and artificially hybridized them to create better varieties. For example, Matsunaga Takamoto, a scientist commissioned by Okinawa Prefecture who was credited with cultivating Okinawa 100, also played an instrumental role in developing other new cultivars like "National Defense" (*Gokoku imo*), "Agricultural Forest 1" (*Nōrin 1*), and "Agricultural Forest 2" (*Nōrin 2*), all of which became popular in wartime and postwar Japan.[46] Incidentally, after Okinawa 100 was introduced by Japan to the Asian mainland, it took the name "Victory 100" (*Shōri 100*). Needless to say, the new name, along with the "National Defense" variety, reflected Japan's desperate desire to win the war.

Nonetheless, Japan lost World War II—the battle for Okinawa was particularly fierce, claiming a large number of lives. In terms of the above sweet potato cultivars, Japanese soldiers and civilians relied heavily on them for survival during the war. Because their country was suffering from widespread starvation after Japan's defeat, they remained highly favored by most Japanese in the immediate postwar years. Indeed, according to Sakai Kenichi's research, sweet potato production spiked to an unprecedented high in 1945 in terms of both sown area and production volume. And through the 1940s, the latter continued to climb up and reached a pinnacle in 1949, whereas the former dipped only slightly in comparison with that of 1945. In the period, Okinawa 100 and its peer "National Defense" were consumed most by the Japanese not only because they were fast ripening but also because their roots, though with a rough and wrinkled skin, were larger in size than all other varieties on the market. Their large size also turned them into favored hog feed among farmers.[47]

However, after a decade or so, Okinawa 100 became strained under its own weight. As the food situation in postwar Japan gradually improved, consumers began to prefer sweet potatoes with a better (sweeter) taste. Despite its large size, Okinawa 100 had a low starch content. People ridiculed it by calling it "radish sweet potato" or "water sweet potato," mocking its bland flavor and sloppy consistency. Instead, the previously mentioned and cultivated "Agriculture Forest 1" and "Agriculture Forest 2" from the early 1940s surpassed the once popular Okinawa 100 beginning in the 1950s. Due to their higher starch content, these two varieties seemed to be more appealing to Japanese palates.[48] However, the knowledge gained from the production of these varieties undoubtedly contributed to the success of the *beni imo* (purple sweet potato), which, with a texture and taste similar to that of "Agriculture Forest 1" and "Agriculture Forest 2," became the sweet potato variety that Okinawa is known for today. There is no question that the color of the *beni imo* is attractive among other sweet potato varieties. But its purple color also makes it easy to confuse with *ube*, a Tagalog term for a purple yam in the Philippines, and the same-colored taro. Both are root crops commonly used in desserts and actually have a similar flavor to *beni imo*. However, their outer skin is coarser and darker, and the inner flesh is less juicy and tender.[49]

In retrospect, it was perhaps predestined that Okinawa 100 would fall out of favor in the 1950s because the overall importance of the sweet potato in Japanese cuisine had begun to diminish around this time. As discussed in the previous chapter, as the country gradually emerged from the postwar food crisis, not only in the rest of the country but also in Okinawa, sweet potato began to lose the critical position in the Japanese diet that it had previously enjoyed. Beginning in 1958, according to Chūman Katsumi, Okinawans gradually replaced sweet potato with rice (largely imported) in their daily meals. In the past, many islanders ate rice only once and sweet potato twice in a day. Now they were eating rice in all three meals. For the first time, therefore, sweet potato lost its traditional status as a staple food on the islands; instead, it was used more as hog feed and for starch making.

Sweet potato farming on Okinawa suffered a setback from the 1960s to the 1980s, at the same time that rice replaced sweet potato as a staple food. The outbreak of sweet potato weevil in the city of Hirara, on Miyako Island, one of the reputed birthplaces of sweet potato cultivation in the region, resulted in the prohibition of sweet potato export from Okinawa to the rest of Japan, in the hopes of controlling the insect's spread to other sweet potato–producing areas. During the 1980s, consumers could only buy sweet potato from a supermarket, not from a farm.[50] All this helped complete the ultimate substitution of sweet potato with rice as the daily staple in the Okinawa archipelago. In a recent article on the evolving of diets in Okinawa, food scientists point out that "traditional Okinawan diet has undergone rapid post-World War II Japanization and Westernization, most notably in terms of increased fat intake. There has also been a decrease in carbohydrate quality, with diversification away from the sweet potato as the staple carbohydrate and toward higher consumption of rice and bread (both mostly white) and noodles as carbohydrate sources."[51]

Regardless, sweet potato is an important traditional food in Okinawa. In recent years, the scientific community has shown a strong interest in assessing the value of traditional Okinawan eating habits, particularly the benefits of consuming sweet potatoes. For example, the authors of the aforementioned article point out that, while Japanese are known for their longevity, Okinawans "have the longest life expectancy within Japan." And the reason for this, they argue, has a lot to do with the archipelago's traditional dietary patterns. Thanks to the ample use of the sweet potato, the Okinawan diet is "low in calories and nutritionally dense, particularly with regard to vitamins, minerals, and phytonutrients in the form of antioxidants and flavonoids." Like most Japanese, Okinawans used to begin their meal with a miso soup in the Okinawan style, followed by sweet potato as the staple carbohydrate and stir-fried vegetables as the main dish, accompanied by "small servings of fish, noodles, and lean meat with herbs, spices, and a little cooking oil." Compared with other Asian cuisines, however, the Okinawan diet is low in salt intake, which is an important health factor. Yet that sweet potato is taken as the staple in the diet is a more beneficial aspect because the root is an "antioxidant-rich and low-GI (glycemic index)" food, contributing to the health and longevity of Okinawans over the years. Their conclusion is echoed by Ishige Naomichi, the aforementioned expert on Japanese food culture.[52]

Okinawans have made notable efforts in recent decades to renew and revive their traditional interest in the root plant as an integral part of Japan's rural revitalization movement. Yomitan Village established a processing center for its local specialty, the purple sweet potato (*beni imo*), in 1992. While fresh sweet potatoes can only be consumed on Okinawa, processed purple sweet potato products can be exported.[53] Every fall, Yomitan also hosts the Beni Imo Hometown Festival as well as the Miss Beni Imo

Figure 5.2 A display of various purple sweet potato products in Okinawa, known for its long tradition of growing the *beni imo* variety. Okinawans pride themselves on their purple sweet potato delicacies, which are exported throughout Japan and beyond.

Contest. On both occasions both villagers and visitors consume ample sweet potatoes and drink plenty of Orion beer, another local specialty brewed in neighboring Urasoe.[54] Determined to promote the purple sweet potato, the Yomitan Village Council issued its "Sweet Potato Day Declaration" in 2001, establishing the sixteenth of each month as "Sweet Potato Day." On this day, schools serve sweet potato dishes and desserts to students in their cafeterias. However, Yomitan Village, designated as the "Sweet Potato Village" by the prefectural government in 2004, is far from the only place on Okinawa that is marketing and promoting the root plant. Visitors can easily find a wide variety of sweet potato products in stores and markets throughout the archipelago, ranging from sweet potato chips and ice cream to sweet potato tarts and packaged sweet potato preserves.[55] To summarize, sweet potato, the root plant whose cultivation in Ryukyu/Okinawa has been intertwined with the archipelago's early modern history, continues to be a significant presence in its culture and society today.

Sweet potato as a gendered crop in Papua New Guinea

In his 1981 article about farming life in the highlands of Papua New Guinea, Paul Sillitoe, an anthropologist then at Manchester University, recorded a tragic incident. Saemom, a husband, killed his wife, Yaelten, in a rage because she dug up sweet potatoes in his field. In Papua New Guinea, women are supposed to farm the sweet potato; male members of the family can only do so when the women are menstruating. Because of this, Saemom had previously obtained the patch from his wife. Yaelten most likely made a mistake by continuing to work in the area after their agreement because, as a woman, she had grown accustomed to planting and tending the sweet potato. In any case, Saemom faced repercussions for the murder of his wife. He paid a large restitution to Yaelten's family for her death, leaving him with no money to marry again. According to Sillitoe, the fallout put Saemom in a difficult position in his community in the years that followed:

> The irony of the situation, and for the Wola the perversely funny side of it, was that he had killed his wife for digging tubers on his patch and as a result he had to plant his own sweet potato from then on. He had no sister to help him and his only daughter was about five years old at the time of the murder. Other female relatives (such as his elder brother's wife and daughter) helped him out, but still he had to do a considerable amount of planting himself until his daughter was old enough to take over responsibility for it. He died some twelve or so years later. This ignominious episode is remembered for Saemom having to get down on his knees and heap up sweet potato mounds. This was ridiculous.[56]

The Wola are a Papua New Guinean Indigenous people who live in the highlands. Like many other groups in the country, they regard sweet potato as a crop farmed solely by women—men can only assist on rare occasions, as demonstrated by Saemom's initial agreement with Yaelten.

Sweet potato is not the only fruit or vegetable to be gendered in many rural communities in Papua New Guinea, and in his article, Sillitoe provides a comprehensive list of crops grown in the country as well as an explanation for their gendered taxonomy; however, he notes that categorization varies by region. Crops grown in Papua New Guinea are generally classified as male or female—"there are some crops which only men may plant and tend, and there are others which only women can cultivate, plus a third category which members of either sex may cultivate." More specifically, Sillitoe finds, plants that are recumbent on the ground are female, or *wiy* in Wola, whereas plants that are vertical to the ground are male, or *hae*. For example, bamboo, banana, yam, and sugarcane are *hae*, and farmed only by men, yet cucumber, dye plant, and sedge (Cyperaceae, a kind of grass) are *wiy*, and farmed only or mainly by women. An important factor is that the sexes of the plants do not necessarily correspond to the sexes of the humans. That is, a male plant can be farmed by women whereas a female plant (e.g., hibiscus green, Chinese cabbage, and watercress) can be farmed by men. In fact, among the plants grown by the Wola, sweet potato is the only *hae*/male plant that is supposedly only farmed by women. To Sillitoe, the categorization of the sweet potato as a male plant is "blatantly wrong," a "misplacement," because the root, with its creeping vines on the ground, does not fit with the *hae* characteristics.[57]

Why is the sweet potato a male crop that can only be farmed by Papua New Guinean women, unlike other male crops? (The Wola are but one example—crop gendering is a widespread practice among many ethnic groups in the country.) To find answers to these questions, we must consider how the American plant was first introduced to Papua New Guinea. Papua New Guinea is an island country in western Oceania, occupying the eastern half of the island of New Guinea as well as some offshore islands in Melanesia. The Solomon Islands are its eastern neighbor, and Indonesia controls the western half of New Guinea. Archaeological evidence suggests that humans arrived in western Oceania and Australia around 50,000 years ago and that agricultural activities began around the same time. The main inhabitants of Papua New Guinea, like many other islands in Oceania, are Austronesian-speaking people who may have colonized the region around 3,500 years ago by replacing earlier settlers. They domesticated pigs, chicken, and dogs and cultivated bananas, taro, and yam. Beginning in the fifteenth century, when Oceania was engulfed in the process of European exploration, Papua New Guinea was no exception. It saw the invasion of American plants, ranging from Bixa, lima beans, cassava, tobacco, and sweet potato. Thanks to the tropical climate, sweet potato became quickly adopted as a main food crop; anyone who comments on the agriculture of Papua New Guinea today usually begins with the statement that "sweet potato is by far the most important food crop" in the country. R. Michael Bourke, an anthropologist at the Australian National University specializing in Papua New Guinean agriculture, and its sweet potato farming in particular, is one such individual, as is Jennifer Woolfe, author of *Sweet Potato: An Untapped Food Resource*. John Bouwkamp, a horticulturalist at University of Maryland, also observes that in Papua New Guinea, sweet potato "is the main staple crop of the population."[58] Bourke explains that prior to the sweet potato's arrival, taro, which originated in Asia, was the staple food among many Papua New Guinean communities. As discussed in chapter 1, taro and its varieties had been a traditional food among most Austronesians when they colonized Oceania. In his well-known study of the diffusion of the sweet potato in Oceania, D. E. Yen states that New Guinea had been a "recipient area" of both taro and yam from Asia whence the two crops were disseminated across the Pacific region. Different from many islands in the region, however, Papua New Guinea has a diverse geography, marked by highlands, wetlands, lowlands, and coasts. The adoption of the sweet potato as a staple food, explains Bourke, improved the diet of Papua New Guineans, for "sweet potato will grow at higher altitude than taro." Indeed, Papua New Guineans grow sweet potato at altitudes ranging from 2,200 to 2,800 meters above sea level, sustaining their permanent settlement in highland areas.[59] The country's tropical weather, needless to say, is also friendly to the root plant. Except in the highlands above 2,100 meters, where night frosts do occur, the daytime temperature is usually above 22 degrees Celsius all year round, which is several degrees higher than the minimum 15 degrees Celsius for sweet potato to flourish. In addition to its efficacy as an easy-to-grow crop, sweet potato has a tender texture and sweet flavor that make it a more appealing food than taro and other root plants.

Though the sweet potato's importance in Papua New Guinea is undeniable, the exact time and method for the root's transfer onto the islands remains a mystery, as discussed in chapter 1.[60] Most researchers believe the plant traveled west to east, entering New Guinea and then its neighboring islands from Indonesia. If this is the case, European explorers, particularly the Portuguese, could be responsible for its introduction into Maritime Southeast Asia. The Portuguese brought the food plant back to Europe after discovering it in the Americas, and then to Africa and Asia. In other words, the sweet potato's transfer to Papua New Guinea might have followed the *batata* line instead of the *camote* line as it did to the Philippines and China, etc. Regardless of how the root plant reached the islands, by the end of the nineteenth century, when Japanese anthropologist Tsuboi Shōgorō conducted his fieldwork on the South Sea islands near New Guinea, he found that the sweet potato had become a staple food among the islanders.[61]

James B. Watson, a twentieth-century cultural anthropologist at the University of Washington, was one of the first scholars to investigate the sweet potato's importance in Papua New Guinean agriculture, focusing on highland regions, prior to the research of R. Michael Bourke and others. As early as 1965, Watson published two major articles revealing his research, in which he pronounced that "The importance of this single crop [sweet potato] in the Highlands is outstanding when measured in almost any meaningful way: by the amount of ground devoted to its production, by the amount of time expended upon its cultivation, or by the amount of the total food intake it constitutes." Though the Portuguese may have introduced the sweet potato to Southeast Asia, it was local traders and hunters of bird of paradise, a beautiful species that lived in the dense forest of Papua New Guinea, who most likely brought the plant from Indonesia to the east coast of New Guinea. The transfer could have occurred in the mid-sixteenth century.[62] In more recent years, others have posited that instead of the Portuguese, it was the Spaniards who first brought the root plant from America to the Philippines, whence it spread to western New Guinea.[63] In sum, similar to the introduction of the sweet potato in Taiwan, Okinawa, China, and Japan, while Europeans may have initially disseminated the American plant throughout Asia, its subsequent spread was carried out by local people.

Since sweet potatoes can withstand drought, they are suitable for growing on highland slopes. In compiling *The Cambridge World History of Food*, Kenneth F. Kiple and Kriemhild Coneè Ornelas note the significance of this for Papua New Guinea: "Among the inhabitants of Wantoat Valley in highland Papua, the sweet potato is the only important cultivated food."[64]

To grow the root in the wetlands of mountain valleys, it requires certain work, such as planting it on mounds or raised garden beds and building ditches around them to drain the extra moisture. In the 1970s, a group of archaeologists examined exactly these conditions at Kuk swamp, in the upper Wahgi Valley in Papua New Guinea, and concluded that sweet potato indeed made landfall in New Guinea around 1665. Their conclusion was based on their observation that "sweet potato can tolerate short periods of waterlogging and as most flood events at Kuk are of short duration—typically a few hours, and only occasionally a day or two—the garden beds would have drained quickly as flood levels subsided." The archaeologists discovered women's houses and pigsties adjacent to the mounds and surrounding ditch network, confirming that the mounds were used for sweet potato cultivation, as Papua New Guinean women tended the plant and the root was used for both human consumption and pig husbandry.[65]

The archaeological find not only confirms the sweet potato's arrival in Papua New Guinea, but it also attests to the root's broad appeal to its people, as it was grown in both mountains and valleys due to its edaphic tolerance and value as a food and fodder source. In fact, the archaeologists began their investigation in the Wahgi Valley to investigate the scope of the so-called Ipomoean Revolution thesis, proposed by James B. Watson two decades ago, in Papua New Guinean society. As mentioned earlier, from the mid-1960s, Watson published his research on the introduction and impact of the American plant in the island country. His anthropological fieldwork in the highlands of New Guinea led him to believe that it was due to the arrival of the sweet potato that the highlanders transited from hunting to horticulture. Moreover, sweet potato farming in the highlands, which had not been densely inhabited for many centuries, attracted migration to the area. Indeed, Watson found that the population density in the highlands had become greater than that of the rest of New Guinea and that "the highest population densities coincide with the areas of largest sweet potato production."[66]

That sweet potato farming propelled population increase, of course, has not been a new and unique story—the root's introduction and subsequent cultivation in East Asia, resulting in demographic expansion and social changes, are well-known cases, covered extensively in chapters 3 and 4 as well as in this chapter. Like the root plant's far-reaching influence in major areas of Asia, the Ipomoean Revolution in Papua New Guinea, put forth by James B. Watson, has also been comprehensive and multifarious, not confined to agriculture or demography.

In fact, as a cultural anthropologist, Watson seemed more interested in researching the sociocultural impact of the sweet potato's adoption in Papua New Guinea's highlands. His account of the Ipomoean Revolution began with observations about how the new crop, after replacing taro, helped improve highlanders' subsistence agriculture. They were doing more tillage and building more houses to plant the root, which led to a sedentary lifestyle. That sweet potato was a good pig feed also aided the transition, as did the construction of fences around communities to prevent theft. Sweet potato farming resulted not in only population growth but also waves of migration into the highlands. This movement of people, described by Watson as "an explosion of population in the Central Highlands," resulted in a noticeable increase in the number of languages spoken in the area. Papua New Guinea is one of the world's most linguistically diverse countries, with hundreds of languages spoken by its various ethnic groups.

Watson argued that as the population grew and subsistence agriculture improved, new changes occurred in gender relations and social structure. One was that, prior to the advent of horticulture, men were primarily responsible for hunting and foraging. The cultivation of sweet potato gardens marked a significant shift, because, as previously stated, the task was traditionally performed by women in Papua New Guinean society. Another was that, while women became tenders of the sweet potato, which was consumed as the people's staple food, thus making women the breadwinners in society, men retained their higher social status because there was a higher frequency of warfare after the sweet potato was adopted as the main food crop. The frequent warfare was caused by two main reasons. First was that the cultivation of sweet potato and the flourishing of pig husbandry led to pressure upon land among the communities. Second was that the improvement of subsistence agriculture, in that women assumed a more important role in the supply of food, led to the "increase of time and manpower available for fighting activities" among men. In a later article, Watson also used the term "the Jones effect" (as in "keeping up with the Jones") to describe the sweet potato's broad appeal (as staple food and pig feed) to Papua New Guineans. Due to the Jones effect, in his opinion, different ethnic groups in the country competed among one another to adopt the root plant in order to reap its multivarious benefits. The competition could at times become hostile and warlike.[67]

James Watson developed his thesis of the Ipomoean Revolution out of his research on the highland societies of Papua New Guinea. Did the sweet potato become a staple food across the country's various regions with different terrains due to the country's diverse topography? As previously stated, sweet potato was cultivated in Papua New Guinea at various elevations and became the country's most cultivated food plant. Watson's thesis, though nearly eighty years old, remains largely plausible in explaining the importance of sweet potato farming in sustaining the island nation. Of course, as subsequent research has revealed, the extent to which the Ipomoean Revolution had an impact was not consistent across the country. Taro and yam are still important food plants in some lowland areas, rivaling sweet potatoes. The presence of the Ipomoean Revolution in the highlands also appeared to be more gradual than Watson's thesis suggested. That is, while sweet potato arrived in Papua New Guinea in the second half of the seventeenth century, its widespread adoption as a major food plant took place over a century and a half. As such, it seems necessary to reconsider and modify Watson's idea of the Jones effect in explaining the quick and revolutionary impact of sweet potato farming in Papua New Guinea. Some have contended that instead of a revolution caused by a single crop, the transformation in the island's agricultural system was evolutionary.[68]

More detailed research has been conducted in recent years to explain how sweet potato, despite its relatively short, three- to four-hundred-year history in Papua New Guinea, has become the most preferred food crop in the country's biodiverse environment. Over time, the plant has developed up to 5,000 cultivars—as Bourke has written, "there may be more cultivars in Papua New Guinea than from any other area of the world."[69] The environment had nurtured a number of plants that are either native or well domesticated by the inhabitants prior to its arrival, including taro, banana, breadfruit, sugarcane, sago, and yam. People can also rely on a variety of vegetables, fruits, and nuts for survival, including Oenanthe, *rungia*, various pandanus, *okari*, and *pitpit*. During and after the cultivation of the sweet potato, Papua New Guineans imported a variety of other crops, vegetables, and fruits, which, due to the warm climate, usually grow quite well on the islands. Cassava, maize, tobacco, peanuts, cabbage, pineapple, mango, pawpaw, and mandarin were among them. Needless to say, some of these food plants and vegetables, such as the sweet potato, can grow at high altitudes. However, altitude influences temperature, which affects both the growth period and the per unit yield of a crop. Despite the low temperature at a high altitude, sweet potato performs better than other food crops such as the more traditional taro. For instance, while it can grow quickly to harvest between three and five months in plain fields, the root also matures between five and eight months at 1600–1800 meters above sea level. By comparison, taro requires eight and twelve months to ripen at the same altitude. Moreover, had it been given a longer ripening time (e.g., eight to twelve months), sweet potato can grow at 2200–2800 meters, whereas taro usually cannot. Lastly, as it is drought-resistant, sweet potato develops well in mountain slopes in the New Guinea highlands, where excessive soil moisture is drained from steep land. In many regions today, the root plant thus monopolizes the field after effectively stamping out "combinations of taro, yam and banana in the highlands."[70]

Valleys in highland regions are also conducive to agricultural development in Papua New Guinea. Because it is widely cultivated in the wetlands of New Guinea's valleys, sweet potato has become the country's most important food crop. However, because sweet potato is sensitive to excessive soil moisture, drainage is essential for wetland agriculture. Sweet potatoes are commonly planted on mounds of varying sizes and heights in Papua New Guinea, with the exception of steep hillsides. The small mounds are 30–80 centimeters in diameter and the large ones are 1–5 meters in diameter, all built on drained beds.[71] If the archaeological excavation in the Kuk swamp of the Wahgi Valley suggested an early cultivation of the plant, this certainly continued in later years and spread across the highlands. According to research, various types of drainages developed among the region's diverse ethnic communities. Chris Ballard of the Australian National University, for example, writes in his detailed article on wetland field systems in the highlands of Papua New Guinea that "there is considerable variation in wetland drain dimensions within and between individual highlands field systems, reflecting local hydrological conditions." He goes on to say that the variations, which demonstrate the sophistication of the drainage systems built by communities, are reflected in the rich vocabularies of the local languages used by the people to describe drain sizes and functions.[72]

Sweet potato's advantage as a food crop is fully demonstrated by ditches and mounds. Anton Ploeg, a Dutch scholar, documents the extent to which taro was supplanted by sweet potato in the region in his survey of wetland agriculture in the central highlands of western Guinea. Though taro used to be a staple and grows well in the moist soil of wetlands, he discovers that it has been replaced by the sweet potato as residents became aware of its numerous benefits. In particular, of the areas he studied, which included Paniai Lakes, Ilaga Valley, the Grand Valley of Baliem, and the Eipo Valley, only Ilaga Valley still grows taro to some extent. The remaining three have all turned to sweet potato farming, which has resulted in population growth and migration inflows. For instance, the Dani people in the Grand Valley, writes Ploeg, "transformed their taro-based economy into a sweet potato one, and the effects of this transition were manifold," for the root grows better at this altitude and on slopes surrounding the valley floor and its root and leaves are a better feed for pig raising. "The way of life," he concludes, "of the Grand Valley Dani in the second half of the 20th century, with great population density and large pig herds, seems easier to sustain in an economy based on sweet potato than in one based on taro."[73]

The abovementioned rapid expansion of sweet potato farming in Papua New Guinea coincided with the island's gradual transformation into a nation. Although Europeans were not directly involved in the introduction of the sweet potato, they were deeply involved in the history of Papua New Guinea. Toward the end of the nineteenth century, the newly unified country Germany increased its overseas expansion. It colonized the northern half of New Guinea in 1884, establishing German New Guinea. Meanwhile, Britain ruled the southern half as British New Guinea. Both European colonies included some nearby Melanesian islands. Germany's rule in Papua New Guinea was challenged and eventually replaced by Australia during and after World War I. During World War II, Papua New Guinea, which was then divided into the Territory of Papua and the Territory of New Guinea, became a contested area among foreign powers once more. Japan's desire to control the Pacific led it to challenge Australia's rule over Papua New Guinea in early 1942. Its invasion of the island as part of its New Guinea campaign was fierce but ultimately unsuccessful. The Allied victory over Japan in 1945 resulted in the union of Papua and New Guinea. Papua New Guinea finally gained independence from Australia in 1975 and became a Commonwealth realm.

Amid foreign competitions and conflicts, sweet potato appears to have made steady progress in Papua New Guinean agriculture. When Europeans first arrived in the highlands in the early twentieth century, they discovered that sweet potato had already become a staple food crop among the local inhabitants. And, as R. Michael Bourke points out, the Ipomoean Revolution has lasted well into the country's postindependence days—so much so that Bourke believes there was a second Ipomoean Revolution in Papua New Guinea in the first half of the twentieth century. As discussed in the previous chapter, Japan's brief control of the islands encouraged its growth, as the root was a major source of food for its army in the occupied region during the war. The Japanese occupying forces tried growing rice on the Gazelle Peninsula in East New Britain and on New Ireland at first but quickly switched to sweet potato after their supply lines became strained and were eventually cut off from the homeland.[74]

The second Ipomoean Revolution was marked by the expansion of sweet potato farming in Papua New Guinea's lowland regions, as well as its cultivation and adoption as a major crop, at varying rates, in neighboring islands such as Admiralty, Bismarck, Solomon, New Britain, New Ireland, and Milne Bay. It was another "revolution" because, prior to 1870, or before European contact with New Guinea, "nowhere was sweet potato an important staple food in PNG [Papua New Guinea] lowlands," according to Bourke. But afterwards, it was gradually accepted in local agricultural systems, eventually displacing yam and taro, among other crops. Aside from lowland areas, sweet potato has made significant inroads into Papua New Guinea's urban food system. Since the late twentieth century, Port Moresby, the country's capital, has seen a significant increase in sweet potato consumption and cultivation within the city limits. All in all, sweet potato appeals to Papua New Guineans because "it has facilitated the intensification of land use, increased food production and enhanced food security. Sweet potato offers a greater yield than taro and yam, particularly under conditions of lower soil fertility as land use is intensified."[75]

Needless to say, the shift from the highlands to the lowlands increased the plant's importance as a food source in Papua New Guinea. During the 1960s, the root accounted for about 45 percent of the staple food crops grown. It increased to 66 percent in 2000, four decades later.[76] This remains more or less the case today. In 2009, when *Food and Agriculture in Papua New Guinea* was published, the editors stated that "Sweet potato dominates food production in PNG. It accounts for almost two thirds (64%) of production of staple crops by weight and 63% of food energy production. No other staple food crop contributes more than 10% of the total national production by weight or food energy." While Papua New Guinea is one of the most ethnically and linguistically diverse countries in the world, sweet potato is farmed by almost all rural villagers (99 percent) throughout the country. Since over 86 percent of Papua New Guineans live in rural areas, sweet potato "is the most important food or an important food for more than 80% of the rural population." Indeed, the root is grown in almost all agricultural systems in the country and among a wide range of environments. And in the Papua New Guinean diet, sweet potato readily occupies the second most important position, following greens, in the proportion of different foods consumed by the country's rural population. Papua New Guineans usually roast sweet potatoes in hot ashes or steam them in a *mumu*, which is an earth oven filled or surrounded with hot coal or stones and covered with leaves.[77] Incidentally, *mumu* is also a common pork dish in the country—many Papua New Guineans consider that an ideal meal must have meat or fish to accompany sweet potato or taro.[78]

Sweet potato has thus established its dominance in the Papua New Guinean food system since its introduction around three to four hundred years ago. Over the past century, "both the white immigrants and local population" attempted to diversify the food supply by importing grain foods from Asia.[79] Rice appears to be the most popular option, accounting for approximately 87 percent of total food consumption by the country's urban population as of today. Domestic rice cultivation was also attempted beginning in the early twentieth century. Domestic rice production had reached 300–400 tonnes by the 1930s, and during World War II, as previously mentioned, the Japanese army experimented with rice growing in New Britain and New Ireland, while Australians sponsored its farming in the territory they controlled. After the war, in places like Bougainville, rice was once cultivated as a cash crop. But crop disease, the recognition of cocoa as a better alternative, and long-term decline in worldwide rice prices quickly ended the venture starting in the mid-1950s. In Bougainville, after the interest in growing sweet potato resumed, the root also superseded taro and yam to become the principal crop. A similar situation also occurred in the Markham Valley, where, "after a failed attempt to grow and sell rice," "the main crops grown and marketed" became sweet potato and peanuts. As a result, domestic rice production in Papua New Guinea today is "negligible compared with that of root crops" and constitutes "less than 1% of the quantity of imported rice."[80]

If the sweet potato's dominance in Papua New Guinean agriculture has remained largely unchanged over the centuries, so has the gender division in its cultivation. In fact, because men and women used to live separately, archaeologists were able to identify evidence of sweet potato farming by looking at nearby women's houses and pigsties, as shown earlier. In Papua

New Guinea today, men do the initial, heavy work of clearing and fencing the areas for cultivation, while women do most of the routine gardening work, such as preparing the soil, tending the crops and vegetables planted, harvesting and transporting them, and finally processing them into foods. Taking care of pigs and processing sweet potato roots and leaves into fodder are also women's jobs.

When did this division of labor take place? Scholars have yet to provide a satisfactory answer, because growing taro, one of the staple foods many islanders cultivated and consumed before the sweet potato, required both sexes. Why not sweet potatoes? Regardless of the answer, it is undeniable that the introduction and cultivation of the sweet potato in the islands strengthened women's roles in Papua New Guinean agriculture. In his analysis of the gender division of both the crops and their laborers in the country, Paul Sillitoe called women "providers" and men "transactors" in Papua New Guinean society, for the former provided sweet potato, or the "bread and butter" in a meal, whereas the latter banana and provided sugarcane, which are more like "appetizers and treats" for the dietary occasion.

The augmentation of women's roles in Papua New Guinean society has had both positive and negative consequences. Among the former, it has been observed that sweet potato cultivation and consumption has increased communication between men and women—the root, to some extent, has been perceived and used as a food exchange that helps bind the otherwise fragmented society together. Furthermore, the increased importance of women's social and economic roles may have resulted in more competition among men for wives, which may have aided in raising women's social status.[81] Indeed, in more recent decades, there seem to be discernible changes regarding women's role in Papua New Guinean agriculture and society. As early as 1972, for example, the British anthropologist Andrew Strathern recorded a lengthy complaint by a Big Man, or tribal leader, in the western highlands about how the women of the new generations differed from the previous ones: "Our women used to rear pigs for us men, I built men's houses. . . . Those women of the old times, they put on girdles at their waist, covered their hair with a head-net, carried sweet potatoes in huge netbags. Today's girl walks out and about and lets the weeds grow over the sweet potato garden, she won't rear pigs anymore."[82]

However, because Papua New Guinean women continue to be the primary agricultural labor force in the country, the heavy workload, certainly heavier than men's, with which women are still burdened on a daily basis has been detrimental to their health and life expectancy, because, despite the onset of modernization, the gender division of labor has remained more or less unchanged.[83] Besides the high risk of death in childbirth, Papua New Guinean women also consume fewer nutrient foods than do men due to restrictions and taboos. Papua New Guinea is actually one of the few countries in the world where men's life expectancy (54.6) is longer than women's (53.5), even though since its independence in 1975, the country has adopted a constitution that "has been most progressive and liberal in employing equal rights for all its citizens." All in all, without doubt, women's agricultural work is "vital for the survival and well-being of the families" in the island nation.[84] This vital importance, as shown above, has been closely associated with the introduction and subsequent cultivation of the sweet potato over the course of four centuries ago.

Sweet potato in Hawaiian agriculture and mythology

Papua New Guinea, in southwestern Oceania, is a great distance from Hawaii, a chain of Polynesian islands far to the northeast. Despite this, they have one thing in common: Sweet potatoes are an important part of their respective populations' diets. Additionally, the plant has been crucial in supporting pig husbandry in Hawaii, as it has in Papua New Guinea. Evidently, there are also significant differences in sweet potato farming between the two ends of Oceania. One is that it is generally accepted, as briefly mentioned in chapter 1, that Hawaii probably saw the sweet potato as early as the eleventh century, or several centuries before the advent of the root in Papua New Guinea. This is due to Hawaii's geographic location to the east of Oceania, which is to say much closer to the Americas, where sweet potatoes originated. The other is that sweet potatoes have been successfully incorporated into Hawaiian culture and religion due to their longer history of cultivation. While some people in Papua New Guinea think that sweet potatoes are "an indigenous crop they have always grown," most scientists think that the crop is relatively new to Papua New Guinean culture. Their evidence is that, as cultural anthropologist James B. Watson put it, "sweet potatoes do not provide a focus for ritual, magic, or folklore, as surrounds taro, bananas, sugar cane, or yams." Interestingly, as seen in the case of Taiwan, some Papua New Guineans also call themselves "men of sweet potato," whereas others are "men of taro." The two different identifications, according to local genealogy, reflect the different timings of the cultivation of one crop versus the other—the "men of sweet potato" generation was about five centuries later than the "men of taro" generation. The epithets, therefore, demonstrated that sweet potato was a much more recent addition to Papua New Guinean horticulture.[85]

In comparison, while sweet potato is also an introduced crop in Papua New Guinea, it is in Hawaii that a rich tapestry of folklore and religious significance has developed around it—making the Hawaiian context particularly compelling from a cultural perspective. For although the sweet potato is native to the Americas, as noted in chapter 1, it was introduced to Hawaii and Polynesia much earlier than the so-called Columbian Exchange. And because of its long history on the island, its cultivation had a significant impact on the religious practices and rituals of native Hawaiians. These important sources allow us to reconstruct another case to demonstrate the agricultural and cultural importance of the root crop to the island society.

Hawaii's location is unique in that it is more remote, both geographically and historically, than the other three regions covered so far in this chapter: Taiwan, Okinawa, and Papua New Guinea. For instance, Okinawa and Taiwan share historical ties with China and Japan in addition to being geographically close to the Asian continent. Papua New Guinea is a populous nation in Southeast Asia that shares the island of New Guinea with Indonesia. It is also close to Australia and Melanesia. The Hawaiian archipelago, on the other hand, though a significant area of Polynesia, is not close to any continent or sizable nation. The isolation and insularity of Hawaii are vividly described by Wade Graham, author of a 2006 environmental history of the archipelago, as being "2,557 miles from Los Angeles, 5,541 miles from Hong Kong, 3,847 miles from Japan, and 5,070 miles from Sydney."[86] As such, it remains somewhat nebulous as to when and how the sweet potato reached the archipelago. Among many hypotheses—reviewed in chapter 1—about the transfer of the root from the Americas to Oceania, it seems that the ones proposed respectively by James Hornell, D. E. Yen, and Roger Green are more probable than others. Having evaluated the linguistic evidence—sweet potato is called *uala* or *uwala*, similar to *kumara* in South America—they hold that the root plant should have come from the continent to Hawaii. Moreover, though differing in their opinion on the timing for the root's arrival to Polynesia, they contend that it was the Polynesians who traveled to America and brought the plant back. In other words, these scholars contend that Polynesians, not Amerindians, made the transoceanic trip because of their superior seafaring ability, which allowed them to colonize the archipelago a few centuries earlier, in contrast to Thor Heyerdahl's belief and his bold attempt to reach Polynesia from South America by raft.[87]

In fact, the scientific community generally agrees that the Austronesians who migrated from Taiwan and Maritime Southeast Asia all the way to Polynesia were the same ones who gave rise to the modern Hawaiian people. Therefore, in a way, Papua New Guineans and Hawaiians are related despite their geographical separation—one is in Near Oceania, while the other is in Remote Oceania. And before the sweet potato was introduced, they had been cultivating taro as their main crop, with yam, banana, coconut, sugarcane, and breadfruit as supplements. The last wave of Austronesian migration in Oceania, the colonization of Hawaii, most likely began in the seventh century. A few centuries later, or between the eleventh and twelfth centuries, the sweet potato was probably first grown in Hawaii. However, as alluded to in chapter 1, the scientific community is divided over whether Hawaii or eastern Polynesia were colonized first. It has been proposed that humans first arrived in Hawaii around 300 BCE whereas sweet potatoes could have grown there around 750 BCE. Some others, however, assert that the "refined dates for the first Polynesian arrival in the Hawaiian Islands" are 1220–161 BCE.[88]

All the same, most people agree that the cultivation of taro and sweet potato represented two distinct phases in Hawaii's agricultural history, despite the differences and challenges in dating the colonization of the island. These phases are also depicted in the Hawaiian myths and folk tales that will be discussed below. Taro is a widespread food crop in Oceania and Asia that was probably brought to Hawaii by the early settlers. It was known as *kalo* in Hawaiian and was grown in windward, wetland areas, which were frequently the first places that colonists settled. In Hawaii and many other islands in Polynesia, where the temperature is usually stable all year round, topography determines levels of precipitation. With only a few exceptions, comments Wade Graham, "windward areas are wet, and leeward ones are dry, irrespective of elevation." Thanks to the rainfall, windward areas are perfect for growing taro because they are sunny and well-watered, which is perfect for the plant that thrives in moist soil. In the course of colonization, new arrivals developed sophisticated methods for farming taro, including raised beds to prevent over-moisturization in swampy coastal valleys, pit cultivation on atolls, and irrigated pond fields surrounded by barrages and ditches in highlands with stream valleys.[89]

As briefly discussed in chapter 1, the importance of taro as a food plant in Hawaii, and in Polynesia generally, predates that of the sweet potato. According to Hawaiian myth, taro plant is the older brother of humankind—there is a senior-junior relationship between the two species—"the name of the ancestor of taro, Hāloa, is the same as that of the ancestor of all men."[90] In today's Hawaii, botanists observe, taro remains "the chief source of starch for the Hawaiians." The common way to cook taro is to mash baked or steamed corms to turn them into *poi*, a staple dish, which, depending on the amount of water added, is a gruel or puree, with the amount of water added corresponding to one's desired level of consistency. Islanders also eat cooked taro corms sliced or grated; the latter can be mixed with coconut milk and/or sweeteners to make a pudding called *kūlolo*. Taro leaves, petioles, and spadices are also eaten as greens.[91]

While the role of taro in Hawaii's development is indisputable, the introduction of the sweet potato, regardless of the precise time it took place, appears to have had a greater influence on the island. In Hawaiian folk belief, there is a hierarchy of gods: Kū, Lono, Kāne, and Kanaloa. Kū, a war god, is on top with multiple manifestations. All activities pertaining to and performed by men, such as war and fishing, are attributed to Kū. Kū's manifestations, or his "bodies," are represented by dogs, hawks, birds, trees, and plants that are related one way or another to fishing and warfare. Many of Kū's "bodies" also evoke virility, states Valerio Valeri, who authored an acclaimed book on religion and society in ancient Hawaii, because they are "straight" and "erect." As such, Kū is worshipped on certain agricultural occasions because some of the tools, such as the digging stick, are "straight."[92]

Yet Lono, immediately below Kū in the hierarchy, is perhaps more a deity of agriculture in Hawaii. With thunder and lightning as its special signs, Lono brings life-giving rains to leeward regions where sweet potato and other dryland crops are planted. Sweet potato, indeed, is Lono's sacred plant. In addition, Lono is the god of medicine and wealth and growth in general, hence more directly associated with human life than is Kū. Kāne, still lower in the hierarchy, is the creator god, playing the role of an assistant to Lono in supporting human life. Responsible for irrigated agriculture, Kāne's sacred plant is taro, and he is associated with flowing waters like streams and springs. Last but not least, Kanaloa, who is associated with death, is the god that presides over the underworld. Kāne and Kanaloa are often paired together, representing two ends of one's life.[93]

As the two most powerful gods, Kū and Lono command the religious rituals dedicated to their worship throughout the year. More specifically, New Year in Hawaii begins with the Makahiki season in honor of Lono, which lasts four lunar months from late fall to early spring. Then the remaining eight months are devoted to rituals that honor Kū, god of war, and hence they become the season for war. During the Makahiki season, however, warfare is forbidden because the festival signals the beginning of the agricultural season, marked by the arrival of Lono from Tahiti (Kahiki), where he supposedly resides during the season of war. Warfare is prohibited during this period so as to ensure Lono's unimpeded passage, or to express the hope and wish for a fertile as well as peaceful season. Indeed, during Makahiki, sweet potato gardens are declared *kapu* (sacred), which, Patrick Vinch Kirch and Clive Ruggles write, is a "ritually inscribed means to assure that nothing adversely affected the new crop." Lono is worshipped as a god of fertility because his arrival is accompanied by southerly rain, commencing the growing season in leeward regions where sweet potato, taro, and other dryland crops are cultivated. "The Hawaiian religious system," argues Kirch, an acclaimed anthropologist born in Hawaii, has "cycled on, year after year, decade after decade, through these alternating seasons of gods. Lono would return from Kahiki, bringing with him the southerly *kona* rains, fertilizing the land. His crops would ripen during the Makahiki days; then the priests would circle the island gathering the *ho'okupu* [offerings] for the king. At the harvest ended, all eyes would turn once more to the flaring-mouthed images of Kū upon the great *luakini* temple platforms."[94]

Like Kū and other deities, Lono also has many manifestations. Besides thunder and lightning, he appears as dark rain clouds signaling the coming of the southerly rain. During the Makahiki season, a chant to Lono's arrival thus goes like this:

Thou art Lonoiki [little Lono],
Thou art Lonolui [great Lono],
[The desire of] my eyes, my love, O Lono.

> Thou art Hiwahiwa [black, sacred, precious],
> Thou art Hamohamo [the anointed].
> The season of fruit, the heavenly season,
> When the heavens are covered by black clouds.
>
> The leaf [offspring] of the *Hiwa* [black hog for sacrifice],
>   ... the *Kukui* ...
> The matured shoot,
> The ripened cloud [*ao o'o*]
> The great foreigner [*haolo*, referring to Lono's coming from Kahiki] with bright eyes;
> Thou Kama [child] of hog excrement [sweet potatoes].
> The cloud shaped like a hog in the sky,
> The hog forms of Kama in the underbrush [wild hogs].[95]

The chant unmistakably highlights Lono's significance for the Makahiki festival by highlighting the rainstorm that the deity brings for the agricultural season, which, the chant also makes clear, is focused on raising hogs and growing sweet potatoes. We'll return to how sweet potatoes are viewed as "hog excrement" in Hawaiian myth shortly, as well as how the root is used as both food and fodder in Hawaii, as it is in most sweet potato–producing regions worldwide. Lono is regarded as a foreigner in this region, which is interesting because the root plant—sacred Lono's crop—arrived later than other crops like taro, yam, banana, sugarcane, and others. It is probably a well-known example of Lono's foreignness that when Captain Cook was welcomed by the Hawaiians—which happened, conveniently, during the Makahiki festival—they referred to him as Lono. Since Makahiki fell during both of Cook's visits to the island, in 1778 and 1779, the islanders took his arrival to be that of Lono, the god of fertility who had just returned from Kahiki to begin the growing season.

> Cook came from the sea, as Lono had promised he would, and Cook's ships had tall masts and white sails, shaped very like the upright sticks and swaths of kapa cloth that were carried in the *makahiki* procession to announce the presence of Lono. Cook's course followed that of the main procession, which always went around the islands in a clockwise direction, and he chose to put in for a long stay at Kealakekua,

the home of the chief whose exploits had become part of the Lono myth and the site of an important *heiau* dedicated to Lono.⁹⁶

Yet despite the higher statuses Kū and Lono occupy in the hierarchy of polytheistic Hawaiian beliefs, Kāne and Kanaloa are deities that were worshipped earlier by the islanders. According to Hawaiian cosmology, Kāne is first (and the highest in some tales), followed by Kanaloa and Kū, while Lono is the last. This sequence, as it were, mirrors the historical order in which food crops were introduced to the islands and subsequently became associated with specific deities in Hawaiian mythology. Kāne goes first not only because it is the creator god but also because taro, its sacred crop, was first grown, along with sugarcane and bamboo, in windward valleys by early settlers. Kanaloa is second not only because death follows life but also because it is associated with banana, ocean, and springs, suggesting the expansion of the settlement. Kū, the god of war, is the third. Its association with coconut, breadfruit, and fishing further attests to lifestyle change on the islands. Finally, Lono, or the new god, comes last because the settlers later opened up dryland areas to grow sweet potato and use it to raise pigs to improve life. Lono's newness or foreign-ness is also shown in its association with gourd, which is another foreign plant probably brought back, together with sweet potato, by Polynesians from the American continent.⁹⁷

Though arriving later, Lono assumes more importance in Hawaiian religious rituals over time. In the Makahiki parades, for instance, the king takes the role of Lono, not Kāne. "Lono is the only god," comments Graham, "who takes human form and has no role in Hawaiian creation myths."⁹⁸ Needless to say, Lono's rise in importance is due to the growing importance of sweet potato farming in Hawaii. In both time and space, the archipelago experienced the aforementioned two phases of agricultural development. The first group of colonizers occupied windward areas in the west to grow taro, Kāne's crop, and developed irrigated agriculture. In low and swampy flats around river mouths and the estuaries, taro grew well in moist clays, sustaining the initial colonization. After the population increased, the settlers found that taro, though an efficient crop, could not be planted extensively because land suitable for it was limited in the archipelago. They hence moved eastward to leeward landscapes to grow dryland crops, of which sweet potato, or *'uala*, Lono's crop, became the principal candidate for broad cultivation. Graham describes the development as follows:

A new agricultural form appeared in Hawai'i around AD 1400: intensive cultivation of 'uala and other crops in unirrigated field systems constructed on broad slopes, mostly in leeward areas. Like the other Hawaiian farming types, this intensive rain-fed, or dryland, cultivation was unevenly distributed across the islands, being possible only where the right conditions existed: rocks old enough to form soils but not so old as to be leached and nutrient-poor, combined with enough rainfall to grow crops of mainly 'uala, supplemented with dryland taro and kō [sugarcane] but not so much as to leach away necessary nutrients.[99]

More specifically, whereas taro is still a significant food crop in West Maui and Kauai, sweet potatoes can be grown quite successfully in East Maui and Hawaii, or the Big Island. The former, which are geologically younger islands, have extensive dryland that is perfect for rain-fed intensive farming but few irrigation areas. Pond-field irrigation predominates in the latter, or older islands, where the topography is marked by "significant valley incisions and permanent streams." In other words, the Hawaiian archipelago is home to two distinct "agro-ecosystems."[100]

Hawaii's agricultural development has mirrored that of Papua New Guinea to some extent. While sweet potato predominates in leeward areas of the Hawaiian archipelago and the highlands of the island of New Guinea, taro is the main crop for wetland areas in both Hawaii and lowland Papua New Guinea. The difference between the two is that in Papua New Guinea, taro and other wetland crops were gradually (though not entirely) replaced as sweet potato spread into lowland areas. This did not take place in the archipelago of Hawaii. The wet/dry dichotomy in Hawaiian agriculture is shaped by the coexistence of taro and sweet potato in two agro-ecosystems throughout the archipelago. Similar to how taro is usually eaten, sweet potatoes are also mashed after roasting to make *poi*. Because of the distinct wet and dry seasons in Polynesia and Oceania as a whole, the two complement one another in the food chain. Islanders have long understood that relying solely on "a single crop in the islands would have been dangerous, since crops were difficult to store and emergency supplies unreliable."[101]

Whatever the case, sweet potato continues to be a better food crop than taro. The strong growth of Hawaii starting in the fifteenth century, after the root was widely cultivated on the islands, is a clear testament to the sweet potato's superior quality. Its well-known edaphic flexibility is the first factor in its widespread acceptance as a food plant. In Hawaii, the root demonstrates its ability to flourish "in less favorable locations with respect to sunlight and soil."[102] Indeed, thanks to sweet potato farming, some areas were opened up whose soil type had not been suitable for the farming of other crops previously. For example, in Kahikinui, on Maui, which is a leeward and young volcanic soil area, "sweet potato was the only crop that would grow." The same could also be said about Kaupō District, next to Kahikinui, where the lava and mudflows were found "highly productive to sweet potato and dryland taro cropping." "Kaupō has been famous for its sweet potatoes," comments Elizabeth S. C. Handy, an ethnologist, "and the greatest continuous dry planting area in the Hawaiian Islands." The Kohala area on the island of Hawaii is also now famous for its sweet potato production. However, prior to the introduction of the sweet potato, anthropologist Thegn Ledefoged and his associates note that most leeward regions in West Hawaii, Kohala in particular, were previously uncultivated. "It was the introduction of the sweet potato into Hawaii that enabled establishment and development of dryland agriculture in areas such as leeward Kohala. Over subsequent centuries this area became one of the most productive zones of dryland agriculture in the archipelago."[103]

Second is sweet potatoes' superior tolerance of high altitudes compared to taro, which has been demonstrated by their successful cultivation in Papua New Guinea and South America, the root's native region. As previously mentioned, the colonization of the Hawaiian archipelago moved from windward, wetland valleys to leeward, dryland landscapes. The study conducted by Thegn Ladefoged and colleagues, mentioned before, found that the elevation of upland regions like Kohala was a major factor in why they had not previously been cultivated. That is to say, even dryland taro, the other important cultivar grown in the drylands, had difficulty growing due to the high altitude. Environmental historian Wade Graham describes the land-reclamation process of dryland regions at high altitudes, comparing dryland taro and sweet potato, in the following words: "farther up the valley on steeper, narrower streamside slopes, partially and intermittently irrigated and dryland, or rain-fed, terraces were built for growing varieties of kalo [taro] requiring less water. On the highest and driest terraces, sweet potato, 'uala, would have been grown in preference to kalo."[104] In other words, even though dryland taro is also grown in Hawaii's drylands, sweet potatoes continue to be a better crop to develop at higher altitudes.

The use of sweet potatoes as food and fodder comes in third. Taro takes longer to mature into food than sweet potatoes do. The root typically matures in three to six months while taro takes nine to eighteen months in Hawaii, where planting is not limited by the seasons. After reaching maturity, sweet potatoes can also be buried for several months without rotting.[105] In regard to fodder, Hawaii is of course not the only place where sweet potato is grown as hog feed. But given the root's earlier cultivation in the archipelago, Hawaiian islanders might well be among the earliest people outside of the Americas to realize the root's multiple utilities as a crop. Indeed, researchers have found that well before European contact, or before Capitan Cook's "discovery" of the islands in 1778, pig herds had been regarded as material assets in traditional Hawaiian society. Captain Charles Clerke, who accompanied Cook on his voyages, remarked that Kauai was "the most extraordinary Hog Island we ever met with, take them for Number and size."[106] Kauai, obviously, is not the only island to exhibit vigorous pig husbandry in the archipelago.

Successful sweet potato farming undoubtedly played a role in sustaining such sizable pig herds. While eighteenth-century European explorers were generally impressed by the number of pig herds, they seemed more amazed by the sweet potatoes they saw on the islands. Lieutenant James King, for example, wrote that in Hawaii, sweet potato "thrived prodigiously, indeed it is such Plenty that the poorest natives would throw them into our Ships for nothing." James Trevenen, who was Capitan Cook's navigator, described how the sweet potatoes in Hawaii were "infinitely superior to any others we ever met at the Society, or Friendly Islands. . . . They are bigger than a Man's head, sweet, and mealy when dressed, & when raw taste something like a chestnut."[107]

Hawaiian mythology does a good job of reflecting the fact that sweet potato farming gave the development of pig husbandry in Hawaii a strong boost. Pigs were a major offering on the altars of Lono at the *heiau* (temple or sacred ground), and Kamapua'a, a demigod or "hog child" with a hog head and human body, has also been closely associated with Lono. In fact, Lono and Kamapua'a are inseparable during religious celebrations because Lono has many bodies, one of which is the dark cloud resembling a hog that heralds the arrival of rain and Hawaii's growing season. In addition to the chant to Lono mentioned above, another prayer describing the food offerings at the Makahiki festival illustrates Lono and Kamapua'a's undivided relationship:

> O Lono of the blue firmament!
> Here are vegetables, here is meat,
> An offering of prayer, a sacrifice,
> An offering of fat things to you, o Lono!
> Let the crop flourish in this *ahu-pua'a* [a subdivision of land]!
> The taro stay in the ground till its top dies down,
> The [sweet] potato lie in its hill till it cracks.
> And here is the pig,
> A pig carved in *kukui* wood for you, o Lono,
> Let it remain in your district-altar
> Until the vegetables rot in the fields.
> Such is thy blessing, o God,
> O Lono, look upon your offspring!
> The burden is lifted! Freedom!

It is noteworthy that in the chant, pig was carved in *kukui* (candlenut) wood for Lono, for in some places in Hawaii, hogs were not supposed to be sacrificed for the god of agriculture because hog was a form of Lono. Some did not even eat fresh pork, as sacred to Lono, during the Makahiki months.[108]

A state tree in Hawaii, *kukui* has been an important plant well integrated into islanders' lives for centuries. Its nuts, for example, were burned by Hawaiians to provide light from very early on. Metonymically, as revealed by anthropological study, the tree is also related to both sweet potato and hog or Kamapua'a. *Kukui* leaves, for instance, resemble "in outline the snout and ears of a pig." Likewise, *kukui* leaves are also comparable to sweet potato leaves in form, "which in turn recalls the form of the pig." Indeed, sweet potato is viewed as a "body" of Kamapua'a.[109] Moreover, in some chants, as mentioned earlier, sweet potato roots are described as Kamapua'a's excrement. One example goes like this:

> O Kamapua'a -kane, and Kamapua'a -wahine,
> I am eating of our food.
> O Kamapua'a -kane, and Kamapua'a -wahine,
> The patch is free for you to eat in,
> To excrete in, to sleep in. Amen.

To regard sweet potatoes as Kamapua'a's excrement, observes Valerio Valeri, represents a "dubious honor" for the plant resulting from its association with the popular demigod. To Hawaiian cultural experts Craighill and Elizabeth Handy, it is an example of the humorous nature of Hawaiian mythology: "It must be remembered always in interpreting prayers and chants of Polynesians in myths, ritual and prayer, the humor of this people played no small part. Plainly these canny farmers did not actually believe that the sweet potatoes which they saw grow on the vine-ends were hog excrement. Rather, here is a lusty touch of humor with which the man-with-the-digging-stick savored his labors by picturing Kamapua'a excreting in the patch as he rooted for tubers."[110]

This humorous, metaphorical association with Kamapua'a ought not surprise us, for the hogman often evokes amusement and delight among people. Notorious for his mischievous and sometimes unruly behavior, Kamapua'a is described as "defiant of all authority, bold and untamed.... Treacherous and tender, he thirsts after the good things in life—adventure, love, and sensual pleasure."[111] Indeed, Kamapua'a is well-known for his relentless pursuit of sensual pleasure—his boisterous relationship with Pele, the fire goddess, is but one example. With respect to his relationship with the sweet potato, there is another humorous image depicted in Hawaiian legend—that of Kamapua'a using his pig-like snout to excavate large tuberous roots from the ground. As such, some Hawaiians also call Kamapua'a "god of the sweet potato,"[112] which further attests to the intrinsic connection between sweet potato farming and pig husbandry in Hawaii.

Evidently, the relationship between sweet potato farming and pig husbandry had a profound impact on the archipelago's economy and the lives of its residents. It altered the traditional three-part economy, which included fishing, dryland agriculture, and wetland agriculture. Major crops in the first two categories were taro and sweet potato. "Since taro and sweet potato," maintains anthropologist Timothy Earle, commenting on the economic development of Hawaii, "are both poor in protein, it was necessary to combine the tuber diet with a good protein source, most often fish."[113] To a degree, one can see how this three-pillared economy is reflected in the evolution of Hawaiian mythology. Kū, though later than Kāne and Kanaloa, occupies the highest status not only because it is the god of war but also because of its association with fishing. Among Kū's other "bodies," such as the canoe, which is related to fishing, the color red is said to belong to Kū, "perhaps because it evokes blood spilled during warfare and fishing."[114]

Yet thanks to the vigorous development of pig husbandry, the islanders were able to diversify the sources for their protein intake, for raising pigs fed by sweet potato vines and roots effectively converted vegetable materials into a stable source of protein in the Hawaiian diet. In the process, states anthropologist Michael Kolb, "pigs act as a repository for cropping surpluses that would otherwise be difficult to store."[115] Indeed, since sweet potatoes contain a good portion of water content, it is usually a challenge for people to store them for a long time or transport them over long distances. But using them to feed animals kills two birds with one stone: It utilizes the root plant to its fullest extent while providing animal protein for humans. As such, sweet potato also demonstrates its advantage over taro, banana, and other plants grown in Polynesia and the Pacific. "It is in the singularly important role," proclaims ethnobotanist D. E. Yen, "of conversion of vegetable material into rich sources of protein and fat—animal husbandry—that the sweet potato attains ascendancy over any other single plant species. Not only is its ability in this direction recognized, but also, in the feeding of pigs, there is a marked preference over other cultigens." The use of sweet potato as hog feed in Hawaii is both similar and dissimilar to other sweet potato–growing regions where the root is a source for pig raising. The islanders not only feed pigs with fresh leaves and vines, but they also use slightly rotted and/or substandard roots from the harvest as fodder. After preparing a meal for themselves, they feed scraps from left over the preparation and consumption to pigs. Moreover, Hawaiians allow pigs to forage sweet potato fields not only after the final harvest, but also before the harvest, out of respect for Kamapuaʻa, since sweet potatoes are his "bodies" or "excrement."[116]

The cultivation of sweet potatoes resulted in combined benefits for the islanders. Researchers have found that the root crop was introduced at the same time that the island's population was growing, a topic discussed in more detail below. By opening up dryland areas, sweet potato farming enriched the island's food system and helped to alleviate increasing population pressure. By supporting pig husbandry, it also generated wealth for the elite class. The consequence is that well before European contact, Hawaii had established a well-developed agriculture as well as a sophisticated, if also highly stratified, society. Among the islands in eastern or Remote Oceania, scholars have maintained, "Hawaii is unequivocally placed at the apex of Polynesian development." In terms of social organization, it reached "a complexity beyond the simple chiefdom and approached the state." With respect to demographic and economic change, the archipelago "experienced a dramatic increase in population, the development of a varied and complex subsistence economy, and the rise of a four-tiered chiefly hierarchy." More specifically, argues Michael Kolb, pig raising supported by sweet potato farming had a two-pronged sociopolitical effect. On the one hand, it undoubtedly improved people's diet and aided population growth. On the other hand, pig husbandry also tended to serve and support the chiefly hierarchy in the swelling society, becoming "chiefly surplus," or a symbol of elites' power and wealth. For example, chiefs gave pigs to their retainers and rewarded loyal subordinates during ritual killings and feasts. As such, pigs became a chiefly commodity, helping individual chiefs to gather more power than their counterparts. All this contributed to social stratification. More than anywhere else in Oceania, argues Timothy Earle, Hawaiian society had a clearer class distinction, chiefs versus commoners, genealogically and economically. And the economic distinction "was defined by differential access to strategic resources."[117]

In addition to its sociopolitical effects, the expansion of sweet potato cultivation had an impact on gender relations in Hawaiian society. In Hawaii, men and women have traditionally lived in separate quarters, somewhat akin to Papua New Guinea. Men and women not only consumed different foods but also prepared them in various *imu*, or ovens. *Imu* is a traditional cooking technique still used in the archipelago on ceremonial occasions. It is usually done by digging a hole and lining it with porous stones. After the fire is lit and dies down, food items such as pig, dog, fish, taro, and sweet potatoes are placed in the cavity and cooked under a cover in the residual heat until they are ready. But because women were usually not allowed to eat pork, bananas, and certain fish, cooking generally became the preserve of men because many foods were *kapu* (sacred), and thus not to be contaminated or consumed by women.[118] In wetland areas, women did not work on taro cultivation, nor were they engaged in fishing. But in dryland regions, such as Maui and the Big Island, after sweet potato was adopted as the principal crop, women joined men to work in the fields. Like their counterparts in Papua New Guinea, the women in dryland regions also tended pig herds. It is believed that women took part in sweet potato farming because dryland agriculture relied on rainfall; to increase the chance for a good harvest, sweet potatoes were planted in various terrains, some of which needed more labor in weeding and mulching. On the other hand, women's work in caring for pigs recalled their traditional role in the production of valuables and material assets, as pigs were symbols of wealth and highly valued gifts.[119]

Undoubtedly, the most noticeable effect of sweet potato cultivation was seen in Hawaii's population expansion, which was fueled by the islands' significant agricultural development. Within a few centuries of the root plant's widespread cultivation in the islands, 80–100 percent of the land below five hundred meters in the archipelago was under cultivation, converting what Wade Graham called "a fragile environment" into "a rich agricultural landscape." That is, "in Hawai'i, both irrigation and dryland farming reached their greatest elaboration and complexity in the Pacific." As a result, inhabitants occupied not only older western islands but also migrated to younger eastern ones.[120] The swelling population probably contributed to more frequent warfare, which, as uncovered in Hawaiian mythology, turned Kū, the god of war, into the most revered as well as the most feared deity. Kamapua'a's courting and pursuit of Pele, the goddess of volcanoes, in Hawaiian legend might have also embedded clues as to the former's—sweet potatoes and pig herds—expansion into young, hot-spot islands built of volcanic basalts and lavas.[121]

All in all, prior to European contact in 1778, there was a notable increase in the Hawaiian population, and the development of sweet potato farming and pig husbandry played a role in relieving some of the pressure that might have resulted. With regard to demographic data, there is no confirmed figure for eighteenth-century Hawaii; estimates range from 242,000 to 800,000.[122] Given the archipelago's size (a total land area of 16,638 square kilometers or 6,423 square miles), the population density was clearly high for a premodern society. Did this population figure put a strain on the food supply, or even cause famine? Scholars have proposed various theories over the years to assess the impact of sweet potato–induced dryland, rain-fed economies in precontact Hawaii. Since there are no textual records, these studies have relied mostly on the methods of ethnography, archaeology, and anthropology in coming up with their conclusions. While the assessments seem to diverge somewhat, a general consensus is that the population upshoot did occasion an apocalyptical disaster in Hawaii, as seemed to be the case also on Easter Island, its neighbor in Polynesia.[123]

If precontact Hawaii largely avoided a Malthusian crisis, its people nonetheless faced a near-death disaster after European explorers made landfall on the archipelago. During his second visit, Captain Cook was killed in a clash with the islanders. His death, however, did not deter other Europeans and, later, Americans from arriving on the shores of the otherwise remote islands, bringing not only advanced weaponry but also lethal diseases. Indeed, while imported firearms may have increased casualties in Hawaiian wars, the infectious diseases carried by Europeans and Americans caused far more devastating damage to the Hawaiian population, according to historical geographer Alfred Crosby. A century after Cook's initial visit, during which Hawaii experienced a series of sociopolitical changes, including the establishment of a unified dynasty—the House of Kamehameha (1795–1874)—the native population suffered a significant loss due to its susceptibility to the diseases brought by Europeans. It was recorded that the population of the islands stood at only 48,000 around 1878, a reduction of over 80 percent, even using the most conservative estimate of 242,000 in precontact Hawaii.[124]

In the decades that followed, before and after it was annexed by the United States in 1898, Hawaii attracted waves of immigrants from around the world who helped rebuild its population. Each immigrant group introduced their preferred foods and food plants, which both enriched and eroded the traditional Hawaiian foodway. Rice, for example (mostly imported, as in Papua New Guinea), has supplanted both taro and sweet potato as the daily staple. Rachel Laudan, author of *The Food of Paradise: Exploring Hawaii's Culinary Heritage*, writes that local food among the people in the archipelago today boasts a unique dietary mixture of many cultures:

> The centerpiece of Local Food is the plate lunch, large quantities of rice and meat covered with gravy and eaten with chopsticks, available from lunch wagons and from numerous small restaurants and restaurant chains. Poke is the local fish dish, raw fish cut in small chunks dressed simply with salt and seaweed or more ambitiously with chili peppers, sesame oil, or soy sauce. Teriyaki is the most popular way of treating all meats, including SPAM, which is an island favorite, and sailor's hardtack (called saloon pilots) remains popular. Soy sauce is the universal condiment. This menu is rounded out by typical snacks (musubi and manapua), soups (saimin), nibbles (crack seed), and sweet things (shave ice and malasadas).[125]

While no longer a staple food, sweet potato remains well represented in Hawaiian diets today, thanks not only to native Hawaiians but also to immigrant groups from Asia, such as Okinawans and Filipinos. Though much smaller in number than before, native Hawaiians still commonly eat *poi* as a staple. As mentioned before, *poi* is usually made of taro but, at times, sweet potato and other starches are also used. From quite early on, David Malo, a nineteenth-century native Hawaiian historian well-known for his works about precontact Hawaiian culture, explained how sweet potato was better than taro because "it is excellent when roasted, a food much to be desired. The body of one who makes his food of the sweet potato is plump and his flesh clean and fair, whereas the flesh of him who feeds on taro-poi is not clear and wholesome."[126] Sweet potato is also indispensable at luau, a traditional Hawaiian party or feast, at which kalua pig (cooked in an *imu*) is the principal meat, supplemented by the root (often roasted or made into a sweet), along with *poi*, poke, and laulau.[127]

Sweet potato is a popular food among Hawaii's sizable Asian immigrant population as well, possibly because it was already well integrated into their foodways prior to their arrival in the late 1800s. Okinawans, for example, are highly accustomed to eating the root prepared in a variety of ways. They and other Japanese immigrants have made sweet potato tempura a popular street food on the islands. Filipinos, on the other hand, are experts at preparing sweet potato greens, which are frequently cooked into soups or mixed in salads with other vegetables. In addition, *bitsu bitsu*, or sweet potato scones, is a well-liked confection attributed to them. It uses grated sweet potatoes as the main ingredient, which are mixed with flour and sugar before being fried into fritters. Filipino *halo-halo*, which translates to "mix-mix," is another Filipino desert, usually available in Filipino restaurants and festivals. It mixes up to a dozen items in shaved ice, including sweetened beans, jackfruit, coconut, sweet potato, sugar palm fruit, and other fruits, covered with a syrup or condensed milk.[128]

In conclusion, Hawaii's population changes from the late eighteenth century onward have had a noticeable impact on the foodways of the archipelago. Sweet potatoes continue to be a common food in Hawaii, just as they are in Taiwan, Okinawa, and Papua New Guinea. They are consumed in a variety of ways that reflect the diverse ethnic and cultural backgrounds of Hawaii's residents. Therefore, the American root plant's use in Hawaii has been similar to that in the other three island groups described in this chapter. Sweet potato cultivation and consumption played a significant role in the evolution of the agronomy and foodways on all four islands. Over the past centuries, and especially in more recent years, the root's dominance as a staple food has somewhat waned—including in Papua New Guinea, even though sweet potato remains the top food crop in that nation. As this chapter has shown, a number of factors, including political/regime changes, economic development, and demographic trends are contributing causes. However, eating sweet potatoes continues to be a pleasant experience for many islanders and evokes tender nostalgia for others, despite no longer being the staple or co-staple food on the islands today. Taiwanese, Okinawans, Papua New Guineans, and Hawaiians are thus, to varying degrees, referred to as "sweet potato folks," not only because the shapes of their respective islands can be compared to that of sweet potato, but also because the root crop, given its historical significance, has contributed to the formation of their memories of the past and played a distinctive role in forming their identities.

CHAPTER VI

From "Asian Crop" to "African Crop"

Recent Global Expansions

Two international symposiums on research into sweet potatoes were organized in Asia in the 1980s. One was funded by the World Vegetable Center, formerly the Asian Vegetable Research and Development Center (AVRDC), which was established in Tainan, Taiwan, in 1971. AVRDC was founded as a global nonprofit organization with assistance from the Asian Development Bank and cooperation from Japan, South Korea, Taiwan, South Vietnam, the Philippines, Thailand, and the United States. Within two decades of its establishment, AVRDC swiftly rose to prominence as the world's leading research center on the production of sweet potatoes. It hosted the First International Symposium on Sweet Potato in 1982 in its original location, attracting scientists and academics from about twenty different nations. The symposium's primary organizer was R. L. Villareal, a Filipino horticulturist who was then employed by the Institute of Plant Breeding at the College of Agriculture, part of the University of the Philippines Los Baños. In his introduction to the conference proceedings, Villareal declared that "Sweet potato is basically an Asian crop" because although the root plant had once been ranked second among all root and tuber crops worldwide, it was, by the time he wrote those words, now the most significant root crop in Asia. "It was grown," he explained, "on about 12.3 million hectares, representing 92% of the world's sweet potato area, and Asia also accounted for 92% of the total production, amounting to 105 million metric tons."[1]

The second worldwide gathering on sweet potato research was held in Leyte, Philippines, some years later, in 1987. The International Research Development Centre of Canada, the Philippine Council for Agriculture, and the Regional Center for Graduate Study and Research in Agriculture of the Southeast Asian Ministers of Education—the latter of which is housed at the University of the Philippines Los Baños —were among the institutions and organizations that sponsored it. In his welcome speech, K. T. Mackay, representing the International Research Development Centre of Canada, echoed Villareal by saying that "Sweet potato is an Asian crop in spite of its new world origin. Over 80 percent of the world's root crop production is now in Asia and Oceania."[2]

It should come as no surprise that the first two international conferences on sweet potato research were held in Taiwan and the Philippines, respectively. As discussed in detail in chapter 5, Taiwan was one of four islands whose people considered the root a staple or co-staple food. During the 1980s, sweet potato production on the island began to decline, which likely encouraged its researchers to help sponsor the conference and reflect on the prospect and potential of the root crop as a source of food for the world population. It was also quite natural to choose the Philippines as the site for the second meeting. Indeed, the second gathering was held at the campus of Visayas State College of Agriculture, which overlooks the Camotes Islands in the Camotes Sea. As discussed in both the introduction and chapter 1, *camote* is one of the names for sweet potato, derived from *camotli* in the Nahuatl language spoken by the Amerindians of central Mexico. The fact that the Philippines has both a series of islands and a sea named after *camote* suggested the robustness of the Spanish influence in the country, which arrived via the historic "Manila galleons" that linked Spain's two major colonies in the Americas and Asia from the mid-sixteenth to the early eighteenth century. In other words, besides East Asian countries like China and Japan, where sweet potato has occupied, in different time periods, an important position in the local food systems, the root plant was also present in the agronomies across Asia and neighboring Oceania. Participants at both conferences presented papers that discussed the cultivation and consumption of the sweet potato not only in well-known places like Taiwan, Papua New Guinea, and the Philippines, but also in India, Sri Lanka, Bangladesh, Vietnam, Indonesia, Western New Guinea, and Micronesia. More notably, the first symposium also included three presentations on sweet potatoes in Liberia, East Africa, and

South Africa, suggesting that the root had already begun to expand its role as a food crop across the African continent at the time. Indeed, from the 1980s to the present, a transitional trend has emerged in which the sweet potato has made significant inroads into shaping African agronomies. Sweet potato was once considered an "Asian crop," but it is now considered an "African crop" as well.

Southeast Asia: The nucleus of sweet potato dissemination

Sweet potato became an "Asian crop" because it has been a major food crop not only in East Asia, as shown in chapters 3 and 4, but also in Southeast Asia. As of today, it has been the third most important sweet potato–producing region in the world, following East Asia and East Africa. According to the research of Dindo Campilan, a Filipino agriculturalist who once worked with the International Potato Center, the region produced 4,158,641 tons (3,772,656 tonnes) of sweet potato in 2006, and Vietnam, the Philippines, and Indonesia are among the world's top fifteen sweet potato–producing countries.[3]

Southeast Asia, both its mainland and maritime parts, was the meeting point for the three routes of the root plant's global journey out of the Americas from the early sixteenth century. First was the *camote* line from Mexico and Central America via the "Manila galleons" to the Philippines. Second was the *batata* line from the Caribbean via Portugal to mainland Southeast Asia. And the third was the *kumara* line from Peru and South America via Polynesia to Maritime Southeast Asia and the islands of the Pacific. In addition, the climate in Southeast Asia is ideal for the tropical sweet potato to grow all year round. Throughout the region, with the exceptions of the mountainous parts of northern Vietnam, Laos, and Myanmar, average temperatures are normally above 15 degrees Celsius even in the winter, which is the minimum temperature required for most varieties of the sweet potato to flourish.

In terms of farmed volume and sown area, the Philippines now pales in comparison to Indonesia and Vietnam. But historically speaking, the Philippines played a key role in the root plant's spread around Southeast Asia as well as to East Asia. In Malaysia, which shares its maritime borders with the Philippines, Indonesia, and Vietnam, for example, sweet potato is simply referred to as "Spanish tuber." Toward the end of the sixteenth century, merchants from the southeast coastal province of Fujian, in China, first brought the root vine back from Luzon. It is also commonly acknowledged that Luzon might be the origin for the sweet potato's dissemination in Vietnam during the same century. In the seventeenth century, sailors from Ryukyu/Okinawa brought the root from China to the archipelago, whence it was later introduced to Japan and Korea. According to Chen Zhenlong, the Chinese merchant credited with introducing the American plant to China, sweet potato was already widely grown on the hillsides of the Philippines when he sojourned there. He discovered that there was no need to use fertilizers because the sweet potato matured quickly in the ground for people to dig out and eat as a cheap and convenient food from summer to autumn, supplementing the country's cereal grain consumption.[4] Chen's observation during the sixteenth century helped reveal how the root plant became an attractive food resource for Filipinos after it was brought by the Spaniards to the island nation. Sweet potato "is a traditional staple food and supplementary carbohydrate source," writes Dindo Campilan, a Filipino official at the International Union for Conservation of Nature, four centuries later in 2009, "in communities where rice is now widely grown, or for households unable to afford rice in their regular diet." Indeed, sweet potato has been cultivated across the Philippines' biodiverse environment for centuries, according to Campilan. It is grown not only along the Pacific coast in typhoon-prone areas, but also in remote uplands and irrigated lowlands.[5]

More specifically, sweet potato is grown across the Philippines' three main geographical divisions: Luzon, Visayas, and Mindanao. Luzon, the country's largest and most populous island, and Eastern Visayas are its main sweet potato–producing regions because, as has been the case in Okinawa, the sweet potato has served an insurance crop when other food plants are damaged by typhoon. Leyte, where the second international conference on the sweet potato was held, and the Camotes Islands are both located in Eastern Visayas. For Filipino farmers value the sweet potato not only for its drought-resistance, but also because it is quite hardy against floods, making it an ideal crop in typhon-prone regions. Moreover, its being an easy-to-grow upland crop also makes it a suitable food plant for the whole country because hills and mountains are common in all three geographical regions. Luzon, for example, has Mount Pulag, which is the highest mountain in the country, and in Visayas and Mindanao, there are, respectively, five and ten mountain ranges. In this regard, the farming experience of the sweet potato in the Philippines is comparable to that in Papua New Guinea—it became a major food plant in upland communities where rice is hard to grow or grow well.

Sweet potato thus is a practical rival or useful supplement to rice, the primary food grain in the Philippines. As a mainly tropical and subtropical food plant, rice has also had a long history as the most important source of food for the islanders. It is widely planted in irrigated lowland fields as well as in terraced planes on hillsides and mountain slopes. The famous rice terraces of the Philippine Cordilleras in Luzon, listed as a UNESCO World Heritage Site since 1995, are just one example. In the centuries since the introduction of the sweet potato into the country, Filipinos have learned that the root is a good candidate for intercropping with rice and other food crops. In 1958, William Henry Scott, an American historian and anthropologist who spent ample years in the Philippines and earned a PhD from the University of Santo Tomas, conducted field research in the upland Kalinga communities of northern Luzon. While his focus was on rice cultivation and its importance in the local food system, Scott found that "only the very rich can eat rice they grow in their own terraces three times a day all year long, . . . the poor subsist mainly on camotes [sweet potatoes] grown in swiddens or rice earned from others by manual labor." To preserve soil fertility, swidden agriculture, or slash-and-burn farming, was practiced, but the locals usually only planted rice once a year. After its harvests, they grew bananas, beans, sweet potatoes, and vegetables the second year. "Only in very fertile areas is the swidden used the second time for rice," explains Scott. Although rice was the more preferred staple food, sweet potato and taro were common "between-meal snacks" during the working season, when most of the farmers went to the fields early to avoid the afternoon heat and only ate two meals a day.[6]

In the Philippine Cordilleras, where the well-known rice terraces have been cultivated and preserved for centuries, researchers find that it is also common for rice farmers to practice intercropping with sweet potatoes and other food plants. For example, in Bayyo, Mountain Province, which has been a popular destination for outside tourists to enjoy the scenery of the rice terraces, three production systems are maintained by the local community: irrigated rice fields (*payew*), permanent swidden (*katualle*), and shifting cultivation (*uma*). As the second staple food in Bayyo, sweet potato is grown in all three systems, along with legumes, peanuts, and beans. By contrast, rice is only grown in the *payew*. And in the *payew* system, the most common practice "is relay cropping of rice and sweet potato." That is, "Immediately after harvesting rice in July and August, the 'baliling' cycle starts, (even if the fields are still muddy). 'Baliling' means raised beds in the payew fields, and sweet potato are [sic] planted in these beds. There must be no slack time between sweet potato and rice." The reason for this age-old practice is that intercropping of rice and sweet potatoes improves soil fertility—"it is not possible to have a rice-rice cropping pattern in payew fields. Sweet potato should be relayed or rotated with the rice crop to ensure good rice growth after sweet potato."[7]

Intercropping has thus been the most common method of growing the sweet potato in the Philippines. In the 1980s, F. G. Villamayor Jr., a professor at Visayas State College of Agriculture, studied sweet potato cultivation in Leyte. He polled the farmers and discovered that 58 percent of them regularly practiced intercropping while growing sweet potatoes. The plants with which the root was intercropped ranged from bananas, coconuts, and yam to corn, peanuts, and cassava. He also explained that intercropping was appealing because not only the local farmers in Leyte but also those in Tarlac, in Central Luzon, a key area of sweet potato farming in the country, seldom applied fertilizers to sweet potato fields. Instead, they devised a total of eight ways, including intercropping, to farm the sweet potato while preserving soil fertility. Specific to the local climate, for example, farmers planted sweet potatoes in the early rainy season; when the rain came, the roots would respond rapidly by forming a canopy, thus helping prevent damage caused by heavy rains in the following months. Staggered planting and harvesting were another way, which helped prevent soil erosion in terraced fields. The farmers also engineered ways to maintain the canopy, in both layered and heavy forms, as well as to develop vegetative terracing with tiger grass, fruit trees, and nitrogen-fixing trees. All of this helped to preserve the fields' fertility while achieving a relatively good yield of the root plant.[8] From time to time, however, the two goals do not always complement one another. When this occurs, observes Dindo Campilan, commenting on the practices of some upland Indigenous communities in more recent years, "yield is intentionally sacrificed in favor of natural resource conserving benefits, e.g. cover crop, erosion control mechanism, green manure."[9]

Recent decades have also seen a new trend in sweet potato farming and intercropping in the Philippines: The root has also been grown as a highly valued commercial crop in the irrigated lowland rice paddies after the cereal grain is harvested. This expansion of sweet potato farming has been due to the increasing demand from agro-industries to process the root into both food and feed. In his aforementioned study, Villamayor stated that both the roots and vines of the sweet potatoes were cultivated principally for human consumption in Leyte. Meanwhile, the locals also used sweet potatoes to feed swine and poultry, especially when they found the roots were "unfit for human consumption" but could safely feed them to the animals in cooked form. Another researcher also suggested that sweet potato is one of the "better forage plants available in the Philippines" to feed livestock and poultry.[10]

The farming of the sweet potato in irrigated lowland fields reflected a marked change in the root plant's utilization in the Philippines. It is not merely consumed in a simple boiled form as before; rather, as seen in other countries, a variety of new ways have been developed to process the root into food. Processing the root industrially also means it can be used to feed not only swine and poultry but also other livestock like cattle, thus producing more beef for consumption. The Filipino government has played a notable role in encouraging the country's farmers to increase sweet potato production and meet the demand for turning it into a cash crop. The government's 2021 *Statguide for Farmers* offers information on the general picture of overall sweet potato production in the country, as well as price fluctuations for the root plant on the seasonal market, in hopes of helping farmers achieve success.[11] Thanks to the incentive and stimulation, major sweet potato regions have seen a marked increase in sown area. In Tarlac Province, for instance, "area grown to sweetpotato reached over 22,000 hectares in 2007, which is more than double the area in the early 2000 [sic]." Meanwhile, production volume has also increased: Thanks to farmers' industriousness and ingenuity, sweet potato can have two cropping seasons in the months after the rice harvest.[12]

Given that the sweet potato has primarily been grown for human consumption in the Philippines for many centuries, it has a prominent place in both historical and contemporary Filipino cuisine. The root has been a chief ingredient in several delicacies traditionally favored by the people. *Lidgid*, particularly popular in Eastern Visayas, is a sweet made of sweet potato or cassava mixed with young coconut and sugar. It is wrapped in banana leaves and boiled before serving. *Ginataan*, which literally means "done with coconut milk," is another. It is a soupy dish that can be cooked with a variety of ingredients. However, its sweet form often uses sweet potato cooked in coconut milk. A similarly made sweet soup is called *tinabudlo*, which uses coconut milk to cook sweet potato cubes. Another, *iraid*, is similar to *lidgid*, and is made from grated sweet potatoes mixed with coconut juice and wrapped in banana leaves before being boiled. In addition, Filipinos customarily eat sweet potato leaves, or "camote tops," as a vegetable.[13] This custom has been present in the country for centuries, though in more recent years, it has increased its appeal among the populace. In the early 1980s, when R. L Vallareal conducted his research on the use of sweet potato leaves with other Asian scholars, they began their report with a memorable story that happened in 1972, when the Philippines suffered a serious flood. Afterward, only two green leafy vegetables were available to the people: sweet potato leaves or tips and water convolvulus (*Ipomoea aquatica*). The latter actually shares the same botanical root with the former.[14]

If these sweet potato–based dishes represent traditional ways of consuming the root in the Philippines, innovations have occurred in recent decades as well, as seen in Japan, China, and Korea. The Department of Agriculture issued *Sweet Potato Production Guide* in 2012, in hopes of improving farm productivity in the country. It states at the outset that "Sweet potato is considered as one of the important crops in the Philippines." Besides being used as flour, starch, and animal feed, the guide continues, "it can be processed into different food products with higher economic value than fresh roots." To that end, the pamphlet recommends a number of recipes, including those for sweet potato bread, sweet potato sponge, sweet potato fritters, sweet potato macarons, sweet potato doughnuts, and sweet potato *polvorón*. If sweet potato doughnuts are influenced by America, sweet potato *polvorón* is inspired by Spanish culinary tradition, as this type of shortbread is a specialty in Andalusia. The former is made by mixing egg, sugar, milk, and margarine into boiled and mashed sweet potato before deep-frying the dough in hot oil until golden brown. The latter includes mashed sweet potato, egg, sugar, whipped cream, margarine, raisons, and nutmeg, and is baked before serving.[15] In addition, new technology has been explored to process sweet potato into beverages in the Philippines. Two types of sweet potato beverages seem especially popular; both are developed from orange-yellow-fleshed variety: One is a fruit juice drink with a colorful look and sweet taste, and the other is a sweet potato–cocoa hot drink made of sweet potato juice and cocoa powder. Their advantage over other similar beverages is that both products are rich in vitamin A, a known propensity of the sweet potato.¹⁶ All in all, these new products, among others, attest to the enduring appeal of the sweet potato as a common and valuable food across the Philippine archipelago.

In Indonesia and Vietnam, where sweet potato has been a major food crop for centuries, its consumption and utilization have also undergone discernible changes in recent decades, albeit at an apparently slower pace. For instance, during the 1980s, when Roberto Soenarjo of the Central Research Institute for Food Crops in Bogor, in West Java, conducted his study, he found that in Indonesia 95 percent of the root was grown for human food, with only 5 percent for animal feed. Up to that time, there were no factories that processed the sweet potato. Due to the high demand for sweet potato as food for Indonesians, found R. L. Villareal, the convenor of the first symposium on the sweet potato in 1982, the island nation witnessed an impressive 21 percent increase in the root's production in the previous decade. Three decades later, in 2019, new research showed that the percentage for human consumption still constituted 89 percent of total production, with the rest for industrial raw materials and animal feed. The high percentage of sweet potato devoted to human food in Indonesia has much to do with the fact that many islanders depend on the root in their day-to-day lives and that it is usually consumed right after harvest by the households that cultivate the plant. "In Indonesia," writes Stanley Levy, a Harvard economist, "sweet potato is a common native food which seldom reaches the market places of the larger cities."¹⁷ This is partly because there is an apparent imbalance of economic and demographic development in the country: Over a half of its population are in Java, making it the world's most populous island. Unsurprisingly, four of the five large cities in Indonesia are located in Java. Java produces roughly 30 percent of the sweet potato cultivated in Indonesia and also has the majority of its production centers that use the root as industrial materials. By comparison, sweet potato is a staple food, or "a familiar daily foodstuff in each household," among those living in the rest of the country, especially in the state of West Papua, an immediate neighbor of Papua New Guinea, as well as the small islands in eastern Indonesia. In those and neighboring areas the root is indispensable for making a subsistence living. Evidence for this was already found in the field study by the Japanese anthropologist Tsuboi Shōgorō in the last years of the nineteenth century; Tsuboi reported that sweet potato was a staple food on the South Sea islands near New Guinea, a large island colonized in part by Dutch and German settlers and then eyed for expansion by the Japanese Empire.[18] After Japan's defeat, the western part of New Guinea would join other islands to its west to form Indonesia. Like everywhere else, Indonesians traditionally prepare sweet potato by boiling, steaming, roasting, and frying it. They also use it to make such desserts as coconut sweet potato cake, a type of *getuk lingri*, to go with evening tea.[19]

Like in Indonesia, sweet potato farming in Vietnam is mostly a family affair, meaning the root is consumed mostly in situ after harvest. Another thing the two Southeast Asian countries have in common is that sweet potato is an important food crop grown across their respective diverse agro-ecological zones. Its utilization, however, varies to a certain extent, both within Vietnam and when compare Indonesia with Vietnam. In the Red River Delta, for example, where most of the sweet potato in Vietnam is farmed, the root is used not only for human consumption but also as a major source of pigfeed, along with its vines. The same can be said about sweet potato production in northern midland and northern central regions, where farmers grow the plant as a winter crop after rice harvest and primarily use the root to raise pigs. By contrast, in coastal areas the root is often a staple food for their residents, and in southern regions, such as the Mekong River Delta, sweet potato is grown as a cash crop and sold at fresh markets. In addition to traditional ways of boiling, roasting, and steaming, famous sweet potato dishes in Vietnam include *banh tom*, in which shrimp and sweet potato are combined in a fried fritter, and *che*, a sweet soup usually served in a glass with ice and eaten with a spoon. To summarize, the fact that sweet potatoes have multiple uses and that a significant portion of sweet potato production is for animal feed causes Vietnam's consumption pattern to resemble that of China rather than Indonesia.[20]

By and large, however, sweet potato farming across Southeast Asia bears more commonalities than differences. Firstly, since rice is without question the dominant food crop, sweet potato and other root crops are regarded as secondary food plants in the region. As will be explained more below, sweet potato was once the most important secondary crop, or *palawija*, after rice, for a significant amount of time in Indonesia. In a later period, this importance was challenged by other American food plants, such as maize, cassava, and the white potato. More specifically, sweet potato became the third most important crop in the early nineteenth century, following rice and maize, and from approximately the mid-nineteenth century, it was ousted by cassava to become the fourth in Indonesia. In Vietnam, by comparison, it managed to maintain the third position, after rice and maize, well into the present times.[21]

Secondly, as shown in the Philippines, sweet potato farming in Southeast Asia is mainly through intercropping with rice, the more preferred grain cereal among its people. Like their counterparts in the Philippines, Indonesian and Vietnamese farmers usually grow sweet potato in irrigated lowland fields after rice is harvested in the dry season. Maize is another major food crop grown in the region, and usually occupies second place, following rice, in terms of its importance as a food. The intercropping system, therefore, involves both rice and maize, following either the rice/sweet potato rotation or the rice/maize/sweet potato rotation, for maize "can substitute for rice as food or as a cash crop in urban areas." In areas where rice and maize are not the main food crops, sweet potato can be intercropped with other plants. "In the lowlands," observed Yan Pieter Karafir, who studied sweet potato farming in Western New Guinea, "mixed cropping is normally practiced. On a piece of land, sweet potato is usually mixed with other crops such as taro, banana, yam, and other vegetables." A similar practice is observed in Vietnam: Sweet potato is cultivated in irrigated flatland, along with rice, maize, groundnut, and vegetables, whereas tea and fruit trees are planted in upland fields. Yet in upland regions of Indonesia, sweet potato is often the sole crop, and people's staple food, thanks to its drought-resistant qualities. Meanwhile, in the rain-fed regions of Indonesia, sweet potato is also a favored crop, as it can be cultivated in mounds alongside other secondary crops such as beans, taro, yam, and jicama.[22]

Thirdly, throughout Southeast Asia, farmers used to grow taro and yam, the experience of which seemed to make them more likely or able to accept sweet potato as a new root crop beginning in the sixteenth century. Indeed, after its introduction to the region by both the Spanish and the Portuguese, sweet potato readily demonstrated its advantage over both taro and yam, which explained its quick ascent to the status of preferred secondary crop in the region. Thanks to the tropical and subtropical climate, farmers in Southeast Asia found that not only was the sweet potato suitable to be planted after rice harvest, often with two crops in a year, but it was also palatable, easy to grow, and offered high yields. Sweet potato can grow year-round across the region, and even in northern Vietnam it can have three seasons if planted as a monocrop. As on many islands across

Oceania, root crops such as taro and yam had been familiar food crops for the people of Southeast Asia. In Indonesia and Vietnam, taro had been well integrated in local dietary traditions for centuries. Yet in Indonesia, taro's dominance was gradually undermined by yam, for though the latter was a bit harder to grow, it stored better, was less susceptible to fungi and insects, and was more suitable for intercropping with rice. The arrival of Islam played a role as well: The disappearance of domesticated pigs reduced the demand for taro, which had been traditionally treated as a pig feed. All the same, this experience in cultivating root crops prepared Southeast Asian farmers to welcome New World crops like sweet potato, cassava, white potato, and jicama, which is called *bengkuang* by Indonesians. In Indonesia and Vietnam, both of which had been European colonies before World War II and briefly thereafter, the advance of sweet potato cultivation was both swift and sequential. According to Peter Boomgaard, a contemporary Dutch expert on Southeast Asia, sweet potato was "the first successful crop" among all foreign food plants introduced to Indonesia. Its development continued for about three centuries, from the early seventeenth to the first half of the nineteenth century in Indonesia. In Java, for example, sweet potato was grown in two-thirds of the land planted to roots and tubers in 1850 and clearly ranked first among all root crops. The sweet potato's popularity was amply recorded by European explorers who visited the archipelago during the period. In addition, Chinese farmers played their part in promoting its cultivation: In the late eighteenth century, they became "explicitly identified as sweet potato growers" around Batavia. It was also quite likely, as explained in chapter 5, that sweet potato spread from Indonesia to Papua New Guinea and became the dominant food plant in the latter, just as it had been in western New Guinea, or Indonesian New Guinea.[23]

Fourthly, thanks to the extended tradition of growing taro and yam in Southeast Asia, sweet potato has been easily integrated into the regional agronomy—so much so that many locals simply consider the American plant to be native to the region. While it was quite clear that the Spaniards brought the sweet potato to the Philippines, there is still some uncertainty regarding how the American plant reached Indonesia and Vietnam. Right after Magellan's circumnavigation voyage, a survivor arrived in what is today Indonesia and found that the locals grew yam, which was called *gumbili*, and which became *batate* in his translation. To this day, though, the Moluccans on the Maluku Islands still use both *gumbili* and/or *batate* to refer to the sweet potato. Nowhere is this linguistic variety on greater display than in Indonesia, where the sweet potato is known by a wealth of different names, testifying to the many peoples who brought the root plant to the country. It reflects not only the history of Maritime Southeast Asia as the meeting point where the root's different routes out of the Americas converged but also the widespread presence of root crops in food culture in general.[24] In the case of Vietnam, there is a distant possibility that the Portuguese might have introduced sweet potato into mainland Southeast Asia. A better likelihood, however, was that the root entered Vietnam via the Philippine island of Luzon. As discussed in chapter 3, some Chinese researchers have identified Vietnam as one of the countries whence sweet potato was introduced to China, suggesting that the root could have had a longer history in Southeast Asia than on the Asian mainland. While presenting his paper on sweet potato farming in Vietnam in the 1980s, Mai Thach Hoanh, who worked at the Food Crop Research Institute under the auspices of the Agriculture and Foodstuffs Industry Ministry of Vietnam, proclaimed at the outset that "sweet potato was first cultivated in Vietnam 2,000 years ago."[25] In his well-documented research, the Dutch historian Peter Boomgaard recorded a similar claim in Indonesia—that "by the 1670s," sweet potato "had become so familiar that its foreign origins were no longer apparent."[26] This perhaps explains why in the above example, the Moluccans (who are by no means alone in Indonesia) use interchangeable terms to refer to both the yam and sweet potato.

South Asia and the Middle East: Sweet potato as vegetable

To the peoples of both South Asia and the Middle East, the sweet potato, in contrast to its image in parts of Southeast Asia, is readily seen as a new and foreign addition to these regions' food systems. Yet there are also marked differences between both with respect to locals' acceptance of American plants. In South Asia, especially along the western coast of India, American food plants were not an unfamiliar presence in general from the early modern period. Due to its successful colonial expansion during the fifteenth and sixteenth centuries, Portugal established the "Estado da India" (State of India), also known as Portuguese India. It stretched from the Cape of Good Hope in Africa via the Persian Gulf all the way to Japan and Timor and was comparable in size with New Spain, the Spanish colonial possession in the Americas. Goa, the capital of the Estado da India, located on the southwestern coast of the Indian subcontinent, was Portugal's first colony in India, established in 1510, and by 1560 it had grown into a city of 225,000, surpassing both Lisbon and Madrid in terms of population. The Portuguese were responsible for introducing a series of American plants into the region and subsequently to the entire subcontinent. These plants included chile pepper, okra, maize, papaya, pineapple, cashew, peanut, guava, tobacco, custard apple, and, of course, both sweet and white potato. In many Indian languages, potatoes are referred to as *alu*, which is a Sanskrit word. But in Maharashtra, a state north of Goa on the western coast, the natives, the Marathi people, use the Portuguese term *batata*, which refers to both sweet and white potato in Portugal, then and now.[27] This usage could well suggest that what they encountered first was the sweet potato instead of the white potato, because the latter came later to Portugal and Europe in general.

While Portuguese were responsible for introducing the American root plant to western India, it was the English who left us the first record of its consumption on the subcontinent. In 1615, the newly founded East India Company sent Sir Thomas Roe, an accomplished diplomat with ample traveling experience, to be the English ambassador to the Agra court of the Mughal Empire. Roe stayed in Surat, a port city in Gujarat, on the western coast, for over three years until 1619. During his term, Roe appointed Edward Terry, an English vicar, to be the chaplain for the English Embassy after the latter's arrival in India in 1616. Terry later wrote a quite detailed account of his voyage to India and his time with the ambassador, offering a kaleidoscopic description of the subcontinent and its inhabitants. In his chapter entitled "Of their Diet and their Cookery in dressing it," Terry complimented local gastronomy and dining habits. He stated that while foods were abundant and versatile, with rice being the staple, people were temperate in their diet because they believed that "full bellies do more [to] oppress than strengthen the body; that too much of the creature doth not the comfort, but destroy nature." But when the Indians treated their guests, it seems they were exceedingly generous in their offerings. One day, the English ambassador and his entourage were invited to dinner by Asaph Chan, a noble in Ajmer, Rajasthan, immediately to the north of Gujarat, who boasted close connections with the Mughal royal family. Asaph Chan treated his guests in a spacious tent, and they all sat on carpeted floors, using dishes of fine silver. They were offered rice as the grain food, dressed in different colors, and several dishes with "flesh of several kinds, and with hens and other sorts of fowl cut in pieces, as before I observed in their Indian cookery." After the main courses, Terry wrote, they were treated with a variety of sweets, made of rice and wheat flour and mingled with almonds and sugar candy. He then offered the following description: "To these potatoes excellently well dressed; and to them divers sallads of the curious fruits of that country, some preserved in sugar, and others raw; and to these many roots candied, almonds blanched, raisons of the sun, prunellas, and I know not what, of all enough to make up the dishes before named; and with these *quelque chose* was that entertainment made up."[28] The "potatoes" mentioned by Terry would likely be sweet potatoes, because of they were in the company of other desserts that completed the banquet.

During the latter part of the seventeenth century, another Englishman recorded his observations of the American root plant in India. In his *A New Account of East India and Persia*, first published in 1698, John Fryer recorded his trip to India between 1672 and 1681. Fryer was an English physician and a fellow of the Royal Society, and he described seeing potatoes planted in gardens with other plants in Surat, such as pumpkins, cucumbers, and gourds. He also observed that "potatoes are their usual Banquet."[29] There is strong reason to believe that what Fryer saw was also sweet potato, for before the eighteenth century, most Europeans, English included, simply called sweet potatoes "potatoes." By the time Fryer traversed the subcontinent, sweet potato had become a common plant cultivated in Kanara, which is the stretch of land along India's southwestern coast.

And the expansion of sweet potato cultivation continued in the following centuries, spreading across the subcontinent. During the second half of the eighteenth century, William Roxburgh, a Scottish physician educated at Edinburgh University, spent many years in Madras (today's Chennai), in southeastern India. Roxburgh developed an intense interest in studying Indian botany, so much so that he later received an official appointment as superintendent of the Calcutta Botanical Garden. He hired local artists to help draw illustrations of the plants he saw in India. After his death, a two-volume *Flora Indica* was published posthumously with the help of his friends. Among the many plants he described was the sweet potato, or *batatas*, in his terminology. Roxburgh described *batatas* as "Root tuberous. Stems creeping, rarely twining. Leaves cordate, angle-lobed. Peduncle many flowered. Segments of the calyx oblong, smooth, acute." He also pointed out that there were two varieties: one red and the other white. And "the red sort," he continued, "is in very general cultivation all over the warmer parts of Asia and very deservedly esteemed one of their most palatable and nutritious roots." He suspected that a similar variety was cultivated in Japan.[30]

Roxburgh's observation about sweet potato's broad appeal was confirmed by his fellow countryman George Watt. Like his predecessor, Watt was also trained as a physician and subsequently acquired an ardent interest in botany. Through his detailed book *The Commercial Products of India*, published in 1908, Watt established himself as a trusted "reporter" on India's economic botany. By the time he wrote about the flora of India, both potatoes had become commonplace throughout the subcontinent, including in what is now Myanmar, Bangladesh, and Pakistan. However, Watt mistook Edward Terry's first mention of "potatoes" for the white potato instead of the sweet potato, and concluded that "It would thus appear that within a remarkably short interval, after the discovery of the potato in America, it had been conveyed to India and was apparently at once taken up by the better class Muhammadans as a desirable addition to the ordinary articles of diet."[31]

Despite his confusion of the two potatoes, Watt's description was perhaps a testament to the quick reception of the sweet potato in India after its introduction by the Portuguese. Meanwhile, his *Commercial Products of India* also provides an extensive description of *batatas*, or sweet potatoes, stating that the root plant "is extensively cultivated in India." More specifically, he wrote,

> The area under the sweet potato in India cannot be definitely ascertained. It is grown all over the country from the Panjáb, the United Provinces, Rajpuntana, Central India, the Central Provinces, Bengal, Assam, Bombay, Madras and Burma. In Bengal, it is more important in the eastern tracts, such as Bogra and Bhagalpur, than in the western and central divisions. Taking India as a whole, it is planted from August to November and reaped from December to May, the variations being a consequence of local climatic conditions and methods of propagation.[32]

Watt's observation confirmed that if sweet potato was first cultivated along India's western coast, by the end of the eighteenth century, it had spread to Bengal, Bombay (Mumbai), and Madras (Chennai) to the east. In fact, he pointed out that the root plant actually became more popular in the eastern part of the country, including what is today Bangladesh, since Bogra is the country's main region, than the west. He also noted that, compared with the white potato, the sweet potato contained more "dry starchy and sugary matter," thus it was "an excellent source of alcohol." More importantly, Watt stressed that sweet potatoes were commonly consumed by the "natives of India" as food, "either cooked in curry or boiled, roasted or fried."[33]

Given its popularity, did sweet potato assume the same status as either a staple or a co-staple food among the peoples of South Asia as it did in East Asia? Sucheta Mazumdar, a Duke University historian mentioned in chapter 3, contributed a chapter to the edited volume *Food in Global History* that compares the reception of the sweet potato, and American food plants in general, in India and China. She states that from the sixteenth century, both sweet potatoes and white potatoes were introduced to India and subsequently integrated into local cuisines as vegetables, but not as staples, as was the case in East Asia. She also believes that given the precedence of the sweet potato over the white potato in Europe at the time, the aforementioned accounts of the "potatoes" by Edward Terry and John Fryer should describe the former rather than the latter. For "the white potato," she writes, "was far from a widespread crop in England at this time and the only potato the Englishmen would have been sufficiently familiar with to recognize straightaway would have been the sweet potato introduced almost a century before." Besides the different patterns of landownership in India and China, Mazumdar also considers certain demographic factors, drawing attention to the slow growth of the Indian population between 1600 and 1850 and the "vast abundance of fertile arable land" on the subcontinent. All of this, she argues, accounted for why the sweet potato, the easy-growing and high-yielding food plant, failed to make quick inroads in the food culture of India.[34] That is, India did not face the same degree of population pressure as China did from the seventeenth century onward.

A game changer occurred in the nineteenth century, acknowledges Mazumdar, when George Watt conducted his study of India's economic botany. Thanks to the active promotion of the British colonial government and the rapid urbanization the country witnessed during the period, both potatoes became integrated into the Indian food system. As of today, she writes, "one can scarcely imagine an urban Indian middle-class meal cooked entirely without using either tomatoes or potatoes." She also cites Watt's study in describing the growing importance of the American roots: "As an article of food, potatoes are now valued by all classes, especially by Hindus on days when forbidden the use of grain. . . . it is now not [an] uncommon circumstance to find cooked potatoes offered for sale at refreshment stalls, in various cold preparations, to be eaten along with so-called sweetmeats that form the midday meal of the city communities."[35] What Watt described here was supposedly related to the white potato. But judging from its context, it could well be a reference to the sweet potato, because the root has been more commonly made into various forms of sweetmeat, either at home or sold on the streets, across the world.

Yet Watt was probably right in saying that sweet potato assumed a more important status as a food crop in the eastern portions of the subcontinent than the western areas beginning in the nineteenth century. The root plant, as pointed out in a recent article on food production in Bangladesh, "plays a significant role in increasing food security and income for the poor farmers" in the country. At present, sweet potato is ranked the fourth most important source of carbohydrate in Bangladesh, following rice, wheat, and white potato.[36] By comparison, a comprehensive book on Indian food, presenting eighty-four recipes in total, offers only one dish that includes sweet potato as an ingredient. The root is also absent in the description of Indian food in the *Food Cultures of the World Encyclopedia*, a massive book that covers foodways and culinary practices around the world in four volumes.[37] In other words, one should probably agree with Sucheta Mazumdar that sweet potato is mostly regarded and treated as a vegetable in South Asia, but not a staple or a co-staple food, as it has been in East Asia.

Meanwhile, it is worth mentioning that there are several traditional dishes that feature the sweet potato as its chief ingredient. And, reflecting perhaps the culinary influence of the Portuguese, these dishes tend to be from western India. The single recipe contained in the aforementioned book on Indian food is called "Gujarati spiced vegetables with coconut." As suggested by its name, it is from the western state of Gujarat, north of Goa, the formal capital of Portuguese rule. The dish is a variation on *undhiyu*, a vegetable mixture from the city of Surat. It is usually cooked in a clay pot called *matla*, in which several layers of assorted vegetables are placed. The *matla* is then fired upside down—as the term *undhiyu* connotes— which gives the vegetables a distinct smoky flavor.

Sweet potato is also a common ingredient in the food culture of Maharashtra, a state sandwiched between Gujarat and Goa. Also called *ratala* in Marathi, sweet potato is particularly desired during fasts. One popular dish with sweet potato as its main ingredient is *ratalyacha khees*, which contains ghee (a clarified butter), ground roasted peanuts, freshly grated coconuts, and various spices to one's taste, offering balanced nutrients of carbohydrate, protein, and fats. If in *ratalyacha khees* and *undhiyu* sweet potato serves as a vegetable, the root can also become an ingredient in making certain Indian grain foods. Sweet potato *puran poli* is an example, which is a sweet, stuffed flatbread with mashed sweet potato and jaggery, prepared in just about every house for every occasion, especially during festivals, in Maharashtra. Sweet potato *roti* is another—a round flatbread or pancake flavored by the sweetness of the root. *Upma* is a thick porridge made of coarse grain flours of rice or semolina, commonly eaten as a breakfast or a snack across the subcontinent. Sweet potato *upma* is the preferred way of eating the porridge because of its distinct flavor. Likewise, one can also make sweet potato *khichdi*, another porridge usually made of rice and lentils.

Sweet potato is a popular ingredient in Goan cuisine. For breakfast, *kangaanchi usli* is a traditional dish consisting of diced sweet potato boiled with grated coconut, chile pepper, salt, sugar, and water. In the Konkani language, sweet potato is called *kanga*, while in Marathi and Hindi it is known as *ratala* and *shakarkand*, respectively. For lunch or supper, one can cook sweet potato curry, or *gajbaje* in Konkani, which is a vegetarian dish—the diced root is cooked with chickpeas in a pot, together with coconut milk, onion, garlic, ginger, and spices. Given its sweetness, sweet potato is of course readily turned into a sweet delicacy. *Kangacheo neureo* are sweet potato dumplings, stuffed with coconut and jaggery, pan-fried in ghee. *Kangachi kheer* is a dessert with sweet potato cubes, coconut milk, and jaggery, usually served cold when it thickens. Finally, as a holiday favorite, bebinca, a common Goan, or Indo-Portuguese, layer cake, made of flour, coconut milk, ghee, egg yolk, and sugar, can also feature sweet potato as one of its layers. In addition to these regional specialties, there are grilled sweet potato, baked sweet potato fries or *shakarkand* finger chips, sweet potato cutlets fried in ghee, and, as sold by roadside street vendors, roasted sweet potato cubes sprinkled with lemon juice and *chaat* masala. Sweet potatoes are also often found in such popular Indian desserts as sweet potato *rabri*, or *shakarkand rabdi*, a condensed-milk dish, sweet potato halva, originally a Persian confection, and sweet potato *daliya*, a sweet porridge made of broken wheat and lentils.[38]

If records about the history of the sweet potato in India are sparse, there is indeed a glaring paucity of information about the dissemination of the American root in the Middle East. Reasons for this are not so hard to find: Nestled in the middle of the East–West interaction on land, the region witnessed the rise of Arabs from the seventh century, followed by that of the Turks from the eleventh century; those strong Muslim powers forced Portuguese, Spaniards, and other Europeans to seek a sea route to Asia from the fifteenth century, and thus to circumvent the Middle East. It was not until the mid-eighteenth century, with the rise of the British Empire and decline of the Ottoman Empire, that parts of the Middle East, such as Syria and Egypt, were opened to European influence. The construction of the Suez Canal between 1859 and 1869 helped return the Middle East to its historical position as the nexus in East–West communication. One could imagine that it was from the nineteenth century on that the sweet potato, the white potato, peanuts, tomato, chile pepper, and other American plants became gradually integrated into Middle Eastern food culture.

Although sweet potatoes have been grown in the Middle East since the nineteenth century, the amount produced in the region has not been sufficient enough to transform the root plant into a major food crop. Egypt may be an exception in that sweet potato cultivation has recently seen a noteworthy increase in sown area due to the country's warm climate and government promotion. As a result, one statistical study reports that Egypt is now ranked third in the world for exporting sweet potatoes, behind the United States and the Netherlands.[39] On Egyptian city streets, like elsewhere, it is not uncommon to see street vendors selling roasted sweet potatoes, while cooking sweet potato casserole in spiced Egyptian sauce made of lean meats such as turkey is often a festive delicacy in many Egyptian homes. The root has frequently been tapped for its sweetness in the preparation of desserts; suggested recipes include Egyptian sweet potato and bechamel pudding, sweet potato crunch, and sweet potato with condensed cream, or *batata bel eshta*.

West of Egypt, nations such as Libya, Algeria, Tunisia, and Morocco frequently prepare sweet potatoes with beef. In the well-known Libyan couscous dish *kushsu*, for instance, chunks of sweet potato are cooked with beef or lamb and other vegetables. It has lately been claimed that using sweet potatoes instead of pumpkin will increase the sweetness of *Kara'a*, a

Libyan pumpkin dip, which is also legendary.[40] *Batata bel kamoun* is a traditional Tunisian beef stew with sweet potato, spiced with cumin and harissa. In Algeria, *chtitha batata* is a common food cooked in most families. Though meat can be used as an ingredient, it is mainly a vegetable dish in which sweet potatoes (or white potatoes) are cooked with a red sauce of onions, tomato puree, paprika, and chickpeas, seasoned with salt, cumin, and pepper. Incidentally, roasted sweet potato seasoned with cumin is popular across the Middle East and the Mediterranean world as a whole.

From the Maghreb to the Levant, sweet potato kibbeh is arguably the Lebanese adaptation of *chtitha batata*. While still a classic meat meal, here the root is substituted for ground beef or lamb and the kibbeh prepared with a range of spices to the individual's preference. Also popular in Lebanon, Syria, and Israel is roasted sweet potato. The Syrians dip their roasted or grilled sweet potatoes in their well-known hot sauce, *muhammara*, while the Israelis often add za'atar. Sweet potato latkes or fritters are a typical Hanukkah dish in Jewish cuisine. With regionally specific sauces and spices, these methods of preparing the sweet potato are also common in Iran and Iraq. In both countries, quince crisp is a popular dessert in which sweet potatoes can be used along with other fruits and vegetables. Overall, even if it is a latecomer, sweet potato is by no means a foreign food in the Middle East. It is true that at present, its appeal in the region is dwarfed by that of the white potato. However, recent studies predict that the market share of sweet potatoes will increase in the years to come.[41]

Becoming an "African crop": Sweet potato's recent advance

When the first and second international symposiums on the sweet potato were held in the 1980s, the majority of the participants agreed that the root plant was de facto an "Asian crop." However, in more recent decades, it has gradually come to be thought of as an "African crop." When Jennifer Woolfe wrote her book *Sweet Potato: An Untapped Food Resource* in the early 1990s, this pattern of development had already emerged. She notes that in the 1960s, sweet potato output in countries like Japan, China, and the United States was already beginning to trend downward, whereas in Africa, production was "more than 80% above its 1960 level."[42] In more recent years, Peter T. Ewell, the International Potato Center's regional representative for Sub-Saharan Africa, points out in his study that the trend of noticeable increase has accelerated and that over the first two decades of the twenty-first century, sweet potato production in Africa has doubled. Indeed, from the mid-twentieth century, Africa is the only continent where sweet potato production has seen a steady increase; this is in contrast to a trend of decline elsewhere in the world. Regarding the current situation of its cultivation, the International Potato Center reports that "sweetpotato is the third most important food crop in seven Eastern and Central African countries—outranking cassava and maize. It ranked fourth in importance in six southern African countries and is number eight in four of those in West Africa."[43]

As of today, sweet potatoes are grown all over Sub-Saharan Africa. While China continues to produce the highest volume of sweet potatoes globally, Malawi, Nigeria, Tanzania, and Uganda are all in the top five. Along with Indonesia, Vietnam, and the United States, the top ten countries on the list also include African nations like Ethiopia and Angola, in the northeastern and southwestern parts of the continent, respectively. Without a doubt, sweet potatoes are a successful food crop in Africa today.[44]

It is worth noting that the history of sweet potato cultivation in Africa actually predates that of Asia. The Portuguese, who had previously succeeded in establishing a route from Europe to Asia by circumnavigating Africa, are widely regarded as the first to introduce the American root plant to West Africa in the sixteenth century. The Portuguese brought a number of new food plants from the Americas to coastal regions in both Africa and Asia as they established the Estado da India, or Portuguese India, which covered the enormous expanses connecting the Atlantic and Indian Oceans. M. Akoroda, a professor at the University of Ibadan and a Nigerian, makes the following observations about the development of sweet potato farming in West Africa:

> Portuguese brought it [sweet potato] to their settlement or trading stations in Africa along [the] Gulf of Guinea: Sierra Leone, Sao Tome and Principe, and Fernando Po [Bioko in Equatorial Guinea]. Sweet-potato cultivation was first mentioned in Sao Tome in 1520. In the 16th century, Portuguese navigators introduced sweetpotato to Africa, Europe, India, and Indonesia; and [sweet potatoes] were widely cultivated in West Africa by [the] end of 17th century.[45]

Nonetheless, as discussed in chapter 1, theories about the potential pre-Columbian movement of the sweet potato from Oceania and Southeast Asia to Africa have begun to circulate in scholarly circles around the world. Some scholars believe that one of the many varieties of the American root plant grown in East Africa made its way there from Asia via the Indian Ocean long before European contact. This theory is bolstered by the fact that the varieties grown in the Tanzanian highlands around Arusha and Papua New Guinea's highlands are genetically related.[46] On the other hand, Harold C. Conklin, who was also mentioned in chapter 1, stated that the most common names for the sweet potato in Africa are *batata*, *tata*, and *mbatata*. This suggests that the root was introduced to Africa via the *batata* line of transfer, with Spaniards and Portuguese bringing it to Europe before transporting it to Africa. More intriguingly, he noted that the sweet potato's second most popular names are *bombe*, *bambai*, *bambaira*, and *bangbe*, indicating that the root plant originated in Bombay (Mumbai), "the large Indian trading and shipping center nearest the Portuguese Goa trade terminus." Conklin's conclusion therefore is that "late fifteenth-and-sixteenth-century Portuguese ships were the first to carry sweet potatoes to Africa, Goa, and parts of Indonesia." The editors of *The Cambridge World History of Food* agree with this conclusion; citing additional research, they add that "The Portuguese ports of Mozambique probably saw the introduction of *batata* from India as well."[47] In a nutshell, the Portuguese are more likely to have introduced the sweet potato to both West and East Africa.

It might make sense to start with sweet potato cultivation in West Africa if we assume that the Portuguese were responsible for bringing the sweet potato to Sub-Saharan Africa. Stanley Alpern provides some details about how the root plant was transferred from Portugal to coastal areas in West Africa as early as the sixteenth century in his study of European crop introductions into the region. He notes that *batatas* were first grown in Madeira in 1526, when the Portuguese brought the root plants to São Tomé, an island country off the coast of Central Africa that became Portugal's first colony in the fifteenth century. Since the locals had become accustomed to growing yam, sweet potato—the yellow-fleshed variety in particular—was readily accepted as a new crop, even though it was not as easy to keep as yam. From São Tomé, sweet potato "went straight to Gold Coast" in the late sixteenth century. The fact that the sweet potato was introduced by Europeans was also shown in the African references to the root: "The modern Akan, Gã, Bini and Igbo use terms for the crop meaning European or white man's yam," Alpern writes. In the seventeenth century, sweet potato was not only grown in the Gold Coast, or today's Ghana, but also throughout neighboring Benin, in such kingdoms as Allada and Whydah.[48] And in Senegal, another country on the coast of West Africa, peanuts are cultivated as the primary crop. But Senegalese also eat couscous, white rice, and sweet potato, which is usually cooked in a stew with meats, vegetables, and spices.

From Benin, sweet potato probably spread to Nigeria, its eastern neighbor, where it quickly took root. Over the past half century, Nigeria has rivaled both Malawi and Uganda for the status of the largest sweet potato producer in Africa. At present, nearly 80 percent of the sweet potato produced in West Africa come from Nigeria. However, unlike these East African countries, where it has become a staple or a co-staple, sweet potato remains by and large a minor crop in Nigeria to this day. Akoroda reveals in his aforementioned study that, in terms of Nigerian farmers' rankings of various crops' "relative importance in diet base[d] on ease of propagation and economic yield," sweet potato is found behind yam, maize, cassava, and cocoyam. Sweet potato is ranked below these crops, some believe, because its roots are "too sweet and too low in dry matter" for the people in some West African countries.[49] Yet sweet potato production has seen a marked growth in Nigeria in recent decades. In 2003, O. O. Tewe, F. E. Ojeniyi, and O. A. Abu, in a collaborative project of the University of Ibadan and the International Potato Center, point out that "sweet potato production, marketing and utilization have expanded beyond the traditional areas of the central and riverine zones to the humid, sub-humid and semi-arid regions in the last two-and-a-half decades. The national production figures reported by the FAO [Food and Agricultural Organization] showed a rapid increase in production and area harvested in the 1990s, surpassing two million tonnes harvested from more than 300,000 hectares annually by the end of the decade." In other words, if sweet potato production saw increased production in Nigeria, it began in the 1990s, as it grew by nearly ten times in the decade. As of today, semiarid states in northern Nigeria, such as Kaduna, Kano, and Bauchi, have become centers of sweet potato production in the country, where the root has also achieved the status of a staple crop. Due to this rapid expansion, there have been contradicting data regarding whether Nigeria is the largest sweet potato producer in Africa. A study conducted in 2014 states that the country follows Uganda in sweet potato production, whereas a newer one, published in 2019, proclaims that Nigeria has become the largest producer in Africa and the second largest globally after China.⁵⁰

Scholars have found various reasons to explain how and why sweet potato farming made a notable advance in Nigeria over the past half century. Firstly, compared with other crops such as cassava, groundnuts, cocoyam, banana, and plantains, sweet potato is more impervious to disease and pest attacks. That is, the root plant proves to be a viable alternative to these traditional crops, one that farmers use to diversify the cropping system when others fail. Secondly, the labor shortage in rural areas as a result of HIV/AIDS infection and migration to urban areas have pushed farmers to select food plants that require less risk, labor, and cost, such as sweet potatoes. And the third factor influencing farmers to switch to sweet potatoes is the decline in government support for maize production in recent years.⁵¹

Despite its impressive growth of late, obstacles still seem to dot the root plant's path from a minor to a major crop in the nation's agronomy. The previously mentioned studies all agree that Nigeria's (and Africa's) per hectare sweet potato yield is still much lower than that of other nations, notwithstanding differences in country standing in terms of production in Africa. While stating that Nigeria produces the highest volume of sweet potatoes on the continent, for example, the 2019 study acknowledges that with respect to levels of production, the country "recorded one of the world's lowest potato yields of less than 3.1 tonnes per hectare." This is approximately less than one-seventh of the yield that countries like the United States and Japan can produce. By quoting the number given by the FAO, Tewe, Ojeniyi, and Abu state that the average yield was between 5 and 8 tonnes per hectare in the country, which was of course much higher than 3.1 tonnes per hectare. However, they also agree that "there is substantial scope for increasing sweetpotato productivity in Nigeria."⁵² Whether the average yield is 3 or 5 tonnes per hectare, it signifies a somewhat worrisome trend because, according to the recent research by a group of scientists at the University of Washington, these figures suggest a significant decline in sweet potato yield from that of the pervious decades in Nigeria. They point out that from 1990 to 2010, the sown area of the root plant had a remarkable increase of 3,233 percent in the country. But in 1990, the yield was 5.1 tonnes per hectare, while three decades earlier, it had actually been 12.4 tonnes per hectare.⁵³

Several reasons contribute to the drop in sweet potato productivity in Nigeria. Lack of improved technology is one, since Nigerian farmers usually use tools like cutlasses and hoes, which demand high human labor input. Failure to plant better cultivars and prevent pest and disease infestation is another, which negatively affects the root's return. Nigerian farmers traditionally grow the white-colored sweet potato variety, which is not only characterized by low yield but also contains less vitamin A than the orange-fleshed variety scientists recommend. In more recent years, like elsewhere in Africa, the Nigerian government has promoted the orange-fleshed sweet potato in an effort to combat vitamin A deficiency, a common health issue among many African children and women, although the change has been slow to materialize. That Nigerian sweet potato growers seldom apply fertilizer is another factor that negatively affects sweet potato production in the country. Worse still, researchers find, because of low yields, farmers are less incentivized to spend time on farming—some chose instead to engage in transport businesses for a better income, leading to a labor shortage on farms. As a result, rather than a high-yield and easy-growing food crop, as seen elsewhere, sweet potato is viewed as a crop that involves high labor costs in Nigeria. The low productivity also leads to high prices for the root as food at market. Across the nation, sweet potatoes are sometimes sold at significantly higher prices than cassava, which limits their use as a raw material for root- and tuber-based industries. In a nutshell, Nigerian investments in the production of sweet potatoes typically yield low returns.[54]

Nigerian scholars have exchanged ideas on how to improve sweet potato farming in the country in light of these difficulties and limitations. For instance, Tewe, Ojeniyi, and Abu list four recommendations in their study. These include conducting a nationwide survey to enable effective planning of sweet potato production and utilization, obtaining promising varieties from the International Potato Center for distribution, investigating the commercial viability of sweet potato production, and wage campaigns to strengthen the connection between sweet potato producers, processors, and consumer industries. To turn sweet potato farming into a profitable industry, I. M. Ahmad and colleagues, who conducted the aforementioned 2014 study, recommend that the Nigerian government assist farmers in gaining better access to improved varieties, develop better technologies for pest and disease control, and apply small amounts of fertilizers for higher productivity; all this, they believe, will help lower the root's price and clear the path for the growth and development of the industry into the future.⁵⁵

Sweet potatoes, despite being a minor food crop in Nigeria, are a common ingredient in the nation's cuisine. In fact, sweet potato consumption has increased dramatically in recent decades; according to FAO, while Nigerians consumed 143,000 tonnes of the root in 1990, that figure had jumped to 2,746,000 tonnes by 2010—a nearly 2,000 percent increase. Sweet potato porridge, for instance, is a staple in the country, where it is usually boiled together with cowpea, lima beans, sesame, millet, and other root or tuber crops. Some choose to cook the porridge with red bell peppers, onions, and tomatoes to enhance its orange/reddish color. An even simpler way to cook sweet potatoes among Nigerians is to boil them with different spices, then add sauce before serving. In addition, sweet potato dough is a popular ingredient in making two staple dishes in Nigeria: *fufu* and *amala*, which are also consumed in other African countries. The former is a swallow made by pounding equal portions of boiled root and tuber crops in a mortar, and the latter is a thick porridge cooked with floured root and tuber crops and served with various kinds of soup. In recent years, sweet potato fries, too, are becoming increasingly popular among the urban population; before frying, some choose to coat them with egg to achieve a golden-brown color.⁵⁶

In neighboring Liberia, it has been a tradition to consume not only sweet potato's tuberous roots but also its leaves. In fact, when M. A. As-Saqui, a Liberian scholar from the country's Central Agricultural Research Institute, wrote his article on the potential use of sweet potato in the early 1980s, he found that "in Liberia, the leaves are more extensively used than the roots when many people grow sweet potato in their backyards mainly for the leaves." Thanks perhaps to the leaf-eating custom, Liberian researchers discovered that the leaves contain three times the amount of actual protein and as much as ten to one hundred times the amount of carotene as the roots. Indeed, sweet potato leaves or vine tips—the latter also includes stems and petioles—contain as much nutrition as the tuberous roots. Also in the 1980s, a group of Asian scientists found that if an active man "eats about 100 g[rams] of tips a day he gets his supply of vitamin A for about two days, one quarter of his vitamin B2 requirement, and more than half of his vitamin C and Fe needs in addition to some quantities of Ca, ash, and fiber."⁵⁷ Moreover, stressed As-Saqui, compared with cassava, another root plant commonly grown and consumed in the nation, sweet potato is more drought-tolerant and better protected from locust attack. However, like in Nigeria and other countries in Africa, sweet potato ranks lower than cassava as a food crop in Liberia, mostly due to the lack of processing industries by which the root can be made for industrial use.[58]

Liberia, however, is by no means the only country where sweet potato leaves are consumed. In fact, given the root plant's ubiquitous presence across the African continent, it is quite common for people to consume sweet potato leaves as a vegetable. The authors of *Unleashing the Potential of Sweetpotato in Sub-Saharan Africa*, a collective project sponsored by the International Potato Center, observe that "Sweetpotato leaves are also consumed in many countries in SSA [Sub-Saharan Africa], with the notable exceptions of Kenya and Uganda where they are principally considered to be animal feed."[59] *Matembela/matembele*, for example, is the word for the leaves in Lingala, which is a major language in Africa, spoken in both the Republic of Congo and the Democratic Republic of Congo, as well as in some parts of Angola and the Central African Republic. Stewed *matembela*, according to Hadia Zebib Khanafer, a Lebanese author of a cookbook based on her extensive living experience in Congo, is "an undiscovered treasure waiting to be found and my favorite way of eating leafy green stews." It is usually cooked with chicken in red palm oil, African piri piri, and plenty of onions.[60] It can serve with white rice and/or *fufu*. As mentioned before, *fufu* is a popular grain meal in West Africa typically made of pounded cassava, plantain, or white potato. However, sweet potato *fufu*, made by mashing the root with yam and other root and tuber crops, is also popular in both the Congos and their neighbors because the root is one of the region's main food plants.

African culinary practices usually mix with local traditions and European influences; together they helped produce regional differences. Indeed, across the continent of Africa, while the root plant is integrated into regional and national culinary cultures, the manner of its consumption typically reflects, to varying degrees, a combination of heritages in African history—namely, colonial influences and local traditions. As mentioned above, sweet potato farming had a long history in West and Central Africa mainly because of the Portuguese and later Belgian and French influences. "Sweet potatoes Congolese" is an example, and is included in a Belgian recipe book. It is made by marinating sliced sweet potatoes in honey, brandy, and lemon peel and dipping them in a batter made of flour and beer before frying.[61] However, in Namibia, a former German colony, white potato has traditionally assumed a bigger role in its foodway. It is the horticultural product consumed most by Namibians. By comparison, many Africans consume couscous, bulgur, yam, plantain, and cocoyam as staple foods, which are more reflective of indigenous foodways.

Looking at South Africa, sweet potato production and consumption have also shown notable differences. First, the root plant was introduced to the country, then a Dutch colony, as early as 1652; its cultivation thus began in approximately the same period as in West Africa. Farmers over time created more cultivars, whereas until recently, sweet potato farming in most other African nations had largely focused on a few varieties.[62] Second, throughout Sub-Saharan Africa, sweet potato is grown mostly by smallholders with a low per hectare yield, as described above, whereas in South Africa, highly mechanized commercial production can generate a yield as high as 40 tonnes per hectare, which is certainly comparable to that of the highest yield of sweet potato production achieved by such countries as Japan. Third, while most of its neighbors in southern Africa, such as Angola, Madagascar, Malawi, and Mozambique, take the sweet potato as a staple food, sweet potato is primarily considered a vegetable in South Africa. It is grown, however, across the country, especially in provinces like Limpopo, Mpumalanga, KwaZulu-Natal, and the Western Cape. It is also consumed in ways similar to that in other parts of the world: Sweet potatoes are boiled, roasted, and made into soups, casseroles, desserts, and breads. Yet some of the country's recipes have also registered, more obviously than in its neighboring countries, the combined heritages (indigenous foodways and diasporic inventions) mentioned above. "Twice-baked sweet potato with ricotta cheese," "slow-cooker vegetarian chili with sweet potatoes," and "sweet potato and kale frittata" are good examples. *Soetpatats*, or caramelized sweet potatoes, are arguably the most distinct sweet potato food from South Africa. A popular Afrikaans side dish, its preparation reflects the country's checkered past: Sweet potato chunks are cooked in a Dutch oven with butter, brown sugar, and cinnamon, and it is often served with grilled snoek, a fish commonly eaten in the country, and other barbecued foods. As one of the favorite plants people grow in their home gardens, sweet potato's popularity has also greatly increased in South Africa because of its multiple health benefits, and particularly its extraordinarily rich vitamin A component. A recent survey reveals that there is a widespread vitamin A deficiency in the country—43.6 percent of children between one and five and 27 percent of women of reproductive age suffer such a deficiency. And South African scientists have found that sweet potato, especially the orange-fleshed cultivar recently being promoted throughout Africa, is "bioavailable and efficacious in improving the vitamin A status in children."[63]

From South Africa to the eastern side of the African continent, there are several countries where the sweet potato assumes the status of a staple food. Malawi is one whose sweet potato production easily ranks among the top five in Africa as well as in the world, even though maize remains the main grain for Malawians. Throughout its history, the country has often been hit by famines due to maize failure, usually caused by droughts. Consequently, the drought-resistant sweet potato and cassava become favored as secondary security crops among farmers. Traditionally, sweet potatoes are grown in the Shire floodplain and cassava on the river shore. Yet in more recent years, both plants are expanding their acreage, thanks to the promotion by several international agencies, such as the newly disbanded USAID (United States Agency for International Development) and the International Potato Center, as well as national offices like the Malawian Ministry of Agriculture and Food Security. For example, from the early twenty-first century, these bodies launched a project to encourage the plantation of orange-fleshed sweet potato, which has produced tangible benefits to Malawian farmers. Gerald Action, a peasant living in southern Malawi, described how the variety benefited his life: "When I adopted this style of farming, planting this crop, it made a big difference to my family." For compared with the other types he used to grow, orange-fleshed sweet potato provides higher yields and is more resilient to drought, pests, and diseases. The Action family now has enough sweet potatoes not only to feed themselves but also to sell in the market. They have also moved from a thatched house to a brick one. Action's improved life is but one example. The initiative for promoting the orange-fleshed sweet potato, according to Putri Ernawati Abidin and Ted Nyekanyeka, who worked for the International Potato Center, has made the cultivar "popular among households, bringing improved food availability and extending food supplies into the 'lean' months."[64] The Malawians usually consume sweet potatoes in a simple way, such as cooking them together with bean and spinach or with *usipa*, a local fish abundant in Lake Malawi, and making sweet potato fritters. Yet in more recent years, *mbatata*, or Malawian sweet potato cookie, has become a pathway not only for family consumption but also for commercializing the root plant.

North of Malawi, Rwanda and Uganda are two East African countries where sweet potato has traditionally been a staple food. In her *Sweet Potato: An Untapped Food Resource*, written in 1992, Jennifer Woolfe chose the two countries as representatives to discuss the advance of sweet potato farming in Africa from the late twentieth century. In fact, the importance of sweet potato in East Africa has drawn attention from scholars over many decades. When the first international symposium on the sweet potato was held in Asia in the 1980s, mentioned at the outset of this chapter, eastern and southern Africa were among the continent's few regions on which scientists presented papers. A decade before, J. D. Acland published *East African Crops*, in which he recognized at the outset that "Sweet potatoes and cassavas are the most important root crops in East Africa. Sweet potatoes are the more widely distributed of the two; they are grown in all areas below 7000 ft (c. 2100 m) provided they do not experience such a long dry period that the vines cannot survive from one growing season to the next."[65] Indeed, though sweet potato was first introduced to West Africa, it become a staple crop in East Africa; in addition to Rwanda and Uganda, Tanzania and Kenya are among the major sweet potato–producing countries on the continent, and indeed around the world. One source states that Tanzania's sweet potato production ranked third in the world in 2020, trailing only China and Malawi.[66]

Sweet potato as a crop thrives in East Africa because the region is blessed by its warm climate: Uganda's mean temperature, for example, is between 21 and 25 degrees Celsius, ideal for farming the root plant all year round. Sweet potatoes were thus cultivated, noted R. K. Jana already in the early 1980s, in "almost all parts of Uganda," and the majority of them "were grown in pure stand," while the remainder were intercropped with other plants. Jana also pointed out that unlike Tanzania and Kenya, where rainfall varies notably due to wide-ranging topography, Uganda's annual rate of rainfall is fairly consistent: The country "receives, on average, between 750 mm and 1250 mm rain four years out five." All of this may help to explain why Uganda was one of the first nations in East Africa where sweet potato took hold. Before 1900, it had already been widely cultivated as an important food plant in southern Uganda, in such places as Buganda, Bwamba, and among the Bakonjo of Ruwenzori. Kigezi, in southwestern Uganda, is the country's center of sweet potato production. That the root assumes the primary importance as a food crop in southern Uganda is also because of its superior propensity; it can grow at high altitudes where cassava, another major root plant in the country, cannot.[67] As a staple food in Uganda, sweet potatoes are usually prepared by wrapping them in banana leaves with beans and steaming them. Roasting them, peeled or unpeeled, on the ashes of a fire is another common method. Given its broad appeal as a staple, second only to plantain in many regions, sweet potato production has seen a steady growth in the country over the past decades. Woolfe cited FAO statistics indicating that Uganda produced 495,000 tonnes in 1961, which rose to 2,002,000 tonnes in 1976. The figure declined a little in the following years but quickly rose back up to 2,000,000 tonnes in 1985.[68] This trend of increased sweet potato production has accelerated in recent years thanks to a nationwide campaign launched by the Ugandan government to promote orange-fleshed sweet potato over the white and yellow varieties that were previously farmed. This is because, as in South Africa, there is widespread vitamin A deficiency among Ugandan women and children. The orange-fleshed sweet potato's high vitamin A content has been hailed as a "miracle" food for combating the deficiency. This campaign has been quite successful thus far, and the proportion of sweet potato farmers growing the orange-fleshed variety increased from 4.2 percent in 2013 to 28.2 percent in 2017.[69]

Thanks to its multiple advantageous propensities and edaphic flexibility, sweet potato is also widely grown in other East African countries. Its main attraction lies in the fact that the plant provides reliable yields because, in the words of Peter Ewell of the International Potato Center, it "propagates and grows with no inputs on degraded soils under a range of rainfall patterns."[70] Granted, according to most research, sweet potato grows best in areas with an average annual rainfall of 30 inches (or 750 millimeters) or more. However, the root plant can also withstand prolonged drought. When there is a drought, sweet potato vines can remain green for a while even though root growth is slowed. During Tanzania's long dry season, for example, farmers typically cultivate sweet potato in swamps, rivers, and seepage areas. Little wonder, then, that although the sweet potato is cultivated throughout Tanzania, its primary production areas are the country's Lake Zone (i.e., Lake Victoria), the Southern Highlands, and the Eastern Zone; in the Lake Zone, where the yields (estimated to be 6.5 tonnes per hectare) are also among the highest in the country, the regions of Mwanza and Shinyanga are the leading producers. More importantly, these relatively high yields have been a factor in the country becoming one of the world's leading sweet potato producers. While Tanzania produced 207,830 tonnes of sweet potatoes in 2000, this jumped to 1,466,120 tonnes in 2002—a 600 percent increase in just two years, though the production stagnated afterwards.[71]

Rwanda is another interesting case, as its geography differs notably from Uganda's. Whereas Uganda's climate is heavily influenced by its many lakes, especially Lake Victoria, the whole country of Rwanda is at a high altitude with varied rainfall patterns. Nonetheless, sweet potato appeals to Rwandan farmers despite the country's mountainous terrain because it can grow all the way from sea level to heights of 7,000 feet (approximately 2,100 meters), making it a superior crop to cassava, which can rarely grow at altitudes of 5,000 ft (approximately 1,500 meters) or higher.[72] As previously stated, Rwanda is another country, in addition to Uganda, where sweet potato cultivation has received considerable scholarly attention. African scientists, for example, conducted experiments to determine which cultivars could produce good or better yields in Rwanda's high-altitude regions. And the plant's average yield in the country is indeed higher than that of its neighbor Tanzania, while trailing that of Kenya.[73]

Legend holds that sweet potatoes were brought back by Rwandan soldiers from Uganda when they invaded their neighbor sometime in the eighteenth century. The soldiers wrapped sweet potato vines around their spears after they found that the plant gives milk (latex) when cut. From then on, sweet potato was quickly adopted to become a major food crop in the country. Unlike yam, a more traditional crop, sweet potatoes do not require as much fertility as yam or as much labor for staking. As elsewhere, the main reason Rwandans turned to sweet potato farming was the country's population growth from the 1960s to the 1980s. In order to meet the rising demand for calories per capita, sweet potato cultivation has expanded steadily since then. According to the same FAO statistics cited above by Woolfe, Rwanda produced 452,000 tonnes of sweet potato in 1961; by 1985, this had increased to 900,000 tonnes. Agriculturalist David Gregory Tardif-Douglin agrees, emphasizing in his own study that sweet potato production in Rwanda expanded at one of the fastest rates of any food plant between 1966 and 1985. He estimates that the root plant grew at a 7 percent annual rate in the period, which was "significantly greater than that for beans, cassava, bananas, and . . . cereals."[74]

At present, about 90 percent of Rwandan farmers grow sweet potato. While the root is planted all over Rwanda, its production center is on the country's central plateau because of the root's adaptability to different climatic conditions and its versatility. According to Jennifer Woolfe, the rapid growth of sweet potato farming in Rwanda, which was once held as a model for other African countries, has been due to three factors. The first is the crop's integration in the intensive cropping system, which exploits its yield potential and reduces inputs of fertilizers, pest control, and weeding. The second is Rwandan farmers' ingenuity in planting the root plant in different seasons and areas. They grow the sweet potato on hillsides during the rainy season, for example, and in the bottoms of valleys during the dry season. And the third is the widespread use of sweet potatoes as both a home garden plant and a field crop, owing to their well-known quality as a security crop capable of producing consistent yields that meet people's nutritional needs.[75] Consequently, not only is sweet potato grown throughout Rwanda, but it also produces several harvests per year, ensuring a year-round supply of food. More specifically, "first season planting is most frequently done on hillside fields between September and January, and second season planting is usually done between February and May, which permits harvesting throughout most of the year."[76]

Rwandans consume sweet potatoes in a manner comparable to that of Ugandans and people in other East African nations. The roots are typically boiled or steamed and served with beans; the former provides calories and the latter protein—like the rest of East Africa, Rwandan food is high in vegetables but low in meat. In some cases, Rwandans peel the roots and mash them before cooking and serving, and at other times, they grill and roast the sweet potato to eat as a snack. Unlike in West and Central Africa, however, sweet potato leaves are seldom consumed as food in East Africa; and whereas Ugandans tend to use them as animal feed, Rwandans hardly use them at all for that purpose.[77] People in northeastern Africa—for example, in Ethiopia, whose sweet potato farming practices will be discussed further below—only recently discovered the benefits of eating sweet potato leaves as a vegetable, thanks to the promotion and education of an American graduate student.[78]

As a predominantly agricultural country, most sweet potatoes in Rwanda are freshly consumed by farmers locally. As such, while we can assume that the sweet potato will remain a staple among the nation's rural population, it has been a challenge to make it an important food for the increasing urban population. Having conducted research on the marketability of the sweet potato in Rwanda in 1991, David Gregory Tardif-Douglin observed that in contrast to other countries and regions in Asia and even some parts of Africa, Rwanda lacked the tradition of sweet potato processing. Most sweet potato farmers in the country rarely traveled more than ten to fifteen kilometers to sell their produce.[79] Yet things have definitely been changing in recent years. Sweet potato fries, called "chips" by the locals, are becoming a favorite snack in urban areas in Rwanda. They are coated in a cornstarch slurry before frying, which makes them crispy instead of soggy. To tap into the urban market for sweet potato consumption, Umugiraneza Regis, a young Rwandan entrepreneur, started a company in 2014 that makes sweet potato bread and biscuits. Regis started the business because "everyone was growing sweet potatoes in the rural areas," he explains, "and eating them boiled, the same way they had been preparing them for decades." He believes his new company can add value to the crop as food among his compatriots since the urban population tend not to enjoy the sweet potato when prepared via traditional methods. Interestingly, Regis's company is not alone; he admits that other big-name bakeries are also processing sweet potatoes in the country.[80] All of this suggests that some Rwandans are changing their sweet potato consumption habits.

Ethiopia, in the Horn of Africa, is one of the top ten sweet potato producers in the world. Cultivated mainly in its southern, southeastern, and eastern regions because of their warm and humid weather, 736,000 tonnes of the root plant were produced in the country in 2010, which was the nineth-highest yield among African countries. In fact, sweet potato production has experienced a steady increase since the early twenty-first century in Ethiopia. The yields of sweet potatoes in Ethiopia were also among the highest in Africa during the decade, hovering around 10 tonnes per hectare. The majority of the sweet potatoes produced are consumed as food by Ethiopians, while its vines are processed as animal feed. But when it comes to its status as a food plant, sweet potato consumption in the Horn of Africa pales in comparison to its standing as a major staple in nations around Lake Victoria in East Africa. In Ethiopia, it is the second most important root crop after *enset*, or Ethiopian banana, which is a flowering plant in the banana family only domesticated in that country. In terms of total food consumption across the country, sweet potatoes make up a comparatively small portion. According to two recent surveys conducted in southern Ethiopia, only 21.1 percent of farmers in one community and 53.3 percent in another named the root as a major staple food in their diet. Ethiopians consume sweet potatoes in a variety of ways, including boiling, roasting, steaming, and frying them into crisps, just like people everywhere else. However, in recent decades, it has not been uncommon to see them cooking sweet potatoes with lentils, a common food in Ethiopia, as a vegetable dish, or with meat and *berbere*, a spice mixture essential to Ethiopian cuisine, to make *wat*, a thick and tasty stew eaten by hand with a sourdough flatbread.[81]

From Asia to Africa: A comparison

In summary, from the late twentieth century, sweet potato has increasingly become an "African crop," its production seeing marked increases throughout the Sub-Saharan regions of the continent. According to the recent FAO data, four of the top five countries producing the highest volumes of sweet potatoes in the world are now in Africa: Malawi, Tanzania, Angola, and Ethiopia, while China remains the leading producer.[82] Sweet potato's recent expansion in Africa bears both similarities and differences to its production in East and Southeast Asia, where the root plant, as discussed at the outset of this chapter, once earned the moniker "Asian crop" during the 1980s. The first similarity is that many sweet potato–producing countries in Africa are located near the equator, or between latitudes 30° north and 30° south, which means they are comparable not only to major sweet potato–producing countries in Asia, but also to countries and regions in the Americas, such as Peru, Ecuador, Columbia, the Caribbean, and southern Mexico, where the root plant was either first domesticated or first propagated. It is not surprising that the sweet potato, as a tropical and semitropical crop, was quickly adopted and easily integrated into the African food system.

The second is that like their Asian counterparts, African farmers were accustomed to growing root crops. That is, while the Chinese and Okinawans were well-known for cultivating Chinese yam (*shanyao*, or *Dioscorea polystachya*) and taro as traditional food crops, many African farmers were also well-versed in yam cultivation. It is no surprise that when American root crops like cassava, sweet potatoes, and white potatoes were introduced to the continent, Africans readily embraced them. And like their Asian counterparts, African farmers also frequently substituted these American crops for the native plant because of the formers' numerous advantages. Nigeria is an example. While the country is the world's largest producer of yam, it has also seen exponential growth in sweet potato farming over the last half century, allowing it to also become one of Africa's leading producers of sweet potato.

Third, with a few exceptions, such as Uganda and Rwanda, where the majority of sweet potatoes are grown in pure stand, most African sweet potato growers, like their Asian counterparts, practice intercropping and plant the root with other food crops. A typical case is African farmers growing cassava and sweet potato, the continent's two most important root crops, on the same farm and in the same region because of the two crops' similar propensities. Indeed, Nigerian farmers find that intercropping is advantageous because when cassava fails due to pest and disease attacks, they can switch to sweet potato farming since the root matures faster than cassava. And this has been a common practice across Sub-Saharan Africa—"when cassava fails, farmers frequently switch to sweetpotato to substitute for the energy delivered by cassava."[83] For in general, sweet potatoes take three to four months to mature compared to seven to twelve months for cassava. Even in Rwanda, where sweet potato is mostly grown individually in stands, farmers also see the benefit of intercropping to increase yields during different planting seasons. Rwandan farmers would grow a portion of sweet potato with beans and maize between September and January, which is their first planting season. They would intercrop sweet potato with beans, sorghum, and cassava during the second planting season, which runs from February to May. In this way, they hope to harvest "a series of crops overlapped on the same field."[84]

Fourth, in Sub-Saharan Africa, as in Asia, particularly Southeast Asia, and Papua New Guinea in western Oceania, women typically play a significant role in sweet potato cultivation. To a degree, therefore, sweet potato farming has been a gendered experience in these places. Back in the 1980s, Asian scholars studied the division of labor among male and female members of sweet potato growers' families. The researchers discovered that, while husbands did most of the land preparation, their wives and daughters did a substantial amount of planting and an even greater amount of harvesting.[85] A similar phenomenon is also observed across Sub-Saharan Africa. According to surveys of sweet potato farmers in Nigeria, for example, the gender breakdown is roughly fifty-fifty—that is, women shoulder as much labor as male members of the family. On the other side of the continent, in Uganda, sweet potato is known as "a women's crop," as women play an important part in cultivating the root. Ugandan men, on the other hand, take on a larger role in transporting and marketing the roots. The situation in Tanzania is very similar: "women, who do the majority of the production and decision-making, largely oversee sweet potato production." Surveys of specific regions in Tanzania show that, while men do up to 50 percent of the work in some places, they are not involved in sweet potato farming at all in others. The only exception perhaps is Ethiopia: Due to the tradition whereby women are not entitled to land inheritance, Ethiopian men are primarily responsible for sweet potato farming. However, while it is not a consistent phenomenon, as shown in the case of Papua New Guinea in the previous chapter, most studies agree that "women predominate in sweetpotato production" in Africa.[86]

Fifth, while it contributes significantly to food supply in both Asia and Africa, sweet potato is mostly regarded as a "poor man's food" on both continents; or, because women play such an important role in its cultivation, some have suggested calling it a "poor person's food" instead. All the same, sweet potato has a low status as a food in both Asia and Africa, in contrast to its image as a festive fare in America and a desired ingredient in European desserts. This contrast suggests that the perception of the sweet potato as a "poor person's food" stems less from its nutritional value and more from its association with poverty. In other words, sweet potato becomes a victim of its own success—its well-known proclivities as a food plant include easy growth, quick ripening, and consistent yield. Poor farmers in Asia and Africa turn to it as a security crop precisely because of these advantageous properties. As a result, the root has a bad reputation as a lowly food. In his foreword to the second international symposium on the root plant in 1987, K. T. Mackay put it succinctly: "Sweet potato is an important staple and emergency food in many countries, especially for the poorest—the resource poor farmers and the urban poor." In the Philippines, for instance, sweet potato is traditionally viewed as "a crop for the poor and the unsuccessful." In Tanzania, sweet potatoes are likewise seen as "an important source of food in the homes of rural and urban poor." Granted, the fact that sweet potatoes cause flatulence may also contribute to their low status. In Bangladesh, the root is regarded as "poor man's food" because urban dwellers avoid it for this very reason. According to Jennifer Woolfe, flatulence, too, is a factor that causes many modern Chinese to avoid eating sweet potatoes. On the other hand, many root crops, including the white potato, and most legumes, have the same effect on the digestive system. In a nutshell, the sweet potato's ingrained reputation as a "poor person's food" is primarily due to its historical associations with poverty in Africa and Asia, rather than any property inherent to the root itself as a food.[87]

Despite the above similarities, the recent advancement of sweet potato farming in Sub-Saharan Africa also differs from that in Asia. And the difference is most noticeable in how the plant is consumed. As mentioned above, while sweet potato leaves or tops are used as a valuable source of feed in Asia, they are also commonly consumed as a popular vegetable, particularly in Southeast Asian countries like the Philippines. Africa's experience, by comparison, is more varied. In South Africa, for example, sweet potato is regarded mostly as a vegetable, which is analogous to how it is viewed in South Asia and the Middle East. And, while it is common for people in West Africa to consume sweet potato leaves, it is uncommon for people in East Africa to do so, even though the root is a staple food in the region. More interestingly, sweet potato leaves are normally used as animal feed in Kenya and Uganda, but not in Rwanda. Only recently have some people in the Horn of Africa learned to cook the leaves as a vegetable. Sweet potato roots, in contrast, tell a more consistent story across Sub-Saharan Africa, with the majority consumed as food and only a tiny minority processed for industrial and other uses. Sweet potatoes were a staple or co-staple food in both China and Japan, as well as in Taiwan, as discussed in previous chapters. However, beginning in the late twentieth century, an increasing number of sweet potato roots were used to produce starch and alcohol for human and medicinal purposes. While the fresh use of the roots has decreased, Jennifer Woolfe notes a trend from China beginning in the 1980s that "consumption of processed food products has increased." She points out that the same pattern had already started in Japan, where sweet potato consumption peaked during World War II, when more than 70 percent of the harvest was intended for human consumption. By the 1980s, it had dropped to just 10 percent.[88] A similar trend has emerged in Southeast Asia in recent years, as discussed earlier in this chapter: People have explored numerous ways to process the sweet potato into a variety of food products and sell them in the market. By contrast, though perceived as a "low-status food" in Tanzania, 95 percent of the sweet potato production in the country is for domestic human consumption and 5 percent is waste. To ensure food supply, Tanzanians process fresh sweet potatoes into either *michembe* or *matobolwa*, two common ways to preserve the root; the former involves cutting them into slices and drying them, and the latter first boiling them before slicing and drying them. A similar picture is found in Uganda, where sweet potatoes "are grown mostly for home consumption." During its dry season, the roots become "a critical food source" and are consumed as sun-dried *amukeke* (sliced) and/or *inginyo* (chunk). However, there are efforts to transform sweet potato roots into more marketable foods, such as the case mentioned earlier in Rwanda, where the roots are used to make bread and biscuits. In Nigeria today, in addition to sweet potato *fufu* and porridge, the root is industrially processed into fried snacks such as sweet potato fries (chips), candy, starch, noodles, and flour.[89] Yet it remains to be seen how these new initiatives will result in a significant change of sweet potato consumption patterns in Africa in the future.

Despite Sub-Saharan Africa's impressive growth in sweet potato production over the last half century, obstacles remain that hinder the root crop from becoming a more significant part of the region's food system. One issue is that sweet potato yields on the continent continue to be much lower than they are elsewhere. In general, sweet potato yields are well below 10 tonnes per hectare across Africa, whereas research has shown that they could be raised to 30 and even 40 tonnes, given favorable climatic conditions where the plant is cultivated on the continent.[90] The other is how to expand sweet potato product markets in urban areas, as most sweet potatoes are currently consumed locally in rural areas.

To address the former issue, governments across the continent are currently urging farmers to grow better cultivars, like the orange-fleshed variety. The orange-fleshed cultivar not only enjoys higher yields than the white, cream, and yellow-fleshed varieties most African farmers grow, but it is also packed with vitamin A, which is crucial for treating the widespread deficiency in many African countries that can lead to health problems such as vision loss, dry and scaly skin, infertility, and stunted growth in children. However, due to the preference for low-sweetness, dry, and bland sweet potatoes among many Africans—that is, for white-, cream-, and yellow-colored varieties—orange-fleshed varieties continue to be "relatively rare" in sweet potato farming throughout the continent. Scientists have recognized the need to create new orange-fleshed sweet potatoes that can replace the traditional orange-fleshed variety's moist and sweet mouthfeel and taste. Finally, the use of fertilizers, either organic or inorganic, may be a desirable approach that, while proven effective, has so far been lacking in the practice of sweet potato farming throughout Africa. J. D. Acland observed in 1972 that "Farmyard manures always gives good responses but it is very seldom applied [sic]." However, little has changed in the intervening decades.[91]

To find solutions for the latter, many scholars and scientists, including those affiliated with the International Potato Center, have conducted extensive research over the last few decades. They have found that, generally speaking, two factors have thus far impeded the marketization of sweet potato as food in Africa. The first has more to do with the root plant itself, such as its water content, weight, and perishability—problems not necessarily confined to Africa, and which present a more universal challenge. "Sweetpotato roots are bulky and perishable," observe Maria Andrade and her associates at the International Potato Center, "unless cured. This limits the distance over which sweet potato can be economically transported. Production areas capable of generating surpluses tend to be relatively localized but dispersed, which leads to a lack of market integration and limits market size." The second factor relates more specifically to the African experience. As previously stated, some efforts have only recently been made in certain African countries to industrially process the sweet potato into foods that are both marketable and transportable. Yet overall, quoting Andrade and her colleagues again, "There is little commercial processing into chips and flour, which could be stored for year round consumption for use in ugali, bread and cakes, or processing into fermented and dried products like fufu."[92] In his analysis of the challenges of expanding the market for sweet potato consumption in Rwanda, agriculturalist David Gregory Tardif-Douglin also notes that, despite the root's importance as a staple food in the country, Rwandans have not yet established a uniform method for sweet potato processing, nor the technique of long- and even medium-term postharvest storage of the roots, both of which are identified as factors inhibiting the root's marketization. As a result, he writes, "Most sweet potato producers are marginally market oriented; what they grow is almost exclusively for consumption on the farm." Meanwhile, Tardif-Douglin cites the success of cassava processing in Rwanda to demonstrate that, given the two root plants' similar proclivities, the same technique is squarely applicable to sweet potato processing.[93] Indeed, notwithstanding the challenges and obstacles, sweet potato farming certainly has an enormous potential for development in Africa. To be sure, addressing these issues is no easy task. However, scientists have already identified strategies for maximizing the potential of sweet potato cultivation and consumption in and for Africa. This will hopefully help

African farmers raise sweet potato yields in the future through, for example, improved breeding, better use of technology, and rationalized crop management. Enhancing sweet potato yields will consequently lead to more initiatives to explore and expand the root's marketization in urban areas and further augment its importance in the existing food systems on the continent. Indeed, when Jennifer Woolfe reported on sweet potato farming in Rwanda and Uganda in 1992, she observed that, unlike in China and Japan, the root was then considered a "major staple food and security crop" in both countries. Over the last two decades, however, a different picture has emerged. According to Jan Low, a scientist at the International Potato Center's Sub-Saharan Africa branch, "Clearly, Sweetpotato is emerging from its traditional role as the classic food security crop to a more commercialized crop, and in the case of orange-fleshed sweetpotato, a healthy food for all" in many parts of Africa in 2015.[94] Given that sweet potato has emerged as an "African crop" over the past half-century, any further transformation in its role or perception will likely have a significant impact on its future global standing as a valuable food crop.

Epilogue

In 2015, something unexpected happened in the world of sweet potatoes. China, the world's largest producer, announced the launch of the "white potato stapleization" (*tudou zhulianghua* 土豆主粮化) campaign. Its goal was to elevate the white potato to the status of one of the nation's staple foods, alongside rice, wheat, and corn. Why not the sweet potato? Naturally, many were curious. After all, sweet potatoes had been a staple food for many Chinese for hundreds of years before the end of the twentieth century. The white potato, by comparison, has been considered primarily a vegetable throughout the country since it first appeared in the seventeenth century. It is grown as a staple only in certain northwestern regions.[1]

The Chinese government did not provide a clear explanation for why it was calling for a switch from sweet potatoes to white potatoes. According to one of the reports, Yu Xinrong, then a deputy minister of agriculture, was the official in charge of the campaign. In 2013, he attended a meeting cosponsored by the Department of Agriculture and the Academy of Agricultural Science with the goal of making white potatoes a staple food, which essentially meant processing the potato into flour and making potato buns, bread, and noodles to be consumed by people in the same way that wheat and rice flour foods are. Yet Yu was not the first to propose the idea; Guo Fenglian, a member of the People's Congress, had proposed it in 2010.[2] Guo, who was born and raised in the northwestern Chinese province of

Shanxi, became known as a model worker during the 1960s, when the country was in the throes of the Cultural Revolution (1966–1976), and Mao Zedong praised her village for its revolutionary spirit. Guo's proposal made the most sense out of the two because white potatoes were grown in Shanxi. Yu was raised instead in Jiangxi, in South China, and grew up eating the tubers as a vegetable. If Yu took the campaign's leadership role, it was more so because he merely carried out a directive from above. Xi Jinping, China's current leader, might be the person behind the campaign to make white potato a staple food because, like Guo Fenglian, Xi grew up in Northwest China, where white potato, instead of sweet potato, was traditionally a staple food in the region. While Xi did not become the top leader until 2012, he became first secretary of the Chinese Communist Party in 2007 and vice president of the People's Republic of China in 2008, respectively. It was in 2008 that the Chinese Academy of Agricultural Sciences was charged with researching the possibility of making white potato the fourth staple food in China. In 2015, after he became China's president, Xi Jinping enjoined in a meeting that the country should worry about its food security because its large population cannot rely on foreign food imports to sustain itself. His warning suggested that he was supportive of the move to diversify China's staple foods.[3]

Though China's specific motivation for launching the "white potato stapleization" remains unknown, there is no doubt that white potato cultivation has made tremendous progress since the post-Mao years. The nation continues to be the world's top producer of sweet potatoes, but since the last decade of the previous century, it has also overtaken other nations as the leading producer of white potatoes. China has contributed about one-fourth of the world's total white potato production in terms of both sown area and total output. While total global production of white potatoes has decreased elsewhere in recent years, it has increased in China. From 1997 to 2006, China more than doubled its white potato production volume and increased its cultivated acreage from 2,860,000 to 5,010,000 hectares. The trend continued between 2007 and 2012, when the sown area increased to 5,200,000 hectares, a figure that has held steady in the years since. In contrast, sweet potato production in the country has declined significantly. During the 1960s and 1970s, its sown area was as large as 10,000,000 hectares, but it is now only about 4,500,000 hectares.[4]

Needless to say, China would continue to expand white potato cultivation throughout the country in order to make it the fourth staple food for its people. In response to a question about why the tuber, rather than the sweet potato, was chosen, a Department of Agriculture spokesperson explained that, while ensuring food security for the populace is a long-term national strategy, the campaign for white potato stapleization was not being launched because the country was facing a food shortage. Rather, it was due to the tuber's high nutritional value as a food crop, as well as its ability to grow well in areas with limited water resources. The spokesperson admitted that in recent decades, Northwest and North China were increasingly concerned about the depletion of their water resources. In addition, since South China has an estimated 100 million *mu* of cultivated land, the department hoped to grow white potatoes there in the winter. The campaign's overall goal was to double the current white potato planting area from 8,000 *mu* to 150 million *mu*, or 10,000,000 hectares.[5]

In other words, if China concentrated on sweet potato production during the Mao era, the shift to white potato production began in the post-Mao Reform and Opening-Up period. The goal of reaching 10,000,000 hectares of white potato cultivation, the same high level once achieved by sweet potato in the country, was a clear indication. Regardless of its changing focus, China has become the world's largest player in potato cultivation, and indeed in global food production in general. Since the 1980s, its government and scientific community have maintained close collaboration with the International Potato Center, the World Potato Congress, the Consultative Group on International Agricultural Research, and the Food and Agriculture Organization (FAO). Qu Dongyu, a Chinese-trained biologist with a PhD from the University of Wageningen in the Netherlands, is the current director general of the FAO, for instance. In fact, no sooner had China entered the Reform and Opening-up period than it began its exchanges with the International Potato Center. A decade or so later, the latter opened its branch office in Beijing. In 2017, the International Potato Center—China Center for Asia Pacific (CCCAP) was formally launched in Yanqing, Beijing.[6]

On the other hand, as mentioned in the introduction, white potato experts such as John Reader and Rebecca Earle have observed that China's increased contact with the outside world during the Reform and Opening-up period helped expose its populace, particularly the younger generation, to white potato products such as french fries. "French fries have accounted for most of the rise," writes Reader, "in potato consumption, as affluence has inspired a taste for convenience food, and it can be no surprise that two fast food giants from the United States, McDonald's and Kentucky Fried Chicken, have been the principal suppliers." And the fact, which was also offered by Reader, that both fast food chains had to import 70 percent of the french fries they sold in China reflected the need for increasing white potato production in the country. As Earle puts it, "As the Chinese state embraced the principles of market economy, interest in the potato has grown markedly."[7] In other words, China, like other Asian countries, has seen some Westernization of its food culture in recent decades. Most Chinese scientists, businessmen, and officials compare data on per capita white potato consumption in China and the West in their explanations for making white potato the fourth staple food. One such report, for example, emphasized that "In 2009, statistical data shows that the annual per capita consumption of potatoes worldwide is 50 kg. The annual per capita consumption in the United States, Britain, Germany, France, and Canada is 56 kg, all much higher than our 37 kg."[8] Citing this data implies that China should catch up with the West by increasing its consumption of white potatoes. Additionally, it may not have been a coincidence that the Chinese Academy of Agricultural Sciences began its research on making white potatoes a staple food in 2008 because FAO designated that year as the "International Potato Year."

However, increasing white potato consumption is not the same as making the tuber a staple food for the entire nation, owing partially to the fact that sweet potato was consumed as a staple food in many parts of the country. A recent report, for example, reveals that farmers in Henan are resisting the campaign to make white potatoes, rather than sweet potatoes, a staple food because the latter has a long tradition of cultivation and consumption in the province.[9] While recalling his collaboration with the International Potato Center, CCCAP's first director, Lu Xiaoping, praised the center's assistance to China in cultivating new white potato varieties. In the meantime, he discusses China's success in sweet potato production and utilization. He acknowledges that 45 percent of the sweet potatoes produced in the country are processed into starch and other industrial materials, compared to only 10 percent of white potatoes. In fact, 70 percent of white potatoes are eaten fresh as vegetables.[10] Since the white potato stapleization campaign wants to get people to eat white potato buns and noodles as if they were made of wheat and rice, it goes without saying that the low percentage of processed white potato shows how difficult it is to make it a staple food.

Almost a decade has passed since the official campaign of white potato stapleization began. How successful has it been? Without a doubt, Chinese consumption of the white potato has increased over the period, as has the area under cultivation. However, both government representatives and agricultural scientists acknowledge that the campaign is far from being a success. First, due to the ingrained habit of eating grain foods, the majority of the Chinese population still primarily views and consumes the tuber as a vegetable, despite the fact that french fries served in international fast-food chains are consumed in large quantities, especially by young people. Even so, those who have grown to enjoy french fries still regard them as a side dish. Second, in the absence of advanced technology, processing white potatoes for use in foods such as buns, bread, and noodles is economically insensible because the cost is several times higher than processing wheat and rice. Scientists discover that because white potatoes lack gluten proteins, they are hard, difficult to shape, and unsuitable for making steamed buns and noodles. As a result, effectively marketing these white potato foods to get people to like them is difficult. Because of the price and taste differences, the volume of white potato foods a bakery typically sells is one-tenth of that made with wheat flour foods, according to surveys. Third, in relation to the previous two factors, as well as the weak international market because of a global decline in white potato consumption, Chinese farmers generally lack incentive to grow more white potatoes in arid regions of the North and during the winter season in the South. The goal of increasing white potato sown area in the country to 10,000,000 hectares has thus yet to be realized.[11]

By contrast, despite a significant decline in sweet potato cultivation in recent decades, the root not only retains its traditional appeal as a favorite snack—roasted sweet potatoes sold on city streets remain commonplace, bought by both old and young on the go—but it has also seen a new global trend of revival in the form of a variety of new food and drink products. Indeed, thanks to centuries of experience in processing sweet potatoes, the root can be processed at a lower cost, though higher than wheat and rice, than the white potato.[12] That is, while sweet potatoes are no longer a staple food, they remain popular among many Chinese, including the younger generations. The success of Kuaile fanshu (Happy Sweet Potato), a rapidly expanding chain store, demonstrates this. Since it is based in Xiamen, Fujian, one of the provinces where the American plant was first grown in China, the company builds on the province's long tradition of eating sweet potatoes. Meanwhile, by combining the sweet potato eating and tea drinking traditions, it specializes in selling a variety of sweet potato bubble tea, with either milk or fruit juice or both. Bubble tea originated in Taiwan, another traditional sweet potato–growing region across from Fujian, and contains not only milk and/or fruit juices, but also tapioca pearls made of cassava or sweet potato starch for people to chew on while drinking the tea. The tea's diversity derives from the various kinds of tapioca pearls and juices, which provide customers with a variety of options. Over the past half century, bubble tea has become markedly popular among young people in China, Asia, and even in many Western countries. In addition, the Happy Sweet Potato company is known for offering a range of sweet potato foods, from traditional roasted sweet potatoes to sweet potato sausage and sweet potato wraps. According to the company's website, since its inception in 2008, it has expanded to nearly 160 franchised stores in cities across the country's twenty-six provinces.[13]

If the Happy Sweet Potato chain's success represents a trend of sweet potato food culture revival in China, the same tendency has also appeared in Japan, only earlier. Jennifer Woolfe observed in her 1992 book *Sweet Potato: An Untapped Food Resource* that, despite being "previously stigmatized as a survival food," sweet potato "is enjoying a comeback, especially among young people, during the present period of prosperity and boom in gourmet food consumption."[14] In both countries, the revival is characterized by the participation of younger generations in sweet potato consumption. However, while the trend in China appears to be propelled by experiments with new products, it seems that the Japanese are driven more by nostalgia. Woolfe based her observation on the fieldwork of Barry R. Duell, an American working at Tokyo International University who had lived in Kawagoe, Japan, for many decades. Despite being a minor sweet potato–growing region in modern Japan, Kawagoe has a history of sweet potato cultivation and consumption, according to Duell's research. Indeed, as mentioned in chapter 4, Akazawa Nihei, a farmer who wrote one of Japan's first sweet potato cultivation manuals, was from Kawagoe. A grassroots organization called Kawagoe Friends of Sweet Potatoes was formed in 1982 with the goal of reviving the tradition of sweet potato consumption in the city and beyond. It organized a special exhibit, *Sweet Potato History in Kawagoe*, in collaboration with the Kawagoe Sweet Potato Research Association, which was formed the same year, to educate residents about the root's cultivation in the city. A year later, it also took the lead in establishing the annual sweet potato festival as a tradition. Sweet potato culture experienced a revival in Kawagoe because of these activities, with many local businesses specializing in making and selling sweet potato foods subsequently resurrected. Kawagoe's sweet potato festival and products drew tourists due to the city's proximity to Tokyo. These tourists are attracted to traditional sweet potato products sold in Kawagoe, such as *imo-senbei*, which are thin-sliced, sugar-coasted sweet potatoes, *imo-natto*, candied sweet potato slices, and *imo-karinto*, deep-fried sweet potatoes coated with sugar. Meanwhile, Kawagoe businesspeople have been experimenting with new products. Imozen Restaurant, which opened in 1982, is one such example. It serves not only a variety of sweet potato dishes, but also a full-course sweet potato dinner, which includes hors d'oeuvres, soup, salad, noodles, and ice cream, all of which contain sweet potato. Another was Tanakaya, a buckwheat noodle shop. It was founded in 1986 and makes its noodles with sweet potato flour. His technique has also been used by others in making different types of sweet potato–based noodles. In addition, sweet potato paste is used in a variety of desserts, such as ice cream, pudding, and pie, which are readily available in bakeries and restaurants throughout the city.[15]

The revival of the sweet potato food culture in Kawagoe is far from unique, but thanks to the onset of globalization, it testifies to a rising global trend of reintroducing the root as a healthier and more nutritious alternative in people's food choices. Roasted sweet potato, as mentioned throughout this book, is arguably the most popular way for people all over the world to enjoy the root's enticing aroma and fibrous yet tender flesh. In Japan, roasted sweet potato is known as *tsubo yaki-imo*, as explained in chapter 4. The Japan Tsubo Yaki-imo Association was established in 2021 in Ōgaki City, Gifu Prefecture, which is located in the center of the country between Tokyo and Kyoto. Its mission, as stated on its website, is "to engage in a variety of activities to preserve the *yaki-imo* food culture that has existed since the Edo period." The association not only teaches people how to cook the root, but it also sells the pot (700 millimeters tall and 450–470 millimeters in diameter) in which it is roasted, which distinguishes *tsubo yaki-imo* from other roasted sweet potatoes. "When sweet potatoes are heated," according to the association's website, "the starch is converted to maltose by the action of enzymes. This change takes place during the heating stage, around 60–70°C, but the general heating method passes through this temperature range too quickly. The *tsubo yaki-imo* is baked slowly by the heat

Figure 7.1 Sweet potato ice cream for sale in Kawagoe, Japan, a town famous for its cherished heritage of sweet potato cultivation and consumption. This illustrates the innovative ways sweet potatoes are being incorporated into modern food products in Japan.

of the charcoal and the reflected heat of the jar, so the temperature range of the transformation to maltose can be maintained for the maximum length of time." It claims that the secret to the deliciousness of *tsubo yaki-imo* is the careful control of the charcoal heat. Apart from offering techniques and equipment, the Japan Tsubo Yaki-imo Association provides ideas and support to anyone interested in starting a business.[16]

Taiwan, one of the "sweet potato islands" discussed in chapter 5, also saw a notable revival of sweet potato culture at the turn of the century.

A former Japanese colony, Taiwan became a major sweet potato–growing region in East Asia during Japan's half-century rule (1895–1945). Furthermore, as in Japan, sweet potato continued to play an important role in the island's food system well into the 1970s. As a result, the island's postwar generation still clearly remembers how they relied on the root during their difficult formative years. While sweet potato is no longer their staple food, they have recalled this collective memory in a variety of writings that have appeared in local newspapers and magazines in recent decades. Their children, some of whom have emigrated from the island, also write about how their parents' generation instilled in them the value of being "sweet potato children" or "sweet potato folks." Some local governments in Taiwan support activities focused on the sweet potato in order to strengthen and promote this cultural identity. In 1999, for example, sweet potato festivals were held in both Kinmen (Quemoy) and Sinhua (Xinhua), one in the North, off China's southeastern coast, and the other in the South, as a suburb of Tainan City. Despite their geographical separation, both prefectures, like the rest of Taiwan, had a long history of sweet potato cultivation and consumption. Thus, it is no accident that Kinmen and Sinhua both chose to start the sweet potato festival, which not only serves foods made with sweet potatoes but also promotes regional culture. These promotional activities are by no means exceptional; indeed, although sweet potatoes are no longer a staple food in Taiwan, there are still many sweet potato–based dishes and beverages that are popular there, such as roasted sweet potatoes, candied sweet potatoes, and dried sweet potato slices, which are, of course, common outside Taiwan. But sweet potato milkshake, sweet potato tapioca balls, and sweet potato cake (*fanshu su* 蕃薯酥—a small square pastry filled with sweet potato paste instead of the island's famous pineapple) could qualify as Taiwanese inventions.[17]

While sweet potato was widely cultivated and consumed as a survival food across most of East Asia, the root was not a staple food among many Koreans, due in part to the peninsula's cold climate, as explained in chapter 4. But that does not mean that sweet potatoes are not welcome there. Emily Kim, a Korean American YouTuber better known as Maangchi for authoring both *Maangchi's Real Korean Cooking* and *Maangchi's Big Book of Korean Cooking*, speaking of her childhood experience growing up in Yeosu, South Korea, an island not far from Tsushima, the island through which the root arrived in Korea, recalls that "the first thing I saw every day was *goguma* [sweet potato]." She remembers fondly how her grandmother prepared sweet potato rice, or *goguma-bap*, for her, which is, of course, a popular way for many East Asians to consume the root. Another sweet potato recipe she learned from her grandmother was dried sweet potato slices, or *goguma-mallaengi*, which are simply steamed sweet potato slices that have been dried for twelve hours until crisp. The fact that both foods were part of her childhood memories attests to the popularity of sweet potatoes among certain segments of the Korean population.[18] While sweet potato rice and dried sweet potato slices are simple and common ways to prepare the sweet potato, Maangchi's cookbooks also include recipes that use sweet potato starch, or *dangmyeon* (glass/cellophane noodle), and sweet potato stems. She demonstrates how to use both in a variety of dishes, including braised pork and stir-fried vegetables. In addition, new methods for making candied sweet potatoes as desserts are being explored.[19]

If Maangchi draws on her childhood experiences to offer sweet potato recipes, her efforts extend what Debbie Wolfe, an English writer sojourning in South Korea, describes as "Korea's Love Affair with Sweet Potatoes." According to her observations, many Koreans consider sweet potato a "favorite." Aside from her own experience with "the epitome of winter comfort eating" by munching a roasted sweet potato, Wolfe offers vivid descriptions of different kinds of sweet potatoes found on the streets of Korean cities as well as how Koreans like to eat the root at home with kimchi, their nation's iconic side dish of spicy fermented napa cabbage. Sweet potato fries are also popular as *anju*, a Korean term for small foods that accompany people's consumption of *soju*, a type of liquor often made from sweet potato starch. Overall, she believes that sweet potatoes are popular among Koreans today because they are a "trendy diet food." Sweet potatoes are not only nutritious, being high in minerals and vitamin A, but they are also filling, which helps people resist "the urge to eat between meals."[20]

The recent growth of Korean sweet potato culture suggests that it is motivated by people's health concerns as well as their nostalgia. Indeed, more and more people around the world are now aware of the nutritional value and health benefits of sweet potatoes. An obvious example is the promotion of orange-fleshed sweet potatoes by governments in Sub-Saharan Africa in an effort to combat vitamin A deficiency in children and women. Sweet potato has also made a strong comeback in Europe and North America, where it began its global journey in modern times—so much so that it has significantly impacted the food industries that the white potato once dominated. Sweet potato fries, for example, have grown in popularity in recent decades. Fried sweet potatoes are a popular dish in many parts of the world. Though not as popular as roasted sweet potatoes, fried sweet potato recipes were included in American cookbooks as early as the eighteenth century, including the one attributed to Martha Washington mentioned in chapter 2. Frying sweet potato has long been accepted as a popular way to consume the root in parts of Asia and Oceania, such as Japan, Korea, India, and Papua New Guinea. Sweet potato tempura, also known as *goguma twigim* in Korean, is a popular street food in both Japan and Korea, as is *Kananga phodi-tawa*, an Indian fried sweet potato dish.

It is worth mentioning again that Martha Washington called the dish "French fried sweet potato" in her *Cook Book*, implying that before white potatoes, sweet potatoes were probably used to make french fries. Needless to say, the most popular french fries today are made of white potatoes. However, in Euro-America, more and more fast-food restaurants are serving sweet potato fries. Though it is difficult to pinpoint when the trend began, fast-food restaurants such as Checkers, Arby's, and Chick-fil-A most likely started it in the American South. Other well-known fast-food restaurants, including Wendy's, Burger King, Jack in the Box, and Carl's Jr., eventually followed suit. McDonald's and KFC are the most recent additions to the list, both of which have added sweet potato fries to their menus in order to attract health-conscious customers. Sweet potato consumption in the United States has increased steadily as a result of the growing popularity of sweet potato fries and other new sweet potato foods. It increased by nearly 42 percent between 2000 and 2016, reaching 7.2 pounds per capita, for example.[21] Moreover, not only has the trend continued into the last decade, but it has also been observed across Europe, and possibly earlier. Sweet potato fries, indeed, have become more popular in various types of restaurants throughout Europe. And, perhaps because they were already available in Dutch restaurants, McDonald's first offered sweet potato fries in Norway and Sweden before anywhere else.[22]

In other words, sweet potato is increasingly gaining a new image as a health food in all corners of the world, in addition to being an "insurance food plant" known for its easy-growing and high-yielding nature, which remains its main attraction in parts of Sub-Saharan Africa, Southeast Asia, and Oceania. The root has been dubbed a "superfood" by such NGOs as

Center for Science in the Public Interest, defined as a food that provides health benefits due to its high nutrient density. Though used primarily as a marketing tool by food companies, the term has undoubtedly played a role in increasing sweet potato consumption in recent decades. It is illustrative that the well-known fast-food restaurants mentioned above have all used sweet potato to make french fries as an addition to traditional white potato french fries. While fried foods are not usually healthy, sweet potato fries carry the veneer of healthiness because they are made with a so-called superfood.[23] White potato fries served in fast-food giants like McDonald's typically use russet potatoes, or Russet Burbank, named after horticulturalist Luther Burbank, who developed it in the 1870s. The variety is "easy to store and has a consistent texture and taste, which makes it the perfect french-fry potato." However, "it consumes a great deal of water, takes a long time to mature, and requires large amounts of pesticides." It also contains more sugar than other breeds.[24] Sweet potato fries, therefore, actually have more health and environmental benefits than white potato fries, despite the fact that both are fried.

Whether or not they qualify as "superfoods," much research has recently been done on the health advantages of sweet potatoes, which are reported in both scientific journals and various media outlets internationally.[25] All of this amounts to an attempt to recast the root's image in relation to that of the white potato and other foods. Some Chinese scientists explained why the white potato was chosen over the sweet potato in the aforementioned stapleization campaign because of the country's recent rapid rise in diabetes cases. The words "sweet" and "potato," as Rachel Meltzer Warren, a nutritionist in New Jersey writing for *Consumer Reports*, puts it, "may conjure up an image of carb and sugar bomb." However, she quickly adds that this is a misconception because it is not scientifically valid.[26] In fact, there seems to be a growing trend among scientists and the media to reclassify sweet potatoes as diabetic-friendly foods because of their capacity to stabilize blood sugar levels and reduce insulin resistance, even though their sugar content is higher than that of white potatoes. It is argued that the root's high content of dietary fiber, which slows down digestion and the release of sugar into the bloodstream, is the cause.[27] This means that to some extent, sweet potato can help combat rather than cause diabetes.

The sweet potato's dense but low-calorie quality is another reason it is popular in health-conscious Western cultures. Though sweet potato is filling, qualifying it as a high-energy food, its content of both calories and carbohydrates are relatively low in comparison with other foods. "A medium sweet potato baked in its skin," according to one media source, "is only about 100 calories, making sweet potatoes an ideal food for weight management." Granted, carbohydrates have a bad reputation among those who are concerned with their health. But the complex carbohydrates in sweet potatoes are "released at a steady pace," according to the same source, so there are fewer chances of sharp changes in blood sugar levels after consumption.[28]

Third, sweet potatoes are high in vitamins A and C, as well as other minerals, according to reported research. As previously covered, the orange-fleshed sweet potato variety, which is currently being promoted by many African countries and is the primary variety consumed in the United States, is well-known for its vitamin A content. More and more research has pointed out how vitamin A, converted from beta-carotene in the sweet potato variety, is beneficial to one's immune system, vision, lung, and bone health.[29] Vitamin C, which is found in both orange- and yellow-fleshed varieties, aids in infection prevention, wound healing, and iron absorption. One report specifies that a cup (200 grams) of sweet potato with skin contains 769 percent of the daily value of vitamin A, 65 percent of the daily volume of vitamin C, 50 percent of the daily volume of manganese, and 27 percent of the daily volume of potassium. Manganese and potassium are believed to help regulate blood sugar and blood pressure.[30]

Fourth, sweet potatoes are thought to contain more antioxidants than many other foods and fruits, which can help reduce the risk of chronic illnesses such as cancer and cardiovascular disease. Sweet potatoes, particularly purple-colored sweet potatoes, contain anthocyanins, an antioxidant proven by medical research to help slow the growth of cancer cells in the stomach, bladder, breast, and colon. Sweet potato is also believed to be an ideal ingredient in an anti-inflammatory diet due to its phytonutrient and flavonoid content; inflammation is blamed for a variety of diseases, including cancer, heart disease, and diabetes, as well as arthritis, Crohn's disease, and Alzheimer's disease. In addition, studies have been carried out to extract dehydroepiandrosterone, or DHEA, from the root, which is an endogenous steroid hormone precursor that is essential to one's health.[31]

Fifth and final, the recent push to rebrand sweet potatoes as healthy foods also involves reevaluating the drawbacks with which the root used to be associated. Jennifer Woolfe and others have noted that consuming sweet potatoes on a regular basis can make people feel cloyed by their sweetness.[32]

This could be one of the reasons the root was surpassed by the white potato and cassava as a more widely consumed staple food in some parts of the world. But according to the research conducted by the School of Public Health at Harvard University, while sweet potato has "a high glycemic index and glycemic load," or carbohydrates and blood sugar in common parlance, "most people don't eat sweet potatoes in the same over-sized quantities as they do white potatoes, which is perhaps why research studies haven't found sweet potatoes to be a major culprit for weight gains and diabetes."[33] Apparently, sweet potatoes' natural sweetness plays a role—they can satiate people when consumed in large quantities.

Sweet potatoes have an "annoying" aspect, which is their propensity to cause flatulence. Because of its high fiber content, which is responsible for the production of flatulence, sweet potato was dubbed a "windy food" in early anglophone texts. Many people have noticed this bodily effect, and as a result, some have developed an aversion to eating the root.[34] According to scientific research, sweet potato is flatulent because it contains both viscous and nonviscous fibers, the former of which absorbs water while the latter does not. Both of these fibers are indigestible and help to keep bowel movements regular. While flatulence can be an annoyance at social gatherings, high fiber diets have been shown to improve gut health and reduce the risk of colon cancer. Besides having a high fiber content, or dietary fiber in scientific terms, sweet potatoes also have a high percentage of resistant starch, which has a similar physiological effect on improving human health. Due to its multivarious nutritional advantages, there is a great deal of recent interest in the scientific community in finding ways to industrialize and market the resistant starch found in sweet potatoes.[35]

In sum, the sweet potato, an important food crop that originated in South and Central America, has acquired divergent reputations over the course of its global expansion beginning in the early sixteenth century. Sweet potato first appealed to Europeans as a favorite ingredient in sweet dishes; some early Europeans mistook the sweet potato for an aphrodisiac. Whereas in Asia and Oceania, the root's efficacy in saving people from starvation earned it the title of "golden spud," among other honors. In Sub-Saharan Africa, the root plant has emerged as a valuable food source in recent decades. Nevertheless, in an ironic, even illogical twist, sweet potatoes have come to be associated with poverty due to their use as an insurance crop to aid people in difficult times, leading to a love-hate relationship with the plant.[36] In many places, the root is thus referred to as "poor person's food," while in others, a gender bias is extended by calling it a "women's crop." Furthermore, the recent resurgence of the sweet potato as a wholesome food or "superfood" in Asia, Europe, and elsewhere suggests the interconnectedness of our globalized world. Meanwhile, it is also interesting to note that the ways in which sweet potatoes are produced as a nutritional superfood remain diverse and localized. In short, the sweet potato's versatility has led to diverse uses and meanings among different peoples and societies. This diversity can be seen, among other things, in the contrast between the global trend to make the sweet potato a healthful food source and the Chinese government's move to make white potatoes instead of sweet potatoes one of the country's staple foods, which is where this epilogue began. That is, the sweet potato has influenced and continued to influence the development of our modern world in manifold ways ever since it began its journey out of the Americas. Not only has it greatly improved food availability worldwide, feeding and fueling population growth, but its many uses have irrevocably shaped the distinctive cultural and political identities of people wherever it is grown around the world.

Notes

Introduction

1. William Shakespeare, *The Merry Wives of Windsor* (London, 1721), 67.
2. John Fletcher, *The Loyal Subject* (London, 1702), 46–47. Also Peter Brears, *Cooking and Dining in Tudor and Early Stuart England* (Prospect Books, 2015), 293.
3. Xu Guangqi, *Nongzheng quanshu jiaozhu* [Complete treatise on agriculture, annotated], vol. 2, ed. Shi Shenghan (Shanghai guji chubanshe, 1977), 688–694.
4. Redcliffe N. Salaman, *History and the Social Influence of the Potato* (Cambridge University Press, 1985), 425.
5. Eloy Terrón, *España, encrucijada de culturas alimentarias: Su papel en la difusión de los cultivos americanos* (Ministerio de Agricultura, Pesca y Alimentación, 1992), 118; Carolyn A. Nadeau, *Food Matters: Alonso Quijano's Diet and the Discourse of Food in Early Modern Spain* (University of Toronto Press, 2016), 95–96.
6. J. R. McNeill, forward to the thirtieth anniversary edition of *The Columbian Exchange: Biological and Cultural Consequences of 1492*, by Alfred Crosby (Praeger, 2003), xii.
7. Crosby, *The Columbian Exchange*, 181.
8. Rebecca Earle, *The Body of the Conquistador: Food, Race and the Colonial Experience in Spanish America, 1492–1700* (Cambridge University Press, 2012), 66; Rachel Laudan, *Cuisine and Empire: Cooking in World History* (University of California Press, 2013), 201–202.

9. George George, *Potatoes: The Poor Man's Own Crop* (Salisbury, 1861); Larry Zuckerman, *The Potato: How the Humble Spud Rescued the Western World* (Faber & Faber, 1998).
10. William H. McNeill, "How the Potato Changed the World's History," *Social Research* 66, no. 1 (April 1999): 67–83.
11. Linda Civitello, *Cuisine and Culture: A History of Food and People*, 2nd ed. (John Wiley & Sons, 2008), 205; Rebecca Earle, *Potato* (Bloomsbury Academic, 2019), 5–6; Andrew F. Smith, *Potato: A Global History* (Reaktion Books, 2011), 98–101.
12. Salaman, *History and the Social Influence of the Potato*, 130–133.
13. Arturo Warman, *Corn and Capitalism: How a Botanical Bastard Grew to Global Dominance*, trans. Nancy L. Westrade (University of North Carolina Press, 2003), xiii, 233–234.
14. See Andrew F. Smith, *Peanuts: The Illustrated History of the Goober Pea* (University of Illinois Press, 2002), and Brian R. Dott, *The Chile Pepper in China: A Cultural Biography* (Columbia University Press, 2020).
15. Following Alfred Crosby, Jack Weatherford also acknowledged the importance of American food plants in world history; his *Indian Givers: How the Indians of the Americas Transformed the World* (Crown Publishers, 1988) stresses that sweet potato was most attractive to Asian and African farmers. For sweet potato farming in today's world, see Maniyam Nedunchezhiyan and Ramesh C. Ray, "Sweetpotato Growth, Development, Production and Utilization: Overview," in *Sweet Potato: Post Harvest Aspects in Food, Feed and Industry*, ed. Ramesh C. Ray and K. I. Tomlins (Nova Science Publishers, 2010), 4. By 2011, sweet potato was being farmed in a total of 117 countries around the world. See W. J. Grüneberg et al., "Advances in Sweetpotato Breeding from 1992 to 2012," in *Potato and Sweetpotato in Africa: Transforming the Value Chains for Food and Nutrition Security*, ed. Jan Low et al. (CABI, 2015), 4. For its support of most people per hectare, see Gad Loebenstein and George Thottappilly, eds., *The Sweetpotato* (Springer, 2009), 504. Sweet potato contains 313 calories per 100 grams of dry weight versus the Irish potato's 306.6 calories. See Wikipedia, "Sweet Potato," last modified March 25, 2025, 18:00 (UTC), https://en.wikipedia.org/wiki/Sweet_potato.
16. Crosby, *The Columbian Exchange*, 200. Also Ping-ti Ho, *Studies on the Population of China, 1368–1953* (Harvard University Press, 1959).
17. Angus Maddison, *China's Economic Performance in the Long Run*, 2nd ed. (OECD, 2007), 24. It seems that Maddison's estimate of the population growth in Europe is lower than that of others. W. W. Rostow states that Europe's population stood at 125 million in 1750 and grew to 208 million in 1850. See his *The Great Population Spike and After: Reflections on the 21st Century* (Oxford University Press, 1998), 25. Andrew Smith also gives a higher figure,

arguing that between 1750 and 1850, Europe's population grew from 140 to 266 million. See his *Potato: A Global History*, 34. Be that as it may, the growth rate still paled in comparison to China's in the same period.
18. Ho, *Studies on the Population of China*, 270–271, 187.
19. "Sweet Potato Production, 2023," Our World in Data, accessed August 3, 2024, https://ourworldindata.org/grapher/sweet-potato-production.
20. R. L. Villareal, "Sweet Potato in the Tropics: Progress and Problems," *Sweet Potato: Proceedings of the First International Symposium*, ed. Ruben L. Villareal and T. D. Griggs (AVRDC, 1982), 3–16, esp. 4.
21. See Low et al., *Potato and Sweetpotato in Africa*; Kenneth T. Mackay et al., eds., *Sweet Potato Research and Development for Small Farmers* (SEAMEO-SEARCA, 1989); Loebenstein and Thottappilly, *The Sweetpotato*.
22. Jennifer Woolfe, *Sweet Potato: An Untapped Food Resource* (Cambridge University Press, 1992), 1, 539–558.
23. Woolfe, *Sweet Potato*, 24; T. P. Smith et al., "Sweetpotato Production in the United States," in Loebenstein and Thottappilly, *The Sweetpotato*, 298.
24. Smith et al., "Sweetpotato Production in the United States," 298.
25. Felipe Fernández-Armesto, *Near a Thousand Tables: A History of Food* (Free Press, 2002), 99.
26. Wikipedia has a "list of sweet potato cultivars" that states that there are over seven thousand. While most cultivars are developed for food, there are also others for their flowering vines. See Wikipedia, "List of Sweet Potato Cultivars," last modified October 9, 2024, 03:56 (UTC), https://en.wikipedia.org/wiki/List_of_sweet_potato_cultivars.
27. Grüneberg et al., "Advances in Sweetpotato Breeding from 1992 to 2012," in Low et al., *Potato and Sweetpotato in Africa*, 15; Nedunchezhiyan and Ray, "Sweetpotato Growth, Development, Production and Utilization," 14.
28. John C. Bouwkamp, ed., *Sweet Potato Products: A Natural Resource for the Tropics* (CRC Press, 1985), 145, 176.
29. Woolfe, *Sweet Potato*, 606–607.
30. The tendency has intensified in recent centuries—at present, two-thirds of the sweet potato production in the world are intended as forage for animal production and the rest for food and industrial use. See Nedunchezhiyan and Ray, "Sweetpotato Growth, Development, Production and Utilization," 4.
31. Woolfe, *Sweet Potato*, 411.
32. See R. Michael Bourke and Tracy Harwood, eds., *Food and Agriculture in Papua New Guinea* (ANU Press, 2009).
33. Woolfe, *Sweet Potato*, 422.
34. Woolfe, 366–370.
35. Bouwkamp, *Sweet Potato Products*, 10–21.

36. Douglas Horton, *Underground Crops: Long-Term Trends of Production in Roots and Tubers* (Winrock International, 1988), 6; S. C. S. Tsou and R. L. Villareal, "Resistance to Eating Sweet Potato," in Villareal and Griggs, *Sweet Potato*, 37–44.
37. Bouwkamp, *Sweet Potato Products*, 3.
38. Vincent Lebot, *Tropical Root and Tuber Crops: Cassava, Sweet Potato, Yams and Aroids* (CABI, 2009), 91.
39. Woolfe, *Sweet Potato*, 121; Thottappilly, "Introductory Remarks," in Loebenstein and Thottappilly, *The Sweetpotato*, 3.
40. Woolfe, *Sweet Potato*, 9, 55, 475, 593.
41. While sweet potato could be regarded by some urbanites in China as a "treat," according to the study of Shiu-ying Hu in her *Food Plants of China* (Chinese University Press, 2005), 126, others disliked it when it was offered to them every meal. See also Zhang Ning, *Tudi de huanghun: Xiangcun jingyan de weiguan quanli fenxi* (The twilight of the earth: Analyses of micro-power based on rural experiences) (Dongfang chubanshe, 2005), 117. Woolfe also observes that a survey in the Dominican Republic found the same among its people. Woolfe, *Sweet Potato*, 593.
42. Stanley J. Kays, while contributing to an international symposium, also states that the hinderance to increasing sweet potato consumption around the world is its intense sweet flavor. See his "Flavor: the Key to Sweetpotato Consumption," in *Proceedings of the 11th International Symposium on Sweetpotato and Cassava: "Innovative Technologies for Commercialization," Kuala Lumpur, Malaysia, 14–17 June 2005* (International Society for Horticultural Science, 2006), 97–105; Woolfe, *Sweet Potato*, 9; Tsou and Villareal, "Resistance to Eating Sweet Potato"; T. H. Yang, "Sweet Potato as a Supplemental Staple Food," in Villareal and Griggs, *Sweet Potato*, 37–45, 31–34; Grüneberg et al., "Advances in Sweetpotato Breeding from 1992 to 2012," 15.
43. Grüneberg et al., "Advances in Sweetpotato Breeding from 1992 to 2012," 15; Thottappilly, "Introductory Remarks," 3. Other sweet potato varieties are discussed in Woolfe, *Sweet Potato*, 57–58, and Nedunchezhiyan and Ray, "Sweetpotato Growth, Development, Production and Utilization," 8–10.
44. The attempts have been seen of late mainly in China and the United States. See Grüneberg et al., "Advances in Sweetpotato Breeding from 1992 to 2012," 8; Ramesh C. Ray et al., "Bio-Processing of Sweet Potato Into Food, Feed and Bio-Ethanol," in Ray and Tomlins, *Sweet Potato*, 163–192.
45. *The Voyage of Christopher Columbus: Columbus' Own Journal of Discovery*, trans. John Cummins (St. Martin's Press, 1992), 93.
46. *The Travels of Marco Polo*, trans. L. F. Benedetto (Routledge, 2004), 263–266.
47. *The Voyage of Christopher Columbus*, 105, 112. For a succinct description of Zayton's, or Quanzhou's, importance and prosperity in the thirteenth and

fourteenth centuries, see Charles C. Mann, *1493: Uncovering the New World Columbus Created* (Alfred A. Knopf, 2013), 405–408.
48. Heike Paul, *The Myths That Made America: An Introduction to American Studies* (Transcript Verlag, 2014), 53.
49. William D. Phillips and Carla Rahn Phillips, *The Worlds of Christopher Columbus* (Cambridge University Press, 1993), 4.
50. Crosby, *The Columbian Exchange*, xxi–xxii.
51. Crosby, 219.
52. Tzvetan Todorov, *The Morals of History*, trans. Alyson Waters (University of Minnesota Press, 1995), 99–100.
53. Ella Shohat and Robert Stam, *Unthinking Eurocentrism: Multiculturalism and the Media*, 2nd ed. (Routledge, 2014), 71–77.
54. Phillips and Phillips, *The Worlds of Christopher Columbus*, 240–241. Another attempt to present a historical understanding of Columbus, using his own writings, can be found in Felipe Fernández-Armesto, *Columbus* (Oxford University Press, 1991), and, more recently, Robert Carle, "Remembering Columbus: Blinded by Politics," *Academic Questions* 32 (2019): 105–113.
55. Stephen J. Summerhill and John Alexander Williams, *Sinking Columbus: Contested History, Cultural Politics, and Mythmaking During the Quincentenary* (University of Florida Press, 2000), 5.
56. Felipe Fernández-Armesto, *1492: The Year Our World Began* (HarperCollins, 2009). Here Fernández-Armesto extends his 1991 *Columbus* to discuss more of Columbus's impact on history. But half of its content nonetheless focuses on the explorer's life and actions. Placing Columbus in a global context, Fernández-Armesto considers his voyages "the Great Leap Forward" in the history of explorations. See also Felipe Fernández-Armesto, *Pathfinders: A Global History of Explorations* (Oxford University Press, 2006), 153–192.
57. Charles C. Mann, *1491: New Revelations of the Americas Before Columbus* (Alfred A. Knopf, 2005).
58. Michael Blake, *Maize for the Gods: Unearthing the 9,000-Year History of Corn* (University of California Press, 2015); Duccio Bonavia, *Maize: Origin, Domestication and Its Role in the Development of Culture*, trans. Javier Flores Espinoza (Cambridge University Press, 2013).
59. Betty Fussell, *The Story of Corn* (Alfred A. Knopf, 1992), 4; Cynthia Clampitt, *Midwest Maize: How Corn Shaped the US Heartland* (University of Illinois Press, 2015); Anthony Boutard, *Beautiful Corn: America's Original Grain from Seed to Plate* (New Society Publishers, 2012); José Antonio Serratos Hernández, *The Origin and Diversity of Maize in the American Continent* (Greenpeace, 2009).
60. Warman, *Corn and Capitalism*, 37–50, 232–242; Elizabeth Fitting, *The Struggle for Maize: Campesinos, Workers, and Transgenic Corn in the Mexican Countryside* (Duke University Press, 2011). See also Michael Owen Jones, *Corn: A*

Global History (Reaktion Books, 2017). However, despite corn's global scope, Jones does not give much attention to corn farming in China, the world's second-largest producer.

61. Smith, *Potato*.
62. John Reader, *Potato: A History of the Propitious Esculent* (Yale University Press, 2008), 265–278.
63. Earle, *Potato*, 60–64; Earle, *Feeding the People: The Politics of the Potato* (Cambridge University Press, 2020).
64. Zhang Jian, *Xindalu nongzuowu de chuanbo he yiyi* (Kexue chubanshe, 2014), especially introduction, 1–11.
65. Sakai Nobuo, *Bunmei o kaeta shokubutsu-tachi: Koronbusu ga nokoshita shushi* (NHK, 2011), especially 38–50, 243–258; Inagaki Hidehiro, *Sekaishi o ōkiku ugokashita shokubutsu* (PHP kenkyūjo, 2018); Yamamoto Norio, *Koronbusu no fubyōdō kōkan sakumotsu dorei ekibyō no sekaishi* (Kadokawa, 2017), especially 223–245.
66. In 1959, several industrial publishers in China published pamphlets on the processing and utilization of the sweet potato. See Fujian Putian xian hongzhuan daxue, ed., *Fanshu zaipeixue* [How to cultivate the sweet potato] (Gaodeng jiaoyu chubanshe, 1959); Nongye ziliao bianji weiyuanhui, ed., *Fanshu de chucang he liyong* [Storage and utilization of the sweet potato] (Nongye chubanshe, 1959); and Qinggongye shipinju, ed., *Fanshu zonghe liyong de jingyan* [The experience in multifarious utilization of the sweet potato] (Qinggongye chubanshe, 1959). See also Zhongguo shehui kexueyuan lishi yanjiusuo Qingshi yanjiushi, ed., *Qingshi ziliao* [Sources of Qing history], vol. 7 (Zhonghua shuju, 1989), which presents historical records on the dissemination of the sweet potato and maize in China.
67. Miyamoto Tsuneichi, *Kansho no rekishi* (Mirai sha, 1962), 2–4.
68. Sakai Kenichi, *Satsuma imo* [The sweet potato] (Hosei daigaku shuppansha, 1999); Chūman Katsumi, *Nihon kansho saibaishi: Kansho no denrai kara denpa, sono saibai-hō no hensen* [Japanese sweet potato chronicles: From introduction and spread to the evolution of cultivation methods] (Taki shobō, 2002); Itō Shōji, *Satsuma imo to Nihon jin: Wasure rareta shoku no ashiato* (PHP Kenkyūjo, 2010).
69. Cai Chenghao and Yang Yunping, *Taiwan fanshu wenhuazhi* (Guoshi chuban, 2004).
70. Mackay et al., *Sweet Potato Research and Development for Small Farmers*, iii.
71. See Villareal and Griggs, *Sweet Potato*; Mackay et al., *Sweet Potato Research and Development for Small Farmers*, iii; Low et al., *Potato and Sweetpotato in Africa*.
72. Ping-ti Ho's *Studies on the Population of China, 1368–1953*, concludes that the crises Qing China faced from the mid-nineteenth century were caused by the overpopulation of the previous century and cautions the government of

the People's Republic of China to learn the historical lesson and design policies for population control (257–278). He essentially reiterates the same caution in concluding his articles on the impact of American plants on Chinese agronomy. See his "Meizhou zuowu de yinjin, chuanbo jiqi dui Zhongguo liangshi zuowu de yingxiang" [The introduction and dissemination of American plants and their influence in Chinese food crops], *Shijie nongye* [World agriculture], 4 (1979): 34–41, 5 (1979): 21–31, 6 (1979): 25–31. Kang Chao expounds Ho's thesis and examines the population pressure in hindering China's economic growth in his *Man and Land in Chinese History: An Economic Analysis* (Stanford University Press, 1986). In both of his studies of Chinese agriculture and rural society of the modern period, Philip C. C. Huang adapts the "involution" concept and argues that the country's high population density contributed to the stagnation of its economic development. See his *The Peasant Economy and Social Change in North China* (Stanford University Press, 1985) and *The Peasant Family and Rural Development in the Lower Yangzi Region, 1350–1988* (Stanford University Press, 1990). Mark Elvin, too, considers China's high population a hindrance to its economic development, causing what he calls "the high-level of equilibrium trap" in the late imperial period. See his *The Pattern of the Chinese Past* (Stanford University Press, 1973), 203–319. In explaining the "European miracle" in industrializing its society from the eighteenth century, Eric L. Jones also considers population control in Europe a factor vis-à-vis the out-of-control situation in Asia, including China. See his *The European Miracle: Environments, Economies and Geopolitics in the History of Europe and Asia*, 3rd ed. (Cambridge University Press, 2003).

73. Adam Smith, *An Inquiry Into the Nature and Causes of the Wealth of Nations*, ed. and intro. Edwin Cannan (University of Chicago Press, 1977), 225, and Thomas Malthus, *An Essay on the Principle of Population and a Summary View of the Principle of Population*, ed. and intro. Antony Flew (Penguin Books, 1970), 115.

74. R. Bin Wong, *China Transformed: Historical Change and the Limits of European Experience* (Cornell University Press, 1997), chaps. 1 and 2; James Z. Lee and Wang Feng, *One Quarter of Humanity: Malthusian Mythologies and Chinese Realities, 1700–2000* (Harvard University Press, 1999); James Z. Lee et al., "Positive Check or Chinese Check?," *Journal of Asian Studies* 61, no. 2 (May 2002): 591–607; Kenneth Pomeranz, *The Great Divergence: China, Europe, and the Making of the Modern World Economy* (Princeton University Press, 2000), 12.

75. See, for example, Kent Deng and Shengming Sun, "China's Extraordinary Population Expansion and Its Determinants During the Qing Period, 1644–1911," *Population Review* 58, no. 1 (2019): 20–77. The period of China's population increase was also closely analyzed in a standard textbook on East Asian

history in the United States. See John K. Fairbank et al., *East Asia: Tradition and Transformation* (Houghton Mifflin, 1989), 241–242.

76. Dwight Perkins stated that despite the tremendous population increase, China was able to feed its people through much of the twentieth century by improving farming technology and introducing new food crops. See his *Rural Development in China* (Johns Hopkins University Press, 1984). Challenging Malthus's thesis, Ester Boserup argued that population increase could lead to better food production, which seems applicable in explaining China's continuous demographic expansion in modern times. See her *The Conditions of Agricultural Growth: The Economics of Agrarian Change under Population Pressure* (1965; repr. Routledge, 2003). About the unchanging standard of living in China and East Asia before the mid-nineteenth century, see Kenneth Pomeranz, "Is There an East Asian Development Path? Long-term Comparisons, Constraints and Continuities," *Journal of the Economic and Social History of the Orient* 44, no. 3 (2001): 322–362, and Pomeranz, *The Great Divergence*, 226. About the connection between the "East Asian miracle" with the "European miracle," see Kaoru Sugihara, "The European Miracle and the East Asian Miracle: Toward a New Global Economic History," in *The Pacific in the Age of Early Industrialization*, ed. Kenneth Pomeranz (Ashgate, 2009), 1–21, and, more generally, Gareth Austin and Kaoru Sugihara, eds., *Labour-Intensive Industrialization in Global History* (Routledge, 2013).

77. Jennifer Woolfe enumerates the various ways in which sweet potato has been received as a food around the world, noting that some take it as a staple or co-staple, whereas others (e.g., in Japan and parts of China and North America) consume it as a treat or as a featured item on special or festive occasions. See Woolfe, *Sweet Potato*, 475–477.

78. Cf. Smith, *Potato*, 102.

1. American Origin and Oceanian Diffusion

1. Thor Heyerdahl, *Kon-Tiki: Across the Pacific in a Raft*, trans. F. H. Lyon (Rand McNally and Company, 1950), 167.
2. Graham E. L. Holton, "Heyerdahl's *Kon-Tiki* Theory and the Denial of the Indigenous Past," *Anthropological Forum* 14, no. 2 (July 2004): 163–181, quote on 177.
3. There is an ever-growing body of literature on the historical possibility of trans-Pacific voyages before the European "Age of Discovery," including such works as Victor H. Mair, ed., *Contact and Exchange in the Ancient World* (University of Hawaii Press, 2005); Terry L. Johns et al., eds., *Polynesians in America:*

Pre-Columbian Contacts with the New World (AltaMira Press, 2010); Alice Beck Kehoe, *Traveling Prehistoric Seas: Critical Thinking on Ancient Transoceanic Voyages* (Routledge, 2015); and Stephen C. Jett, *Ancient Ocean Crossings: Reconsidering the Case for Contacts with the Pre-Columbian Americas* (University of Alabama Press, 2017). Andrea Ballesteros Daniel's PhD dissertation, "Trans-Pacific Contact: A History of Ideas on the Oceanian-Americas Connection" (Australian National University, 2020), offers a good and recent synthesis of existing works on the subject. The theories about the sweet potato's diffusion in the Pacific receive a concise discussion on 194–195 and 271–279.

4. O. F. Cook, "Quichua Names of Sweet Potatoes," *Journal of the Washington Academy of Sciences* 6, no. 4 (February 1916): 86–90. Also see D. E. Yen, *The Sweet Potato and Oceania: An Essay in Ethnobotany* (Bishop Museum Press, 1974), 335–345.
5. Heyerdahl, *Kon-Tiki*, caption of the photos of the boat's kitchen, 64–65.
6. Thor Heyerdahl, *American Indians in the Pacific: The Theory Behind the Kon-Tiki Expedition* (Gyldendal Norsk Forlag, 1952), 428.
7. Crosby, *The Columbian Exchange*, 5.
8. Jared Diamond, *Guns, Germs and Steel: The Fates of Human Societies* (W. W. Norton, 1997), 161.
9. Donald D. Brand, "The Origin and Early Distribution of New World Cultivated Plants," *Agricultural History* 13, no. 1 (April 1939): 112–113.
10. J. G. Hather and P. V. Kirch, "Prehistoric Sweet Potato (*Ipomoea batatas*) from Mangaia Island, Central Polynesia," *Antiquity*, no. 65 (1991): 887–893; T. N. Ladeforged and M. W. Graves, "The Introduction of Sweet Potato in Polynesia: Early Remains in Hawaii," *Journal of Polynesian Society* 114, no. 4 (2005): 359–374; M. Horrocks and R. B. Rechtman, "Sweet Potato (*Ipomoea batatas*) and Banana (*Musa sp.*) Microfossils in Deposits from the Kona Field System, Island of Hawaii," *Journal of Archaeological Science* 36, no. 5 (2009): 1115–1126.
11. Alphonso de Candolle, *Origin of Cultivated Plants* (London, 1884), 53–58.
12. Fernández-Armesto, *Near A Thousand Tables*, 99.
13. Lebot, *Tropical Root and Tuber Crops*, 91–93.
14. Estelle Levetin and Karen McMahan, *Plants and Society*, 7th ed. (McGraw-Hill, 2016), 226; Washington Irving, *The Life and Voyages of Christopher Columbus*, vol. 1 (London, 1850), 242; William D. Phillips Jr. and Carla Rahn Phillips, *The Worlds of Christopher Columbus*, 266.
15. Roland B. Dixon, "The Problem of the Sweet Potato in Polynesia," *American Anthropologist* 34, no. 1 (January–March 1932): 40–66; Robert J. Gustafson, Derral R. Herbst, and Philip W. Rundel, *Hawaiian Plant Life: Vegetation and Flora* (University of Hawaii Press, 2014), 226; Patrick V. Kirch, John Holson, and Alexander Baer, "Intensive Dryland Agriculture in Kaupō, Maui,

Hawaiian Islands," *Asian Perspectives* 48, no. 2 (Fall 2009): 265–290; Patricia J. O'Brien, "The Sweet Potato: Its Origin and Dispersal," *American Anthropologist* 74, no. 3 (June 1972): 342–365.
16. Dixon, "The Problem of the Sweet Potato"; D. E. Yen, "Sweet Potato Variation and Its Relation to Human Migration in the Pacific," in *Plants and the Migrations of Pacific Peoples: A Symposium*, ed. J. Barrau (Bishop Museum Press, 1963), 112.
17. Bourke and Harwood, *Food and Agriculture in Papua New Guinea*, 133.
18. Woolfe, *Sweet Potato*, 19–21.
19. See Uchibayashi Masao, "Koronbusu izen kara porineshia ni atta satsuma imo—gaikan" [Sweet potato in pre-Columbian Polynesia: An overview], *Yakugaku zasshi* [Pharmaceutical Society of Japan] 126, no. 12 (2006): 1347.
20. H. B. Guppy, *The Observations of a Naturalist in the Pacific between 1896 and 1899* (Macmillan, 1906), 532; Yen, *The Sweet Potato and Oceania*, 2.
21. Uchibayashi, "Koronbusu izen kara porineshia ni atta satsuma imo—gaikan," 1347.
22. Chris Ballard, Paula Brown, R. Michael Bourke, and Tracy Harwood, eds., *The Sweet Potato in Oceania: A Reappraisal* (University of Sydney, 2005).
23. Yen, *The Sweet Potato and Oceania*, 20. Also more specifically, D. E. Yen, "The Sweet Potato in the Pacific: The Propagation of the Plant in Relation to Its Distribution," *Journal of Polynesian Society* 69, no. 4 (1960): 368–375.
24. Yen, *The Sweet Potato and Oceania*, 2–5; Harold C. Conklin, "The Oceanian-African Hypotheses and the Sweet Potato," in Barrau, *Plants and the Migrations of Pacific Peoples*, 129–136.
25. Yen, *The Sweet Potato and Oceania*, 2.
26. Elmer D. Merrill, *The Botany of Cook's Voyages and Its Unexpected Significance in Relation to Anthropology, Biogeography, and History* (Chronica Botanica Co., 1954).
27. Robert Langdon, "Bamboo Raft as a Key to the Introduction of the Sweet Potato in Prehistoric Polynesia," *Journal of Pacific History* 36, no. 1 (June 2001): 51–76.
28. Yen, *The Sweet Potato and Oceania*, 2–3; J. W. Purseglove, "Some Problems of the Origin and Distribution of Tropical Crops," *Genetica Agraria* 17 (1963): 105–122, especially 110–111.
29. Harold M. Ross, for example, states that the origin of the root's presence in the Solomons remains uncertain. See his "The Sweet Potato in the South-Eastern Solomons," *Journal of Polynesian Society* 86, no. 4 (December 1977): 521–530.
30. R. C. Green, "Sweet Potato Transfers in Polynesian Prehistory," in Ballard et al., *The Sweet Potato in Oceania*, 43–62.
31. Pablo Muñoz-Rodríguez et al., "Reconciling Conflicting Phylogenies in the Origin of Sweet Potato and Dispersal in Polynesia," *Current Biology* 28, no. 8

(April 2018): 1253; Tom Carruthers et al., "The Temporal Dynamics of Evolutionary Diversification in Ipomoea," *Molecular Phylogenetics & Evolution* 146 (May 2020): 1–16.

32. Dan Garisto, "Polynesian Sailings to America Doubted," *Science News*, April 28, 2018.
33. Heyerdahl, *Kon-Tiki*, 14; Heyerdahl, *American Indians in the Pacific*; Heyerdahl, *The White Gods: Caucasian Elements in Pre-Inca Peru* (Allen and Unwin, 1952).
34. Heyerdahl, *American Indians in the Pacific*, 1–68.
35. Heyerdahl, 76.
36. Cf. Wikipedia, "Kon-Tiki Expedition," last modified February 7, 2025, 20:56 (UTC), https://en.wikipedia.org/wiki/Kon-Tiki_expedition.
37. See Denis McLellan, "Thor Heyerdahl of 'Kon-Tiki' Fame Dies," *Los Angeles Times*, April 19, 2002. Ai, Axel Andersson has written a biography entitled *A Hero for the Atomic Age: Thor Heyerdahl and the Kon-Tiki Expedition* (Peter Long, 2010), which duly notes that Heyerdahl was motivated by a racist intent to launch the adventure. See also Holton, "Heyerdahl's *Kon-Tiki* Theory and the Denial of the Indigenous Past."
38. One example is Garry W. Trompf's "Kon-Tiki and the Critics" in *Melbourne Historical Journal* 4 (January 1964): 52–65.
39. Uchibayashi, "Koronbusu izen kara porineshia ni atta satsuma imo—gaikan," 1342.
40. See Wikipedia, "Polynesia," last modified March 18, 2025, 19:15 (UTC), https://en.wikipedia.org/wiki/Polynesia.
41. Erik Thorsby, "The Polynesian Gene Pool: An Early Contribution by Amerindians to Easter Island," *Philosophical Transactions: Biological Sciences* 367, no. 1590 (March 2012): 812.
42. Thorsby, 818.
43. Yen, *The Sweet Potato and Oceania*, 259–267, 340–348. A more recent study, using genetic analysis, seems to confirm the thesis concerning the *kumara* line of transfer, whereas the other two lines remain a bit inconclusive due to the variety of local variants appearing in modern western Oceania and Southeast Asia. See Caroline Roullier, Laurie Benoit, Doyle B. McKey, and Vincent Lebot, "Historical Collection Reveal Patterns of Diffusion of Sweet Potato in Oceania Obscured by Modern Plant Movements and Recombination," *Proceedings of the National Academy of Sciences of United States of America* 110, no. 6 (February 2013): 2205–2210.
44. Bourke and Harwood, *Food and Agriculture in Papua New Guinea*, 17.
45. Yen, *The Sweet Potato and Oceania*, 3. See also Yen, "Reflection, Refraction and Recombination," in Ballard et al., *Sweet Potato in Oceania*, 181–188.

46. Christopher Columbus, *Journals and Other Documents on the Life and Voyages of Christopher Columbus*, trans. and ed. Samuel Eliot Morison (Heritage Press, 1963), 88–89, 234; Salaman, *History and the Social Influence of the Potato*, 131; De Candolle, *Origin of Cultivated Plants*, 54–56.
47. Yen, *The Sweet Potato and Oceania*, 338.
48. Peter Martyr d'Anghiera, *De Orbe Novo: The Eight Decades of Peter Martyr D'Anghera*, trans. Francis Augustus MacNutt (Putnam's Sons, 1912), 63–64, 385. In this translation of Peter Martyr's book, the sweet potato was also referred to as "agoes" as well as "ages."
49. Gonzalo Fernández de Oviedo, *Natural History of the West Indies*, trans., Sterling Stoudemire (University of North Carolina Press, 1959), 98; Salaman, *History and the Social Influence of the Potato*, 131.
50. Salaman, *History and the Social Influence of the Potato*, 132.
51. De Candolle, *Origin of Cultivated Plants*, 55; Conklin, "Oceanian-African Hypothesis and the Sweet Potato," in Barrau, *Plants and the Migrations of Pacific Peoples*, 129–136.
52. Conklin, "Oceanian-African Hypothesis and the Sweet Potato," 132–141; Reader, *Potato*, 22–23.
53. Brand, "Origin and Early Distribution of New World Cultivated Plants," 111. Also, Salaman, *History and the Social Influence of the Potato*, xx.
54. Yen, *The Sweet Potato and Oceania*, 347–348.
55. *The Voyages of Pedro Fernandez de Quiros, 1595–1606*, vol. 1, trans. Sir Clements Markham (London, 1804), 154.
56. Henry Kamen, *Spain's Road to Empire: The Making of a World Power, 1492–1763* (Penguin Books, 2003), especially chap. 5.
57. Jacob Roggeveen, *Extracts from the Official Log of Mynheer J. Roggeveen (1721–22)*, 2nd sers, vol. 13 (Hakluyt Society, 1908), 120; Patricia J. O'Brien, "The Sweet Potato: Its Origin and Dispersal," *American Anthropologist* 74, no. 3 (June 1972): 342–365.
58. *The Journals of Captain Cook*, ed. Philip Edwards (Penguin Books, 1999), 81.
59. *Journals of Captain Cook*, 336.
60. De Candolle, *Origin of Cultivated Plants*, 55; Robert Langdon, "The Bamboo Raft as a Key to the Introduction of the Sweet Potato in Prehistoric Polynesia," *Journal of Pacific History* 36, no. 1 (June 2001): 51–76; Reader, *Potato*, 79.
61. Georg Friederici, "Zu den vorkolumbischen Verbindungen der Südsee-Völker mit Amerika," *Anthropos* 24, nos. 3–4 (May–August 1929): 441–487, quotes on 469–470.
62. Berthold Laufer, "The American Plant Migration," *Scientific Monthly* 28, no. 3 (March 1929): 230–251, quote on 241.

63. Ronald B. Dixon, "The Problem of the Sweet Potato in Polynesia," *American Anthropologist* 34, no. 1 (January–March 1932): 54.
64. Peter H. Buck, *Vikings of the Sunrise* (Frederick A. Stokes Company, 1938), 316.
65. Uchibayashi, "Koronbusu izen kara porineshia ni atta satsuma imo—gaikan," 1347.
66. James Hornell, "How Did the Sweet Potato Reach Oceania?," *Journal of the Linnean Society of London* 53 (1946): 41–62; James Hornell, "Was There Pre-Columbian Contact Between the Peoples of Oceania and South America?," *Journal of the Polynesian Society* 54, no. 4 (December 1945): 167–191.
67. Roullier et al., "Historical Collection Reveal Patterns of Diffusion of Sweet Potato in Oceania Obscured by Modern Plant Movements and Recombination."
68. Barbara Watson Andaya and Leonard Y. Andaya, *A History of Early Modern Southeast Asia, 1400–1830* (Cambridge University Press, 2015), 20.
69. Wang Dayuan, *Daoyi zhilue jiaoshi*, annotated by Su Jiqing (Zhonghua shuju, 1981), 38, 89, 126, 172, 190.
70. Ross, "The Sweet Potato in the South-Eastern Solomons," 525–526.
71. Green, "Sweet Potato Transfers in Polynesian Prehistory," 42; J. W. Macnab, "Sweet Potatoes and Settlement in the Pacific," *Journal of Polynesian Society* 76, no. 2 (June 1967): 219–221.
72. Helen Leach, "*Ufi Kumara*, the Sweet Potato as Yam," in Ballard et al., *The Sweet Potato in Oceania*, 63–70.
73. Leach, 63.
74. Yen, *The Sweet Potato and Oceania*, 266.
75. James Coil and Patrick V. Kirch, "An Ipomoean Landscape: Archaeology and the Sweet Potato in Kahikinui, Maui, Hawaiian Islands," and Paul Wallin, Christopher Stevenson, and Thegn Ladefoged, "Sweet Potato Production on Rapa Nui," in Ballard et al., *The Sweet Potato in Oceania: A Reappraisal*, 71–84, 85–88.
76. Yen, *The Sweet Potato and Oceania*, 18–19, 337, 339; John F. G. Stokes, "Spaniards and the Sweet Potato in Hawaii and Hawaiian-American Contact," *American Anthropologist* 34, no. 4 (October–December 1932): 599.
77. Hornell, "How Did the Sweet Potato Reach Oceania?," 43; Yen, *The Sweet Potato and Oceania*, 18–19.
78. Green, "Sweet Potato Transfers in Polynesian Prehistory," 43.
79. Antonio Pigafetta, *The First Voyage Round the World by Magellan: Translated from the Accounts of Pigafetta and Other Contemporary Writers*, ed. Henry Edward John Stanley (London, 1874; Cambridge University Press, 2010), 70.
80. J. C. Beaglehole, *The Voyage and Resolution of the Discovery 1776–1780: The Journals of James Cook on His Voyage of Discovery*, vol. 3, pt. 1 (Cambridge University Press, 1967), 278.

81. William Ellis, *Polynesian Researches, During a Residence of Nearly Six Years in the South Sea Islands Including Descriptions of the Natural History and Scenery of the Islands with Remarks on the History, Mythology, Traditions Government, Arts, Manners, and Customs of the Inhabitants*, vol. 1 (London, 1829), 34, 360.
82. Yen, *The Sweet Potato and Oceania*, 291.
83. Cf. Ross, "The Sweet Potato in the South-Eastern Solomons."
84. E. S. Craighill Handy and Elizabeth Green Handy, *Native Planters in Old Hawaii: Their Life, Lore and Environment* (Bishop Museum, 1978), 123.
85. E. S. Craighill Handy, *The Hawaiian Planter*, vol. 1 (Bishop Museum), 7, 143, 152.
86. Wallin et al., "Sweet Potato Production on Rapa Nui."
87. Elsden Best, *Maori Agriculture* (Dominion Museum, 1925), bulletin 9, 99.
88. Yen, *The Sweet Potato and Oceania*, 34.
89. Yen, 256–257; Ross, "The Sweet Potato in the South-Eastern Solomons."
90. Hornell, "How Did the Sweet Potato Reach Oceania?," 48; Christopher Ehret, *History and the Testimony of Language* (University of California Press, 2011), 237, 243; Maria Andrade et al., *Unleashing the Potential of Sweetpotato in Sub-Saharan Africa: Current Challenges and War Forward* (International Potato Center, 2009), 26.
91. Conklin, "Oceanian-African Hypothesis and the Sweet Potato," 129–133.
92. Yen, *The Sweet Potato and Oceania*, 249–259.
93. Bourke and Harwood, *Food and Agriculture in Papua New Guinea*, 15–17.
94. Hornell, "How Did the Sweet Potato Reach Oceania?"; Yen, *The Sweet Potato and Oceania*, 341–347.
95. Matthew G. Allen, "The Evidence for Sweet Potato in Island Melanesia," in Ballard et al., *The Sweet Potato in Oceania*, 99–108; Yen, *The Sweet Potato and Oceania*, 129; Ross, "The Sweet Potato in the South-Eastern Solomons."
96. Yen, *The Sweet Potato and Oceania*, 343–345.
97. See Villareal and Griggs, *Sweet Potato: Proceedings of the First International Symposium* and Mackay et al., *Sweet Potato Research and Development for Small Farmers*.
98. Bourke and Harwood, *Food and Agriculture in Papua New Guinea*, 1.
99. Chris Ballard, "Still Good to Think With: The Sweet Potato in Oceania," in Ballard et al., *The Sweet Potato in Oceania*, 2.

2. A Sweet Connection Between Europe and America

1. Columbus, *Journals and Other Documents on the Life and Voyages of Christopher Columbus*, 88–89. The text here originally used *mames* instead of *niames*,

whereas on page 128, it used *niames* to refer to the same species. According to I. H. Burkill, Columbus himself used *niames* in his original journal. "Aji and Batata as Group Names Within the Species *Ipomoea batatas,*" *Ceiba* 4 (1954): 227.

2. Diego Álvarez Chanca, *Letter of Dr. Chanca on the Second Voyage of Columbus* (American Journeys Collection, Document No. AJ-065, Wisconsin Historical Society, Digital Library and Archives, 2003), 308n4.

3. Candice Goucher, *Congotay! Congotay! A Global History of Caribbean Food* (M. E. Sharpe, 2014), 19.

4. Columbus, *Journals and Other Documents on the Life and Voyages of Christopher Columbus*, 91.

5. Columbus, 122, 128.

6. Columbus, 234–235.

7. Diego Álvarez Chanca, *Letter of Dr. Chanca on the Second Voyage of Columbus*, 308–309. Note 1 on page 309 says that "By the Indians Dr. Chanca means the Tainos, the native inhabitants of Española."

8. Gerard Paul, "History of the Sweet Potato: From Mystery Migration to Thanksgiving Staple," ManyEats, November 14, 2020, https://manyeats.com/history-of-the-sweet-potato/.

9. Columbus, *Journals and Other Documents on the Life and Voyages of Christopher Columbus*, 356.

10. Columbus, 88, 89n6, 96, 137, 139n1, 216, 218n14, 223, 225n19.

11. Francisco López de Gómara, *Historia general de las Indias, y vida de Hernán Cortés* (Biblioteca Ayacucho, 1977), 32.

12. Salaman, *History and the Social Influence of the Potato*, 131. Recent scholars like Carolyn A. Nadeau agree with Salaman that Columbus brought the sweet potato back to Spain after his first voyage. See Nadeau, *Food Matters: Alonso Quijano's Diet and the Discourse of Food in Early Modern Spain*, 95. The same is stated by Bethany Aram and Bartolomé Yun-Casalilla, eds., *Global Goods and the Spanish Empire, 1492–1824* (Palgrace Macmillan, 2014), 23.

13. In his detailed study of the sweet potato and its names, I. H. Burkill wrote that, while López de Gómara might have some living experience in the New World, we should not so trustful of his records. With regard to whether Columbus returned with the sweet potato to Spain after his first voyage, Burkill stated that "It may be assumed that both ships started the voyage [Columbus's first return trip back to Spain] abundantly provided with sweet potato tubers as they were, in Columbus' words, good and plentiful." Burkill, "Aji and Batatas as Group Names Within the Species *Ipomoea batatas*," 228–229. In other words, it remains an assumption that Columbus brought the plant back to Spain in early 1493, after his first voyage to the New World.

14. Burkill, 233. On Peter Martyr, see Henry R. Wagner, "Peter Martyr and His Works," *American Antiquarian Society* 56, no. 2 (October 1946): 239–288.
15. Peter Martyr d'Anghiera, *De Orbe Novo: The Eight Decades of Peter Martyr d'Anghera*, vol. 1, 63–64.
16. d'Anghiera, 1:382–383. Here the quote is modified by consulting Richard Eden's translation of the first three chapters of Peter Martyr's book because Francis A. MacNutt used *agoes* rather than *ages*, as he did before in the last quote. By comparison, Richard Eden wrote that "of the setting the roots of *Maizium, Ages, Iucca, Battatas*, and such other, being their common food and of the vse of the same, we haue spoken sufficiently before." *De nouo orbe, or, The historie of the West Indies*, trans. R. Eden and edited by M. Lok (London, 1612), 142a.
17. d'Anghiera, *De Orbe Novo*, 1:385. The quote here is again modified by consulting Richard Eden's translation on page 143a, which consistently used *ages* rather than *agoes*, as in MacNutt's translation.
18. Columbus, *Journals and Other Documents on the Life and Voyages of Christopher Columbus*, 89n6. Also see Bartolomé de Las Casas, *Historia de las Indias*, vol. 1 (Fondo de Cultura Economica, 1951), 229.
19. Bartolomé de Las Casas, *Apologética história de las Índias*, ed. M. Serrano y Sanz (Dailly, Bailliére é Dijos, 1909), 28.
20. Las Casas, 29.
21. López de Gómara, *Historia general de las Indias*, 109.
22. Gonzalo Fernández de Oviedo, *Sumario de la natural historia de las Indias*, ed. Álvaro Baraibar (Iberoamericana-Vervuert, Universidad de Navarra, 2010), 138, 313. Oviedo's book has an English translation, which can be consulted for the translated quotes here: *National History of the West Indies*, trans. Sterling A. Stoudemire (University of North Carolina Press, 1959), 41, 98.
23. Sophie D. Coe, *America's First Cuisines* (University of Texas Press, 1994), 19–20.
24. Salaman, *History and the Social Influence of the Potato*, 132.
25. Bernal Díaz del Castillo, *Historia verdadera de la conquista de la Nueva España*, trans. and ed. Miguel León-PortillaCover (1632; HimaliDigital, 2018), 106. See also Bernal Díaz del Castillo, *The True History of the Conquest of New Spain*, trans. Alfred Percival Maudslay (Hakluyt Society, 1908), 41. The English translator translated both *boniatos* and *batatas* as sweet potatoes.
26. Nicolás Monardes, *Joyfull Newes Out of the Newe Founde Worlde*, trans. John Frampton (London, 1577), fol. 104. The spellings and some terms in the quote have been modified for modern readers.
27. Caroli Clvsii Atrebat [Carolus Clusius], *Rariorum aliquot stirpium per Hispaniam observatorum Historia* (Antwerp, 1576), 297–299.
28. Francisco Hernández, *Historia de las plantas de Nueva España* (Imprenta Universitaria, 1943), 520–521.

29. Bernardino de Sahagún, *General History of the Things of New Spain, Book 11—Early Things*, trans. Charles E. Dibble and Arthur J. O. Anderson (University of Utah Press, 1963), 125, 139–140. In his *The War of Conquest: How It Was Waged Here in Mexico*, another volume of his *General History of the Things of New Spain*, Sahagún recorded that Spaniards ate sweet potato while in Mexico: "In due course the Spaniards ate; they had white tortillas, degrained corn, eggs, turkey, various kinds of sweet potato, manioc, avocado, acacia beans, and *jícama*, and ended with a choice of custard apple, mamey, sapota, plum, jobo, guava, *cuajilote, tejocote*, American cherry, blackberry, prickly pear, and *pitahaya*." Trans. Arthur O. Anderson and Charles E. Dibble (University of Utah Press, 1978), 19–20.
30. Salaman, *History and the Social Influence of the Potato*, 131.
31. Terrón, *España, Encrucijada de culturas alimentarias*, 74, 83.
32. Cook, "Quichua Names of Sweet Potatoes"; Yen, *The Sweet Potato and Oceania*, 335–345.
33. Rachel Laudan, *Cuisine and Empire: Cooking in World History* (University of California Press, 2013), 187.
34. Joseph de Acosta, *The Natural and Moral History of the Indies*, vol. 1, ed. Clements R. Markham (Cambridge University Press, 2009), 235.
35. Oviedo, *Sumario de la natural historia de las Indias*, 138nn351–352; Burkill, "Aji and Batatas as Group Names Within the Species *Ipomoea batatas*": Douglas Taylor, "Aji and Batata," *American Anthropologist* 59, no. 4 (August 1957): 704–705; Yen, *Sweet Potato and Oceania*, 338, 15.
36. Columbus, *Journals and Other Documents on the Life and Voyages of Christopher Columbus*, 154–155n6.
37. López de Gómara, *Historia general de las Indias*, 33.
38. Francisco Hernández described chile pepper and its popularity as follows: "The chile or Mexican pepper is the plant that gives those pods called by the Haitians *ajíes*, by the ancients peppers, and by the Spaniards Indian pepper. And although it has long since been brought to Spain, where it is highly esteemed and is planted in gardens and pots as an ornament for their use and as a useful plant. However, as there are among the Indians many other genera used daily to excite the appetite and spice meals, so that no table is found without chile, and its properties are therefore well known by the daily experience." *Historia de las plantas de Nueva España*, 428–429.
39. Burkill, "Aji and Batata as Group Names Within the Species *Ipomoea batatas*," 235–237.
40. Woolfe, *Sweet Potato*, 9.
41. Woolfe, 55, 121, 475. According to Bouwkamp, sweet potato was ranked sixth in world production as a food crop back in the 1970s. *Sweet Potato Products*, 3.

42. Burkill, "Aji and Batatas as Group Names Within the Species *Ipomoea batatas*," 235–239.
43. Jack Goody, *Cooking, Cuisine and Class: A Study in Comparative Sociology* (Cambridge University Press, 1982), especially chap. 4, "The High and the Low: Culinary Culture in Asia and Europe," 97–153.
44. Jodi Campbell, *At First Table: Food and Social Identity in Early Modern Spain* (University of Nebraska Press, 2017), 109.
45. Laudan, *Cuisine and Empire*, 176–177, 196.
46. Rebecca Earle, *The Body of the Conquistador: Food, Race and the Colonial Experience in Spanish America, 1492–1700* (Cambridge University Press, 2012), 5–8, 26–46, 49–66, quote on 8 (italics in the original); Goucher, *Congotay! Congotay!*, 29.
47. David Gentilcore, *Food and Health in Early Modern Europe: Diet, Medicine and Society, 1450–1800* (Bloomsbury Academic, 2016), 144, 150. Also Salaman, *History and the Social Influence of the Potato*, 108–116.
48. Andrew F. Smith, *Potato: A Global History*, 27–29.
49. Zuckerman, *The Potato*, xi.
50. Earle, *The Body of the Conquistador*, 127.
51. Marcy Norton, "Tasting Empire: Chocolate and the European Internalization of Mesoamerican Aesthetics," *American Historical Review* 111, no. 3 (June 2006): 684.
52. Quoted in Ken Albala, *Eating Right in the Renaissance* (University of California Press, 2002), 66.
53. Alan K. Outram, "Hunter-Gatherers and the First Farmers: The Evolution of Taste in Prehistory," in *Food: The History of Taste*, ed. Paul Freedman (University of California Press, 2007), 47.
54. Sydney W. Mintz, *Sweetness and Power: The Place of Sugar in Modern History* (Penguin Books, 1985), 9–14.
55. Mintz, 5.
56. Jeri Quinzio, *Dessert: A Tale of Happy Endings* (Reaktion Books, 2018), 26.
57. Brian Cowan, "New Worlds, New Tastes: Food Fashions After the Renaissance," in Freedman, *Food: The History of Taste*, 219–220.
58. Quinzio, *Dessert*, 21.
59. Andrew F. Smith, *Sugar: A Global History* (Reaction Books,), chap. 1.
60. Elizabeth Thompson Newman, "A Critical Edition of an Early Portuguese Cookbook" (PhD diss., University of North Carolina at Chapel Hill, 1964). On page xxi, for instance, Newman writes that "sugar is plentifully used in meat recipes."
61. In commenting on Messisbugo's cooking, Ken Albala notes that "practically everything is seasoned with sugar and spices." See his *Food in Early Modern Europe* (Greenwood Press, 2003), 127.

62. Terry Breverton, *Tudor Kitchen: What the Tudors Ate and Drank* (Amberley Publishing, 2015), 44.
63. Albala, *Eating Right in the Renaissance*, 173.
64. Earle, *The Body of the Conquistador*, 129.
65. Mintz, *Sweetness and Power*, 16.
66. Goucher, *Congotay! Congotay!*, 112–124.
67. Smith, *Sugar*, chap. 2.
68. López de Gómara, *Historia general de las Indias*, 55.
69. Laudan, *Cuisine and Empire*, 187, 197–198; Gentilcore, *Food and Health in Early Modern Europe*, 172–174.
70. Albala, *Food in Early Modern Europe*, 85. John A. West states that it was in 1824, with the advent of hydraulic presses, that people began to "extract cacao butter, a development that brought cocoa powder into extensive use for the first time." See his "A Brief History and Botany of Cacao," in *Chilies to Chocolate: Food from the Americas Gave the World*, ed. Nelson Foster and Linda S. Cordell (University of Arizona Press, 1992), 114.
71. Columbus, *Journals and Other Documents on the Life and Voyages of Christopher Columbus*, 131.
72. Acosta, *The Natural and Moral History of the Indies*, 1:244–245.
73. Laudan, *Cuisine and Empire*, 198.
74. Smith, *Sugar*, chap. 4. A more detailed study of the ways Spaniards and Europeans prepared chocolate drinks from the sixteenth century can be found in Irene Fattacciu, "The Resistance and Boomerang Effect of Chocolate: A Product's Globalization and Commodification," in Aram and Yun-Casalilla, *Global Goods and the Spanish Empire*, 255–276.
75. Gentilcore, *Food and Health in Early Modern Europe*, 175.
76. Mintz, *Sweetness and Power*, 109.
77. Fattacciu, "The Resistance and Boomerang Effect of Chocolate," 261–266.
78. Thomas Gage, *The English-American: A New Survey of the West Indies, 1648* (RoutledgeCurzon, 1928), 23.
79. Hernando Cortés, *Five Letters 1519–1526* (The Argonaut Series), ed., E. D. Ross and E. Power, trans. J. B. Morros (Robert M. McBride and Co. 1929); Bernal Díaz del Castillo, *Historia verdadera de la conquista de la Nueva España*, chaps. 39, 61, 68, and 71.
80. Earle, *The Body of the Conquistador*, 136.
81. Sergio Ramírez, "What the Palate Knows: Nicaragua's Culinary Cultures," in *Food, Text and Cultures in Latin America and Spain*, ed. Rafael Climent-Espino and Ana M. Gomez-Bravo (Vanderbilt University Press, 2020), 273.
82. Norton, "Tasting Empire," 677.
83. Sarah Moss and Alexander Badenoch, *Chocolate: A Global History* (Reaktion Books, 2009), 20–21.

84. Cited in Francisco Zamora Rodríguez, "Interest and Curiosity: American Products, Information, and Exotica in Tuscany," in Aram and Yun-Casalilla, *Global Goods and the Spanish Empire*, 183.
85. Gentilcore, *Food and Health in Early Modern Europe*, 173–174.
86. Bernal Díaz del Castillo, *Historia verdadera de la conquista de la Nueva España*, 426.
87. Moss and Badenoch, *Chocolate*, 20, 26.
88. Sophie D. Coe and Michael D. Coe, *The True History of Chocolate* (Thames and Hudson, 2007), 88, 30.
89. William Shakespeare, *Troilus and Cressida* (William Heinemann, 1904), 117–118.
90. John Fletcher, *The Elder Brother* (London, 1637), h2 (act 4, scene 4), and Fletcher, *The Loyal Subject*, 46–47.
91. Terrón, *España, Encrucijada de culturas alimentarias*, 118.
92. Earle, *The Body of the Conquistador*, 139.
93. Thomas Dawson, *The Good Huswife's Jewell* (London, 1596), 20–21.
94. William Harrison, *Elizabethan England*, ed. Lothrop Withington, intro. F. J. Furnivall (Walter Scott Publishing, 1945), 92.
95. Quoted in Salaman, *History and the Social Influence of the Potato*, 104.
96. J. H. [in non-Latin alphabet], *A Description of the Last Voyage to Bermudas in the Ship Marygold, S. P. Commander* (London, 1671), 10.
97. Fernández-Armesto, *Near A Thousand Tables*, 32–33.
98. Coe and Coe, *The True History of Chocolate*, 77.
99. Gentilcore, *Food and Health in Early Modern Europe*, 136.
100. Irene Fattacciu, "Cacao: From an Exotic Curiosity to a Spanish Commodity. The Diffusion of New Patterns of Consumption in Eighteenth-century Spain," in *Food History: Critical and Primary Sources*, vol. 3, *Global Contact and Early Industrialization*, ed., Jeffrey M. Pilcher (Bloomsbury, 2014), 444.
101. Salaman, *History and the Social Influence of the Potato*, 425.
102. Albala, *Eating Right in the Renaissance*, 150.
103. Albala, 101–150, quotes on 145 and 148. Also, Amy I. Aronson, "Aphrodisiac Texts in Medieval Iberian Texts," in *Forging Communities: Food and Representation in Medieval and Early Modern Southeastern Europe*, ed. Montserrat Piera (University of Arkansas Press, 2018), 221–232; reference to Galen is on 223.
104. Laudan, *Cuisine and Empire*, 200–201.
105. Earle, *The Body of the Conquistador*, 66, 151–152.
106. Gregorio Saldarriaga, "Taste and Taxonomy of Native Food in Hispanic America, 1492–1640," in Climent-Espino and Ana M. Gomez-Bravo, *Food, Text and Cultures in Latin America and Spain*, 86, 91. Italics added.
107. Laudan, *Cuisine and Empire*, 200–201.

108. Dawson, *The Good Huswife's Jewell*, 20–21.
109. See Albala, *Food in Early Modern Europe*, 169.
110. Jeffrey L. Forgeng, *Daily Life in Elizabethan England*, 2nd ed. (ABC-CLIO, 2010), 167–168.
111. See Albala, *Eating Right in the Renaissance*, 41.
112. Kenelm Digby, *The closet of the eminently learned Sir Kenelme Digbie Kt. opened: Whereby is discovered several ways for making of metheglin, sider, cherry-wine, &c. : Together with excellent directions for cookery, as also for preserving, conserving, candying, &c. / published by his son's consent* (London, 1669).
113. Harrison, *Elizabethan England*, 92.
114. John Gerard, *Gerard's Herball—Or, Generall Historie of Plantes* (London, 1597), 780–782; Alison Sim, *Food and Feast in Tudor England* (Sutton Publishing, 1997), 124–126.
115. Gerard, *Gerard's Herball*, 780–781.
116. Sim, *Food and Feast in Tudor England*, 125.
117. John Parkinson, *Paradisi in sole paradisus terrestris. Or A garden of all sorts of pleasant flowers which our English ayre will permitt to be noursed vp with a kitchen garden of all manner of herbes, rootes, & fruites, for meate or sause vsed with vs, and an orchard of all sorte of fruitbearing trees and shrubbes fit for our land together with the right orderinge planting & preseruing of them and their vses & vertues collected by Iohn Parkinson apothecary of London 1629* (London, 1629), 516–517.
118. Gerard, *Gerard's Herball*, 781.
119. Parkinson, *Paradisi in sole paradisus terrestris*, 516–517.
120. John Forster, *Englands Happinesse Increased, Or a Sure and Easie Remedy against All Succeeding Dear Years by A Plantation of the Roots Called Potatoes*, Early English Books Online, Michigan Library Digital Collections, accessed March 14, 2022, https://quod.lib.umich.edu/e/eebo/A40002.0001.001?view=toc.
121. Sim, *Food and Feast in Tudor England*, 125.
122. Clarissa Dickson Wright, *A History of English Food* (Random House, 2011), 136.
123. Salaman, *The History and Social Influence of the Potato*, 188–221.
124. Sim, *Food and Feast in Tudor England*, 126. See also Breverton, *The Tudor Kitchen*, 182–183.
125. Coe, *America's First Cuisines*, 20.
126. Nicolás Bautista Monardes, *Primera Y Segunda 1580* (Ministero de Sanidad Y Consumo, 1989), fols. 94v–95v.
127. Saldarriaga, "Taste and Taxonomy of Native Food in Hispanic America, 1492–1640," 92.
128. Albala, *Food in Early Modern Europe*, 142.
129. Diego Granado, *Libro del arte de cozina* (Lérida, 1614), 279, 283.

130. Quotes are in Albala, *Eating Right in the Renaissance*, 233n52 (translation is mine).
131. See prologue in Francisco Martínez Montiño, *Arte de cocina, pastelería, bizcochería y conservería* (Barcelona, 1763).
132. Albala, *Eating Right in the Renaissance*, 174–181.
133. Terrón, *España, Encrucijada de culturas alimentarias*, 99, 103.
134. About the arrival of the white potato and sweet potato in Europe, see J. G. Hawkes and J. Francisco-Ortega, "The Potato in Spain During the Late 16th Century," *Economic Botany* 46, no. 1 (March 1992): 86–97, and J. G. Hawkes and J. Francisco-Ortega, "The Early History of the Potato in Europe," *Euphytica* 70, nos. 1–2 (1993): 1–7.
135. See María de los Ángeles Pérez Samper, "The Early Modern Food Revolution: A Perspective from the Iberian Atlantic," in Aram and Yun-Casalilla, *Global Goods and the Spanish Empire*, 22, 24.
136. Granado, *Libro del arte de cozina*, 48–49.
137. Robert May, *The Accomplisht Cook, or the Art & Mystery of Cookery* (Londo, 1685), 1–2, 92.
138. Giovanni Domenico Sala, *De alimentis et eorum recta administratione* (Padua, 1622), 12, cited in Albala, *Eating Right in the Renaissance*, 238n73.
139. See Wikipedia, "Juan de la Mata," last modified July 30, 2024, 21:05 (UTC), https://es.wikipedia.org/wiki/Juan_de_la_Mata.
140. Juan de la Mata, *Arte de repostería* (Madrid, 1791), 48, 67.
141. See Wikipedia, "Juan de Altamiras," last modified December 6, 2024, 21:53 (UTC), https://es.wikipedia.org/wiki/Juan_de_Altamiras, and also Vicky Hayward's preface and introduction in *New Art of Cookery: A Spanish Friar's Kitchen Notebook by Juan de Altamira* (Rowman and Littlefield, 2017), x, xx.
142. Hayward, *New Art of Cookery*, 28–30.
143. Maria Paz Moreno, *Madrid: A Culinary History* (Rowman and Littlefield, 2018), 144, 57–61.
144. Hayward, *New Art of Cookery*, 249n18.
145. Forster, *Englands Happinesse Increased*.
146. Breverton, *The Tudor Kitchen*, 78–79, 90.
147. Domingos Rodrigues, *Arte de cozinha* (Lisboa, 1693), 86–87.
148. François Pierre de la Varenne, *La Varenne's Cookery*, trans. and ed. Terence Scully (Prospect Books, 2006).
149. Barbara Ketcham Wheaton, *Savoring the Past: The French Kitchen and Table from 1300 to 1789* (Touchstone, 1983), 84–85.
150. See Wright, *A History of English Food*, 95.
151. Thomas Muffet, *Healths Improvement, or Rules Comprizing and Discovering the Nature, Method and Manner of Preparing All Sorts of Food in This Nation*, corrected and enlarged by Christopher Bennet (London, 1655), 226.

152. Tobias Venner, *Via recta ad vitam longam* (London, 1638), 185–186.
153. Gervase Markham, *The English Huswife: Containing the Inward and Outward Virtues Which Ought to Be in a Complete Woman* (London, 1623), 74–75.
154. Markham, 94–95.
155. Gentilcore, *Food and Health in Early Modern Europe*, 150.
156. Albala, *Eating Right in the Renaissance*, 239.
157. David Davidson, "European's Wary Encounter with Tomatoes, Potatoes, and Other New World Foods," in Nelson and Cordell, *Chilies to Chocolate*, 11.
158. Albala, *Eating Right in the Renaissance*, 239.
159. Gerard, *Gerard's Herball*, 780.
160. Albala, *Eating Right in the Renaissance*, 94.
161. Gerard, *Gerard's Herball*, 871.
162. Salaman, *History and the Social Influence of the Potato*, 427–428.
163. Gerard, *Gerard's Herball*, 781.
164. See Wikipedia, "*Sium sisarum*," last modified March 30, 2025, 23:57 (UTC), https://en.wikipedia.org/wiki/Sium_sisarum#Culinary_use.
165. Peter Brears, *Cooking and Dining in Medieval England* (Prospect Books, 2012), 268.
166. May, *The Accomplisht Cook*.
167. E. E. Rich and C. H. Wilson, eds., *The Cambridge Economic History of Europe*, vol. 4 (Cambridge University Press, 1967), 279.
168. See Albala, *Eating Right in the Renaissance*, 233–236, quotes on 233.
169. William Rabisha, *The Whole Body of Cookery dissected taught, and fully manifested, methodically, artificially, and according to the best tradition of the English, French, Italian, Dutch* (London, 1661), 150. Peter Brears provides a modernized description of the sweet potato pie in his *Cooking and Dinning in Tudor and Early Stuart England*, 165.
170. Brears, *Cooking and Dining in Medieval England*, 268.
171. Brears, 293–294.
172. *El libro de las familias: Novísimo manual práctico de cocina Española, Francesa y Americana, higiene y economía doméstica* (Décimasexta Edición), 26th ed. (Madrid, 1874), 620, 637, 648.
173. See John Gerard, "Potato's of Virginia," in which he said that the potato was grown "naturally in America," *Gerard's Herball*, 781–782.
174. Reader, *Potato*, 84.
175. Wright, *A History of English Food*, 136–137; Reader, *Potato*, 87.
176. Peter J. Hatch, *Thomas Jefferson's Revolutionary Garden at Monticello* (Yale University Press, 2012), 193, and Ruth Salvaggio, "Eating Poetry in New Orleans," *Writing in the Kitchen: Essays on Southern Literature and Foodways*, ed. David A. Davis, Tara Powell, and Jessica B. Harris (University Press of Mississippi, 2014), 105–123.

177. Lebot, *Tropical Root and Tuber Crops*, 96. George Milligen Johnston wrote in his *A Short Description of the Province or South-Carolina, with An Account of the Air, Weather, and Disease at Charles-Town, Written in the Year 1763* (London, 1770) that sweet potato, corn, and pumpkin were foods people relied on in those days.
178. Hatch, *Thomas Jefferson's Revolutionary Garden at Monticello*, 192–193.
179. Dave DeWitt, *The Founding Foodies: How Washington, Jefferson and Franklin Revolutionized American Cuisines* (Sourcebooks, 2010), 107.
180. Mary Randolph, *The Virginia House-Wife* (Washington, 1824), 132–133.
181. Though Martha Washington's cookbook was attributed to her, it was kept within the Lewis family until 1892, when it was given to the Historical Society of Pennsylvania and printed several decades later. Eliza Warren's recipe of french-fried potatoes was in her *The Economical Cookery Book: For Housewives, Cooks, and Maids-of-All-Work* (London, 1858), 88.
182. Hatch, *Thomas Jefferson's Revolutionary Garden at Monticello*, 183.
183. DeWitt, *The Founding Foodies*, 75, 90, 138.
184. Richard Bradley, *The Country Housewife and Lady's Director* (London, 1732), 31–32, 153–154.
185. Eliza Smith, *The Complete Housewife*, 5th ed. (London, 1758), 157–158.
186. Peter Brears included a "(Sweet) Potato Pie" recipe in his *Cooking and Dinning in Tudor and Early Stuart England*, suggesting that perhaps in the sixteenth century the sweet potato would be its chief ingredient, 165.
187. Smith, *The Complete Housewife*, 156–157.
188. Hannah Glasse, *The Art of Cookery Made Plain and Easy*, 5th ed. (London, 1755), 134, 192–193, 197, 206–207, 224, 247.
189. Elizabeth Raffald, *The Experienced English House-Keeper* (Manchester, 1769), 263.
190. Raffald, 349–360.
191. Glasse, *The Art of Cookery Made Plain and Easy*, 327–328.
192. Gilbert White, *The Natural History and Antiquities of Selborne* (London, 1837), 129.
193. Richard Briggs, *The English Art of Cookery* (London, 1791), 184, 189, 319–320, 331, 343–344, 426, 447, 602–611.
194. Briggs, 79–80, 98, 100.
195. Amelia Simmons, *American Cookery* (1796; Hudson and Goodwin, 1963), 34–35, 65, 78.
196. Maria Eliza Rundell, *A New System of Domestic Cookery*, new ed. (London, 1808), 135, 145, 153, 155, 164–165, 167, 174–175, and Eliza Acton, *Modern Cookery for Private Families*, new ed. (London, 1897).
197. R. H. Price, *Sweet Potato Culture for Profit* (Dallas, 1896), 106–107.

198. Jessica B. Harris, "Three Is a Magic Number," *Southern Quarterly* 44, no. 2 (2007): 10. Also see Jessica B. Harris, *High on the Hog: A Culinary Journey from Africa to America* (Bloomsbury, 2011), chap. 3, "The Power of Three," 41–60.
199. Margaret Jones Bolsterli, Toni Tipton-Martin, and John T. Edge, "Soul Food," in *The Encyclopedia of Southern Culture*, vol. 7, *Foodways*, ed. John T. Edge (University of North Carolina Press, 2007), 104.
200. Edward Bleier, *The Thanksgiving Ceremony: New Traditions for America's Family Feast* (Crown Publishers, 2003), 37. See also, Kathryn K. Blue and Anthony Dias Blue, *Thanksgiving Dinner* (HarperPerennial, 1990), xiii.
201. Bleier, *The Thanksgiving Ceremony*, 39–41. Washington's proclamation was excerpted in Ralph Linton and Adelin Linton, *We Gather Together: Story of Thanksgiving* (Henry Schuman, 1949), 84–86. That Thanksgiving is observed on the fourth Thursday of November was decided by an act of Congress in 1942, under President Franklin Roosevelt.
202. Cf. Elizabeth Pleck, "The Making of the Domestic Occasion: The History of Thanksgiving in the United States," *Journal of Social History* 32, no. 4 (1999): 773–789, especially 775–776.
203. Sarah Josepha Hale, *The Good Housekeeper* (Boston, 1839), 66.
204. Sarah Josepha Hale, *The New Household Recipe-Book* (New York, 1853), 322, 336.
205. See Linton and Linton, *We Gather Together*, 97.
206. Lebot, *Tropical Root and Tuber Crops*, 96.
207. *Housekeeper's Annual & Ladies Register* (Boston, 1839), 39–40.
208. James W. Baker, *Thanksgiving: The Biography of an American Holiday* (University of New Hampshire Press, 2009), 56.
209. Michael W. Twitty, "The African Virginian Roots of Edna Lewis," in *Edna Lewis: At the Table with an American Original*, ed. Sarah B. Franklin (University of North Carolina Press, 2018), 130. Also see Judith A. Carney and Richard Nicholas Rosomoff, *In the Shadow of Slavery: Africa's Botanical Legacy in the Atlantic World* (University of California Press, 2009), 89; Goucher, *Congotay! Congotay!*, 56.
210. Frederick Douglass Opie, *Hog and Hominy: Soul Food from Africa to America* (New York: Columbia University Press, 2008), 32.
211. Harris, *High on the Hog*, 10.
212. Carney and Rosomoff, *In the Shadow of Slavery*, 113.
213. Harris, *High on the Hog*, 84.
214. Gary R. Kremer, *George Washington Carver: A Biography* (Greenwood, 2011), 1–76. The quote is from Linda O. Hines, "George W. Carver and the Tuskegee Agricultural Experiment Station," *Agricultural History* 53, no. 1 (January 1979): 71.

215. Barry Mackintosh, "George Washington Carver: The Making of a Myth," *Journal of Southern History* 42, no. 4 (November 1976): 511. For the incomplete list of Carver's experiments with peanuts and sweet potatoes, see Christina Vella, *George Washington Carver: A Life* (Louisiana State University Press, 2015), app. 2.
216. Kremer, *George Washington Carver*, 98–99; Mackintosh, "George Washington Carver," 518–519.
217. Twitty, "The African Virginian Roots of Edna Lewis," 129.
218. Edna Lewis, *The Taste of Country Cooking* (Alfred A. Knopf, 2006), 116–144.
219. Edna Lewis and Scott Peacock, *The Gift of Southern Cooking: Recipes and Revelations from Two Great Southern Cooks* (Alfred A. Knopf, 2003), 163–167.
220. Baker, *Thanksgiving*, 52–53, and Diana Karter Applebaum, *Thanksgiving: An American Holiday, an American History* (Facts on File Publications, 1984), 266–272.
221. Hale, *The Good Housekeeper*, 40–41, 72–73.
222. Opie, *Hog and Hominy*, 33.
223. Opie, 36.
224. Randolph, *The Virginia Housewife*, 34–35, 62–63, 95–96, 135–138.
225. Opie, *Hog and Hominy*, 36.
226. Frederick Douglass Opie, "Influence, Sources, and African American Foodways," in *Food in Time and Place: The American Historical Association Companion to Food History*, ed. Paul Freedman, Joyce E. Chaplin, and Ken Albala (University of California Press, 2014), 202–203. Also see William C. Witt, "Soul Food as Cultural Creation," and Anne Yentsch, "Excavating the South's African American Food History," in *African American Foodways: Explorations of History and Culture*, ed. Anne L. Bower (University of Illinois Press, 2009), 45–98.
227. Jessica B. Harris, "African American Foodways," in Edge, *The Encyclopedia of Southern Culture*, 7:17.
228. Harris, *High on the Hog*, 178.
229. Opie, *Hog and Hominy*, 69.
230. N. M. Nesbit, *Farmer's Bulletin*, no. 129, in "Sweet Potatoes," *Everyday Housekeeping* 18, no. 2 (1902): 60, Nineteenth Century Collections Online database. M. F. Maury, in his *Travels in the New South I & II*, also noted that sweet potato was produced from New Jersey to North Carolina along the East Coast of the United States. M. F. Maury, *Travels in the New South I & II* (Wheeling, WV, 1876), 72, Archives Unbound database.
231. Opie, *Hog and Hominy*, 64.
232. The Dutrieuilles family catering business is featured in episode 4 of *High on the Hog: How African American Cuisine Transformed America*, a documentary

series inspired by Jessica B. Harris's *High on the Hog* book. Made by Netflix in 2021, the documentary was directed by Roger Ross Williams.
233. Pleck, "The Making of the Domestic Occasion," 781.
234. Blue, *Thanksgiving Dinner*, 118–121.
235. Joe Kissell, *Take Control of Thanksgiving Dinner* (TidBITS Publisher, 2012).
236. A candied sweet potato dish is often one of the top ten classic dishes for Thanksgiving on many U.S. websites.
237. Bruce Kraig and Colleen Taylor Sen, eds., *Street Food Around the World: An Encyclopedia of Food and Culture* (ABC-CLIO, 2013), 233, 238. See also Karen Hursh Graber, "Mexican Sweet Potatoes, from Soup to Dessert: Los Camotes," MexConnect, accessed September 23, 2021, https://www.mexconnect.com/articles/2145-mexican-sweet-potatoes-from-soup-to-dessert-los-camotes/.
238. Cf. Ken Albala, ed., *Food Cultures of the World Encyclopedia*, vol. 2 (Greenwood, 2011), 126, 171, 5.
239. Goucher, *Congotay! Congotay!*, 19.
240. Albala, *Food Cultures of the World Encyclopedia*, 2:288.
241. Albala, 2:211, 244, 157.
242. Woolfe, *Sweet Potato*, 558–560.
243. The introduction of the policy for promoting native plants in Peru has resulted in several theses at both the doctoral and master's levels. See, for example, Daniel Tobin, "Interactions Between Livelihoods and Pro-Poor Value Chains: A Case Study of Native Potatoes in the Central Highlands of Peru" (PhD diss., Pennsylvania State University, 2014); Mary Katherine Weeler, "The Uncertain Promise of Agriculture: Two Essays on Climate Change, Agriculture and Nutrition in the Andean Highlands of Peru" (master's thesis, Cornell University, 2017); and Michel Ian Collins, "Economic Analysis of Wholesale Demand for Sweet Potatoes in Lima, Peru" (master's thesis, University of Florida, 1989).
244. See the website of the International Potato Center at https://cipotato.org/.
245. Cf. Judith E. Fan, "Can Ideas About Food Inspire Real Social Change? The Case of Peruvian Gastronomy," *Gastronomica* 13, no. 2 (Summer 2013): 29–40, quote on 32.
246. Kraig and Sen, *Street Food Around the World*, 13–14, and Albala, *Food Cultures of the World Encyclopedia*, 2:266.
247. Albala, *Food Cultures of the World Encyclopedia*, 2:268.
248. Collins, "Economic Analysis of Wholesale Demand for Sweet Potatoes in Lima, Peru," 60.
249. Cf. Laura Zanotti, *Radical Territories in the Brazilian Amazon* (University of Arizona Press, 2016), 126–162.

250. Tom Kime, *Street Food: Exploring the World's Most Authentic Tastes* (Dorling Kindersley, 2007), 98.
251. Woolfe, *Sweet Potato*, 564–572. Also see Collins, "Economic Analysis of Wholesale Demand for Sweet Potatoes in Lima, Peru," 8–9, 15–46.

3. Mundane or Miracle?

1. Ho, *Studies on the Population of China*, 187.
2. L. Zhang, Q. Wang, Q. Liu & Q. Wang, "Sweetpotato in China," in Loebenstein and Thottappilly, *The Sweetpotato*, 327.
3. Wang Jiaqi, "Luetan ganshu he *Ganshu lu*" [Some discussions of the sweet potato and the *Sweet Potato Records*], *Wenwu* [Cultural relics] 3 (1961): 27–29; Xia Nai, "Luetan fanshu he shuyu" [Brief discussions of sweet potato and Chinese yam], *Wenwu* [Cultural relics] 8 (1961): 58–59; Wu Deduo, "Guanyu ganshu he *Jinshu chuanxilu*" [On the sweet potato and *The Cultivation of the Sweet Potato*], *Wenwu* [Cultural relics] 8 (1961): 60–61; Wang Jiaqi, "'Lutan fanshu he shuyu' erwen duhou" [My comments on Xia Nai and Du Deduo's essays], *Wenwu* [Cultural relics] 8 (1961): 61. There were others who also argued that the sweet potato entered China prior to the sixteenth century. Li Tianxi believes that it was introduced from Sulu in the early Ming. See his "Huaqiao yinjin fanshu xinkao" [New study of the introduction of the sweet potato by overseas Chinese], *Zhongguo nongshi* [Agricultural history of China] 1 (1998): 107–112.
4. Manuel Perez Garcia, "Challenging National Narratives: On the Origins of Sweet Potato in China as Global Commodity During the Early Modern Period," in *Global History and New Polycentric Approaches: Europe, Asia and the Americas in the World Network System*, ed. Manuel Perez Garcia and Lucio de Sousa (Springer Nature, 2018), 53–80.
5. Zhou Yuanhe, "Ganshu de lishi dili: Ganshu de tusheng, chuanru, chuanbo yu renkou" [Historical geography of the sweet potato: The origin, transfer, diffusion of the sweet potato and population], *Zhongguo nongshi* [Agricultural history of China] 3 (1983): 75–88. While extending his support of Wang Jiaqi, Zhou also discusses those who disagreed with Wang in his article.
6. "Sweet Potato Production, 2023," Our World in Data, accessed August 3, 2024, https://ourworldindata.org/grapher/sweet-potato-production. See also Zhang et al., "Sweetpotato in China," 327.
7. Crosby, *The Columbian Exchange*, 198.
8. Zhou, "Ganshu de lishi dili," 75–76. Wu Deduo in his "Dui *Jinshu chuanxilu* de zairenshi" [New understanding of *The Cultivation of the Sweet Potato*]

also mentioned that it was after 1976 that scholars in mainland China began to notice Ping-ti Ho's works of the 1950s. See Chen Shiyuan et al., *Jinshu chuanxilu zhongshupu hekan* [Cultivation of the sweet potato and manual for growing the sweet potato, combined edition] (Nongye chubanshe, 1982), 175–176.

9. Quan Hansheng, "Meizhou faxian duiyu Zhongguo nongye de yingxiang" [Influence of the discovery of the Americas in Chinese agriculture], *Zhongguo jingjishi yanjiu*, vol. 2 (Zhonghua shuju, 2011), 183–191. The article was originally published in 1966.

10. Ping-ti Ho, "The Introduction of American Food Plants into China," *American Anthropologist* 57, no. 2, pt. 1 (April 1955): 191–201.

11. Chen Shuping, "Yumi he fanshu zai Zhongguo chuanbo qingkuang yanjiu" [Study of the diffusion of maize and sweet potato in China], *Zhongguo shehui kexue* [Chinese social sciences] 3 (1980): 187–204; Liang Jiamian and Qi Jingwen, "Fanshu yinzhong kao" [Study of the introduction of the sweet potato], *Huanan nongxueyuan xuebao* [Journal of South China Agricultural College] 3 (1980): 74–79; Cao Shuji, "Yumi he fanshu chuanru Zhongguo luxian xintan" [New exploration of the transfer routes of maize and sweet potato into China], *Zhongguo shehui jingjishi yanjiu* [Journal of Chinese social and economic history] 4 (1988): 62–74.

12. See, for example, Zhang Jian, *Xindalu nongzuowu de chuanbo he yiyi*; Song Junling, "MingQing shiqi Meizhou nongzuowu zai Zhongguo de chuanzhong jiqi yingxiang yanjiu" [Spread and influence of American plants in China during the Ming and Qing periods] (PhD diss., Henan University, 2007); Cao Ling, "Meizhou liangshi zuowu de chuanru, chuabo jiqi yingxiang yanjiu" [Introduction, diffusion and influence of American food plants] (master's thesis, Nanjing Agricultural University, 2003); Huang Fuming, "MingQing shiqi fanshu yinjin Zhongguo yanjiu" [Introduction of the sweet potato in China during the Ming and Qing periods] (master's thesis, Shandong Normal University, 2010).

13. Cf. Cao Ling, "MingQing Meizhou liangshi zuowu chuanru Zhongguo yanjiu zongshu" [Review of the studies of the introduction of American food plants into China during the Ming and Qing periods], *Gujin nongye* [Ancient and modern agriculture] 2 (2004), 95–103, especially 99.

14. The story about Chen Yi was originally recorded in his family history and later entered the county's gazetteer. Excerpted in Chen Shuping, ed., *MingQing nongyeshi ziliao, 1368–1911* [Historical sources of Ming and Qing agriculture, 1368–1911], vol. 1 (Shehui kexue wenxian chubanshe, 2013), 281–282, and cited in Chen Xuewen, "Lun fanshu de yinjing Zhongguo jiqi yiyi" [On the introduction of the sweet potato into China and its significance], *Si yu yan* [Ideas and words] 30, no. 1 (1992): 4–5.

15. The story was recorded in both *Dianbai xianzhi* [Gazetteer of Dianbai County] and *Guiping xianzhi* [Gazetteer of Guiping County], excerpted in Chen, *MingQing nongyeshi ziliao*, 1:282.
16. Yang Baolin, "Woguo yinjin fanshi de zuizao zhiren he yinjin fanshu de zuizao zhidi" [First person and first place of the introduction of the sweet potato into China], *Nongye kaogu* [Agricultural archaeology] 2 (1982): 79–83.
17. L. C. Goodrich, "The Introduction of the Sweet Potato into China," *China Journal* 27, no. 4 (October 1937): 206–208; Chen et al., *Jinshu chuanxilu zhongshupu hekan*, 17.
18. Cf. Wan Guoding, *Wugu shihua* [Stories of the five grains] (Zhonghua shuju, 1961), 36; Li Yukun, "Xian Shu Ci yu fanshu de chuanru" [Sweet potato ancestral shrine and the introduction of the sweet potato], *Gujin nongye* [Ancient and modern agriculture] 2 (1992): 61–62.
19. Min Zongdian, "Haiwai nongzuowu de chuanru he dui Zhongguo nongye shengchan de yingxiang" [Introduction of agricultural crops and its influence on China's agricultural production], *Gujin nongye* [Ancient and modern agriculture] 1 (1991): 5–6; Wu Deduo in his aforementioned "Dui *Jinshu chuanxilu* de zairenshi" and Chen Shuping in his "Yumi he fanshu zai Zhongguo chuanbo qingkuang yanjiu" believed that besides Guangdong and Fujian, Taiwan could be another place where sweet potato was cultivated in the sixteenth century. Wan Guoding pointed out that in addition to other routes, the root's transference to Fujian could be multiple as there were many Fujianese merchants traveling back and forth between China and the Philippines in the time. See his *Wugu shihua*, 35–37.
20. It is commonly believed that sweet potato spread to Vietnam from the Philippines in the late sixteenth century, which was not so much earlier than China. But the stories of both Chen Yi and Lin Huailan seemed to imply that when the two visited Vietnam in the period, sweet potato had already become a well-accepted food plant in the country.
21. See Chen, "Lun fanshu de yinjin Zhongguo jiqi yiyi," 6.
22. Rich and Wilson, *Cambridge Economic History of Europe*, 4:286; Sucheta Mazumdar, "The Impact of New World Food Crops on the Diet and Economy of China and India, 1600–1900," in *Food in Global History*, ed. Raymond Grew (Westview, 2000), 58–78.
23. See Quan Hansheng, *MingQing jingjishi yanjiu* [Studies of economic history in the Ming and Qing[(Lianjing chubangongsi, 1987), 66; Yang Guozhen et al., *MingQing Zhongguo yanhai shehui yu haiwai yimin* [Chinese coastal society and overseas emigration in the Ming and Qing periods] (Gaodeng jiaoyu chubanshe, 1997), 40.
24. Chen, *Jinshu chuanxilu zhongshupu hekan*.

25. See Zheng Nan, "Cong yumi, fanshu, malingshu de chuanru kan wailai nongzuowu chuanru dui Zhongguo shehui de yingxiang" [Impact of foreign food crops on Chinese society: Examples of corn, sweet potato and white potato], *Liuzhu zuxian canzhuo de jiyi: Yazhou shixue luntan* [Keep memories of ancestors' tables: The proceedings of 2011 Asian Food Cultural Heritage Forum] (Yunnan renmin chubanshe, 2011), 450. According to Wu Liqing's research, it was first in *Puning Xianzhi* [Gazetteer of Puning County] in the year 1745 that sweet potato was first placed on a par with grain crops. See his "Fanshu zai Chaozhou diqu de chuanbo yu nongye tixi biandong" [Diffusion of the sweet potato in Chaozhou and the change of agricultural system], *Nongye kaogu* [Agricultural archaeology] 4 (2012): 31–38.
26. Wu, "Dui *Jinshu chuanxilu* de zairenshi," 178–180; Xiao Kezhi, "*Jinshu chuanxilu* banbenshuo" [Discussions of the editions of *The Cultivation of the Sweet Potato*], *Gujin nongye* [Ancient and modern agriculture] 3 (2001): 64–65.
27. Wang Guozhong, "Xu Guangqi de *Ganshu shu*" [Xu Guangqi's *Explanations of the Sweet Potato*], *Zhongguo nongshi* [Agricultural history of China] 3 (1983): 71–74.
28. Xu Guangqi, *Nongzheng quanshu jiaozhu* [Complete treatise on agriculture, annotated], vol. 2, ed. Shi Shenghan (Shanghai guji chubanshe, 1977), 688–694.
29. Xu, *Nongzheng quanshu jiaozhu*. Shinoda Osamu's 1944 article translated into Chinese in Chen Shiyuan, *Jinshu chuanxilu zhongshupu hekan*, 253–274.
30. Xu, *Nongzheng quanshu*, 2:689.
31. See Wang, "Luetan ganshu he *Ganshu lu*," 28–29.
32. Seo Yugu, *Jongjeobo*, in Shiyuan, *Jinshu chuanxilu zhongshupu hekan*, 201–252. Xu Guangqi's discussion of the "thirteen advantageous utilities" of the sweet potato was in his *Nongzheng quanshu*, 2:694.
33. See Bouwkamp, *Sweet Potato Products*, 4–6; Woolfe, *Sweet Potato*, 25.
34. The report, which appeared across major news websites in China, compares Chen Zhenlong's contribution to Yuan Longping's cultivation of high-yield rice variety in recent decades. See "He Brought Back a Vine from Abroad and Saved 300 Million People from Famine," Baidu, January 15, 2025, https://baijiahao.baidu.com/s?id=1821294736192856437&wfr=spider&for=pc, and Ling Yun, "Ancient Version of Yuan Longping, Who Stole a Vine from Abroad and Increased the Population of the Qing Dynasty by 300 Million," Sohu.com, April 2, 2025, https://www.sohu.com/a/878449827_121199304. However, this was not the first time that Chen Zhenlong was credited with the introduction of the sweet potato by Chinese media. Under an equally sensational title, a major newspaper in Henan Province, which has been one of the major sweet potato–producing regions in China, reported a decade

before that "Fanshu de chuanru chedi gaibian le Zhongguo lishi jincheng" [Introduction of the sweet potato thoroughly changed the course of Chinese history], *Henan ribao* [Henan daily], February 15, 2011, 3.
35. Xu, *Nongzheng quanshu*, 2:689. See the previous note for modern variations of the story.
36. Eugene Anderson, *The Food of China* (Yale University Press, 1988), 80.
37. Quoted in Mark Elvin, *The Pattern of the Chinese Past* (Stanford University Press, 1973), 206, with slight modifications.
38. Cf. Liu Xitao, *Fujian lishi dili yanjiu* [Study of Fujian historical geography] (Fujian jiaoyu chubanshe, 2017), 81–84.
39. Chen Ying, *Haicheng Xianzhi—fengtu* [Gazetteer of Haicheng County: Local customs] (n.p., 1762), *juan* 15, 140–141.
40. Cf. Xiamen Daxue Lishi Yanjiusuo and Zhongguo Shehui Jingjishi Yanjiushi, eds., *Fujian jingji fazhan jianzhi* [Concise history of economic development in Fujian] (Xiamen daxue chubanshe, 1989), 25–70.
41. Ouyang Chunlin, "Fanshu de yinzhong yu MingQing Fujian yanhai shehui, 1594–1911" [Introduction of the sweet potato and the coastal society of Fujian in the Ming and Qing periods, 1594–1911] (master's thesis, Fujian Normal University, 2012), 9–20.
42. Zhou Lianggong, *Min xiaoji* [Brief history of Fujian], excerpted in Chen, *MingQing nongyeshi ziliao*, 1:280.
43. Chen Hong, *Pubian xiaocheng* [Analysis of the incidents in Putian] and Yang Siqian, *Quanzhou fuzhi* [Gazetteer of Quanzhou Prefecture], excerpted in Zhongguo shehui kexueyuan lishi yanjiusuo Qingshi yanjiushi [hereafter ZSKLYQY], ed., *Qingshi ziliao* [Sources of Qing history], vol. 7 (Zhonghua shuju, 1989), 369.
44. Quoted in Shiba Yoshinobu, *Commerce and Society in Sung China*, trans. Mark Elvin (Center for Chinese Studies, University of Michigan), 61.
45. Chen, *Jinshu chuanxilu zhongshupu hekan*, 36–39.
46. From the end of the seventeenth through the eighteenth century, Fujian's prefectural gazetteers recorded sweet potato as either a vegetable or a root plant, showing that it had been planted across the province. See ZSKLYQY, *Qingshi ziliao*, 7:358–384. About Jin Xuezeng's contribution, see Ouyang Chunlin, "Jin Xuezeng tuijin fanshu zai Min zhongzhikao" [Study of Jin Xuezeng's promotion of sweet potato cultivation in Fujian], *Qingdao nongye daxue xuebao* [Journal of Qingdao Agricultural University] 1 (2012): 79–82.
47. See, for example, Chen Shuping, "Yumi he fanshu zai Zhongguo chuanbo qingkuang yanjiu," 194.
48. Yang Baolin, "Woguo yinjin fanshi de zuizao zhiren he yinjin fanshu de zuizao zhidi."

49. Wu Zhenfang, *Lingnan zaiji* [Miscellaneous notes on Guangdong] and Wang Yongming, *Huaxian zhi* [Gazetteer of Hua County], excerpted in ZSKLYQY, *Qingshi ziliao*, 7:300, 303.
50. See Yoshinobu, *Commerce and Society in Sung China*, 62.
51. Wu Liqing, "Fanshu zai Chaozhou diqu de chuanbo yu nongye tixi biandong," 34–37.
52. Wan, *Wugu shihua*, 37.
53. Chen Shuping, "Yumi he fanshu zai Zhongguo chuanbo qingkuang yanjiu," 193; Gao Gongqian, *Taiwan fuzhi* [Gazetteer of Taiwan Prefecture], excerpted in ZSKLYQY, *Qingshi ziliao*, 7:384; Ouyang Chunlin, "Fanshu de yinzhong yu MingQing Fujian yanhai shehui, 1594–1911," 37.
54. Wan, *Wugu shihua*, 37; Chen Shuping, "Yumi he fanshu zai Zhongguo chuanbo qingkuang yanjiu," 193. Wan dated sweet potato cultivation in Shandong to 1752, whereas Chen Shuping stated that it began in 1742.
55. Chen, *Jinshu chuanxilu zhongshu pu hekan*, 32–35, 48–54. The gazetteer of Jiaozhou also acknowledged Chen Shiyuan's role in transferring the sweet potato from Fujian to Shandong. See Cao Ling, "Meizhou liangshi zuowu de chuanru, chuabo jiqi yingxiang yanjiu," 44.
56. Chen, *Jinshu chuanxilu zhongshu pu hekan*, 36–48.
57. Cf. William T. Rowe, *Saving the World: Chen Hongmou and Elite Consciousness in Eighteenth-Century China* (Stanford University Press, 2001).
58. Rowe, 234–235; Pan Guisheng, *Chen Hongmou zhuan* [Biography of Chen Hongmou] (Guangxi renmin chubanshe, 2007), 102–107.
59. The severity of the disaster of 1743–1744 has caught the attention of Pierre-Etienne Will, who uses it as a case study to examine and evaluate the success of Qing relief measures. See his *Bureaucracy and Famine in Eighteenth-Century China*, trans. Elborg Forster (Stanford University Press, 1990), 149–175.
60. Chen Hongmou, *Peiyuan tang oucun gao* [Chen Hongmou's scattered manuscripts], excerpted in ZSKLYQY, *Qingshi ziliao*, 7:260–261.
61. Cf. Cao Ling, "Meizhou liangshi zuowu de chuanru, chuabo jiqi yingxiang yanjiu," 46–47; Laura Murray, "New World Food Crops in China: Farms, Food, and Families in the Wei River Valley, 1650–1910" (PhD diss., University of Pennsylvania, 1985), 274–345.
62. Rowe, *Saving the World*, 234–235.
63. The following discussion of the trend of sweet potato development draws on the studies of Cao Shuji, Liu Pubing, and Cao Ling, among others. See Cao Shuji, "Qingdai yumi fanshu fenbu de dili tezheng" [Geographic characteristics of the dissemination of maize and sweet potato in the Qing period], *Lishi dili yanjiu* [Studies of historical geography] 2 (1992): 287–303; Liu Pubing, "Fanshu de yinjin he chuanbo" [Introduction and diffusion of the sweet

potato], *Zhonghua yinshi wenhua jijinhui huixun* [Newsletter of the foundation of Chinese food and drink culture] 17, no. 4 (November 2011): 35–40; Cao Ling, "Meizhou liangshi zuowu de chuanru, chuabo jiqi yingxiang yanjiu," 47.

64. Chen Dongsheng, "Ganshu zai Shandong chuanbo zhongzhi shilue" [Brief history of the sweet potato's cultivation in Shandong], *Nongye kaogu* [Agricultural archaeology] 1 (1991): 219–220.

65. Zhang et al., "Sweetpotato in China," 333–334.

66. See Li Wei's instructions in Shiyuan, *Jinshu chuanxilu zhongshu pu hekan*, 36–48.

67. According to Huang Kerun (?–1763), it was in Dezhou, Shandong, where he saw that the farmers had mastered the technique to preserve storage roots through the winter. See Chen Dongsheng, "Ganshu zai Shandong chuanbo zhongzhi shilue," 220; Cao Ling, "Meizhou liangshi zuowu de chuanru, chuabo jiqi yingxiang yanjiu," 50.

68. Murray, "New World Food Crops in China," 274–345.

69. Zhang et al., "Sweetpotato in China," 329–333.

70. Anderson, *Food of China*, 65.

71. Ho, *Studies on the Population of China*, 183–184.

72. Chinese officials from Xu Guangqi and Li Wei to Chen Hongmou and Weng Ruomei (fl. 1760–1780), a magistrate in Sichuan, all used the phrase *jiuhuang diyiyi* to describe the quality of the sweet potato in dealing with a famine. See Song Junling, "MingQing shiqi Meizhou nongzuowu zai Zhongguo de chuanzhong jiqi yingxiang yanjiu," 168–181.

73. Ho, *Studies on the Population of China*, 187.

74. Ho, 191, 206.

75. Susan Mann and Philip Kuhn, "Dynastic Decline and Roots of Rebellion," *Cambridge History of China*, vol. 10, *Late Ch'ing, 1800–1911, Part 1*, ed. John K. Fairbank (Cambridge University Press, 1978), 108–112; Susan Naquin and Evelyn Rawski, *Chinese Society in the Eighteenth Century* (Yale University Press, 1987), 106; William T. Rowe, "Social Stability and Social Change," in *Cambridge History of China*, vol. 9, Part 1, the Ch'ing Empire to 1800, ed. Willard J. Peterson (Cambridge University Press, 1978), 474–475.

76. Wang Ruiping, "Mingdai renkou kao" [A study of the Ming population], *Huanghuai xuekan* [Journal of Huanghuai] 4 (1992): 48–52.

77. Jonathan Spence, "The K'ang-hsi reign," in Twitchett and Fairbank, *Cambridge History of China*, 9:178.

78. Ge Jianxiong, *Zhongguo renkou fazhanshi* [A history of the development of Chinese population] (Fujian renmin chubanshe, 1991), 63. Dwight Perkins also believes that China's demographic data of the early Ming and the high Qing were more reliable. See Perkins, *Agricultural Development in China, 1368–1968* (Aldine Publishing Company, 1969) 7–8.

79. John K. Fairbank and Edwin O. Reischauer, *China: Tradition and Transformation* (Houghton Mifflin, 1978), 241.
80. Ho, *Studies on the Population of China*, 277–278; Ge, *Zhongguo renkou fazhanshi*, 235, 240; Zhao Wenlin and Xie Shujun, *Zhongguo renkoushi* [History of Chinese population] (Renmin chubanshe, 1988), 356–357; Cao Shuji, *Zhongguo renkoushi—Ming shiqi* [History of Chinese population—the Ming period], vol. 4 (Fudan daxue chubanshe, 2000), 34; Cao Shuji, *Zhongguo renkoushi—Qing shiqi* [History of Chinese population—the Qing period], vol. 5 (Fudan daxue chubanshe, 2001), 833; Martin Heijdra, "The Social-Economic Development of Rural China During the Ming," in *Cambridge History of China*, vol. 8, *The Ming Dynasty, 1368–1644, Part 2*, ed. Denis C. Twitchett and Frederick W. Mote (Cambridge University Press, 1998), 437–439. Wang Ruiping in his "Mingdai renkou kao" also believed that the peak of the Ming population should be above 200 million.
81. Heijdra, "The Social-Economic Development of Rural China During the Ming," in Twitchett and Fairbank, *Cambridge History of China*, 8:439. Among others, Wang Ruiping in his "Mingdai renkou kao" also argued that China's population expansion had started in the Ming.
82. About the population decline in the Ming–Qing transition, Jiang Tao has offered some vivid descriptions in his *Zhongguo jindai renkoushi* [Modern history of Chinese population] (Zhejiang renmin chubanshe, 1993), 12–15.
83. Ge, *Zhongguo renkou fazhanshi*, 249–250; Zhao and Xie, *Zhongguo renkoushi*, 390; Cao, *Zhongguo renkoushi—Ming shiqi*, 4:452; Cao, *Zhongguo renkoushi—Qing shiqi*, 5:833; Jiang, *Zhongguo jindai renkoushi*, 15. Incidentally, Ping-ti Ho decided not to give a specific number for the population reduction during the Ming–Qing dynastic transition; he only wrote that "the nation suffered severe losses in population." See his *Studies on the Population of China*, 277.
84. Rowe, "Social Stability and Social Change," 475.
85. Examples were found in the TV series like the *San'guo yanyi* [Romance of the Three Kingdoms], made in 1994, and *Shuihui* [Outlaws of the marsh] in 1983.
86. In fact, the uniform resembling scrambled eggs with tomato was worn by Chinese athletes more than once at the Olympics, which was joked about by Bai Yansong, a popular TV anchor. See "The 'Scrambled Eggs and Tomatoes' Uniform Is a Hit! The Chinese Team Actually Wore It for Three Olympic Games," *China Daily*, August 11, 2016, https://cn.chinadaily.com.cn/2016olympics/2016-08/11/content_26433023.htm.
87. Cf. Brian Dott, *The Chile Pepper in China: A Culinary Biography* (Columbia University Press, 2020).
88. ZSKLYQY, *Qingshi ziliao*, 7:369, 384.
89. ZSKLYQY, 340, 343, 382.

90. ZSKLYQY, 316, 367–368, 373,
91. ZSKLYQY, 263, 382, 255, 257, 295, 351, 364, 368, 371–374, 380.
92. Ho, *Studies on the Population of China*, 184–189; Zheng Nan, "Cong yumi, fanshu, malingshu de chuanru kan wailai nongzuowu chuanru dui Zhongguo shehui de yingxiang," 433.
93. Zhao Gang, *Qingdai liangshi* mu *chanliang yanjiu* [Studies of acreage productivity during the Qing] (Zhongguo nongye chubanshe, 1995), 62. See also Zheng Nan, "Cong yumi, fanshu, malingshu de chuanru kan wailai nongzuowu chuanru dui Zhongguo shehui de yingxiang," 434, 441.
94. Lu Yao, *Ganshu lu*, quoted in Zheng Nan, "Cong yumi, fanshu, malingshu de chuanru kan wailai nongzuowu chuanru dui Zhongguo shehui de yingxiang," 439.
95. Cf. Perkins, *Agricultural Development in China, 1368–1968*, 184–185.
96. George Macartney, *An Embassy to China: Being the Journal Kept by Lord Macartney During His Embassy to the Emperor Ch'ien-lung, 1793–1794*, ed. J. L. Cranmer-Byng (1963; repr., Longmans, 1972), 245–247.
97. George Staunton, *An Authentic Account of an Embassy from the King of Great Britain to the Empire of China*, vol. 2 (London, 1797), 544–545.
98. John Barrow, *Travels in China: Containing Descriptions, Observations and Comparisons, Made and Collected in the Course of a Short Residence at the Imperial Palace of Yuen-min-yuen, and on a Subsequent Journey Through the Country from Pekin to Canton* (Cambridge University Press, 2010), 574–586.
99. Macartney, *An Embassy to China*, 244.
100. Aeneas Anderson, *An Accurate Account of Lord Macartney's Embassy to China* (London, 1797), 27; Barrow, *Travels in China*, 67.
101. Staunton, *An Authentic Account of an Embassy from the King of Great Britain to the Empire of China*, 2:69; Barrow, *Travels in China*, 86.
102. Barrow, *Travels in China*, 83, 109, 500, 508, 539, 549, 554, 556, 563; Staunton, *An Authentic Account of an Embassy from the King of Great Britain to the Empire of China*, 2:6–7, 107; Anderson, *An Accurate Account of Lord Macartney's Embassy to China*, 41.
103. Samuel Holmes, *The Journal of Mr. Samuel Holmes, One of the Guard on Lord Macartney's Embassy to China and Tartary* (Cambridge University Press, 2010), iv, 84, 137, 166.
104. Malthus, *An Essay on the Principle of Population and a Summary View of the Principle of Population*, 93–103.
105. Malthus, 195, 87.
106. Malthus, 115.
107. Malthus, 243–261.
108. There have been some articles comparing the ideas of Thomas Malthus and Hong Liangji. See Gan Qinghua, "Hong Liangji yu Maersasi de renkoulun

bijiao" [Comparison of the demographic theories of Hong Liangji and Thomas Malthus], *Lingling xueyuan xuebao* [Journal of Lingling University] 2 (2003): 72–74, and Ren Weiling, "Hong Liangji renkou sixiang he Maersasi renkou lilun de bijiao ji qishi) [Comparison of the demographic ideas of Hong Liangji and Thomas Malthus and its heuristic meaning], *Changzhou gongxueyuan xuebao* [Journal of Changzhou Institute of Technology] 1 (2019): 1–8.

109. Hong Longji, "Shengji pian" [On livelihood], in *Hong Liangji ji* [Hong Longji's works], vol. 1, ed. Liu Dequan (Zhonghua shuju, 2001), 15–16.

110. Hong Liangji, "Zhiping pian" [On the rule of peace], in *Hong Liangji ji*, 1:14–15.

111. Hong, 14–15.

112. See Wu Shenyuan, "Dui 'Qingdai renkou yanjiu' yiwen de liangdian yijian" [Two criticisms of the article "A Study of the Qing population"], *Zhongguo shehui kexue* [Chinese social sciences] 4 (1982): 221–222.

113. Wu, 221–222.

114. Xie Zhongliang, "Hong Liangji de renkou lilun jiqi chansheng de shidai tiaoji shulue" [Discussion of Hong Liangji's population theory in relation with the circumstances under which it was developed], *Guizhou shifan daxue xuebao* [Journal of Guizhou Normal University] 3 (1979): 44–48.

115. Liu Ming, "Wang Shiduo renkou sixiang chutan" [Preliminary exploration of Wang Shiduo's demographic theory], *Anhui shixue* [Historical research in Anhui] 3 (1989): 32–36; Yang Pengcheng and Tan Yangfang, "Wang Shiduo renkou sixiang tanxi" [Analysis of Wang Shiduo's demographic ideas], *Anhui daxue xuebao* [Journal of Anhui University] 1 (2002): 17–21.

116. Wang Shiduo, *Wang Huiweng (Shiduo) yibing riji* [Wang Shiduo's selected diary], ed. Deng Zhicheng (Wenhai chubanshe, 1967), 148–154.

117. Dwight Perkins believes that the estimated total loss of population during the Taiping Rebellion was over 60 million. Perkins, *Agricultural Development in China, 1368–1968*, 210–211. This seems a very conservative estimate, for Chinese demographers have given it a much higher figure. For example, Lu Yu and Teng Zezhi in their *Zhongguo renkou tongshi* [General history of Chinese population] (Shandong renmin chubanshe, 1999) argue that the loss was nearly 73 million (796), whereas Ge Jianxiong in his *Zhongguo renkou fazhanshi* estimates that the loss was as high as 112 million (253). Jiang Tao also states that the reduction was between 60 million and 100 million (*Zhongguo jindai renkoushi*, 74). Also see Li Bozhong, "'Renkou yali' yu 'zuidi shenghuo shuizhun' zhiyi" [Query on "population pressure" and "minimum subsistence level of living"], *Zhongguo shehui jingjishi yanjiu* [Studies on Chinese socio-economic history] 3 (1996): 31–37.

118. Ho, *Studies on the Population of China*, 238, 246, 256.

119. Ho, 257–278.
120. Jiang Tao, *Renkou yu lishi: Zhongguo chuantong renkou jiegou yanjiuu* [Population and history: A study of traditional demographic structure in China] (Renmin chubanshe, 1998), 372–403. In her dissertation, Laura Murray also observes that the family size in the Wei River Valley remained around five people from the eighteenth to the early twentieth century. Murray, "New World Crops in China," 203–207.
121. William Lavely and R. Bin Wong, "Revising the Malthusian Narrative: The Comparative Study of Population Dynamics in Late Imperial China," *Journal of Asian Studies* 57, no. 3 (August 1998): 714–748. See also R. Bin Wong, *China Transformed: Historical Change and the Limits of European Experience* (Cornell University Press, 1997), chaps. 1 and 2.
122. James Z. Lee and Wang Feng, *One Quarter of Humanity: Malthusian Mythologies and Chinese Realities, 1700–2000* (Harvard University Press, 1999), 7–9; James Z. Lee et al., "Positive Check or Chinese Check?," *Journal of Asian Studies* 61, no. 2 (May 2002): 591–607.
123. Ann Waltner, "Infanticide and Dowry in Ming and Early Qing China," in *Chinese Views of Childhood*, ed. Anne B. Kenney (University of Hawaii Press, 1995), 193–217.
124. James Z. Lee and Cameron Campbell, *Fate and Fortune in Rural China: Social Organization and Population Behavior in Liaoning 1774–1873* (Cambridge University Press, 1997), 66.
125. Lin Liyue, "Fengsu yu zuiqian: Mingdai de ninü jixu jiqi wenhua yihan" [Customs and crimes: The narrative and cultural meaning of female infanticide in the Ming], in *Wusheng zhisheng: Jindai Zhongguo de funü yu shehui 1600–1950* [Unheard voices: Women and society in modern China], vol. 2, ed. You Jianming (Zhongyang yanjiuyuan jindaishi yanjiusuo, 2003), 1–24; Chang Jianhua, "Mingdai niying wenti chutan" [Preliminary study of infanticide during the Ming], in *Zhongguo shehui lishi pinglun* [Review of Chinese social and economic history], ed. Zhang Guogang (Shangwu yinshuguan, 2002), 121–136, and Chang Jianhua, "Qingdai niying wenti xintan" [New research on infanticide during the Qing], in *Hunyin jiating yu renkou xingwei* [Marriage, family and population behavior], ed. Li Zhongqing (James Z. Lee) et al. (Beijing daxue chubanshe, 1999), 197–219.
126. Ogawa Yoshiyuki, "Shindai Kōsei, Fukken ni okeru 'dekijo' shūzoku to hō ni tsuite: 'Atsu yome' 'ton yan si' nado no shūzoku to no kankei o megutte" [Custom and law of female infanticide in Jiangxi and Fujian during the Qing: The customs of "expensive dowry" and "child bride" and their relations with infanticide], in *Chūgoku kinsei no kihan to chitsujo* [Rule and order in early modern China], ed. Yamamoto Eishi (Tōyō bunko, 2014), 247–278; Waltner, "Infanticide and Dowry in Ming and Early Qing China."

127. Yi Ruolan, "Qingdai Taiwan ninü wenti chutan" [Preliminary study of female infanticide in Qing Taiwan], *Jianzhong xuebao* [Journal of Jianguo High School] 3 (1997): 203–214.
128. In response to James Z. Lee and Wang Feng's critique, Cao Shuji and Chen Yixin published a long article arguing that while refusing Malthus and his followers, what Lee and his collaborators attempted has created a "new myth" about China's demographic behavior, as though it was immune to Malthusian "positive checks" such as wars and epidemics. They maintain that the revolts in the mid-nineteenth century were examples of these "positive checks" because China's population had reached a level of crisis in the early years of the century. Cao Shuji and Chen Yixin, "Maer sasi lilun he Qingdai yilai de Zhongguo renkou: Ping Meiguo xuezhe jinnian laide xiangguan yanjiu" [Malthusian theory and China's population from the Qing: A review of related works by American scholars], *Lishi yanjiu* [Historical research] 1 (2002): 41–54.
129. Ester Boserup, *The Conditions of Agricultural Growth: The Economics of Agrarian Change Under Population Pressure* (1965; repr. Routledge, 2003), 11; Ester Boserup, *Population and Technological Change: A Study of Long-Term Trends* (University of Chicago Press, 1981).
130. Murray, "New World Crops in China," 292, 362–368; Wang Siming, "Meizhou yuanchan zuowu de yinzhong zaipei jiqi dui Zhongguo nongye shengchan jiegou de yingxiang" [Introduction and cultivation of American plants and its influence in Chinese agricultural production structure], *Zhongguo nongshi* [Chinese agricultural history] 2 (2004): 16–27; Wang Siming, "Meizhou zuowu de chuanbo jiqi dui Zhongguo yinshi yuanliao shengchan de yingxiang" [Dispersal of American plants and its impact on China's food supply], *Di 8 jie Zhongguo yinshi wenhua xueshu yantaohui lunwenji* [Proceedings of the 8th Academic Symposium on Chinese Food Culture) (Zhongguo yinshi wenhua jijinhui, 2004), 141–161. Li Xinsheng, "Meizhou zuowu yu renkou zengzhang: jianlun 'Meizhou zuowu juedinglun' de lailong qumai" [American plants and population increase: A discussion of the origin and development of "American crop determinism"], *Zhongguo jingjishi yanjiu* [Studies on Chinese economic history] 3 (2020): 157–173; Li Xinsheng and Wang Siming, "Qing zhi Min'guo Meizhou zuowu shengchan zhibiao guji" [Estimates of American plants' per unit production in the Qing and Republican periods], *Qingshi yanjiu* [Qing history journal] 3 (2017): 126–139. Dwight Perkins also observes that the increase of maize farming in China only began after 1918. Perkins, *Agricultural Development in China, 1368–1968*, 47–48.
131. The discussion of the five major sweet potato farming regions in the following few paragraphs is based on Zhang et al., "Sweetpotato in China," 329–333. While using their study, I have reversed their order of the five zones by

following the south–north route, which mimicked the spread of the plant in Chinese history.

132. Chen Hong, *Pubian xiaocheng*, in ZSKLYQY, *Qingshi ziliao*, 7:369.
133. Wu Liqing, "Fanshu zai Chaozhou diqu de chuanbo yu nongye tixi biandong," 33.
134. Zheng Weikuan, "Qingdai yumi he fanshu zai Guangxi chuanbo wenti xintan" [New explorations of the cultivation of maize and sweet potato in Guangxi], *Guangxi minzu daxue xuebao* [Journal of Guangxi University for Nationalities] 6 (2009): 114–121; Li Xinsheng and Wang Siming, "Qingdai yumi he fanshu zai Guangxi chuanbo wenti zaitan" [New study of the cultivation of maize and sweet potato in Guangxi], *Zhongguo lishi dili luncong* [Journal of Chinese historical geography] 4 (2018): 78–86.
135. Gong Shengsheng, "Qingdai lianghu diqu de yumi he fanshu" [Maize and sweet potato in Hunan and Hubei in the Qing period], *Zhongguo nongshi* [Chinese agricultural history] 3 (1993): 47–57.
136. Zhang Qian, "Ganshu zai Sichuan de chuanbo jidui Sichuan yinshi wenhua de yingxiang" [Spread of the sweet potato in Sichuan and its influence in Sichuanese food culture], *Nongye kaogu* [Agricultural archaeology] 3 (2013): 181–185.
137. Chen Dongsheng, "Ganshu zai Shandong chuanbo zhongzhi shilue."
138. Cf. Zheng Nan, "Meizhou zouwu fanshu de chuanru ji zai Heilongjiang diqu de yinzhong he zaipei" [Introduction of the American plant sweet potato and its cultivation in Heilongjiang], *Chuxiong shifan xueyuan xuebao* [Journal of Chuxiong Normal University] 8 (2013): 4–11. Despite the title, the article only mentions the sweet potato cultivation in Heilongjiang very briefly at the end.
139. Perkins, *Agricultural Development in China, 1368–1968*, 31.
140. Shi Zhihong, *Agricultural Development in Qing China: A Quantitative Study, 1661–1911* (Brill, 2018), 34–35, 77–78.
141. Wang Siming, "Meizhou yuanchan zuowu de yinzhong zaipei jiqi dui Zhongguo nongye shengchan jiegou de yingxiang," 22.
142. Li and Wang, "Qing zhi Min'guo Meizhou zuowu shengchan zhibiao guji," 136–137. They also quote Dwight Perkins, who estimated that sweet potato acreage was roughly 1.7 percent of total cultivated acreage in China between 1914 and 1957. Perkins, *Agricultural Development in China, 1368–1968*, 47. However, Li and Wang believe that these estimates are likely to be significantly lower than the actual acreage.
143. Shi, *Agricultural Development in Qing China*, 164.
144. Li and Wang, "Qing zhi Min'guo Meizhou zuowu shengchan zhibiao guji," 137.
145. Sucheta Mazumdar, *Sugar and Society in China: Peasants, Technology and the World Market* (Harvard University Press, 1998), 259.

146. Philip C. C. Huang, *The Peasant Economy and Social Change in North China* (Stanford University Press, 1985), 116–117.
147. This poem was collected in Seo Yugu's *Jongjeobo*, in Shiyuan, *Jinshu chuanxilu zhongshupu hekan*, 250.
148. Bozhong Li, *Agricultural Development in Jiangnan, 1620–1850* (St. Martin's Press, 1998), 66. Also see Guo Songyi, "Fanshu zai Zhejiang de yinzhong he tuiguang" [Introduction and promotion of the sweet potato in Zhejiang], *Zhejiang xuekan* [Zhejiang journal] 3 (1986): 45–49.
149. Bozhong Li, 45–49.
150. See local gazetteers in Fujian, Jiangxi, and Hunan excerpted in Chen, *MingQing nongyeshi ziliao, 1368–1911*, 1:288, 294–295, 305.
151. Lavely and Wong, "Revising the Malthusian Narrative," 726.
152. See local gazetteers and official memorials excerpted in Chen, *MingQing nongyeshi ziliao, 1368–1911*, 1:253–254, 260, 262, and in ZSKLYQY, *Qingshi ziliao*, 7:157, 159, 177–180,
153. Naquin and Rawski, *Chinese Society in the Eighteenth Century*, 132.
154. Naquin and Rawski, 130.
155. Cf. Ge Jianxiong et al., *Jianming Zhongguo yiminshi* [Concise history of Chinese immigration] (Fujian renmin chubanshe, 1993), 405–428.
156. Zhou Yuanhe, "Ganshu de lishi dili," 85–86.
157. See *Yushan Xianzhi* [Gazetteer of Yushan County], excerpted in ZSKLYQY, *Qingshi ziliao*, 7:352.
158. Chen Quanqing, "Qingdai de renkou zengzhang yu weiji" [Population growth and crisis in the Qing], *Hunan shifan daxue xuebao* [Journal of Hunan Normal University] 6 (1991): 57.
159. Lin Renchuan and Wang Puhua, "Qingdai Fujian renkou xiang Taiwan de liudong" [Fujianese emigration to Taiwan during the Qing], *Lishi yanjiu* [Historical research] 3 (1983): 130–141; Li Wei, *Qingdai liangshi duanque yu dongbei tudi kaifa* [Food shortage and the northeastern immigration in the Qing] (Jilin renmin chubanshe, 2011).
160. Rewi Alley, *Travels in China, 1966–71* (New World Press, 1973), 34.
161. Alley, 28, 182–183.
162. In terms of sown areas, sweet potato reached its peak in 1971. See Jia Lian Sheng et al., "Sweetpotato Breeding, Production and Utilization in China," in *Sweetpotato Technology for the 21st Century*, ed. Walter A. Hill et al. (Tuskegee University, 1992), 154.
163. John Lossing Buck, *Land Utilization in China* (University of Chicago Press, 1937), 407; Hao Hu et al., eds., *Chinese Agriculture in the 1930s: Investigation Into John Lossing Buck's Rediscovered "Land Utilization in China" Microdata* (Palgrave Macmillan, 2019), 265, 280.

164. Martin C. Yang, *A Chinese Village, Taitou, Shantung Province* (Columbia University Press, 1945), 16, 34.
165. Chen Dongsheng, "Ganshu zai Shandong chuanbo zhongzhi shilue," 221.
166. Alley, *Travels in China*, 321.
167. Hanchao Lu, "The Tastes of Chairman Mao: The Quotidian as Statecraft in the Great Leap Forward and Its Aftermath," *Modern China* 41, no. 5 (September 2015): 546–547.
168. Fujian Putian xian hongzhuan daxue, ed., *Fanshu zaipeixue* [How to cultivate the sweet potato] (Gaodeng jiaoyu chubanshe, 1959); Nongye ziliao bianji weiyuanhui, ed., *Fanshu de chucang he liyong* [Storage and utilization of the sweet potato] (Nongye chubanshe, 1959); Qinggongye shipinju, ed., *Fanshu zonghe liyong de jingyan* [Experience in multifarious utilization of the sweet potato] (Qinggongye chubanshe, 1959).
169. Zhongguo nongye yanjiuyuan ganshu yanjiusuo, ed., *Ganshu zaipeixue jiangyi* [Lecture notes on sweet potato farming] (Nongye chubanshe,1960). The first comprehensive work on the sweet potato, claimed by its authors, was published in the year when the Chinese Academy of Agricultural Science was founded. See Sheng Jialian et al., *Ganshu* [Sweet potato] (Kexue chubanshe, 1957). It was arguably the first by modern Chinese scientists.
170. Lu, "The Tastes of Chairman Mao," 548.
171. Isabel Crook and David Crook, *The First Years of Yangyi Commune* (1966; Routledge, 2003), 27n3, 117.
172. John Lossing Buck et al., *Food and Agriculture in Communist China* (Frederick A. Praeger, 1966), 56–60.
173. The figure of sweet potato production was indeed inflated during the Great Leap Forward. See Zhou Xun, *Forgotten Voices of Mao's Great Famine, 1958–1962: An Oral History* (Yale University Press, 2013), 45, 167.
174. Bruce Stone, "An Analysis of Chinese Data on Root and Tuber Crop Production," *China Quarterly*, no. 99 (September 1984): 594–630. For the death estimates (30–45 million) in the wake of the Great Leap Forward, see Frank Dikötter, *Mao's Great Famine: The History of China's Most Devastating Catastrophe, 1958–62* (Bloomsbury, 2010), preface, xi–xviii, and Zhou Xun, *The Great Famine in China, 1958–1962: A Documentary History* (Yale University Press, 2012), introduction, ix–xv.
175. Dikötter, *Mao's Great Famine*, 136; Zhou, *The Great Famine in China, 1958–1962*, 84, 94.
176. Zhou, *The Great Famine in China, 1958–1962*, 84–85.
177. Crook and Crook, *The First Years of Yangyi Commune*, 52, 113.
178. Crook and Crook, 152.
179. Zhou, *Forgotten Voices of Mao's Great Famine, 1958–1962*, 48, 253, 255–257, 276; Zhou, *The Great Famine in China, 1958–1962*, 57, 122, 130–131.

180. Buck, *Land Utilization in China*, 146–147.
181. Kimberley Ens Manning and Felix Wemheuer, eds., *Eating Bitterness: New Perspectives on China's Great Leap Forward and Famine* (UBC Press, 2011), 209, 261–262; Ellen Oxfeld, *Bitter and Sweet: Food, Meaning and Modernity in Rural China* (University of California Press, 2017), 82–86.
182. Reporting in 1977 to the U.S. Congress, the Committee on Science and Technology concludes that food production in China must include non-grain food crops and that these non-grain foods helped the country to keep pace of its population growth. Committee on Science and Technology, U.S. House of Representatives, Ninety-Fifth Congress, *The Role of Science and Technology in China's Population/Food Balance* (Science Policy Research Division, Congressional Research Service, Library of Congress, 1977), 55–56. Dwight Perkins also maintains that food crops in China should include roots and tubers. See his *Agricultural Development in China*, chap. 2.
183. Gao Hua, "Food Augmentation Methods and Food Substitutes During the Great Famine," and Richard King, "Romancing the Leap: Euphoria in the Moment Before Disaster," in Manning and Wemheuer, *Eating Bitterness*, 62–64, 175–188; Crook and Crook, *The First Years of Yangyi Commune*, 152–153.
184. Anderson, *Food of China*, 122.
185. Lu, "The Tastes of Chairman Mao," 549–550.
186. Yang Jisheng, *Tombstone: The Great Chinese Famine, 1958–1962* (Farrar, Straus and Giroux, 2012), 329.
187. Oxfeld, *Bitter and Sweet*, 10; Lu, "The Tastes of Chairman Mao," 550.
188. Stone, "An Analysis of Chinese Data on Root and Tuber Crop Production," 596, 612. By comparing the numbers between 1971 and 1989, the sown area of sweet potato in China declined by as much as 85 percent. See Jia Lian Sheng et al., "Sweetpotato Breeding, Production and Utilization in China," 154.
189. Oxfeld, *Bitter and Sweet*, 209n3.
190. Bouwkamp, *Sweet Potato Products*, 3.
191. Oxfeld, *Bitter and Sweet*, 82–87.
192. Woolfe, *Sweet Potato*, 411. In the 1990s, the percentage of sweet potato for animal feed was 45 percent, compared to 15 percent for human food and over 30 percent for industrial uses. See Jia Lian Sheng et al., "Sweetpotato Breeding, Production and Utilization in China," 155.
193. Shiu-ying Hu, *Food Plants of China* (Chinese University of Hong Kong, 2005), 12, 126.
194. Frederick Simoons, *Food in China: A Cultural and Historical Inquiry* (CRC Press, 1991), 123.
195. Cf. Lu, "The Tastes of Chairman Mao," 561–562.

4. Hunger Food? Daily Meals?

1. Seo Yugu, *Jongjeobo*, in Chen Shiyuan, *Jinshu chuanxilu zhongshupu hekan*, 201–252; No Seonghwan, "Chaoxian tongxinshi chuanru fanshu kao" [A study of how the Korean emissaries brought back the sweet potato], in *Dongya wenhua jiande bisai: Chaoxian furi tongxinshi wenxian de yiyi* [The race among Asian cultures: The significance of the sources by Korean emissaries to Japan], ed., Fudan daxue wenshi yanjiuyuan (Advanced Humanistic Studies, Fudan University) (Zhonghua shuju, 2019), 217–244; Felix Siegmund, "Tubers in a Grain Culture: The Introduction of Sweet and White Potatoes to Chōson Korean and Cultural Implications," *Korean Histories* 2, no. 2 (2010): 61–62.
2. Basil Hall Chamberlain, "The Luchu Islands and Their Inhabitants I: Introductory Remarks," *Geographical Journal* 5, no. 4 (April 1895): 301–302.
3. See Edmund Simon, "Auf welchem Wege kam dis Süße Kartoffel nach Japan?," *Anthropos* 8, no. 1 (January–February 1913): 135–137, and Edmund Simon, "The Introduction of the Sweet Potato into the Far East," *Transactions of the Asiatic Society of Japan* 42 (1914): 711–724; Mock Joya, *Japan and Things Japanese* (Kegan Paul, 2006), 281.
4. Sakai, *Satsuma imo*, 34–35.
5. Yamada Syoji [Shōzi], *Satsuma imo: Tenrai to bunka* [Sweet potato: Its spread and culture] (Shun'endō shuppan, 1994), 62–65. He also states that since Nokuni Sōkan brought sweet potato back from China, it was called *fanshu* by the people in Ryukyu, the same name used in Chinese. See his "Kansho no Nantō he no fukyū" [How sweet potatoes were introduced to Nansei-Shotō Islands], *Nantō shigaku* [Journal of Ryukyuan studies] 31 (April 1988): 79–81.
6. The first story is recorded by Sakai Kenkichi in his *Satsuma imo* (see note 2), and, more briefly, in Chūman Katsumi's *Nihon kansho saibaishi*, 12–13, but not in Miyamoto Tsuneichi's *Kansho no rekishi*. Yamada Syoji's *Satsuma imo*, which is a more detailed study of the sweet potato's early dissemination in Ryukyu and Japan, only mentions it in passing, 73–74. Sakai, too, acknowledges that the second story about Nokuni Sōkan and Gima Shinjō is better remembered in Okinawa. In English, it has been retold by George H. Kerr, who had ample living experience in both Taiwan and Okinawa, in his *Okinawa: The History of an Island People*, rev. ed. (Tuttle Publishing, 2000), 183–184.
7. Miyamoto, *Kansho no rekishi*, 36–37; Yamada, *Satsuma imo*, 143–144.
8. Richard Cocks, *Diary of Richard Cocks: Cape-Merchant in the English Factory in Japan, 1615–1622*, vol. 1, ed. Edward Maunde Thompson (Cambridge University Press, 2010), 2, 11, 13, 59. Also see Kerr, *Okinawa*, 183–184.

9. Miyamoto, *Kansho no rekishi*, 39.
10. Barry Duell, "Anthropological Problems Connected with the Introduction and Diffusion of the Sweet Potato in Japan (1)," *Kokusai shōka daigaku ronsō* [Forum of International College of Commerce] 28 (September 1983): 55–56. The quote here is excerpted from the English translation of Miyazaki's description of the sweet potato by Duell in his "Anthropological Problems Connected with the Introduction and Diffusion of the Sweet Potato in Japan (2)," *Kokusai shōka daigaku ronsō* 29 (March 1984): 66–67. The Japanese version is available in Matsuoka Jo'an's *Bansho roku* [Sweet potato book] (n.p., 1717, available in the digital collections of the National Diet Library, Tokyo).
11. Yamada, "Kansho no Nantō he no fukyū," 88–89, and Yamada, *Satsuma imo*, 200–205.
12. Miyamoto, *Kansho no rekishi*, 78–80; Yamada, *Satsuma imo*, 146–149.
13. Yamada, "Kansho no Nantō he no fukyū," 82; Wikipedia, "Maeda Riemon" [in Japanese], last modified March 11, 2024, 03:06 (UTC), https://ja.wikipedia.org/wiki/%E5%89%8D%E7%94%B0%E5%88%A9%E5%8F%B3%E8%A1%9B%E9%96%80.
14. Miyamoto, *Kansho no rekishi*, 135–136; Itō Shōji, *Satsumaimo to Nihonjin*, 84–85; and Yamada Syoji, *Satsuma imo*, 170–176, which discusses different versions of Aoki Konyō's *Bansho kō*. Barry Duell writes that it was during the 1720s that sweet potatoes were sown on Izu Islands. See his "Anthropological Problems Connected with the Introduction and Diffusion of the Sweet Potato in Japan (1)," 53.
15. Matsuoka, *Bansho roku*, n.p.
16. Miyamoto, *Kansho no rekishi*, 135.
17. Barry Duell offers some details about how Tokugawa Yoshimune realized that sweet potato was not poisonous and hoped to educate his subjects about this. See his "Anthropological Problems Connected with the Introduction and Diffusion of the Sweet Potato in Japan" (1), 52–54.
18. Aoki Konyō, *Bansho kō* [Sweet potato study] (n.p., 1735, available in the digital collections of the National Diet Library, Tokyo), n.p. Also, Miyamoto, *Kansho no rekishi*, 137. Xu's thirteen "advantages" of the sweet potato were discussed in his *Nongzheng quanshu*, 2:694.
19. Mikiso Hane, *Peasants, Rebels, Women and Outcasts: The Underside of Modern Japan* (Rowman and Littlefield, 2003), 7.
20. Cf. Jean-Pascal Bassino, "Market Integration and Famines in Early Modern Japan, 1717–1857," World Economic History Congress, August 2009, Utrecht, Netherlands; Kalland and Pedersen, "Famine and Population in Fukuoka Domain During the Tokugawa Period," 51.
21. Hane, *Peasants, Rebels, Women and Outcasts*, 8.

22. Naomichi Ishige, *The History and Culture of Japanese Food* (Routledge, 2001), 95.
23. Susan B. Hanley, "Tokugawa Society: Material Culture, Standard of Living and Life-Styles," in *The Cambridge History of Japan*, vol. 4, *Early Modern Japan*, ed. John Whitney Hall (Cambridge University Press, 1991), 682.
24. Japanese economic historian Matao Miyamoto sets "the 1600 population at 12 million and the 1721 population at 31.28 million." "Quantitative Aspects of Tokugawa Economy," in *The Economic History of Japan: 1600–1990*, vol. 1, *Emergence of Economic Society in Japan, 1600–1859*, ed. Akira Hayami et al. (Oxford University Press, 1999), 37.
25. Hanley, "Tokugawa Society," 663.
26. Kalland and Pedersen, "Famine and Population in Fukuoka Domain During the Tokugawa Period," 31–72. Hayami Akira's estimates are in his "Population Changes," in *Japan in Transition: from Tokugawa to Meiji*, ed. Marius Jansen and Gilbert Rozman (Princeton University Press, 1986), 287.
27. Miyamoto, "Quantitative Aspects of Tokugawa Economy," 45.
28. Akira Hayami and Hiroshi Kitō, "Demography and Living Standards," in Hayami et al., *Economic History of Japan*, 1:222.
29. Akira Hayami, *Japan's Industrious Revolution: Economic and Social Transformations in the Early Modern Period* (Springer Japan, 2015), 81–90.
30. Yamada, *Satsuma imo*, 205.
31. It should be noted that Arizono's comparison here uses raw sweet potatoes instead of dried ones, for dried sweet potatoes have roughly the same amount of protein content as does rice (7.0 vs. 8.1 grams per 100 grams dry weight). See Wikipedia, "Sweet Potato," last modified April 15, 2025, 06:41 (UTC), https://en.wikipedia.org/wiki/Sweet_potato.
32. Arizono Shōichirō, "Seinan Nihon 3 chiiki ni okeru satsumaimo no fukyū to jinkō zōka to no kakawari" [Relationship between the spread of sweet potatoes and population growth in the three regions of southwest Japan), Aichi Daigaku Bugakubu Rekishi, Chiri-gakka hen [Departments of History and Geography, College of Arts, Aichi University], ed., *Aidai shigaku, Nihon shigaku, seikai shigaku, chirigaku* [History, Japanese history, world history, and geography Aichi University] 14 (2005): 173–196.
33. Joya, *Japan and Things Japanese*, 282.
34. I have, however, not been able to locate Suyama Don'ō's *Kansho seshi*. That Suyama told the "filial yam/taro" story in it is stated by No Seonghwan in his "Chaoxian tongxinshi chuanru fanshu kao," 224–225. Miyamoto also describes it in his *Kansho no rekishi* (96–98), which is the basis of my discussion.
35. Woolfe, *Sweet Potato*, 45.

36. Suyama Don'ō's description is excerpted in Miyamoto, *Kansho no rekishi*, 98. A more detailed instruction on making sweet potato *sen* (starch) practiced in Tsushima is given in Itō Shōji's *Satsuma imo to Nihon jin*, 114–115. In modern-day Japan, *sen* is made by following a more sophisticated procedure. The Japanese Wikipedia entry on *sen* describes the process as follows: "In November and December, when the harvest is done and the weather is getting cold, sweet potatoes are washed and ground in a mortar. In the past, a foot-operated mortar was used, which is now being replaced by a mechanical mortar; in some cases, a slicer is used to cut the sweet potatoes into plates. The potatoes are placed in a coarse colander on top of a vat, and soaked in water for several days to ferment, changing the water. The starch precipitates at the bottom of the vat, which is washed several times with water and then dried to produce 'raw *sen*' or 'white *sen*.' The potatoes left in the colander are spread out on a flat board to a thickness of 7–8 mm and left to ferment outdoors for another 10–40 days. After a few days on a flat board, the outside of the potatoes becomes dry and cracked, while the inside turns yellowish-brown due to fermentation by lactic acid bacteria and emits an odor similar to that of earthen walls. In some areas, they are placed in baskets lined with straw and fermented for 20–30 days. After a month or so, the surface becomes hard, dry, cracked, and dark brown. Put them in a vat filled with water and soak them for three to five days, changing the water every day to remove the lye. When the water is clear, crush them with a mortar, soak them in the water in the vat, and rub them well to release the starch. Remove the skins and fibers of the floating potatoes from the supernatant water and pass it through a coarse colander to remove the fine fibers. When only white sediment remains at the bottom, decant the water and transfer the sediment to a container lined with bleach. After a night, the sediment becomes as hard as clay, so one should pinch it with fingers to make a nose-shaped dumpling that dries easily outdoors to finish the process. The entire process is completed between January and March, and the fully dried *sen* can be stored for several years." Wikipedia, "Sen (food)" [in Japanese], last modified November 27, 2023, 17:09 (UTC), https://ja.wikipedia.org/wiki/%E3%81%9B%E3%82%93 _(%E9%A3%9F%E5%93%81).
37. Quoted in Felix Siegmund, "Die Einführung von Süßkartoffel und Kartoffel in Korea: Verbreitung und Nutzung von agronomischen Wissen," *Bochumer Jahrbuch zur Ostasienforschung*, vol. 33, ed. Fakultät für Ostasienwissenschaften der Ruhr-Universität Bochum (Iudicium Verlag, 2009), 225–226.
38. No Seonghwan, "Chaoxian tongxinshi chuanru fanshu kao," 218–224.
39. Felix Siegmund, "Einführung und Verbreitung der Süßkartoffel und Kartoffel in Korea und China" (master's thesis, Ruhr-Universität Bochum, 2009),

81–84. No Seonghwan, "Chaoxian tongxinshi chuanru fanshu kao," 218–224, 230–231.
40. Quoted in Felix Siegmund, "Tubers in a Grain Culture: The Introduction of Sweet and White Potatoes to Chōson Korean and Cultural Implications," *Korean Histories* 2, no. 2 (2010): 63.
41. Seo Yeongbo's memorial presented to the Joseon court is translated in Siegmund's "Tubers in a Grain Culture," 70–71.
42. Siegmund, "Die Einführung von Süßkartoffel und Kartoffel in Korea," 232–241, and Siegmund, "Tubers in a Grain Culture," 64–66.
43. Siegmund has made the argument in the works cited above, especially "Tubers in a Grain Culture."
44. Jeong Yakyong's *Mongmin Simseo* has an English translation: Chong Yagyong (Jeong Yakyong), *Admonitions on Governing the People*, trans. Choi Byonghyon (University of California Press, 2010). His discussion on how to deal with famines are in book 11, *Famine Relief*, 893–974.
45. Marius B. Jansen, "Introduction," in *The Cambridge History of Japan*, vol. 5, *The Nineteenth Century*, ed. Marius B. Jansen (Cambridge University Press, 1989), 15.
46. Harold Bolitho, "The Tempō Crisis," in Jansen, *Cambridge History of Japan*, 5:118; Takeshi Horie et al., "Effects of Elevated CO_2 and Global Climate Change on Rice Yield in Japan," in *Climate Change and Plants in East Asia*, eds., K. Omasa et al. (Springer Japan, 1996), 41.
47. Bolitho, "The Tempō Crisis," 118–119
48. Bolitho, 161.
49. Yamada, *Satsuma imo*, 205.
50. Chūman, *Nihon kansho saibaishi*, 48–49, 39.
51. Chūman, 79.
52. Tsuboi Shōgorō, "Izu nii jima no dozoku" [Local customs on the Izu New Island], *Tokyo jinrui gakukai zasshi* [Journal of Tokyo Anthropology Association] 10, no. 113 (1895): 421–430.
53. Duell, "Anthropological Problems Connected with the Introduction and Diffusion of the Sweet Potato in Japan" (2), 55.
54. Cf. Basil Hall Chamberlain, *Things Japanese: Being Notes on Various Subjects Connected with Japan* (1905; repr., Stone Bridge Press, 2007), 20. Chamberlain noted that the enumeration of the five grains in both China and Japan "differed slightly according to time and space."
55. Charlotte von Verschuer, "Agriculture and Food Production," in *Routledge Handbook of Premodern Japanese History*, ed., Karl F. Friday (Routledge, 2017), 380–385.
56. Emiko Ohnuki-Tierney, *Rice as Self: Japanese Identities Through Time* (Princeton University Press, 1993), 4–5, 34–36.

57. Vaclav Smil and Kazuhiko Kobayashi, *Japan's Dietary Transition and Its Impact* (MIT Press, 2012), 10–11.
58. *Lust, Commerce, and Corruption: An Account of What I Have Seen and Heard, by an Edo Samurai*, trans. Mark Teeuwen et al. (Columbia University Press, 2014), 278.
59. Eric C. Rath, *Food and Fantasy in Early Modern Japan* (University of California Press, 2010), 91, 149, 177. The sweet potato dish on the daimyo's banquet might have been a version of *hikado*, which was common in Nagasaki. It was made by dicing tuna, daikon, carrots, and sweet potatoes and simmering them in soy sauce. See Ishige, *History and Culture of Japanese Food*, 93.
60. Robert Fortune, *Yedo and Peking: A Narrative of a Journey to the Capitals of Japan and China* (London, 1863), 19, 207, 273–280.
61. Charlotte von Verschuer, *Rice, Agriculture and Food Supply in Premodern Japan*, trans. Wendy Cobcroft (Routledge, 2016), chap. 4.
62. Chamberlain, *Things Japanese*, 22.
63. Arizono Shōichirō, *Kinsei shomin on nichijōshoku: Hyakushō wa kome o taberarenakatta* [Daily foods of the modern commoners: Did ordinary people eat rice?] (Kaiseisha, 2007), especially 31–32, 86. Shunsaku Nishikawa, "Grain Consumption: The Case of Chūshū," and Susan B. Hanley, "The Material Culture: Stability in Transition," in Jansen and Rozman, *Japan in Transition*, 422, 455.
64. Son Chin-t'ae, "Gamjeo jeonpa ko" [A study of the diffusion of the sweet potato], *Chin-tan hakpo* [Journal of Chin-tan Society] 13 (1941): 86–109.
65. Cf. Siegmund, "Einführung und Verbreitung der Süßkartoffel und Kartoffel in Korea und China," 98–107; Kim Hee-sun and Kim Sook-Hee, "Joseonhugi kigeun manseonghoa'oa guhoangsigpum kaepal'ui sahoi·gieongje'jeog kochar" [Social-economic approach to the chronic state of famine and exploration of famine relief food in the late half of the Joseon period], *Hankuk shik munhwa hakhochi* [Korean journal of dietary culture] 2, no. 1 (1987): 81–92.
66. E. Sydney Crawcour, "Economic Change in the Nineteenth Century," in Jansen, *Cambridge History of Japan*, 5:613; Hane, *Peasants, Rebels, Women and Outcasts*, 31
67. Chūman, *Nihon kansho saibaishi*, 93–94.
68. Ann Waswo, "The Transformation of Rural Society, 1900–1950," in *The Cambridge History of Japan*, vol. 6, *The Twentieth Century*, ed. Peter Duus (Cambridge University Press, 1988), 542–543.
69. Hane, *Peasants, Rebels, Women and Outcasts*, 24–29.
70. Hane, 20.
71. E. Sydney Crawcour, "Industrialization and Technological Change, 1885–1920," in Duus, *Cambridge History of Japan*, 6:388–389.
72. Hane, *Peasants, Rebels, Women and Outcasts*, 40.

73. Chūman, *Nihon kansho saibaishi*, 96–97, 99–100, 103–105, 108–109, 111–113, 116–117.
74. Sakai, *Satsuma imo*, 90; Itō, *Satsuma imo to Nihon jin*, 151.
75. Crawcour, "Industrialization and Technological Change, 1885–1920," 436, 389.
76. Hane, *Peasants, Rebels, Women and Outcasts*, 40. Here the "yam" could be the sweet potato instead, because in Japanese yam is *sato-imo* and sweet potato is *Satsuma-imo*. In everyday conversation, however, *imo* is often used to refer to any root. At that time, sweet potato was also more common than yam.
77. Crawcour, "Industrialization and Technological Change, 1885–1920," 438–439. Chūman, *Nihon kansho saibaishi*, 118, 140.
78. Hane, *Peasants, Rebels, Women and Outcasts*, 115–116, 129, 133–134. Judging from the contexts, the "potatoes" mentioned here were more likely to be sweet potatoes due to their much higher popularity than the white potato in Japan at the time. Potato is called *jaga-imo* whereas sweet potato is *Satsuma-imo*. Both are simply called *imo* in everyday language in Japan, as explained before.
79. Lizzie Collingham, *The Taste of War: World War II and the Battle for Food* (Penguin Press, 2012), 56–57.
80. Collingham, 52.
81. Itō, *Satsuma imo to Nihon jin*, 159–174.
82. Chūman, *Nihon kansho saibaishi*, 142; Itō, *Satsuma imo to Nihon jin*, 57–58.
83. Collingham, *The Taste of War*, 232; Takafusa Nakamura, "Depression, Recovery, and War, 1920–1945," in Duus, *Cambridge History of Japan*, 6:491.
84. Cited in David J. Lu, ed., *Japan: A Documentary History: The Late Tokugawa Period to the Present*, vol. 2 (Routledge, 2015), 429.
85. Itō, *Satsuma imo to Nihon jin*, 58–63.
86. Siegmund, "Einführung und Verbreitung der Süßkartoffel und Kartoffel in Korea und China," 105–106.
87. Chūman, *Nihon kansho saibaishi*, 159–160, 164–165, 168, 173–174.
88. Itō, *Satsuma imo to Nihon jin*, 151; Ishige, *History and Culture of Japanese Food*, 161.
89. Collingham, *The Taste of War*, 316.
90. Itō, *Satsuma imo to Nihon jin*, 151–152.
91. Yamakawa Osamu, *Satsuma imo no sekai, sekai no satsuma imo* (Gendai shokan, 2020), 12.
92. Itō, *Satsuma imo to Nihon jin*, 151; Chūman, *Nihon kansho saibaishi*, 187–188, 195–196, 200–201, 205, 210–211.
93. Smil and Kobayashi, *Japan's Dietary Transition and Its Impact*, 1–2, 23.
94. Woolfe, *Sweet Potato*, 505–507.
95. Joya, *Japan and Things Japanese*, 85–86.
96. Yamada, *Satsuma imo*, 214.

97. Itō, *Satsuma imo to Nihon jin*, 179–184.
98. Russell Thomas, "The Endurance of Japan's Simple Street Snack," BBC, May 18, 2022, https://www.bbc.com/travel/article/20220518-japans-beloved-sweet-potato-vendors?ocid=ww.social.link.email.
99. Duell, "Anthropological Problems Connected with the Introduction and Diffusion of the Sweet Potato in Japan" (2), 51–55.
100. Barry Duell, "Consumption and Utilization of Sweet Potato in Japan," paper presented at the 2nd Annual UPWARD International Conference, Sweet Potato Cultures of Asia and South Pacific, Los Banos, Philippines (April 1991), 473–474, https://www.researchgate.net/publication/341179391_CONSUMPTION_AND_UTILIZATION_OF_SWEET_POTATO_IN_JAPAN.
101. Siegmund, "Tubers in a Grain Culture," 68–69; Michael J. Pettid, *Korean Cuisine: An Illustrated History* (Reaktion Books, 2008), 164–165.
102. Duell, "Consumption and Utilization of Sweet Potato in Japan," 471.
103. See Wikipedia, "Daikaku imo (University sweet potato)" [In Japanese], last modified April 22, 2025, 11:33 (UTC), https://ja.wikipedia.org/wiki/%E5%A4%A7%E5%AD%A6%E8%8A%8B.
104. The recipe is available from the Japanese version of Wikipedia on *"suito poteto:"* https://ja.wikipedia.org/wiki/%E3%82%B9%E3%82%A4%E3%83%BC%E3%83%88%E3%83%9D%E3%83%86%E3%83%88, accessed on June 30, 2022.
105. See Samantha Chew, "10 Sweet Potato Recipes Besides Your Staple Steamed Goguma," Smart Local, June 17, 2020, https://thesmartlocal.com/korea/sweet-potato-recipes/.
106. Yamakawa, *Satsuma imo no sekai, sekai no satsuma imo*, 15.
107. Woolfe, *Sweet Potato*, 48.
108. See, for example, *Maangchi's Big Book of Korean Cooking: From Everyday Meals to Celebration Cuisine* (Houghton Mifflin Harcourt, 2019), which offers a variety of recipes that feature sweet potato starch noodles, or *dangmyeon*.
109. Yamakawa, *Satsuma imo no sekai, sekai no satsuma imo*, 99–100.

5. Sweet Potato Islands, Sweet Potato Peoples

1. Lung Ying-tai, "Fanshu" [Sweet potato], in her *Kan shijimo xiang ni zoulai* [Behold the fin de siècle approaching you] (Shanghai wenyi chubanshe, 1996), 8–10.
2. Lin Haiying, "Fanshu ren" [Sweet potato folks], in *Lao Beijing Taiwanren de gushi* [Stories of Taiwanese by old Beijingers], ed. He Biao (Taihai chubanshe, 2009), 99.

3. Chang Kwang-chih, *Fanshuren de gushi* [A story of a sweet potato guy] (Lianjing chuban shiye gongsi, 1998).
4. Lin, "Fanshu ren," 99.
5. Cai Chenghao and Yang Yunping, *Taiwan fanshu wenhuazhi* [A cultural history of the sweet potato in Taiwan] (Guoshi chuban, 2004), 27.
6. Cai and Yang, 36–53.
7. Cai and Yang, 67–70; Hui-tun Chuang, "Fabricating Authentic National Cuisine: Identity and Culinary Practice in Taiwan" (PhD diss., New School for Social Research, 2011), 42.
8. For instance, Lin Qingxuan, a Taiwanese writer, recalls that in his childhood, every meal was either rice cooked with sweet potatoes (usually two to one in proportion) or rice porridge with sweet potatoes, often causing him to complain to his father. See his "Hongxin fanshu" [Red-heart sweet potato] in *Xianzai jiushi zuihaode shiguang* [The present is the best time] (Jiangxi jiaoyu chubanshe, 2016), 187–192. Cai Lan (a.k.a. Chua Lam), a gourmet of southern Chinese cuisine, remarks that rice porridge with sweet potatoes is best prepared in Fujian and Taiwan. See his *Cai Lan de cailanzi: Shijie shicai jingxuan* [Cai Lan's food basket: Best food ingredients around the world] (Guangdong lüyou chubanshe, 2013), 82–83. T. H. Yang of National Taiwan University observed that in Taiwan, "Sweet potato was cooked alone or cooked with rice and served as a staple food." See his "Sweet Potato as a Supplemental Staple Food," in Villareal and Griggs, *Sweet Potato*, 31.
9. Cf. Chuang, "Fabricating Authentic National Cuisine," 54–61.
10. Cai and Yang, *Taiwan fanshu wenhuazhi*, 80–84.
11. Cai and Yang, 86–88.
12. H. Wan, "Cropping Systems Involving Sweet Potato in Taiwan," in Villareal and Griggs, *Sweet Potato*, 225–232.
13. Cai and Yang, *Taiwan fanshu wenhuazhi*, 112; Asian Vegetable Research and Development Center, *Farmers' Viewpoint of Sweet Potato Production in Taiwan*, Technical Bulletin No. 4 (Asian Vegetable Research and Development Center, December 1977), 3–44, quotes on 3.
14. Asian Vegetable Research and Development Center, *Farmers' Viewpoint of Sweet Potato Production in Taiwan*, 3. H. Y. Chen identifies three similar reasons for the decline of sweet potato from 1970: (1) the improvement of living standards among the Taiwanese; (2) the hog-raising industry turning more to corn instead of sweet potato for feed; and (3) better irrigation systems for growing more rice. See Chen, "Marking of Sweet Potato in Taiwan," in Villareal and Griggs, *Sweet Potato*, 413–419. And, about the association of sweet potato consumption with poverty, Denny Roy, *Taiwan: A Political History* (Cornell University Press, 2003), 103.
15. Cai and Yang, *Taiwan fanshu wenhuazhi*, 112–113.

16. Chuang, "Fabricating Authentic National Cuisine," 54–55.
17. Cai and Yang, *Taiwan fanshu wenhuazhi*, 11.
18. See Wang Jie, "Taiwan dangdai yinshi sanwen yanjiu" [A study of essays on food and drinks by contemporary Taiwanese writers] (master's thesis, Fujian Shifan Daxue [Fujian Normal University], 2014), 43–44.
19. In reporting the sweet potato festival in Xinhua, Taiwan, Wen Xiujiao wrote that Chen Tangshan, the mayor, recalled in his opening remarks that when he grew up, kids all ate rice cooked with sweet potatoes daily—and they often ate the rice in it and let their parents eat the sweet potatoes. "Guzao wei, fanshu qing: You Taiwan wei de Xinhua fanshujie" [Nostalgic taste and sweet potato affection: The sweet potato festival in Xinhua steeped in Taiwanese culture], *Fengnian* [Harvest] 49, no. 9 (1999): 20–21.
20. George H. Kerr, *Formosa Betrayed* (Houghton Mifflin, 1965), 168–169, 257. Besides identifying the mainlanders as "taro folks," Taiwanese also referred to them derogatively as "pigs." Lin Qingxuan also recalls that taro and sweet potato were used to distinguish the two groups in schools, and that he ate the latter on a daily basis when he grew up. See his "Hongxin fanshu," 187–192.
21. Cf. Zhang Jiaxing, "Cong fangyan shuyu kan MinTai wenhua chayi: Yi 'shebei' wenhua, 'shefan' wenhua yu 'fanshu wenhua qingjie' weili" [Cultural differences between Fujian and Taiwan in vernacular and folk languages: Examples of "northern related" culture, "foreign related" culture and "sweet potato cultural complex"], in *Nanfang yuyanxue* [Studies of southern languages], ed. Gan Yuen (Ji'nan daxue chubanshe, 2011), 37–43.
22. Kerr, *Okinawa*, 83–90.
23. Yamada, *Satsuma imo*, 113.
24. Mamoru Akamine, *The Ryukyu Kingdom: Cornerstone of East Asia*, trans. Lina Terrell; ed. Robert Huey (University of Hawaii Press, 2017), 68–70.
25. Akamine, 68.
26. Ishige, *History and Culture of Japanese Food*, 136–137. Also, Sakai, *Satsuma imo*, 113–114. He points out that at present, sweet potato is grown in 90 percent of the fields in Okinawa.
27. Joya, *Japan and Things Japanese*, 385.
28. At the beginning of the twentieth century, Japanese anthropologist Yoshiwara Shigeyasu visited Ryukyu and observed that sweet potato had become the staple among the islanders and that the root was eaten four times a day—at eight, eleven, three, and six o'clock. See his "Ryukyu ryokō no oboegaki" [Ryukyu trip memo], *Tokyo jinrui gakukai zasshi* [Journal of the Tokyo Anthropology Association] 15, no. 170 (1900): 325, and "Ryukyu shima ryokō banashi" [Remarks on my trip to Ryukyu], *Chigagu zasshi* [Journal of geography] 12, no. 8 (1900): 484–489. Yoshiwara's descriptions of Ryukyuan

dietary habits are largely reiterated by Chūman Katsumi in his *Nihon kansho saibaishi*, except the latter states that the islanders had three meals a day (28–29).

29. Cf. T. H. Yang, "Sweet Potato as a Supplemental Staple Food," in Villareal and Griggs, *Sweet Potato*, 31–33.
30. Yamada, *Satsuma imo*, 65–68; Sakai, *Satsuma imo*, 115–116.
31. Chūman, *Nihon kansho saibaishi*, 28–29.
32. Gregory Smits, *Visions of Ryukyu: Identity and Ideology in Early-Modern Thought and Politics* (University of Hawaii Press, 1999), 44–49. Original emphasis.
33. Yamada, *Satsuma imo*, 70–71; Kerr, *Okinawa*, 181.
34. Cai and Yang, *Taiwan fanshu wenhuazhi*, 26.
35. Kerr, *Okinawa*, 420–458.
36. Staunton, *An Authentic Account of an Embassy from the King of Great Britain to the Empire of China*, 2:459–460.
37. Kerr, *Okinawa*, 231–232.
38. Cf. Megumi Chibana, "Till the Soil and Fill the Soul: Indigenous Resurgence and Everyday Practices of Farming in Okinawa" (PhD diss., University of Hawaii at Manoa, 2014), 35, 102.
39. Katō Sango, "Ryukyu zakki (san)" [Okinawa miscellany, 3], *Tokyo jinrui gakukai zasshi* [Journal of the Tokyo Anthropology Association] 18, no. 209 (1903): 448–457.
40. Robert Walker, *Okinawa and the Ryukyu Islands* (Tuttle Publishing, 2014), 114.
41. Yamakawa, *Satsuma imo no sekai, sekai no satsuma imo*, 33.
42. Itō, *Satsumaimo to Nihonjin*, 44–46.
43. Itō, 28–32.
44. Chūman, *Nihon kansho saibaishi*, 148–175.
45. Yamakawa, *Satsuma imo no sekai, sekai no satsuma imo*, 29.
46. Sakai, *Satsuma imo*, 126–130; Itō, *Satsumaimo to Nihonjin*, 44–53.
47. Sakai, *Satsuma imo*, 90–95; Chūman, *Nihon kansho saibaishi*, 185.
48. Sakai, *Satsuma imo*, 134.
49. Cf. Samantha Kwok, "Which Is Actually a Purple Potato? Ube vs. Taro vs. Beni Imo," Kokoro Care, June 25, 2022, https://kokorocares.com/blogs/blog/purple-potatoes-ube-vs-taro-vs-beni-imo.
50. Chūman, *Nihon kansho saibaishi*, 219, 258.
51. D. Craig Willcox, Bradley J. Willcox, Hidemi Todoriki, and Makoto Suzuki, "The Okinawan Diet: Health Implications of a Low Calorie, Nutrient Dense, Antioxidant-Rich Dietary Pattern Low in Glycemic Load," *Journal of the American College of Nutrition* 28, no. 4 (2009): 503s–504s.
52. Willcox et al., 501s–506s. See also Donald Craig Willcox, Giovanni Scapagnini, and Bradley J. Willcox, "Healthy Aging Diet Other than the Mediterranean: A Focus on Okinawan Diet," *Mechanisms of Ageing and Development*,

nos. 136–137 (2014): 148–162. Ishige believes that the Okinawans' longevity had something to do with their diet being different from that of the Japanese because of "high consumption of sweet potatoes, seaweed, and the locally produced brown sugar." See his *History and Culture of Japanese Food*, 139.

53. Chūman, *Nihon kansho saibaishi*, 258.
54. Walker, *Okinawa and the Ryukyu Islands*, 114.
55. Itō, *Satsumaimo to Nihonjin*, 222–225; Chibana, "Till the Soil and Fill the Soul," 131–134.
56. Paul Sillitoe, "The Gender of Crops in the Papua New Guinea Highlands," *Ethnology* 80, no. 1 (January 1981): 4.
57. Sillitoe, 4–11.
58. R. M. Bourke, "Sweet Potato in Papua New Guinea," in Villareal and Griggs, *Sweet Potato*, 45; Woolfe, *Sweet Potato*, 539; Bouwkamp, *Sweet Potato Products*, 154.
59. Yen, *The Sweet Potato and Oceania*, 271; Bourke and Harwood, *Food and Agriculture in Papua New Guinea*, 17.
60. A. J. Kimber, "The Sweet Potato in Subsistence Agriculture," *Papua New Guinea Agricultural Journal* 23, nos. 3–4 (1972): 80–83; Woolfe, *Sweet Potato*, 539.
61. Tsuboi Shōgorō, "Nanyō shotō ni okana hareta tabuu sei no hanashi" [Stories about the taboo system practiced on the Southern Islands], *Tokyo jinrui gakukai zasshi* [Journal of the Tokyo Anthropology Association] 8, no. 86 (1893): 319–328.
62. James B. Watson, "The Significance of a Recent Ecological Change in the Central Highlands of New Guinea," *Journal of the Polynesian Society* 74, no. 4 (December 1965): 438–440, and "From Hunting to Horticulture in the New Guinea Highlands," *Ethnology* 4, no. 3 (July 1965): 299–300. Watson also expressed his belief that the root was transferred westward on the island of New Guinea by citing the work of Leonard J. Brass (1900–1971).
63. Donald Denoon, "Human Settlement," in *The Cambridge History of the Pacific Islanders*, ed. Donald Denoon, Malama Meleisea, Stewart Firth, Jocelyn Linnekin, and Karen Neo (Cambridge University Press, 1997), 68.
64. Kenneth F. Kiple and Kriemhild Coneè Ornelas, eds., *The Cambridge World History of Food*, vol. 1 (Cambridge University Press, 2000), 211.
65. Tim Bayliss-Smith et al., "Archaeological Evidence for the Ipomoean Revolution at Kuk Swamp, Upper Wahgi Valley, Papua New Guinea," in Ballard et al., *The Sweet Potato in Oceania*, 109–120, and Tim Bayliss-Smith, Jack Golson, and Philip Hughes, "Phase 6: Impact of the Sweet Potato on Swamp Landuse, Pig Rearing and Exchange Relations," in *Ten Thousand Years of Cultivation at Kuk Swamp in the Highlands of Papua New Guinea*, ed. Jack Golsom,

Tim Denham, Philip Hughes, Pamela Swadling, and John Muke (Australian National University Press, 2017), 297–323.
66. Watson, "From Hunting to Horticulture in the New Guinea Highlands," 296.
67. Watson, "The Significance of a Recent Ecological Change in the Central Highlands of New Guinea," 442–448, and "Pigs, Fodder, and the Jones Effect in Postipomoean New Guinea," *Ethnology* 16, no. 1 (January 1977): 57–70. Also, Paula Brown and Harold Brookfield, "Sweet Potato, Pigs and the Chimbu of Papua New Guinea Highlands," in Ballard et al., *The Sweet Potato in Oceania*, 131–136. Nicholas Modjeska examines the impact of the "Ipomoean Revolution" on social relations in Papua New Guinean society and concludes that men remain socially and economically superior to women. See his "Production and Inequality: Perspectives from Central New Guinea," in *Inequality in New Guinea Highlands Societies*, ed. Andrew Strathern (Cambridge University Press, 1982), 50–108.
68. It must be noted that James Watson himself recognized that after the perceived "Ipomoean Revolution," some elements of the pre-Ipomoean culture persisted in Papua New Guinea. See his "The Significance of a Recent Ecological Change in the Central Highlands of New Guinea," 448–449. About the controversy and criticisms of the "Ipomoean Revolution," see H. C. Brookfield and J. Peter White, "Revolution or Evolution in the Prehistory of the New Guinean Highlands: A Seminar Report," *Ethnology* 7, no. 1 (January 1968): 43–52; David J. Boyd, "Beyond the Ipomoean Revolution: Sweet Potato on the 'Fringe' of the Papua New Guinea Highlands," and Anton Ploeg, "Sweet Potato in the Central Highlands of West New Guinea," in Ballard et al., *The Sweet Potato in Oceania*, 137–162.
69. Woolfe, *Sweet Potato*, 540–541; R. Michael Bourke, "Recent Research on Sweetpotato and Cassava in Papua New Guinea," in *Proceedings of the 2nd International Symposium on Sweetpotato and Cassava: Innovative Technologies for Commercialization*, ed. Abd Shukor bin Abd Rahman (International Society for Horticultural Science, 2006), 241–246.
70. R. Michael Bourke, "Environment and Food Production in Papua New Guinea," in Golsom et al., *Ten Thousand Years of Cultivation at Kuk Swamp*, 51–64.
71. Bourke, "Sweet Potato in Papua New Guinea," 46–47.
72. Chris Ballard, "The Wetland Field Systems of the New Guinea Highlands," in Golsom et al., *Ten Thousand Years of Cultivation at Kuk Swamp*, 65–83.
73. Ploeg, "Sweet Potato in the Central Highlands of West New Guinea," 149–162.
74. Bourke and Harwood, *Food and Agriculture in Papua New Guinea*, 168–169.

75. R. Michael Bourke, "The Continuing Ipomoean Revolution in Papua New Guinea," in Ballard et al., *The Sweet Potato in Oceania*, 171–180.
76. Bourke, 171.
77. Bourke and Harwood, *Food and Agriculture in Papua New Guinea*, 139, 132, 194. See also Alison Orr-Ewing, "Papua New Guinea: What People Eat," *Nutrition and Health* 2 (1983): 26–32.
78. Albala, *Food Cultures of the World Encyclopedia*, 3:204–205.
79. Anton Poleg, "Food Imports into Papua New Guinea," *Bijdragen tot de Taal-, Land- en Volkenkunde* 141, nos. 2–3 (1985): 303–322; Philip W. Harvey and Peter F. Heywood, "Twenty-Five Years of Dietary Change in Simbu Province, Papua New Guinea," *Ecology of Food and Nutrition* 13, no. 1 (1983): 27–35.
80. Scott MacWilliam, *Securing Village Life: Development in Late Colonial Papua New Guinea* (Australian University Press, 2013), 66–71; Woolfe, *Sweet Potato*, 539; Bourke and Harwood, *Food and Agriculture in Papua New Guinea*, 21, 133.
81. Leopold Pospisil, *The Kapauku Papuans of West New Guinea* (Holt, Reinhart and Winston, 1963); Watson, "Pigs, Fodder, and the Jones Effect in Postipomoean New Guinea," 67–68.
82. Donald Denoon, "Pacific Edens? Myths and Realities of Primitive Affluence," in Denoon et al., *Cambridge History of the Pacific Islanders*, 99.
83. John Connell, *Papua New Guinea: The Struggle for Development* (Routledge, 1997), 46; Denoon, "Human Settlement," 52–53.
84. David Kavanamur, Charles Yala, and Quinton Clements, *Building a Nation in Papua New Guinea: Views of the Post-Independence Generation* (Pandanus Books, Australian National University, 2003), 271–279. It should be noted that recent data indicate that female life expectancy in Papua New Guinea now surpasses that of males.
85. Sillitoe, "The Gender of Crops in the Papua New Guinea Highlands," 11; Watson, "From Hunting to Horticulture in the New Guinea Highlands," 300.
86. Wade Graham, *Braided Waters: Environment and Society in Molokai, Hawaii* (University of California Press, 2006), 15.
87. Hornell, "How Did the Sweet Potato Reach Oceania?," 41–62; Yen, *The Sweet Potato and Oceania*, 18–19, 337, 339; Green, "Sweet Potato Transfers in Polynesian Prehistory," 43–62.
88. Cf. Denoon, "Human Settlement," 64. There have been different dates regarding the diffusion of the sweet potato in Polynesia. Denoon suggests vaguely that it began in the "initial period of settlement" of the region, whereas Roger Green and D. E. Yen believe that it happened later, between the eleventh and twelfth centuries. Archaeological digs have, however, revealed that cultivation did not take place in Hawaii until 1425. See Yen, *The Sweet Potato and Oceania*, 18–19; Green, "Sweet Potato Transfers in

Polynesian Prehistory," 43–62; Michael W. Kaschko and Melinda S. Allen, "The Impact of the Sweet Potato on Prehistoric Hawaiian Cultural Development," in *Proceedings of the Second Conference in Natural Sciences: Hawaii Volcanoes National Parks*, ed. C. W. Smith (Cooperative National Parks Resources Studies Unit and the University of Hawaii at Manoa, 1978), 178; Macnab, "Sweet Potatoes and Settlement in the Pacific," 219–221; Robert J. Gustafson, Derral R. Herbst, Philip W. Rundel, *Hawaiian Plant Life: Vegetation and Flora* (University of Hawaii Press, 2014), 225.
89. Graham, *Braided Waters*, 16, 14.
90. Jocelyn Linnekin, "Contending Approaches," in Denoon et al., *Cambridge History of the Pacific Islanders*, 14; Valerio Valeri, *Kingship and Sacrifice: Ritual and Society in Ancient Hawaii*, trans. Paula Wissing (University of Chicago Press, 1985), 80.
91. Gustafson et al., *Hawaiian Plant Life*, 226.
92. Valeri, *Kingship and Sacrifice*, 13–15.
93. Patrick Vinton Kirch, *Kua'aina Kahiko: Life and Land in Ancient Kahikinui, Maui* (University of Hawaii Press, 2014), 192–194.
94. Patrick Vinton Kirch and Clive Ruggles, *Heiau, 'Āina Lani: The Hawaii Temple System in Kahikinui and Kaupō, Maui* (University of Hawaii Press, 2019), 30, 17–19; Kirch, *Kua'aina Kahiko*, 194.
95. Handy and Handy, *Native Planters in Old Hawaii*, 340.
96. See Marshall Sahlins, *Islands of History* (University of Chicago Press, 1985), 104–110; Malama Meleisea and Penelope Schoeffel, "Discovering Outsiders," in Denoon et al., *Cambridge History of the Pacific Islanders*, 133. After his botched attempt to kidnap the Hawaiian king, Captain Cook was killed by Hawaiians a few weeks later, though he remained Lono to them afterwards.
97. Handy and Handy, *Native Planters in Old Hawaii*, 13–17; Graham, *Braided Waters*, 36–37.
98. Graham, *Braided Waters*, 37.
99. Graham, 14–31; quotes on 31.
100. Patrick V. Kirch, John Holson, and Alexander Baer, "Intensive Dryland Agriculture in Kaupō, Maui, Hawaiian Islands," *Asian Perspectives* 48, no. 2 (Fall 2009): 265.
101. Graham, *Braided Waters*, 36–43; Denoon, "Pacific Edens?," 86.
102. Kaschko and Allen, "The Impact of the Sweet Potato on Prehistoric Hawaiian Cultural Development," 178.
103. Viliamu Iese et al., "Facing Food Security Risks: The Rise and Rise of the Sweet Potato in the Pacific Islands," *Global Food Security* 18 (2018): 50; Alexander Baer, "Ceremonial Agriculture and the Spatial Proscription of Community: Location Versus Form and Function in Kaupō, Maui, Hawaiian Islands," *Journal of Polynesian Society* 125, no. 3 (September 2016): 301, 292;

Patrick V. Kirch, "Hawaii as a Model System for Human Ecodynamics," *American Anthropologist* 109, no. 1 (March 2007): 8–26; Thegn N. Ladefoged and Michael W. Graves, "Evolutionary Theory and the Historical Development of Dry-Land in North Kohala, Hawai'i," *American Antiquity* 65, no. 3 (July 2000): 423–448; Thegn N. Ladefoged, Michael W. Graves, and James H. Coil, "The Introduction of Sweet Potato in Polynesia: Early Remains in Hawai'i," *Journal of the Polynesian Society* 114, no. 4 (December 2005): 369.

104. Ladefoged and Graves, "Evolutionary Theory and the Historical Development of Dry-Land in North Kohala, Hawai'i," 441; Graham, *Braided Waters*, 21.
105. Kaschko and Allen, "The Impact of the Sweet Potato on Prehistoric Hawaiian Cultural Development," 178.
106. See Thomas S. Dye, "Wealth in Old Hawai'i: Good Year Economics and the Rise of Pristine States," *Archaeology in Oceania* 49 (2014): 61.
107. Dye, 61.
108. Handy and Handy, *Native Planters in Old Hawaii*, 356–357, 349–350.
109. Valeri, *Kingship and Sacrifice*, 10.
110. Valeri, 123; Handy and Handy, *Native Planters in Old Hawaii*, 138.
111. Lilikalā K. Kameʻeleihiwa, *Moʻolelo kaʻao o Kamapuaʻa: A Legendary Tradition of Kamapuaʻa, the Hawaiian Pig-God* (Bishop Museum Press, 1996), x–xiii. See also Harvey Mindess, "Humor in Hawai'i: Past and Present," *Hawaiian Journal of History* 40 (2006): 177–199.
112. Gustafson et al., *Hawaiian Plant Life*, 226.
113. Timothy Earle, *Economic and Social Organization of a Complex Chiefdom: The Halelea District, Kaua'i, Hawaii* (Museum of Anthropology, University of Michigan, 1978), 17.
114. Valeri, *Kingship and Sacrifice*, 11–12.
115. Michael J. Kolb, "Ritual Activity and Chiefly Economy at an Upland Religious Site on Maui, Hawai'i," *Journal of Field Archaeology* 21, no. 4 (Winter 1994): 431.
116. Yen, *The Sweet Potato and Oceania*, 52; Dye, "Wealth in Old Hawai'i," 62.
117. Kolb, "Ritual Activity and Chiefly Economy at an Upland Religious Site on Maui, Hawai'i," 417, 431–432; Earle, *Economic and Social Organization of a Complex Chiefdom*, 11–13.
118. James L. Flexner and Mark D. McCoy, "After the Missionaries: Historical Archaeology and Traditional Religious Sites in the Hawaiian Islands," *Journal of Polynesian Society* 125, no. 3 (September 2016): 308–309; Albala, *Food Cultures of the World Encyclopedia*, 2:178.
119. Dye, "Wealth in Old Hawai'i," 65; Graham, *Braided Waters*, 27–35.
120. Graham, *Braided Waters*, 27, 37–41.
121. Cf. Christina Bacchilega, *Legendary Hawai'i and the Politics of Place: Tradition, Translation and Tourism* (University of Pennsylvania Press, 2006), 162, 182n28.

122. Alfred W. Crosby, "Hawaiian Depopulation as a Model for the Amerindian Experience," in *Epidemics and Ideas: Essays on the Historical Perception on Pestilence*, ed. Terrence Ranger and Paul Slack (Cambridge University Press, 1992), 176.
123. Thomas Dye and others have exchanged their opinions on the demographic and economic changes in precontact Hawaii. See Dye, "Wealth in Old Hawai'i," 59–85.
124. Crosby, "Hawaiian Depopulation as a Model for the Amerindian Experience," 176.
125. Rachel Laudan, *The Food of Paradise: Exploring Hawaii's Culinary Heritage* (University of Hawaii Press, 1996), 5.
126. Quoted in John F. G. Stokes, "Spaniards and the Sweet Potato in Hawaii and Hawaiian-American Contact," *American Anthropologist* 34, no. 4 (October–December 1932): 598.
127. Albala, *Food Cultures of the World Encyclopedia*, 2:180–181.
128. Laudan, *The Food of Paradise*, 78–79.

6. From "Asian Crop" to "African Crop"

1. R. L. Villareal, "Sweet Potato in the Tropics: Progress and Problems," in Villareal and Griggs, *Sweet Potato*, 4.
2. K. T. Mackay, "Foreword," in Mackay et al., *Sweet Potato Research and Development for Small Farmers*, iii.
3. D. Campilan, "Sweetpotato in Southeast Asia: Assessing the Primary Functions of a Secondary Crop," in Loebenstein and Thottappilly, *The Sweetpotato*, 469.
4. Chen, *Jinshu chuanxilu zhongshupu hekan*, 22–23.
5. Campilan, "Sweetpotato in Southeast Asia," 473.
6. William Henry Scott, "A Preliminary Report on Upland Rice in Northern Luzon," *Southern Western Journal of Anthropology* 14, no. 1 (Spring 1958): 87–105.
7. Damasa Magcale-Macandog and Lovereal Joy M. Ocampo, "Indigenous Strategies of Sustainable Farming Systems in the Highlands of Northern Philippines," *Journal of Sustainable Agriculture* 26, no. 2 (2005): 117–138.
8. F. G. Villamayor Jr., "Indigenous Technologies in Sweet Potato Production and Utilization," in Mackay et al., *Sweet Potato Research and Development for Small Farmers*, 329.
9. Campilan, "Sweetpotato in Southeast Asia," 473.

10. Villamayor, "Indigenous Technologies in Sweet Potato Production and Utilization," 331–332; Eetafanio C. Farinas, "The Better Forage Plants Available in the Philippines for Feeding Livestock and Poultry," *Araneta Journal of Agriculture* 14, no. 4 (1967): 234–244.
11. Philippine Statistic Authority, *Statguide for Farmers: Production and Marketing Strategies on Sweet Potato* (Philippine Statistic Authority, May 2021).
12. Campilan, "Sweetpotato in Southeast Asia," 473.
13. Villamayor, "Indigenous Technologies in Sweet Potato Production and Utilization," 331.
14. R. L Vallareal, S. C. Tsou, H. F. Lo, and S. C. Chiu, "Sweet Potato Vine Tips as Vegetables," in Bouwkamp, *Sweet Potato Products*, 176.
15. Department of Agriculture, Regional Field Office No. 02, High Value Crops Development Program (Philippines), *Sweet Potato Production Guide* (Department of Agriculture, November 2012).
16. Truong Van Den, "Sweetpotato Beverages: Product Development and Technology Transfer," in *Sweetpotato Technology for the 21st Century*, ed. Walter A. Hill, Conrad K. Bonsi, and Philip A. Loretan (Tuskegee University, 1992), 389–399.
17. Roberto Soenarjo, "Indigenous Technologies and Recent Advances in Sweet Potato Production, Processing, Utilization, and Marketing in Indonesia," in Mackay et al., *Sweet Potato Research and Development for Small Farmers*, 313–315; Villareal, "Sweet Potato in the Tropics," 5; Yoesti Silvana Arianti and Yos Wahyu Harinta, "Sweet Potatoes: Development and Potential as Alternative Food Ingredients in Karanganyar Regency, Indonesia," *E3S Web of Conferences*, no. 226 (2021): 1–6; Stanley Levy, "Agriculture and Economic Development in Indonesia," *Economic Botany* 11, no. 1 (January–March 1957): 3–39, quote on 14; and Campilan, "Sweetpotato in Southeast Asia," 471–472.
18. Tsuboi, "Nanyō shotō ni okana hareta tabuu sei no hanashi," 319–328.
19. Yan Pieter Karafir, "Sweet Potato in Irian Jaya," in Mackay et al., *Sweet Potato Research and Development for Small Farmers*, 317–322; Arianti and Harinta, "Sweet Potatoes," 1–6; Campilan, "Sweetpotato in Southeast Asia: Assessing the Primary Functions of a Secondary Crop," 471–472.
20. Cf. Campilan, "Sweetpotato in Southeast Asia," 472.
21. Peter Boomgaard, "In the Shadow of Rice: Roots and Tubers in Indonesian History, 1500–1950," *Agricultural History* 77, no. 4 (Autumn 2003), 593–594.
22. J. Wagiono, E. Tuherkih, and D. Pasaribu, "Sweetpotato Cropping Systems at Production Center Areas in Indonesia," in Hil et al., *Sweetpotato Technology for the 21st Century*, 185–194; Karafir, "Sweet Potato in Irian Jaya," 320; Bui Tan Yen, Saskia M. Visser, Chu Thai Hoanh, and Leo Stroosnijder, "Constraints on Agricultural Production in the Northern Uplands of

Vietnam," *Mountain Research and Development* 33, no. 4 (November 2013): 404–415; Campilan, "Sweetpotato in Southeast Asia," 471–472.
23. Boomgaard, "In the Shadow of Rice," 583–595.
24. Boomgaard, 593; Yen, *Sweet Potato and Oceania*, 346
25. Mai Thach Hoanh, "Sweet Potato Development, Selection, and Creation from Seeds in Vietnam," in Mackay et al., *Sweet Potato Research and Development for Small Farmers*, 343.
26. Boomgaard, "In the Shadow of Rice," 585.
27. Colleen Taylor Sen, *Feasts and Fasts: A History of Food in India* (Reaktion Books, 2015), 212–213.
28. Edward Terry, *A Voyage to East-India* (London, 1777), 195–197.
29. John Fryer, *A New Account of East-India and Persia* (London, 1698), 104, 179.
30. William Roxburgh, *Flora Indica, or, Descriptions of Indian Plants*, vol. 1 (Serampore, 1832), 483.
31. George Watt, *The Commercial Products of India* (John Murray, 1908), 1028.
32. Watt, 687.
33. Watt, 688.
34. Sucheta Mazumdar, "The Impact of the New World Food Crops on the Diet and Economy of China and India, 1600–1900," in *Food in Global History*, ed. Raymond Grew (Westview Press, 1984), 61, 70–71.
35. Mazumdar, 71–72. Also see Watt, *The Commercial Products of India*, 1030.
36. Mahe Alam Sorwar, Tanvir Ahmed, Sudhir Chandra Nath, Harum-or-Rashid, and Chris Wheatley, "Analysis of Value Chain of Sweet Potato in Two Districts of Bangladesh," *International Journal of Agricultural Marketing* 2, no. 3 (August 2015): 78–83.
37. Brinder Narula, Vijendra Singh, Sanjay Mulkani, and Thomas John, *The Food of India* (Periplus Editions, 2004), 51; Albala, *Food Cultures of the World Encyclopedia*, 3:93–101.
38. Cf. "Uses of Sweet Potatoes, Shakarkand," Tarladalal.com, accessed May 3, 2025, https://www.tarladalal.com/article/article-uses-of-sweet-potatoes-shakarkand-350/.
39. Cf. "Egyptian Sweet Potato," Mosader.com, December 15, 2020, https://mosader.com/egyptian-sweet-potato/. See also "Sweet Potato," Tridge, accessed April 25, 2025, https://www.tridge.com/intelligences/sweet-potato/EG.
40. Tom Kline, *Street Food: Exploring the World's Most Authentic Tastes* (Dorling Kindersley, 2007), 182.
41. See "Middle East and Africa Sweet Potatoes Market—Industry Trends and Forecast to 2029," Data Bridge Market Research, September 2020, https://www.databridgemarketresearch.com/reports/middle-east-and-africa-sweet

-potatoes-market. Also, Sally Butcher, *New Middle Eastern Street Food: Snacks, Comfort Food, Mezze from Snackistan* (Interlink Books, 2013), 164–189.

42. Woolfe, *Sweet Potato*, 19.

43. Peter T. Ewell, "Sweetpotato Production in Sub-Saharan Africa: Patterns and Key Issues," International Potato Center, accessed May 3, 2025, http://www.sweetpotatoknowledge.org/wp-content/uploads/2016/04/Sweetpotato-Production-in-Sub-Saharan-Africa-Patterns-and-Key-Issues.pdf; "Sweetpotato in Sub-Saharan Africa," International Potato Center, accessed April 25, 2025, https://cipotato.org/sweetpotato-in-sub-saharan-africa/. See also W. J. Grüneberg et al., "Advances in Sweetpotato Breeding from 1992 to 2012," in Low et al., *Potato and Sweetpotato in Africa*, 5.

44. "10 World's Biggest Sweet Potatoes-Producing Countries," Science Agriculture, November 14, 2024, https://www.scienceagri.com/2023/06/10-worlds-biggest-sweet-potatoes.html.

45. M. Akoroda, "Sweetpotato in West Africa," in Loebenstein and Thottappilly, *The Sweetpotato*, 442.

46. See Christopher Ehret, *History and the Testimony of Language* (University of California Press, 2011), 237, 243; Maria Andrade et al., *Unleashing the Potential of Sweetpotato in Sub-Saharan Africa: Current Challenges and War Forward* (International Potato Center, 2009), 26. Ehret, Andrade, and her associates acknowledge that it was during the nineteenth century that most sweet potato varieties reached East Africa.

47. Conklin, "Oceanian-African Hypothesis and the Sweet Potato," 129–133; Kiple and Ornelas, *The Cambridge World History of Food*, 1:246.

48. Stanley B. Alpern, "The European Introduction of Crops into West Africa in Precolonial Times," *History in Africa* 19 (1992): 26.

49. Akoroda, "Sweetpotato in West Africa," 448, 443; Andrade et al., *Unleashing the Potential of Sweetpotato in Sub-Saharan Africa*, 33.

50. O. O. Tewe, F. E. Ojeniyi, and O. A. Abu, *Sweetpotato Production, Utilization and Marketing in Nigeria* (International Potato Center, 2003), viii, 4; I. M. Ahmad, S. A. Makama, V. R. Kiresur, and B. S. Amina, "Efficiency of Sweet Potato Farmers in Nigeria: Potentials for Food Security and Poverty Alleviation," *IOSR Journal of Agriculture and Veterinary Science* 7, no. 9 (September 2014): 1; Abigail Adeyonu, Olubunmi Balogun, Babatunde Ajiboye, Issac Oluwatayo, and Abiodun Otunaiya, "Sweet Potato Production Efficiency in Nigeria: Application of Data Envelopment Analysis," *AIMS Agriculture and Food* 4, no. 3 (August 2019): 673.

51. Kathryn Bergh, Patricia Orozco, Mary Kay Gugerty, and C. Leigh Anderson, "Sweet Potato Value Chain: Nigeria," Evans School of Public Affairs, University of Washington, December 14, 2012, 10.

52. Adeyonu et al., "Sweet Potato Production Efficiency in Nigeria," 673, and Tewe et al., *Sweetpotato Production, Utilization and Marketing in Nigeria*, viii.
53. Bergh et al., "Sweet Potato Value Chain: Nigeria," 2–3.
54. Bergh et al., 14–17; Adeyonu et al., "Sweet Potato Production Efficiency in Nigeria," 672–682.
55. Tewe et al., *Sweetpotato Production, Utilization and Marketing in Nigeria*, viii–ix; Ahmad et al., "Efficiency of Sweet Potato Farmers in Nigeria," 1–5.
56. Cf. Bergh et al., "Sweet Potato Value Chain: Nigeria," 6.
57. Vallareal et al., "Sweet Potato Vine Tips as Vegetables," in Bouwkamp, *Sweet Potato Products*, 176.
58. M. A. As-Saqui, "Sweet Potato and Its Potential Impact in Liberia," in Villareal and Griggs, *Sweet Potato*, 59–62.
59. Andrade et al., *Unleashing the Potential of Sweetpotato in Sub-Saharan Africa*, 75.
60. Hadia Zebib Khanafer presents her cookbook online, including the recipe on sweet potato leave stew in Congo. See Hadia Zebib, "Sweet Potato Leaf Stew, Matembele," Hadia's Lebanese Cuisine, October 13, 2015, https://hadiaslebanesecuisine.com/blog/?p=10504.
61. For the recipe, see "Sweet Potatoes Congolese," Food.com, accessed April 25, 2025, https://www.food.com/recipe/sweet-potatoes-congolese-456144.
62. J. T. Meynhardt and T. G. Joubert, "The Development of Sweet Potato in South Africa," in Villareal and Griggs, *Sweet Potato*, 285–290.
63. J. Low et al., "Sweetpotato in Sub-Saharan Africa," in Loebenstein and Thottappilly, *The Sweetpotato*, 376; Andrade et al., *Unleashing the Potential of Sweetpotato in Sub-Saharan Africa*, 74; "Sweet Potatoes Grow like Weeds," Gardening in South Africa, accessed April 25, 2025, https://www.gardeningin southafrica.co.za/sweet-potatoes-grow-like-weeds.
64. Putri Ernawati Abidin and Ted Nyekanyeka, "Less Hunger, Better Health and More Wealth: the Benefits of Knowledge Sharing in Malawi's Orange-fleshed Sweet Potato Project," Mary Robinson Foundation—Climate Justice, April 16, 2013, https://www.mrfcj.org/pdf/case-studies/2013-04-16-Malawi-OFSP.pdf; "Sweet Returns for Sweet Potato Farmers in Malawi," Consultative Group on International Agricultural Research, September 19, 2020, https://www.cgiar.org/food-security-impact/photo_stories/sweet-returns-for-sweet-potato-farmers-in-malawi/.
65. J. D. Acland, *East African Crops: An Introduction to the Production of Field and Plantation Crops in Kenya, Tanzania, and Uganda* (Longman/FAO, 1971), 204.
66. See Wikipedia, "Sweet Potato," last modified April 15, 2025, 06:41 (UTC), https://en.wikipedia.org/wiki/Sweet_potato.
67. R. K. Jana, "Status of Sweet Potato Cultivation in East Africa and Its Future," in Villareal and Griggs, *Sweet Potato*, 63–68.

68. Woolfe, *Sweet Potato*, 549, 553–557.
69. Amy Fallon, "In Uganda, Finding Out Just How 'Miraculous' Sweet Potatoes Can Be," News Deeply, February 21, 2018, https://deeply.thenewhumanitarian.org/malnutrition/articles/2018/02/21/in-uganda-finding-out-just-how-miraculous-sweet-potatoes-can-be.
70. Ewell, "Sweetpotato Production in Sub-Saharan Africa."
71. Bergh et al., "Sweet Potato Value Chain: Nigeria," 2–3.
72. Acland, *East African Crops*, 204–205.
73. G. Ndamage, "Sweet-Potato Production Potential in Rwanda," *Tropical Root Crops: Production and Uses in Africa*, ed. E. R. Terry, E. V. Doku, O. B. Arene, and N. M. Mahungu (Proceedings of the Second Triennial Symposium of the International Society for Tropical Root Crops—Africa Branch held in Douala, Cameron, August 14–19, 1983), 189–192; Bergh et al., "Sweet Potato Value Chain: Nigeria," 3.
74. Woolfe, *Sweet Potato*, 549; David Gregory Tardif-Douglin, *The Role of Sweet Potato in Rwanda's Food System: The Transition from Subsistence Orientation to Market Orientation* (DAI, 1991), 5–7.
75. Woolfe, *Sweet Potato*, 550–551.
76. Tardif-Douglin, *The Role of Sweet Potato in Rwanda's Food System*, 8.
77. Woolfe, *Sweet Potato*, 549.
78. Brenda Dawson, "How to Cook Sweet Potato Leaves, with Thanks from Ethiopia," Feed the Future Innovation Lab for Horticulture, University of California, Davis, November 28, 2018, https://horticulture.ucdavis.edu/blog/cooking-sweet-potato-leaves-for-nutrition.
79. Tardif-Douglin, *The Role of Sweet Potato in Rwanda's Food System*, 37.
80. "Rwandan Sweet Potato Fries (Chips)," International Cuisine, March 28, 2019, https://www.internationalcuisine.com/rwandan-sweet-potato-fries/, and Nelly Murungi, "Rwandan Company Making Bread and Biscuits from Sweet Potatoes," How We Made It in Africa, December 22, 2020, https://www.howwemadeitinafrica.com/rwandan-company-making-bread-and-biscuits-from-sweet-potatoes/83857/.
81. Dan Jones, Mary Kay Gugerty, and C. Leigh Anderson, "Sweet Potato Value Chain: Ethiopia," Evans School of Public Affairs, University of Washington, January 27, 2013, 1–6; Tinsae Abrham, Hussien Mohammed Beshir, and Ashenafi Haile, "Sweetpotato Production Practices, Constraints, and Variety Evaluation under Different Storage Types," *Food and Energy Security* 10, no. 1 (2021):e263, https://doi.org/10.1002/fes3.263.
82. This data is cited in Wikipedia, "Sweet Potato," last modified April 15, 2025, 06:41 (UTC), https://en.wikipedia.org/wiki/Sweet_potato.
83. Bergh et al., "Sweet Potato Value Chain: Nigeria," 10; Low et al., "Sweetpotato in Sub-Saharan Africa," 361.

84. Tardif-Douglin, *The Role of Sweet Potato in Rwanda's Food System*, 8. Also, Ndamage, "Sweet-Potato Production Potential in Rwanda,", 189–192.
85. Dolores I. Alcober and Leonila S. Parrilla, "Gender Roles in Sweet Potato Production, Processing, and Utilization in Eastern Visayas, the Philippines," in Mackay et al., *Sweet Potato Research and Development for Small Farmers*, 227–236.
86. Bergh et al., "Sweet Potato Value Chain: Nigeria," 11–12; Jones et al., "Sweet Potato Value Chain: Tanzania," 8; Claire Kpaka, Mary Kay Gugerty, and C. Leigh Anderson, "Sweet Potato Value Chain: Uganda," Evans School of Public Affairs, University of Washington, January 23, 2013, 5–6; Low et al., "Sweetpotato in Sub-Saharan Africa," 359.
87. Low et al., "Sweetpotato in Sub-Saharan Africa," 359; Mackay, "Foreword," iii; Nerelito P. Pascual, Antonio P. Abamo, and Ma Salome G. Binongo, "Economic Tests for Profitability, Marketability, and Alternative Uses of Sweet Potato in the Philippines," in Mackay et al., *Sweet Potato Research and Development for Small Farmers*, 255; M. M. Rashid, "Indigenous Technologies and Recent Advances in Sweet Potato Production, Processing, Utilization, and Marketing in Bangladeshi," in Mackay et al., *Sweet Potato Research and Development for Small Farmers*, 287; Woolfe, *Sweet Potato*, 491.
88. Woolfe, *Sweet Potato*, 503, 505.
89. Jones et al., "Sweet Potato Value Chain: Tanzania," 4–5; Kpaka et al., "Sweet Potato Value Chain: Uganda," 3–4; Bergh et al., "Sweet Potato Value Chain: Nigeria," 6.
90. When Bede N. Okigbo gave his presidential address at a symposium held by the International Society for Tropical Root Crops—Africa Branch in Cameron in 1983, he recognized that root crops grown in Africa had low yields. Terry et al., *Tropical Root Crops*, 16. The issue, however, has not been addressed effectively since then.
91. Ewell, "Sweetpotato Production in Sub-Saharan Africa"; Grüneberg et al., "Advances in Sweetpotato Breeding from 1992 to 2012," in Low et al., *Potato and Sweetpotato in Africa*, 21; Acland, *East African Crops*, 206. An experiment conducted in Rwanda demonstrated that the use of manure increased sweet potato yields; even still, the method is not widely practiced. See Ndamage, "Sweet-Potato Production Potential in Rwanda," Terry et al., *Tropical Root Crops*, 189–192.
92. Andrade et al., *Unleashing the Potential of Sweetpotato in Sub-Saharan Africa*, 106.
93. Tardif-Douglin, *The Role of Sweet Potato in Rwanda's Food System*, 87–99.
94. Andrade et al., *Unleashing the Potential of Sweetpotato in Sub-Saharan Africa*, 1–42; Ahmad et al., "Efficiency of Sweet Potato Farmers in Nigeria," 1–5; Woolfe, *Sweet Potato*, 549; Jan Low, "Preface," in Low et al., *Potato and Sweetpotato in Africa*, xxiii.

Epilogue

1. See Fu Lili, "Woguo jiang qidong tudou zhulianghua zhanlue" [China to launch white potato stapleization strategy], *Jinri keyuan* [Modern science] 2 (2015): 14–17. About the white potato's introduction to China and its reception, see Zhang, *Xindalu nongzuowu de chuanbo he yiyi*, and Ping-ti Ho, "The Introduction of American Food Plants Into China."
2. See "Biena tudou budang liangshi, tudou zhulianghua de guojia zhanlue" [Don't treat white potatoes as insignificant, the national strategy behind the stapleization of white potatoes], Tech-food.com, accessed March 18, 2023, https://www.tech-food.com/news/detail/n1175944.htm.
3. See Sun Junmao, "Zhongguo nongkeyuan juban malingshu zhulianghua fazhan zhanlue yantaohui" [Chinese Academy of Agricultural Sciences held a seminar on white potato stapleization development strategy], Institute of Food Science and Technology, Chinese Academy of Agricultural Sciences, January 16, 2025, https://ifst.caas.cn/xwzx/mtbd/208097.htm; "Zhongguo shixian tudou 'zhulianghua' xuyao shijian" [China's realization of stapleization of the white potato needs time], *Cankao xiaoxi* [Reference news], February 6, 2015.
4. Lu Xiaoping and Xie Kaiyun, *Guoji malingshu zhongxin zai zhongguo: 30nian youyi, hezuo yu chengjiu* [The International Potato Center in China: Friendship, collaboration and accomplishments over the past thirty years] (Zhongguo nongye kexue jishu chubanshe, 2014), 16–21. In their article, Jia Lian Sheng, Qi Han Xue, and Da Peng Zhang state that sweet potato cultivation reached its peak in China in 1971, with its sown area expanded to as much as 12,000,000 hectares. "Sweetpotato Breeding, Production and Utilization in China," in Hill et al., *Sweetpotato Technology for the 21st Century*, 153–154.
5. Feng Hua, "Rang tudou dang zhuliang, bushi yinwei zan queliang" [Treat white potatoes as staple food not because we lack food], *Renmin ribao* [People's daily], January 16, 2015.
6. See "New CCCAP Facilities in China Mark New Era for Root and Tuber Research in Asia," International Potato Center, accessed March 17, 2023, https://cipotato.org/pressreleases/new-cccap-facilities-china-mark-new-era-root-tuber-research-asia/.
7. Reader, *Potato*, 267; Earle, *Potato*, 62.
8. "Biena tudou budang liangshi, tudou zhulianghua de guojia zhanlue."
9. Zhang Gengchang, "Henan: Nancheng tudou zhulianghua tuiguang zhongdian quyu" [It is difficult to make Henan a principal region for making the stapleization of white potatoes], *Zhongguo shipin* [China food] 3 (2015): 46.
10. Lu and Xie, *Guoji malingshu zhongxin zai zhongguo*, 23–24.

11. Wang Jinqiu and Wu Shunchen, "Malingshu zhulianghua zhanlue de dongli, zhang'ai yu qianjing" [The motive, obstacle, and prospect of the stapleization of the white potato], *Nongye jingji* [Agricultural economy] 4 (2018): 17–19; "Tudou dang zhuliang, weihe chichi buneng shixian? zhendeshi yinwei bei chengwei 'pinmin shiwu?'" (Why has the stapleization of the white potato taken so long to realize? Is it really because of its being "poor people's food?"), NetEase, October 15, 2020, https://www.163.com/dy/article/FOVRD2P205327JMJ.html.
12. Sun Jie, "Tudhou zhulianghua, qianlu reng manchang" [It is a long way to go to make white potato a staple food], *Zhongguo nongcun keji* [China rural science and technology] 244 (September 2015): 34–37.
13. See the Happy Sweet Potato company website at https://www.klfspp.com/index.html.
14. Woolfe, *Sweet Potato*, 119.
15. Barry R. Duell, "Sweetpotato Product Innovations by Small Businesses in Kawagoe, Japan," in Hill et al., *Sweetpotato Technology for the 21st Century*, 381–388.
16. See the website of the Japan Tsubo Yaki-imo Association at https://tsubo-yakiimo.or.jp/.
17. See Zheng Xinling, "Nanyi fanshu" [How I miss the sweet potato!], *Yecheng* (Haikou) 10 (2008): 52; Wen Xiujiao, "Guzaowei, fanshu qing: You 'Taiwan' wei de Xinhua fanshujie" [The nostalgia and emotion of sweet potatoes: The sweet potato festival in Sinhua], *Fengnian* [Harvest] 49, no. 9 (May 1999): 20–21; Jilu, "Jinmen wenhuajie, fanshu dang zhujue" [Sweet potatoes are the main actor at Kinmen's cultural festival], *Beijing dang'an* [Beijing archives] 2 (1999): 42; Liu Shuangzhang, "Meinong fanshu, yiku sitian" [Meinong's sweet potato, a sweet and sour memory], *Kejia* (Hakka) 209 (November 2011): 73–74; and Yang Potato company website at https://yangpotato.com.tw/.
18. "Korean Recipes Made with Sweet Potato," Maangchi, accessed April 30, 2025, https://www.maangchi.com/made-with/sweet-potato.
19. Maangchi, *Maangchi's Real Korean Cooking: Authentic Dishes for the Home Cook* (Houghton Mifflin Harcourt, 2015), and *Maangchi's Big Book of Korean Cooking*.
20. Debbie Wolfe, "Korea's Love Affair with Sweet Potatoes," Koreabridge.net, October 9, 2015, https://koreabridge.net/post/korea%E2%80%99s-love-affair-sweet-potatoes-crazykoreancooking.
21. Mark Bittman, "Fast, Good and Good for You," *New York Times Magazine*, April 7, 2013; "Sweet Potatoes," Agricultural Marketing Resource Center, November 2021, https://www.agmrc.org/commodities-products/vegetables/sweet-potatoes.

22. See "What Fast Food Restaurants Sell Sweet Potato Fries?," Quora, accessed May 5, 2025, https://www.quora.com/What-fast-food-restaurants-sell-sweet-potato-fries; "You Can Now Order Sweet Potato Fries at McDonald's," *Men's Health*, June 13, 2018, https://www.menshealth.com/uk/nutrition/a759191/you-can-now-order-sweet-potato-fries-at-mcdonalds/.
23. Bill Krueger, "Can Sweet Potatoes Save the World?," NC State News, April 2019, https://news.ncsu.edu/2019/04/sweet-potato-industry-research-yencho/; "What Fast Food Restaurants Sell Sweet Potato Fries?," Quora, accessed May 5, 2025, https://www.quora.com/What-fast-food-restaurants-sell-sweet-potato-fries.
24. From 2009, the fast-food giant has been seeking to use "a greener potato," but it remains a challenge. See Bruce Watson, "McDonald's Prepares to Switch Its Fries to a Greener Potato," *Daily Finance*, September 23, 2009, https://web.archive.org/web/20151127081449/http://www.dailyfinance.com/2009/09/23/mcdonalds-prepares-to-switch-its-fries-to-a-greener-potato/.
25. See, for example, Mohammad Khairul Alam, "A Comprehensive Review of Sweet Potato (Ipomoea batatas [L.] Lam): Revisiting the Associated Health Benefits," *Trends in Food Science & Technology* 115 (September 2021): 512–529.
26. Rachel Meltzer Warren, "Are Sweet Potatoes Good for You?," *Consumer Reports*, October 18, 2024, https://www.consumerreports.org/nutrition-healthy-eating/are-sweet-potatoes-good-for-you-a9373907274/; and "Hongshu weishenme buneng zhulianghua?" [Why cannot sweet potatoes become a staple food?], Sohu.com, September 2, 2022, https://www.sohu.com/a/582049480_593212.
27. "Sweet Potato Nutrition and Health Benefits," Mouthpower.org, accessed April 30, 2025, https://www.mouthpower.org/nutrition/sweet-potato-nutrition-and-health-benefits/.
28. Stephanie Booth, "Health Benefits of Sweet Potatoes," WebMD, July 17, 2023, https://www.webmd.com/food-recipes/benefits-sweet-potatoes.
29. Gloria Peace Lamaro et al., "Essential Mineral Elements and Potentially Toxic Elements in Orange-Fleshed Sweet Potato Cultivated in Northern Ethiopia," *Biology* 12, no. 2 (February 2023): 266–275.
30. "Sweet Potatoes: Six Health Benefits," Helen G. Nassif Community Cancer Center, August 28, 2019, https://www.communitycancercenter.org/nutrition/sweet-potatoes-six-health-benefits/; Warren, "Are Sweet Potatoes Good for You?"
31. Alexandra A. Bennett et al., "Untargeted Metabolomics of Purple and Orange-Fleshed Sweet Potatoes Reveal a Large Structural Diversity of Anthocyanins and Flavonoids," *Scientific Reports* 11, no. 1 (August 12, 2021), 1–13; Mouthpower.org, "Sweet Potato Nutrition and Health Benefits";

Warren, "Are Sweet Potatoes Good for You?"; Junjian Ran et al. "Optimization of DHEA Extraction from Sweet Potato Pomace by Ultrasonic-Microwave Synergistic Employing Response Surface Methodology," *Journal of AOAC International* 102, no. 2 (March 2019): 680–682.

32. Woolfe, *Sweet Potato*, 9, 55, 121, 475.
33. "Sweet Potatoes," Nutrition Source, Harvard T. H. Chan School of Public Health, accessed April 30, 2025, https://www.hsph.harvard.edu/nutrition source/food-features/sweet-potatoes/.
34. S. C. S. Tsou and R. L. Villareal, "Resistance to Eating Sweet Potato," in Griggs and Villareal, *Sweet Potato: Proceedings of the First International Symposium*, 37–44; Woolfe, *Sweet Potato*, 491.
35. David M. Bodjreno, Xin Li, Xiaodan Lu, Suzhen Lei, Baodong Zheng, and Honliang Zeng, "Resistant Starch from Sweet Potatoes: Recent Advancements and Applications in the Food Sector," *International Journal of Biological Macromolecules* 225 (January 2023): 13–26.
36. Esther Ngumbi, "Sweet Potato Takes: 3 Odes, and 1 Hit Piece, on the Iconic Tuber by African Writers," *Goats and Soda* (blog), NPR, November 25, 2021, https://www.npr.org/sections/goatsandsoda/2018/11/21/669608296/sweet-potatoes-its-a-thin-line-between-love-and-hate. This complex attitude has by no means developed only in Africa; it also appeared across many parts of Asia where the root plant was/is a staple food.

Selected Bibliography

Acland, J D. *East African Crops: An Introduction to the Production of Field and Plantation Crops in Kenya*. Longman/FAO, 1971.

Acosta, Joseph de. *The Natural and Moral History of the Indies*. Edited by Clements R. Markham. Cambridge University Press, 2009.

Acton, Eliza. *Modern Cookery for Private Families*. New ed. London, 1897.

Albala, Ken. *Eating Right in the Renaissance*. University of California Press, 2002.

Albala, Ken, ed. *Food Cultures of the World Encyclopedia*. Greenwood, 2011.

Anderson, Aeneas. *An Accurate Account of Lord Macartney's Embassy to China*. London, 1797.

Anderson, Eugene. *The Food of China*. Yale University Press, 1988.

Andrade, Maria, Ian Barker, Donald Cole et al., eds. *Unleashing the Potential of Sweetpotato in Sub-Saharan Africa: Current Challenges and War Forward*. International Potato Center, 2009.

Aoki, Konyō. *Bansho kō* [Sweet potato study]. N.p., 1735. Available in the digital collections of the National Diet Library, Tokyo.

Aram, Bethany, and Yun-Casalilla, Bartolomé. *Global Goods and the Spanish Empire, 1492–1824*. Palgrave Macmillan, 2014.

Arizono, Shōichirō. *Kinsei shomin on nichijōshoku: Hyakushō wa kome o taberarenakatta* [Daily foods of the modern commoners: Did ordinary people eat rice?]. Kaiseisha, 2007.

Baker, James W. *Thanksgiving: The Biography of an American Holiday*. University of New Hampshire Press, 2009.

Ballard, Chris, Paula Brown, R. Michael Bourke, and Tracy Harwood, eds. *The Sweet Potato in Oceania: A Reappraisal.* University of Sydney, 2005.

Barrow, John. *Travels in China: Containing Descriptions, Observations and Comparisons, Made and Collected in the Course of a Short Residence at the Imperial Palace of Yuen-min-yuen, and on a Subsequent Journey Through the Country from Pekin to Canton.* Cambridge University Press, 2010.

Beaglehole, J. C. *The Voyage and Resolution of the Discovery 1776–1780: The Journals of James Cook on His Voyage of Discovery.* Cambridge University Press, 1967.

Best, Eldsen. *Maori Agriculture.* Dominion Museum, 1925.

Bleier, Edward. *The Thanksgiving Ceremony: New Traditions for America's Family Feast.* Crown Publishers, 2003.

Blue, Kathryn K., and Anthony Dias Blue. *Thanksgiving Dinner.* HarperPerennial, 1990.

Boserup, Ester. *The Conditions of Agricultural Growth: The Economics of Agrarian Change Under Population Pressure.* Routledge, 2003. Originally published in 1965 by Cambridge University Press.

Boserup, Ester. *Population and Technological Change: A Study of Long-Term Trends.* University of Chicago Press, 1981.

Bourke, R. Michael, and Tracy Harwood, eds. *Food and Agriculture in Papua New Guinea.* ANU Press, 2009.

Bouwkamp, John C. *Sweet Potato Products: A Natural Resource for the Tropics.* 2nd ed. CRC Press, 2018.

Bradley, Richard. *The Country Housewife and Lady's Director.* London, 1732.

Brears, Peter. *Cooking and Dining in Tudor and Early Stuart England.* Prospect Books, 2015.

Breverton, Terry. *Tudor Kitchen: What the Tudors Ate and Drank.* Amberley Publishing, 2015.

Briggs, Richard. *The English Art of Cookery.* London, 1791.

Buck, John Lossing. *Land Utilization in China.* University of Chicago Press, 1937.

Butcher, Sally. *Snackistan: Street Food, Comfort Food, Meze.* Pavilion, 2013.

Cai, Chenghao, and Yunping Yang. 2004. *Taiwan fanshu wenhuazhi* [Cultural atlas of the sweet potato in Taiwan]. Guoshi chuban, 2004.

Campbell, Jodi. *At First Table: Food and Social Identity in Early Modern Spain.* University of Nebraska Press, 2017.

Cao, Shuji. *Zhongguo renkoushi—Ming shiqi* [A history of Chinese population—the Ming period], vols. 4–5. Fudan daxue chubanshe, 2000.

Chamberlain, Basil Hall. *Things Japanese: Being Notes on Various Subjects Connected with Japan.* Stone Bridge Press, 2007.

Chanca, Diego Álvarez. *Letter of Dr. Chanca on the Second Voyage of Columbus*. Document No. AJ-065, American Journeys Collection, Wisconsin Historical Society, Digital Library and Archives, 2003.

Chen, Shiyuan, ed. *Jinshu chuanxilu zhongshupu hekan* [On the cultivation of the sweet potato and manuals for sweet potato cultivation]. Nongye chubanshe, 1982.

Chen, Shuping, ed. *MingQing nongyeshi ziliao, 1368–1911* [Historical sources of Ming and Qing agriculture, 1368–1911], vol. 1. Shehui kexue wenxian chubanshe, 2013.

Chen, Ying. *Haicheng Xianzhi—fengtu* [Gazetteer of Haicheng County—local customs]. N.p., 1762.

Chong, Yagyong [Jeong Yakyong]. *Admonitions on Governing the People*. Translated by Choi Byonghyon. University of California Press, 2010.

Chūman, Katsumi. *Nihon kansho saibaishi: Kansho no denrai kara denpa, sono saibaihō no hensen* [Japanese sweet potato chronicles: From introduction and spread to the evolution of cultivation methods]. Taki shobō, 2002.

Civitello, Linda. *Cuisine and Culture: A History of Food and People*. 2nd ed. John Wiley and Sons, 2008.

Climent-Espino, Rafael, and Ana M. Gomez-Bravo, eds. *Food, Text and Cultures in Latin America and Spain*. Vanderbilt University Press, 2020.

Clusius, Carolus [Caroli Clvsii Atrebat]. *Rariorum aliquot stirpium per Hispaniam observatorum Historia*. Antwerp, 1576.

Cocks, Richard. *Diary of Richard Cocks: Cape-Merchant in the English Factory in Japan, 1615–1622*. Edited by Edward Maunde Thompson. Cambridge University Press, 2010.

Coe, Sophie D. *America's First Cuisines*. University of Texas Press, 1994.

Collingham, Lizzie. *The Taste of War: World War II and the Battle for Food*. Penguin Press, 2012.

Columbus, Christopher. *The Voyage of Christopher Columbus: Columbus' Own Journal of Discovery*. Translated by John Cummins. St. Martins, 1992.

Crosby, Alfred. 2003. *The Columbian Exchange: Biological and Cultural Consequences of 1492*. 30th anniv. ed. Praeger, 2003.

d'Anghiera, Peter Martyr. *De Orbe Novo: The Eight Decades of Peter Martyr d'Anghiera*. G. P. Putnam's Sons, 1912.

Dawson, Thomas. *The Good Huswifes Jewell*. London, 1596.

DeWitt, Dave. *The Founding Foodies: How Washington, Jefferson and Franklin Revolutionized American Cuisines*. Sourcebooks, 2010.

DeWitt, Dave. *Precious Cargo: How Foods from the Americas Changed the World*. Counterpoint, 2014.

Diamond, Jared. *Guns, Germs and Steel: The Fates of Human Societies*. W. W. Norton, 1997.

Díaz del Castillo, Bernal. *Historia verdadera de la conquista de la Nueva España*. Introduction and notes by Miguel Le.n-Portilla. Himali, 2018. Originally published in 1632.

Dott, Brian. *The Chile Pepper in China: A Culinary Biography*. Columbia University Press, 2020.

Earle, Rebecca. *The Body of the Conquistador: Food, Race and the Colonial Experience in Spanish America, 1492–1700*. Cambridge University Press, 2012.

Earle, Rebecca. *Potato*. Bloomsbury Academic, 2019.

Ehret, Christopher. *History and the Testimony of Language*. University of California Press, 2011.

Ellis, William. *Polynesian Researches, During a Residence of Nearly Six Years in the South Sea Islands Including Descriptions of the Natural History and Scenery of the Islands with Remarks on the History, Mythology, Traditions, Government, Arts, Manners, and Customs of the Inhabitants*. London, 1829.

Elvin, Mark. *The Pattern of the Chinese Past*. Stanford University Press, 1973.

Fairbank, John K., and Edwin O. Reischauer. *China: Tradition and Transformation*. Houghton Mifflin, 1978.

Fernández-Armesto, Felipe. *Near A Thousand Tables: A History of Food*. Free Press, 2002.

Fernández-Armesto, Felipe. *Pathfinders: A Global History of Explorations*. Oxford University Press, 2006.

Forster, John. 1664. *Englands Happinesse Increased, Or a Sure and Easie Remedy against All Succeeding Dear Years by A Plantation of the Roots Called Potatoes*. 1664. Early English Books Online, Michigan Library Digital Collections. Accessed March 14, 2022. https://quod.lib.umich.edu/e/eebo/A40002.0001.001?view=toc.

Fortune, Robert. *Yedo and Peking: A Narrative of a Journey to the Capitals of Japan and China*. London, 1863.

Fryer, John. *A New Account of East-India and Persia*. London, 1698.

Ge, Jianxiong. *Zhongguo renkou fazhanshi* [A history of the development of Chinese population]. Fujian renmin chubanshe, 1991.

Ge, Jianxiong, Shuji Cao, and Songdi Wu. *Jianming Zhongguo yiminshi* [Concise history of Chinese immigration]. Fujian renmin chubanshe, 1993.

Gentilcore, David. *Food and Health in Early Modern Europe: Diet, Medicine and Society, 1450–1800*. Bloomsbury Academic, 2016.

Gerard, John. *Gerard's Herball—Or, Generall Historie Of Plantes*. London, 1597.

Glasse, Hannah. *The Art of Cookery Made Plain and Easy*. 5th ed. London, 1755.

Goody, Jack. *Cooking, Cuisine and Class: A Study in Comparative Sociology*. Cambridge University Press, 1982.

Goucher, Candice. *Congotay! Congotay! A Global History of Caribbean Food*. M. E. Sharpe, 2014.

Granado, Diego. *Libro del arte de cozina*. Lérida, 1614.

Guppy, H. B. *The Observations of a Naturalist in the Pacific Between 1896 and 1899.* Macmillan, 1906.

Hale, Sarah Josepha. *The Good Housekeeper.* Boston, 1839.

Handy, E. S. Craighill, and Elizabeth Green Handy. *Native Planters in Old Hawaii: Their Life, Lore and Environment.* Bishop Museum Press, 1978.

Hane, Mikiso. *Peasants, Rebels, Women and Outcasts: The Underside of Modern Japan.* Rowman and Littlefield, 2003.

Harris, Jessica B. *High on the Hog: A Culinary Journey from Africa to America.* Bloomsbury, 2011.

Harrison, William. *Elizabethan England.* Edited by Lothrop Withington. Introduction by F. J. Furnivall. Walter Scott Publishing, 1945. Originally published in 1577.

Hatch, Peter J. *Thomas Jefferson's Revolutionary Garden at Monticello.* Yale University Press, 2012.

Hayami, Akira. *Japan's Industrious Revolution: Economic and Social Transformations in the Early Modern Period.* Springer Japan, 2015.

Hayward, Vicky. *New Art of Cookery: A Spanish Friar's Kitchen Notebook by Juan de Altamiras.* Rowman and Littlefield, 2017.

Hernández, Francisco. *Historia de las Plantas de Nueva España.* Imprenta Universitaria, 1943.

Heyerdahl, Thor. *Kon-Tiki: Across the Pacific in a Raft.* Translated by F. H. Lyon. Rand McNally and Company, 1950.

Ho, Ping-ti. "The Introduction of American Food Plants Into China." *American Anthropologist* 57, no. 2 (1955): 191–201.

Ho, Ping-ti. *Studies on the Population of China, 1368–1953.* Harvard University Press, 1959.

Holmes, Samuel. *The Journal of Mr. Samuel Holmes, One of the Guard on Lord Macartney's Embassy to China and Tartary.* Cambridge University Press, 2010.

Horton, Douglas. *Underground Crops: Long-Term Trends of Production in Roots and Tubers.* Winrock International, 1988.

Huang, Philip C. C. *The Peasant Economy and Social Change in North China.* Stanford University Press, 1985.

Ishige, Naomichi. *The History and Culture of Japanese Food.* Routledge, 2001.

Itō, Shōji. *Satsuma imo to Nihon jin: Wasure rareta shoku no ashiato* [Sweet potato and the Japanese: Forgotten footprints of food]. PHP Kenkyūjo, 2010.

Jett, Stephen C. *Ancient Ocean Crossings: Reconsidering the Case for Contacts with the Pre-Columbian Americas.* University of Alabama Press, 2017.

Jiang, Tao. *Renkou yu lishi: Zhongguo chuantong renkou jiegou yanjiuu* [Population and history: A study of traditional demographic structure in China]. Renmin chubanshe, 1998.

Jiang, Tao. *Zhongguo jindai renkoushi* [Modern history of Chinese population]. Zhejiang renmin chubanshe, 1993.

Johns, Terry L., and Alice A. Storey. *Polynesians in America: Pre-Columbian Contacts with the New World.* AltaMira Press, 2011.

Jones, Eric L. *The European Miracle: Environments, Economies and Geopolitics in the History of Europe and Asia.* 3rd ed. Cambridge University Press, 2003.

Jones, Michael Owen. *Corn: A Global History.* Reaktion Books, 2017.

Joya, Mock. *Japan and Things Japanese.* Kegan Paul, 2006.

Kamen, Henry. *Spain's Road to Empire: The Making of a World Power, 1492–1763.* Penguin Books, 2003.

Kehoe, Alice Beck. *Traveling Prehistoric Seas: Critical Thinking on Ancient Transoceanic Voyages.* Routledge, 2015.

Kerr, George H. *Okinawa: The History of an Island People.* Rev. ed. Tuttle Publishing, 2000.

Kime, Tom. *Street Food: Exploring the World's Most Authentic Tastes.* Dorling Kindersley, 2007.

Kraig, Bruce, and Colleen Taylor Sen, eds. *Street Food Around the World: An Encyclopedia of Food and Culture.* ABE-CLIO, 2013.

Kremer, Gary R. *George Washington Carver: A Biography.* Greenwood, 2011.

Las Casas, Bartolomé de. *Historia de las Indias.* Fondo de Cultura Económica, 1951.

Laudan, Rachel. *Cuisine and Empire: Cooking in World History.* University of California Press, 2013.

Lebot, Vincent. *Tropical Root and Tuber Crops: Cassava, Sweet Potato, Yams and Aroids.* CABI, 2009.

Lee, James Z., and Cameron D. Campbell. *Fate and Fortune in Rural China: Social Organization and Population Behavior in Liaoning, 1774–1873.* Cambridge University Press, 1997.

Lee, James Z., and Feng Wang. *One Quarter of Humanity: Malthusian Mythologies and Chinese Realities, 1700–2000.* Harvard University Press, 1999.

Lewis, Edna. *The Taste of Country Cooking.* Alfred A. Knopf, 2006.

Lewis, Edna, and Scott Peacock. *The Gift of Southern Cooking: Recipes and Revelations from Two Great Southern Cooks.* Alfred A. Knopf, 2003.

Li, Bozhong. *Agricultural Development in Jiangnan, 1620–1850.* St. Martin's Press, 1998.

Li, Wei. *Qingdai liangshi duanque yu dongbei tudi kaifa* [Food shortage and the northeastern immigration in the Qing]. Jilin renmin chubanshe, 2011.

Lin, Liyue. "Fengsu yu zuiqian: Mingdai de ninü jixu jiqi wenhua yihan" [Customs and crimes: The narrative and cultural meaning of female infanticide in the Ming]. In *Wusheng zhisheng: Jindai Zhongguo de funü yu shehui 1600–1950* [Unheard voices: Women and society in modern China, 1600–1950], vol. 2, edited by You Jianming. Zhongyang yanjiuyuan jindaishi yanjiusuo, 2003.

Liu, Xitao. *Fujian lishi dili yanjiu* [Study of Fujian historical geography]. Fujian jiaoyu chubanshe, 2017.

Loebenstein, Gad, and George Thottappilly, eds. *The Sweetpotato*. Springer, 2009.

López de Gómara, Francisco. *Historia general de las Indias, y vida de Hernán Cortés*. Biblioteca Ayacucho, 1977.

Low, Jan, Moses Nyongesa, Sara Quinn, and Monica Parker, eds. *Potato and Sweetpotato in Africa: Transforming the Value Chains for Food and Nutrition Security*. CABI, 2015.

Lu, Xiaoping, and Kaiyun Xie. *Guoji malingshu zhongxin zai zhongguo: 30nian youyi, hezuo yu chengjiu* [International Potato Center in China: Friendship, collaboration, and accomplishments over the past thirty years]. Zhongguo nongye kexue jishu chubanshe, 2014.

Maangchi. *Maangchi's Real Korean Cooking: Authentic Dishes for the Home Cook*. Houghton Mifflin Harcourt, 2015.

Macartney, George. *An Embassy to China: Being the Journal Kept by Lord Macartney during his Embassy to the Emperor Ch'ien-lung, 1793–1794*. Edited by J. L. Cranmer-Byng. Longmans, 1972.

Mackay, Kenneth T., Manuel K. Palomar, and Rolinda T. Sanico, eds. *Sweet Potato Research and Development for Small Farmers*. SEAMEO-SEARCA, 1989.

Maddison, Angus. *China's Economic Performance in the Long Run*. 2nd ed. OECD, 2007.

Mair, Victor H. *Contact and Exchange in the Ancient World*. University of Hawaii Press, 2005.

Mann, Charles C. *1493: Uncovering the New World Columbus Created*. Alfred A. Knopf, 2013.

Markham, Gervase. *The English Huswife: Containing the Inward and Outward Virtues Which Ought to Be in a Complete Woman*. London, 1623.

Martínez Montiño, Francisco. *Arte de cocina, pastelería, vizcochería y conservería*. Barcelona, 1763.

Mata, Juan de la. *Arte de repostería*. Madrid, 1791.

Matsuoka, Jo'an. *Bansho roku* [Sweet potato book]. N.p., 1717. Available in the digital collections of the National Diet Library, Tokyo.

Mazumdar, Sucheta. "The Impact of New World Food Crops on the Diet and Economy of China and India, 1600–1900." In *Food in Global History*, edited by Raymond Grew. Westview, 2000.

Mazumdar, Sucheta. *Sugar and Society in China: Peasants, Technology and the World Market*. Harvard University Press, 1998.

Merrill, Elmer D. *The Botany of Cook's Voyages and Its Unexpected Significance in Relation to Anthropology, Biogeography, and History*. Chronica Botanica Co., 1954.

Mintz, Sydney W. *Sweetness and Power: The Place of Sugar in Modern History*. Penguin Books, 1985.

Miyamoto, Tsuneichi. *Kansho no rekishi* [History of the sweet potato]. Mirai sha, 1962.

Monardes, Nicolás. *Joyfull Newes out of the Newe Founde Worlde.* London, 1577.

Monardes, Nicolás Bautista. *Primera y segunda 1580.* Ministero de Sanidad y Consumo, 1989. Originally published in 1580.

Moreno, Maria Paz. *Madrid: A Culinary History.* Rowman and Littlefield, 2018.

Moss, Sarah, and Alexander Badenoch. *Chocolate: A Global History.* Reaktion Books, 2009.

Muffet, Thomas. *Healths Improvement, or Rules Comprising and Discovering the Nature, Method and Manner of Preparing All Sorts of Food in This Nation.* Corrected and enlarged by Christopher Bennet. London, 1655.

Nadeau, Carolyn. *Food Matters: Alonso Quijano's Diet and the Discourse of Food in Early Modern Spain.* University of Toronto Press, 2016.

Narula, Brinder, Vijendra Singh, Sanjay Mulkani, and Thomas John. *The Food of India.* Periplus Editions, 2004.

Ogawa, Yoshiyuki. "Shindai Kōsei, Fukken ni okeru 'dekijo' shūzoku to hō ni tsuite: 'Atsu yome' 'ton yan si' nado no shūzoku to no kankei o megutte" [Custom and law of female infanticide in Jiangxi and Fujian during the Qing: The customs of "expensive dowry" and "child bride" and their relations with infanticide]. In *Chūgoku kinsei no kihan to chitsujo* [Rule and order in early modern China], edited by Yamamoto Eishi. Tōyō bunko, 2014.

Ohnuki-Tierney, Emiko. *Rice as Self: Japanese Identities Through Time.* Princeton University Press, 1993.

Opie, Frederick Douglass. *Hog and Hominy: Soul Food from Africa to America.* Columbia University Press, 2008.

Oviedo, Fernandez de Gonzalo. *Natural History of the West Indies.* Translated by Sterling Stoudemire. University of North Carolina Press, 1959.

Oviedo, Fernández de Gonzalo. *Sumario de la natural historia de las Indias.* Edited by Álvaro Baraibar. Iberoamericana-Vervuert, Universidad de Navarra, 2010.

Pan, Guisheng. *Chen Hongmou zhuan* [Biography of Chen Hongmou]. Guangxi renmin chubanshe, 2007.

Parkinson, John. *Paradisi in sole paradisus terrestris. or A garden of all sorts of pleasant flowers which our English ayre will permitt to be noursed vp with a kitchen garden of all manner of herbes, rootes, & fruites, for meate or sause vsed with vs, and an orchard of all sorte of fruitbearing trees and shrubbes fit for our land together with the right orderinge planting & preseruing of them and their vses & vertues collected by Iohn Parkinson apothecary of London 1629.* London, 1629.

Perez Garcia, Manuel. "Challenging National Narratives: On the Origins of Sweet Potato in China as Global Commodity during the Early Modern Period." In *Global History and New Polycentric Approaches: Europe, Asia and the Americas in the World Network System,* edited by Manuel Perez Garcia and Lucio de Sousa. Springer Nature, 2018.

Perkins, Dwight. *Agricultural Development in China, 1368–1968.* Aldine Publishing Company, 1969.
Pettid, Michael J. *Korean Cuisine: An Illustrated History.* Reaktion Books, 2008.
Phillips, William D. and Carla Rahn Phillips. *The Worlds of Christopher Columbus.* Cambridge University Press, 1993.
Pigafetta, Antonio. *The First Voyage round the World by Magellan.* London, 1874.
Pomeranz, Kenneth. *The Great Divergence: China, Europe and the Making of the Modern World Economy.* Princeton University Press, 2000.
Price, R. H. *Sweet Potato Culture for Profit.* Dallas, 1896.
Quan, Hansheng. *MingQing jingjishi yanjiu* [Studies of economic history in the Ming and Qing]. Lianjing chubangongsi, 1987.
Quijano, Alonso. *Diet and the Discourse of Food in Early Modern Spain.* University of Toronto Press, 2016.
Quinzio, Jeri. *Dessert: A Tale of Happy Endings.* Reaktion Books, 2018.
Raffald, Elizabeth. *The Experienced English House Keeper.* Manchester, 1769.
Randolph, Mary. *The Virginia House-Wife.* Washington, DC, 1824.
Rath, Eric C. *Food and Fantasy in Early Modern Japan.* University of California Press, 2010.
Ray, Ramesh C., and K. I. Tomlins, eds. *Sweet Potato: Post Harvest Aspects in Food, Feed and Industry.* Nova Science Publishers, 2010.
Reader, John. *Potato: A History of the Propitious Esculent.* Yale University Press, 2009.
Rodrigues, Domingos. *Arte de cozinha.* Lisbon, 1693.
Rostow, W. W. *The Great Population Spike and After: Reflections on the 21st Century.* Oxford University Press, 1998.
Rowe, T. William. *Saving the World: Chen Hongmou and Elite Consciousness in Eighteenth-Century China.* Stanford University Press, 2001.
Roxburgh, William. *Flora Indica, or, Descriptions of Indian Plants.* Serampore, 1832.
Rundell, Maria Eliza. *A New System of Domestic Cookery.* New ed. London, 1808.
Sahagún, Bernardino de. *General History of the Things of New Spain, Book 11—Early Things.* University of Utah Press, 1963.
Sakai, Kenichi. *Satsuma imo* [The sweet potato]. Hosei daigaku shuppansha, 1999.
Salaman, Redcliffe N. *History and the Social Influence of the Potato.* Cambridge University Press, 1985.
Seo, Yugu. *Jongjeobo* [Manuals for sweet potato cultivation]. In *Jinshu chuanxilu zhongshupu hekan* [On the cultivation of the sweet potato and manuals for sweet potato cultivation]. Nongye chubanshe, 1982.
Shi, Zhihong. *Agricultural Development in Qing China: A Quantitative Study, 1661–1911.* Brill, 2018.
Siegmund, Felix. "Einführung und Verbreitung der Süßkartoffel und Kartoffel in Korea und China." Master's thesis, Ruhr-Universität Bochum, 2009.
Sim, Alison. *Food and Feast in Tudor England.* Sutton Publishing, 1997.

Simmons, Amelia. *American Cookery.* Hudson & Goodwin, 1963. Originally published in 1796.

Frederick Simoons. *Food in China: A Cultural and Historical Inquiry.* CRC Press, 1991.

Smil, Vaclav, and Kazuhiko Kobayashi. *Japan's Dietary Transition and Its Impact.* MIT Press, 2012.

Smith, Adam. *An Inquiry into the Nature and Causes of the Wealth of Nations.* Edited and introduced by Edwin Cannan. University of Chicago Press, 1977.

Smith, Eliza Smith. *The Complete Housewife.* 5th ed. London, 1758.

Smith, F. Andrew. *Potato: A Global History.* Reaktion Books, 2011.

Staunton, George. *An Authentic Account of an Embassy from the King of Great Britain to the Empire of China,* vol. 2. London, 1797.

Summerhill, Stephen J., and John Alexander Williams. *Sinking Columbus: Contested History, Cultural Politics, and Mythmaking During the Quincentenary.* University of Florida Press, 2000.

Tardif-Douglin, David Gregory. *The Role of Sweet Potato in Rwanda's Food System: The Transition from Subsistence Orientation to Market Orientation.* DAI, 1991.

Taylor Sen, Colleen. *Feasts and Fasts: A History of Food in India.* Reaktion Books, 2015.

Terrón, Eloy. *España, encrucijada de culturas alimentarias: Su papel en la difusión de los cultivos americanos.* Ministerio de Agricultura, Pesca y Alimentación, 1992.

Terry, Edward. *A Voyage to East-India.* London, 1777.

Tewe, O. O., F. E. Ojeniyi, and O. A. Abu. *Sweetpotato Production, Utilization and Marketing in Nigeria.* International Potato Center, 2003.

Varenne, François Pierre de la. *La Varenne's Cookery.* Translated and annotated by Terence Scully. Prospect Books, 2006.

Vella, Christina. *George Washington Carver: A Life.* Louisiana State University Press, 2015.

Venner, Tobias. *Via recta ad vitam longam.* London, 1638.

Villareal, Ruben L., and T. D. Griggs, eds. *Sweet Potato: Proceedings of the First International Symposium.* AVRDC, 1982.

von Verschuer, Charlotte. *Rice, Agriculture and Food Supply in Premodern Japan.* Translated by Wendy Cobcroft. Routledge, 2016.

Wan, Guoding. *Wugu shihua* [Stories of the five grains]. Zhonghua shuju, 1961.

Warman, Arturo. *Corn and Capitalism: How a Botanical Bastard Grew to Global Dominance.* Translated by Nancy L. Westrade. University of North Carolina Press, 2003.

Watson, Barbara, and Leonard Y. Andaya. *A History of Early Modern Southeast Asia, 1400–1830.* Cambridge University Press, 2015.

Watt, George. *The Commercial Products of India.* John Murray, 1908.

Weatherford, Jack. *Indian Givers: How the Indians of the Americas Transformed the World.* Crown Publishers, 1988.

Wheaton, Barbara Ketcham. *Savoring the Past: The French Kitchen and Table from 1300 to 1789*. Touchstone, 1983.

Will, Pierre-Etienne. *Bureaucracy and Famine in Eighteenth-Century China*. Translated by Elborg Forster. Stanford University Press, 1990.

Wong, R. Bin. *China Transformed: Historical Change and the Limits of European Experience*. Cornell University Press, 1997.

Woolfe, Jennifer. *Sweet Potato: An Untapped Food Resource*. Cambridge University Press, 1992.

Wright, Clarissa Dickson. *A History of English Food*. Random House, 2011.

Xiamen Daxue Lishi Yanjiusuo, Zhongguo Shehui Jingjishi Yanjiushi [History Research Institute, Chinese Social History Unit, Xiamen University], eds. *Fujian jingji fazhan jianzhi* [Concise history of economic development in Fujian]. Xiamen daxue chubanshe, 1989.

Xu, Guangqi. *Nongzheng quanshu jiaozhu* [Complete treatise on agriculture], vol. 2. Annotated by Shi Shenghan. Shanghai guji chubanshe, 1977.

Yamada, Syoji [Shōzi]. *Satsuma imo: Tenrai to bunka* [Sweet potato: Its spread and culture]. Shun'endō shuppan, 1994.

Yamakawa Osamu. *Satsumo imo no sekai, sekai no satsumo imo* (Sweet potato's world and the world of the sweet potato). Gendai shokan, 2020.

Yang, Guozhen, Puhong Zheng, and Qian Sun. *MingQing Zhongguo yanhai shehui yu haiwai yimin* [Chinese coastal society and overseas emigration in the Ming and Qing periods]. Gaodeng jiaoyu chubanshe, 1997.

Yang, Jisheng. *Tombstone: The Great Chinese Famine, 1958–1962*. Farrar, Straus and Giroux, 2012.

Yen, D. E. *The Sweet Potato and Oceania: An Essay in Ethnobotany*. Bishop Museum Press, 1974.

Zanotti, Laura. *Radical Territories in the Brazilian Amazon*. University of Arizona Press, 2016.

Zhang, Jian. *Xindalu nongzuowu de chuanbo he yiyi* [Diffusion and significance of New World crops]. Kexue chubanshe, 2014.

Zhao, Gang. *Qingdai liangshi mu chanliang yanjiu* [Studies of acreage productivity during the Qing]. Zhongguo nongye chubanshe, 1995.

Zhao, Wenlin, and Shujun Xie. *Zhongguo renkoushi* [History of Chinese population]. Renmin chubanshe, 1988.

ZSKLYQY (Zhongguo shehui kexueyuan lishi yanjiusuo Qingshi yanjiushi), ed. *Qingshi ziliao* [Sources of Qing history]. Zhonghua shuju, 1989.

Zuckerman, Larry. *The Potato: How the Humble Spud Rescued the Western World*. North Point Press, 1998.

Index

1421: China Discovered the World (Gavin Menzies), 17

1491: New Revelations of the Americas Before Columbus (Charles Mann), 21

1492: The Year Our World Began (Felipe Fernández-Armesto), 21, 391n56

1493: Uncovering the New World Columbus Created (Charles Mann), 21

Abu, O. A., 352–354, 449n50
Accomplisht Cook, The (Robert May), 115, 122–123, 126, 130
Acland, J. D., 359, 368, 450n65, 451n72, 452n91
Acosta, José de, 85, 90, 94, 97, 403n34, 405n72
Acton, Eliza, 183, 187
Adams, William, 223–226
Africa: Its People and Their Culture History (George Murdock), 40
Agricultural Development in Qing China (Shi Zhihong), 203, 426n140

aha, 119
ahi, 119
Ahmad, I. M., 354, 449n50, 450n55, 452n94
Aiauhtona, 83
ajes/ages, 48–49, 72, 74–80, 82–86, 103, 119, 166, 178, 241
aji/ají, 85–88, 92, 110, 151, 403n38
ajies/ajíes, 48, 74–75, 77, 79
akaimo, 227, 241, 288
Akamine, Mamoru, 285
Akazawa Nihei, 256, 376
Albala, Ken, 92, 100–101, 110, 113, 123, 126, 404n61
Allen, Matthew G., 69
Alley, Rewi, 208–210
Alpern, Stanley, 351–352
Altamiras, Juan de, 116–117, 119
alu, 342
Álvarez Chanca, Diego, 73, 96
amala, 355
Amarillo verdadero (type of sweet potato), 150

[469]

America's First Cuisines (Sophie D. Coe), 78
American Cookery (Amelia Simmons), 134
American Indians in the Pacific: The Theory Behind the Kon-Tiki Expedition (Thor Heyerdahl), 43
amukeke, 368
Andalusia, 80, 99, 114, 129, 336
Andaya, Barbara Watson, 58
Andaya, Leonard Y., 58
Anderson, Aeneas, 188–190
Anderson, Eugene N., 177, 216
Andrade, Maria, 67, 369, 449n46
Anghiera, Peter Martyr d', 49, 75–76, 398, 402n15, n17
Anju, 380
Aoki Konyō, 229–235, 240–242, 245, 249, 256, 431n18
aphrodisiac, 29–30, 93, 97–101, 124, 384
Apologética historia summaria de las gentes destas Indias (Bartolomé de Las Casas), 76
Arizono Shōichirō, 238–240, 253–254, 290, 432n32, 435n63
Art of Cookery Made Plain and Easy, The (Hannah Glasse), 132–133
Arte de cocina, pastelera, bizcochera y conservera (Francisco Martnez Montiño), 113
Arte de Cozinha (Domingues Rodriguez), 119
Arte de repostería (Juan de la Mata), 116
As-Saqui, M. A, 355, 450n58
Asagao, 227
Asami Kichijūrō, 230
Asian Vegetable Research and Development Center, 280, 329
asses, 48, 72–75
Atienza, Pedro de, 93
Ayauhtona, 83

Badenoch, Alexander, 96, 405n83, 406n87
baijiu, 10
Baker, James W, 138, 142, 411n208, 412n220
Bakkeljauw, 147
Ballard, Chris, 69–70, 306, 396–397, 399n72, 400n95, 441n65, 442n67, 443n75
Ballestero, Miguel, 93
bam goguma, 267
bambai, 351
bambaira, 351
Banchetti, 92
Bancroft, George, 17
Bangbe, 351
banh tom, 338
Banquete de nobles caballeros (Luis Lobera de Ávila), 111
Bansho kō (Aoki Konyō), 231–232
Bansho roku (Matsuoka Jo'an), 231
bansho, 227. *See* sweet potato
Barrow, John, 187–191, 422n98, n102
basi fanshu, 202
Batata bel eshta, 348
Batata bel kamoun, 349
batata(s), 30, 47–50, 52, 54, 57, 67–69, 74–88, 91–92, 95, 102, 110–111, 114, 128, 151, 302, 331, 342–344, 348–349, 351, 401–404. *See* sweet potato
batate, 63, 68, 340
Bauhin, Gaspard, 99
bengkuang, 340
beni imo, 292–294, 296–298
Benincasa hispida, 59
beonseo, 243
bitsu bitsu, 327
Blaff de Poisson Blancs, 147
Blake, Michael, 21, 391
Blue, Anthony, 145, 411n200, 413n234

Blue, Kathryn, 145, 411n200, 413n234
bombe, 351
Bonavia, Duccio, 21, 391n58
Boniatillo seco, ó paota, 128
Boniato asado, 128
Boniato salcochado sin agua, 128
boniatos, 79
Boserup, Ester, 198–199, 202, 394n76, 425n129
botato, 48
Bourke, R. Michael, 69, 301–302, 305, 307–308, 389n32, 396n17, 397n44, 400n93, n98, 441n58, 442n70, n71, n74, 443n75, n76, n77
Bouwkamp, John C, 165, 301, 389n28, n35, 90n37, 403n41, 417n33, 429n190, 441n58, 447n14, 450n57
Bradley, Richard, 130–131, 410n184
Brand, Donald D, 36, 49, 395n9, 398n53
Brears, Peter, 126, 387n2, 409n170, 410n186
Breverton, Terry, 109, 118, 405n62, 407n124, 408n146
Brevísima relación de la destrucción de las Indias (Bartolomé de Las Casas), 76
Briggs, Richard, 133–134, 410n193, n194
Brookes, Joshua, 130
Broughton, William Robert, 290–291
Brown, Paula, 69, 396n22, 442n67
Buck, John Lossing, 209–210, 212, 215, 427n163, 428n172
Buck, Peter (Hīroa, Te Rangi), 55, 59
bukgamjeo, 254
Bunmei o kaeta shokubutsu-tachi (Sakai Nobuo), 23
Buñuelo de boniato, 128
Burkill, Isaac H, 75, 85–87, 110, 401n1, 402n14, 403n39, 404n42
Buyō Inshi, 252

Cai Chenghao, 25, 274, 277, 282, 392n69, 438n6, n7, n10, n13
Cambridge History of China, The (John K. Fairbank, et al), 178, 180
Cambridge History of Japan, The (John Whitney Hall, et al), 234, 257
Cambridge World History of Food (Kenneth F. Kiple and Kriemhild C. Ornelas), 302, 351
camote, 47–48, 50, 53–54, 57, 67–69, 80, 83–86, 128, 146, 148, 302, 330–333, 336. See sweet potato
camotli, 50, 74, 82–83, 330
camoxalli, 83
Campbell, Cameron, 197, 424n124
Campbell, Jodi, 88, 404n44
Campilan, Dindo, 331–332, 334, 446n3, n5, n9, 448n22
Cao Shuji, 155, 415n11, 419n63, 421n80, 425n128
Carne de limon y batatas, 111
Carney, Judith, 139, 411n212
Carruthers, Tom, 41–42, 397n31
Carver, George Washington, 6, 140–141, 411n214, 412n215, n216
casabe, 146
Central Research Institute for Food Crops, 337
Centro Internacional de la Papa. See International Potato Center
ceviche, 149–150
Chakin Shibori, 268
Chamberlain, Basil Hall, 221, 253, 430n2, 434n54, 435n62
chancaca, 151
Chang Jianhua, 197, 424n125
Chang, Kwang-chih, 272, 438n3
Chen Di, 274–275
Chen Dongsheng, 210, 420n64, n67, 426n137, 428n165
Chen gong shu, 175

Chen Hongmou, 173–174, 176, 184, 419n57, n–60, 420n72
Chen Jinglun, 159
Chen Shiyuan, 157–159, 163, 165, 168–169, 172, 175, 201, 415n8, 419n55, n56, 430n1
Chen Shu, 159
Chen Shuping, 155, 171–172, 415n14, 416n19, 419n53, n54
Chen Yi, 156–157, 415n14, 416n20
Chen Yun, 159
Chen Zhenlong, 157–159, 165–166, 169, 221–222, 332, 417n34
Chengnan jiushi (Lin Haiying), 272
Chiang Kai-shek, 210, 279, 282
Chinese Society in the Eighteenth Century (Susan Naquin and Evelyn Rawski), 178
chiqing, 215
chocolate, 90–91, 93–102, 140
Chōshin Sunagawashinya, 222
Chow-ta-zhin, 187
chtitha batata, 349
Chūman Katsumi, 249–250, 257, 263, 294, 296, 392n68, 430n6, 434n50, n51, 435n67, 436n73, n77, n82, n87, n92, 440n28, n31, n44, n47, n50, 441n53
Cidra y patata, 111
Cieza de León, Pedro, 114
Civitello, Linda, 4, 388n11
Closet of the Eminently Learned Sir Kenelme Digbie Kt, Opened, The (Kenelme Digby), 104
Clusius, Carolus, 80–81, 83, 108, 120, 123, 128, 402n27
cocina criolla, 149
Cocos nucifera L, 36
Coe, Michael D, 97, 406n88
Coe, Sophie D, 78, 97–98, 110, 402n23, 406n88, 407n125
Coil, James, 61, 399n75, 445n103

Collingham, Lizzie, 259, 436n79, n83, n89
Colocasia esculenta. *See* taro
Columbian Exchange, 16, 19–21, 24, 29, 88–89, 101, 312
Columbian Exchange: Biological and Cultural Consequences of 1492, The (Alfred Crosby), ix, 2–3, 6, 19, 35, 154
Columbus, Christopher, 4, 53, 93–94, 96, 112, 146, 151, 155, 167; receptions of, 15–21, 23–24, 30; voyages to Americas, 37, 48–49, 71–76, 78, 84–86, 88–89
Columbus: His Enterprise Exploding the Myth (Hans Koning), 19
Coma, Guillermo, 72, 74
Commercial Products of India (George Watt), 344
Complete Housewife, The (Eliza Smith), 131–132
Conditions of Agricultural Growth: The Economics of Agrarian Change under Population Pressure, The (Ester Boserup), 199
Conklin, Harold C, 67–68, 351, 396n24, 398n52, 400n91, 449n47
Conquest of America: How the Indian Nations Lost Their Continent, The (Hans Koning), 20
Conquest of America: the Question of the Other (Tzvetan Todorov), 19
conquistadors, 3, 30, 49, 76, 78, 87, 93, 97, 114
Consultative Group for International Agricultural Research, 149
Consumer Reports, 382
Cook Book (Martha Washington), 130, 381
Cookery for Maids of All Work (Eliza Warren), 130

Cooking, Cuisine and Class (Jack Goody), 87
Corn and Capitalism: How a Botanical Bastard Grew to Global Dominance (Arturo Warman), 5
Cortés, Hernán, 78, 96–97, 405n79
couch potato, 90
Country Housewife and Lady's Director, The (Richard Bradley), 130
Crawcour, E, Sydney, 257–258, 435n66, n71, 436n75, n77
Crawfurd, John, 34
Crosby, Alfred, ix, 2–3, 6, 154, 387n6, n7, 388n15, n16, 391n50, n51, 395n7, 414n7, and "Columbian Exchange," 16, 19, 21, 23–24, 35, 88, 101; and Hawaiian population, 326, 446n122
Cuisine and Culture (Linda Civitello), 4
Cuisine and Empire (Rachel Laudan), 89
cumar. See kumar/kumara

daigaku imo, 268
Daily Life in Elizabethan England (Jeffrey Forgeng), 103
daliya, 348
dangmyeon, 220, 269, 380
dango, 242, 249
Daoyi zhilue (Wang Dayuan), 58, 399n69
Dawson, Thomas, 99, 103–104, 115, 120–121, 127, 165, 406n93, 407n108
De alimentis et eorum recta administratione (Giovanni Domenico Sala), 115
De honesta voluptate et valetudine (Bartolomeo Sacchi), 92
De Orbe Novo (Peter Martyr d'Anghiera), 49, 75
Deng Xiaoping, 22, 213
Description of England (William Harrison), 99
Dessert: A Tale of Happy Endings (Jeri Quinzio), 91

Diamond, Jared, 35, 395
Díaz del Castillo, Bernal, 78, 96–97, 402n25, 405n79, 406n86
Digby, Kenelme, 104, 407n112
digua, 162, 183–184, 202, 243, 281
Dikötter, Frank, 214, 216, 428n175
ding (male adults), 179
Dioscorea alata. See yam
Dioscorea esculenta. See yam
Dioscorea polystachya. See *shanyao*/ Chinese yam
Directions for Health both Naturall and Artificial (William Vaughan), 104
Dixon, Ronald B, 55, 395n15, 396n16, 399n63
Dongfanji (Chen Di), 274–275
Drake, Francis, 4, 223
Duell, Barry, 227, 266–268, 376, 431n10, n17, 434n53, 437n99, n102, 454n15

Earle, Rebecca, 3–4, 22, 89, 91, 93, 96, 102, 373, 387n8, 388n11, 392n63, 404n46, n50, 405n64, n80, 406n92, n105, 453n7
Earle, Timothy, 322, 324, 445n113, n117
East African Crops (J. D. Acland), 359
Eating Right in the Renaissance (Ken Albala), 123
Echinochloa crus-galli, 254
Ehret, Christopher, 67, 400n90, 449n46
Ein New Kochbuch (Marx Rumpolt), 119
El Libro de Las Familias: Novísimo Manual Práctico de Cocina Española, Francesa y Americana, Higiene y Economía Doméstica (n.a.), 127–128
Ellis, William, 63, 400n81
Ellison, Ralph, 144
Emperor Jiaqing, 160
Emperor Kangxi, 179
Emperor Qianlong, 160, 173, 179, 222, 245

Englands Happinesse Increased, Or a Sure and Easie Remedy against All Succeeding Dear Years by A Plantation of the Roots Called Potatoes (John Forster), 108

English Art of Cookery, The (Richard Briggs), 133–134

English Huswife, The (Gervase Markham), 121–123

Erikson, Leif, 17

España, Encrucijada de Culturas Alimentarias: Su papel en la difusion de los cultivos americanos (Eloy Terrón), 2, 84, 113

Essay on the Principle of Population, An (Thomas Malthus), 190

Estebanillo González, 99

Experienced English Housekeeper, The (Elizabeth Raffald), 133–134

Fairbank, John K, 180, 394n75, 420n75, 421n79

fan (cooked rice), 217

fan (foreign, barbarian), 156

fanshu, 25, 156–157, 163, 183–184, 201, 219, 227, 243, 272, 274, 277, 281, 288, 375, 379. See sweet potato

Fanshu de haizi (Xiao Xiao), 282

fanshu qian, 277

fanshu. See sweet potato

fanshulao, 25

Fanshuren de gushi (Chang Kwang-chih), 272

FAO (Food and Agriculture Organization), 352–353, 355, 360–361, 364, 373–374

Farmer's Bulletin (D. M. Nesbit), 144

Fattacciu, Irene, 100, 405n77, 406n100

Fensi/fentiao, 10, 201, 220

Fernández Oviedo y Valdés, Gonzalo, 49, 78, 83, 86, 90, 110, 398n49, 402n22, 403n35

Fernández-Armesto, Felipe, 9, 21, 99, 389n25, 391n56, 395n12, 406n97

Fitting, Elizabeth, 22, 391n60

Fletcher, John, 1–2, 98, 387n2, 406n90

Flora Indica (William Roxburgh), 343

Food and Agriculture in Communist China (John Lossing Buck, et al), 212

Food and Agriculture in Papua New Guinea (Michael Bourke and Tracy Harwood), 69

Food and Feast in Tudor England (Alison Sim), 109

Food Cultures of the World Encyclopedia (Ken Albala), 346

Food in China (Frederick Simoons), 218

Food in Global History (Raymond Grew), 345

Food of China, The (Eugene Anderson), 177, 216

Food of Paradise: Exploring Hawaii's Culinary Heritage, The (Rachel Laudan), 326

Food Plants of China (Shiu-ying Hu), 218

Forgeng, Jeffrey L, 103, 407n110

Forme of Cury, The (n. a.), 126

Forster, John, 108–109, 118, 123, 127, 131, 407n120, 408n145

Fortune, Robert, 252–253, 435n60

Frampton, John, 79, 402n26

Friederici, Georg, 53–56, 398n61

Fryer, John, 343–345, 448n29

fufu, 355–356, 368–369

fukushoku, 287

Fussell, Betty, 22, 391n59

Gage, Thomas, 95, 405n78

gajbaje, 347

Galen of Pergamon, 101, 406n103

Gama, Vasco Da, 16, 49

gamjeo, 221, 243–245, 254

Gamjeobo, 221, 243

Gang Pilri, 220, 243–245
Gansho saibai ho (Akazawa Nihei), 256
Ganshu lu (Lu Yao), 153, 163, 173
Ganshu shu (Xu Guangqi), 160–163, 165, 168, 174
ganshu, 153–154, 156, 163, 243–244. See sweet potato
Gao Hua, 215, 429n183
gaoliang. See sorghum
Ge Jianxiong, 179, 182, 420n78, 423n117, 427n155
Genovese, Eugene D, 143
Gentilcore, David, 90, 95, 97, 100, 122–123, 404n47, 405n75, 406n85, 409n155
George, George, 3, 388n9
Gerard, John, 104–105, 108–109, 115, 122–125, 128, 407n114, n115, n118, 409n159, n161, n163
getuk lingri, 338
Gift of Southern Cooking, The (Edna Lewis and Scott Peacock), 142
Gima Shinjō, 222, 225, 283, 287, 291, 430n6
Ginataan, 335
Glasse, Hannah, 132–134, 410n188
Godey's Lady's Book (Sarah Hale), 137
Goguma mattang, 268
goguma twigim, 381
goguma-bap, 380
goguma-mallaengi, 380
goguma, 243–244, 267–268, 379–381. See sweet potato
Gokoku imo, 295
gokoku, 250
Gómez, Raimundo, 116
Good Housekeeper, The (Sarah Josepha Hale), 137
Good Huswife's Jewell, The (Thomas Dawson), 99, 103, 120
Goodrich, L.C, 155, 157, 416n17

Goody, Jack, 87–88, 404n43
Graham, Wade, 312–313, 317, 319, 325, 443n86, 444n89, n97, n98, n99, n101, 445n104, n119, n120
Granado, Diego, 111–112, 114–116, 122, 407n129, 408n136
Great Kyōhō Famine, 229, 232, 234, 237, 249
Great Leap Forward, 211–213, 216, 218
Great Tenmei Famine, 233, 236–237, 248–249
Green, Roger, 59–62, 312, 443
Grüneberg, Wolfgang, 67, 388n15, 389n27, 390n42, n43, n44, 449n43, 452n91
gu, 160, 184
gumbili, 340
gun gogkuma, 267
Guns, Germs, and Steel (Jared Diamond), 35
Guo Fenglian, 371–372

Habichuelas con Dulce, 146
hae, 300
hage, 73, 85,
haje, 85
Hale, Sarah Josepha, 137, 142, 411n203, n204, 412n221
Han-tsi-a-kiann in Hokkien, 272
Handy, Craighill, 65, 322, 400n85, 444n95, n97, 445n108, n110
Handy, Elizabeth S,C, 65, 319, 322, 444n95, n97, 445n108, n110
Hane, Mikiso, 233, 256–257, 431n19, n21, 435n66, n69, n70, n72, 436n76, n78
hansu-umu, 288
hansu, 288
Harada Saburōemon, 241
Harris, Jessica B, 135, 139, 143–144, 409n176, 411n198, n211, n213, 412n227, n228, 413n232

Harrison, William, 99, 102, 104, 406n94, 407n113

Harwood, Tracy, 69, 389n32, 396n17, n22, 397n44, 400n93, n98, 441n59, 442n74, 443n77, n80

Hawaiian Planter, The (Craighill Handy), 65

Hayami Akira, 235–237, 290, 432n28, n29

Hayward, Vicky, 117–118, 408n142, n144

Health's Improvement (Thomas Muffet), 120, 123

heiau, 317, 320

Heijdra, Martin, 180–181, 421n80, n81

Her'heri, 147

Herball (John Gerard), 104, 106, 108–109, 122–123, 125, 128

Hernández de Maceras, Domingo, 111, 115, 122

Hernández, Francisco, 82–83, 90, 97–98, 402n28, 403n38

Heyerdahl, Thor, 29, 33–36, 39, 41–44, 46, 48, 56, 62, 313, 394n1, 395n5, n6, 397n33, n34, n35, n37

High on the Hog (Jessica B. Harris), 143

Hīroa, Te Rangi. *See* Peter Buck

Historia de las Indias (Bartolomé de Las Casas), 49, 76

Historia de las Plantas de Nueva España (Francisco Hernández), 82, 98

Historia general de las cosas de la Nueva España (Bernardino de Sahagún), 83

Historia general de las Indias (López de Gómara), 77, 93

Historia medicinal de las cosas que se traen de nuestras Indias Occidentales (Nicolás Monardes), 79–80, 123

Historia natural de las Indias (Gonzalo Fernández de Oviedo), 78

Historia natural y moral de las Indias (José de Acosta), 85, 94, 97

Historia verdadera de la conquista de la Nueva España (Bernal Diáz del Castillo), 79, 96–97

History and Social Influence of the Potato (Redcliffe Salaman), 2

History of Early Modern Southeast Asia, A (Barbara Watson Andaya and Leonard Y. Andaya), 58

History of English Food, A (Clarissa Dickinson Wright), 109

History of the Indian Archipelago (John Crawfurd), 34

History of the United States (George Bancroft), 17

Ho, Ping-ti, ix–x, 6–7, 23, 154, 388n16, 392n72, 415n8, n10, 421n83, 453n1; and American food plants in China, 155–158; China's population and American plants, 176, 178, 180–181, 185–186, 195, 199

ho'okupu, 315

Hog and Hominy (Frederick Douglass Opie), 144

Holmes, Samuel, 189–190, 422n103

Hong Liangji, 191–194, 422n108, 423n109, n110, n111

Horeki Famine, 233

Hornell, James, 56–57, 62, 67, 312, 399n66, n77, 400n94, 443n87

Housekeeper's Annual & Ladies Register, 137

hu (households), 179

Hu, Shiu-ying, 218, 390, 429n193

huan-tsi-a, 283

Huang, Philip C. C, 205, 393n72, 427n146

Ido Masaakira, 230

idōhanbai, 266

Imo hyakuchin, 270
imo kui, 24
imo shgatsu, 251
imo ufusu, 222
Imo-karinto, 377
Imo-natto, 377
Imo-senbei, 377
imomoshi, 294
imu, 324, 327
Inagaki Hidehiro, 24, 392n65
inginyo, 368
inhame, 71
International Potato Center (*Centro Internacional de la Papa*), 8, 67, 149, 331, 349–350, 352, 354, 356, 358, 360, 369–370, 373–374
International Research Development Centre of Canada, 330
Invisible Man, 144
Ipomoea aquatica, 9–10, 336
Ipomoea batatas. See sweet potato
Ipomoea trifida, 41–42
iraid, 336
irori, 264–265
Irving, Washington, 17–18, 395n14
ishi yaki-imo, 266
Ishige Naomichi, 234, 262, 286, 297, 432n22, 435n59, 436n88, 439n26, 441n52
Itō Jinsei, 229
Itō Shōji, 25, 294, 392n68, 431n14, 433n36, 436n74, n81, n82, n85, n90, n92, 437n97, 440n42, n43, n46, 441n55
Itō Tōgai, 229, 231
iztac camotli, 83

Jana, R. K, 359, 450n67
Japan and Things Japanese (Mock Joya), 221, 264
Jeong Yakyong, 246, 434n44

jigwa, 243
Jin gong shu, 157
Jin Xuezeng, 157–159, 165, 169–170, 172–173, 175, 200, 222
Jinshu chuanxilu (Chen Shiyuan), 157–161, 163, 165, 168–169, 172, 201
jinshu, 12, 157
Jo Eom, 220, 242–245
Jongjeobo, 161, 163–165, 220, 243, 246, 255
Joseon Wangjo Sillok, 245–247
Joya, Mock, 221, 240, 264, 430n3
Joyfull Newes out of the Newe Founde Worlde (Nicolás Monardes), 79
junxiang you fa, 193

Kaibara Ekken, 226
Kaibara Rakuken, 226
kalo, 313, 319
kamute, 68
kamuti, 68
Kananga phodi-tawa, 381
kanbai, 294
kangaanchi usli, 347
Kangacheo neureo, 347
Kangachi kheer, 347
Kangkong, 10
Kansho no rekishi (Miyamoto Tsuneichi), 24
Kansho seshi, 241
kanshoō, 228
kao digua, 281
kapu, 315, 324
Karafir, Yan Pieter, 339, 447n19
Karaimo jinja, 229
Karaimo, 220–221, 229, 288
Karaimoden, 229
kashi, 232
Katō Sango, 291–292, 440n39
katualle, 333
Kawagoe, 256, 266–267, 295, 376–377

Kerr, George H, 282–285, 290, 430n6, 439n20, 440n37
Khanafer, Hadia Zebib, 356, 450n60
khichdi, 347
khumara, 85
kichō na, 258
Kinoshita Jun'an, 240
Kiple, Kenneth F, 302, 441n64, 449n47
Kirch, Patrick V, 61, 315, 395n10, n15, 399n75, 444n93, n94, n100, 445n103
Koishikawa yakuen (Koishikawa Herbal Garden), 230
kōkōimo, 241, 243
kokuō, 285
kokushi, 285
Kolb, Michael, 323–324, 445n115, n117
Kon-Tiki, 33–36, 42–44, 46
Koning, Hans, 19–20
Koronbusu no fubyōdō kōkan sakumotsu dorei ekibyō no sekaishi (Yamamoto Norio), 24
Kuaile fanshu, 219, 375
kudoku, 231
Kuhn, Philip, 178, 420n75
kukui, 316, 321
kūlolo, 314
kuma'a, 62
kumal, 62
kumala, 61–62
kumar, 34, 39, 42, 47–49
kumara, 34, 42, 46, 48–50, 53–54, 57, 60–63, 66, 69, 85, 148, 312, 331. See sweet potato
kushsu, 348

l'Écluse, Charles, 80
La Varenne's Cookery, 119
Ladefoged, Thegn, 61, 319, 445n103, n104
Lagenaria vulgaris Ser., 36
Land Utilization in China (John Lossing Buck), 212
Langdon, Robert, 41, 53, 396n27, 398n60
Las Casas, Bartolomé de, 18, 76–77, 83, 86, 110, 402n19, n20
Laudan, Rachel, 3, 85, 89, 101–102, 326, 387n8, 403n33, 404n45, 405n69, n73, 406n104, n107, 446n125, n128
Laufer, Berthold, 4, 54–55, 398n62
Lavely, William, 195–196, 206, 424n121, 427n151
Le Cuisinier françois (François Pierra de la Verenne), 119
Leach, Helen, 59–60, 399n72, n73
Lebot, Vincent, 129, 390n38, 395n13, 397n43, 410n177, 411n206
Lee, James Z, 27, 196–197, 393n74, 424n122, n124, 425n128
Levy, Stanley, 337, 447n17
Lewis, Edna, 141–142, 412n218, n219
Li Dingyuan, 289
Li Wei, 159, 172–173, 176, 420n66, 427n159
Li Xinsheng, 199, 203–204, 425n130, 426n134
Li, Bozhong, 205, 423n117, 427n149
Liang Jiamian, 155–156, 415n11
Libro del arte de cozina (Diego Granado), 111
lidgid, 335–336
Life and Voyages of Christopher Columbus, The (Washington Irving), 17
Lin Haiying, 272
Lin Huailan (Lin Huaizhi), 156–158, 416n20
Lin Liyue, 197, 424n125
Lingnan zaiji, 171
Lister, Martin, 119
liumin, 206

Livro de Cozinha da Infanta Maria (Infanta Maria), 118
Llibre del Coch (Robert de Nola), 110
Lobera de Ávila, Luis, 111
longue durée, 21
López de Gómara, Francisco, 74–75, 77, 86, 93, 110, 401n11, 403n37, 405n68
Low, Jan, 370, 388n15, 452n87
Lu Xiaoping, 374, 453n4
Lu Yao, 163, 173, 176, 422n94
Lu, Hanchao, 211, 428n167, n170
Lung Ying-tai, 271–272, 437n1
Luzon, 50, 157–159, 165, 168, 221, 332–334, 341

Maangchi's Big Book of Korean Cooking (Emily Kim), 379
Maangchi's Real Korean Cooking (Emily Kim), 379
Macartney, George, 187–190, 196, 290, 422n96, n99
Mackay, K.T, 330, 366, 389n21, 392n70, n71, 400n97, 446n2, n8, 447n17, n19, 448n25, 452n85, n87
Macnab, J.W, 59, 399n71, 444n88
Maddison, Angus, 6, 388n17
Madrid: A Culinary History (Maria Paz Moreno), 118–119
Maeda Riemon, 228–229
Magellan, Ferdinand, 16, 50, 63, 112, 340
Mai Thach Hoanh, 341, 448n25
Maize for the Gods (Michael Blake), 21
Maize: Origin, Domestication, and Its Role in the Development of Culture (Duccio Bonavia), 21
Makahiki, 65–66, 315–317, 321
Málaga, 79–80, 99, 110, 116, 128–129
Malthus, Thomas, 27–28, 190–198, 393n73, 394n76, 422n104, n105, n106, n107, n108, 425n128

Mann, Charles C, 21, 391n57
Manila galleons, 25, 50, 52–53, 158, 330–331
Mao Zedong, 210, 212, 216, 372
Marco Polo, 15, 167, 390n46
Markham, Gervase, 121–123, 409n153
Martnez Montiño, Francisco, 113
Mata, Juan de la, 116, 408n140
Matembela, 356
matla, 346
matobolwa, 367
Matsuoka Jo'an, 231, 431n10, n15
May, Robert, 115, 122–123, 126, 130, 408n137
Mazumdar, Sucheta, 204, 345–346, 416n22, 426n145, 448n34
mbatata, 351, 358
McNeill, John R, 2, 387n6
McNeill, William H, 3, 6, 27, 388n10
Menzies, Gavin, 17
Merrill, Elmer D, 40, 396n26
Merry Wives of Windsor, The (William Shakespeare), 1
Messisbugo, Cristoforo di, 92, 404n61
michembe, 367
Mintz, Sydney W, 91, 93, 95, 404n54, n55, 405n65
Miyamoto Tsuneichi, 24, 225, 392n67, 430n7, 431n9, n12, n14, n16, 432n27, 433n36
Miyazaki Yasusada, 226–229, 231, 431n10
mochi, 251, 265, 292
Modern Cookery for Private Families (Eliza Acton), 134, 137
Monardes, Nicolás, 79–80, 82–83, 110, 114–115, 123, 402n26, 407n126
Mongmin Simseo, 246
Morado legitimo (type of sweet potato), 150
Morals of History, The (Tzvetan Todorov), 19

Moreno, Maria Paz, 117–118, 408n143
Moss, Sarah, 96, 405n83, 406n87
muchi, 292
Muffet, Thomas, 120–121, 123, 408n151
mugi, 254
Mul goguma, 267
mumu, 309
Muñoz-Rodríguez, Pablo, 41–42, 396n31
Murdock, George, 40

ñame, 71
nanban ryōri, 234, 252
napi, 73, 85, 147
Naquin, Susan, 178, 207, 420n75, 427n153, n154
National Sweetpotato Collaborators Group, 8
Natural History and Antiquities of Selborne, The (Gilbert White), 133
Natural History of the West Indies (Fernandez de Gonzalo Oviedo), 71, 110
Navagiero, Andrea, 114
Nesbit, D. M, 144, 412n230
New Account of East India and Persia, A (John Fryer), 343
New Household Recipe-Book, The (Sarah Josepha Hale), 137
New System of Domestic Cookery, A (Maria Eliza Rundell), 134
niame(s), 71–72, 76
Nishimura Tokinori, 228
No Seonghwan, 243–244, 430n1, 432n34, 433n38
Nōgyō zensho, 226–228, 241
Nokuni Sōkan, 222, 225, 283, 287, 291, 430n5
Nola, Robert de, 110
Nongsang jiyao (Sinongsi), 173

Nongzheng quanshu (Xu Guangqi), 161–163, 226–227, 229, 231, 242, 243, 245
Norton, Marcy, 91, 96, 404n51, 405n82
Nuevo arte de cocina: sacado de la escuela de la experiencia económica (Juan de Altamiras), 116–117
Núñez de Oria, Francisco, 112

Ogawa Yoshiyuki, 197–198, 424n126
Ogura Shinpei, 243
Ōhara Magozaburō, 260
Ohnuki-Tierney, Emiko, 250–251, 254, 434n56
Ojeniyi, F,E, 352–354, 449n50
okari, 305
Okinawa: The History of an Island People (George Kerr), 283
Olepotrige, 121–122
Olha podrida, 119
Olio Podrida, 115, 122
Olla podrida, 114–115, 118–119, 122
Olmsted, Frederick Law, 143
oni, 292
onimochi, 292
Ōoka Tadasuke, 231
oomara, 61
Opera dell'arte del cucinare, 111
Opie, Frederick Douglass, 139, 142–144, 411–412
Ornelas, Kriemhild Coneè, 302, 441n64, 449n47
Oryza glaberrima. See rice
Outram, Alan K, 91, 404n53
Oxfeld, Ellen, 215–218, 429n181, n187, n189, n191

pain patate, 146–147
papa, 84–85, 114
Paradisi in sole paradisus terrestris (John Parkinson), 105, 107

Parkinson, John, 105, 107–109, 123, 407n117, n119
patata(s), 48–49, 84–85, 111, 117
Paul, Heike, 17, 391n48
payew, 333–334
Peacock, Scott, 142, 412n219
pengmin, 205
Perez Garcia, Manuel, 153, 414n4
Perkins, Dwight H, 202–203, 394n76, 420n78, 422n95, 423n117, 425n130, 426n139, n142, 429n182
Phillips, Carla Rahn, 17, 20, 391n49, n54, 395n14
Phillips, William D, 17, 20, 391n49, n54, 395n14
Pigafetta, Antonio, 63, 68, 399n79
pitpit, 305
Pleck, Elizabeth, 145, 411n202, 413n233
Ploeg, Anton, 306–307, 442n68
poi, 314, 318, 327
Potato and Sweetpotato in Africa: Transforming Value Chains for Food and Nutrition Security (Jan Low, Moses Nyongesa, Sara Quinn, and Monica Parker), 26
potato head, 90
potato. *See* white potato
Potato: How the Humble Spud Rescued the Western World, The (Larry Zukerman), 3, 90
Potatoes: The Poor Man's Own Crop (George George), 3
Powell, Adam Clayton Jr, 144
poxcauhcamotli, 82, 83
Price, R. H, 135, 410
Prodromos, The, 99
puran poli, 347
Purseglove, Jeremy W, 41–42, 396n28

Qi Jingwen, 155–156, 415n11
Quan Hansheng, 154, 415n9, 416n23

Quechua, 34, 39, 61, 84–85, 148
Queirós, Pedro, 51–52, 54–55, 63, 69
Quinzio, Jeri, 91, 404n56

Rabisha, William, 127, 409n169
rabri, 347
Raffald, Elizabeth, 133–134, 410n189, n190
Raleigh, Sir Walter, 4, 109, 128–129
Ramírez, Sergio, 96, 405n81
Randolph, Mary, 129–130, 139, 142–143, 410n180, 412n224
Rariorum aliquot stirpium per Hispaniam observatorum Historia (Carolus Clusius), 80–81
ratala, 347
ratalyacha khees, 347
Rawski, Evelyn, 178, 207, 420n75, 427n153, n154
Reader, John, 22, 53, 128, 373–374, 392n62, 398n52, 409n174, 453n7
Reform and Opening-up (China), 22, 213, 217, 373
Rice as Self: Japanese Identities through Time (Emiko Ohnuki-Tierney), 251
rice, 6–7, 11–13, 27, 32, 69, 87, 91, 308–309, 326, 374–375, 380; and Africa, 352, 356, 371; and Americas, 139, 143, 146–151; and China, 164–167, 169–171, 174, 177–178, 181–186, 188–189, 191, 193, 200–201, 203–205, 207, 209, 212, 215–217; and Japan and Korea, 221, 224, 227–230, 233–234, 237–239, 242, 246–248, 250–264, 267–268; and Ryukyu/Okinawa, 286–289, 291–292, 296–297; and Taiwan, 276–280, 282; and South and Southeast Asia, 332–335, 338–340, 342–343, 346–347
Rodrigues, Domingos, 119, 122, 408n147

Roe, Thomas, 342
Roggeveen, Jacob, 52, 63, 398n57
Rōken mantō, 260
rokoko, 231
Rosomuff, Richard N, 139
Roullier, Caroline, 42, 57, 397n43, 399n67
Rowe, William T, 178, 182, 419n57, n58, 421n84
Roxburgh, William, 343–344, 448n30
Ruggles, Clive, 315, 444n94
Rumpolt, Marx, 119
Rundell, Maria Eliza, 134, 410n196
Ryukyu-imo/Ryukyuimo, 221, 227, 241

Sacchi, Bartolomeo, 92
Sahagún, Bernardino de, 83, 101, 403n29
Sakai Nobuo, 23–24, 392n65
sakoku, 226
Sala, Giovanni Domenico, 115, 408n138
Salaman, Redcliffe, 2, 4, 49, 74, 84, 98, 100, 109, 111, 387n4, 388n12, 398n46, n49, n50, n53, 401n12, 402n24, 403n30, 404n47, 406n95, n101, 407n123, 409n162
Saldarriaga, Gregorio, 102, 110, 114, 406n106, 407n127
Satsuma, 161, 221–223, 227–230, 232, 234, 236–239, 241, 248–249, 252–253, 283–286, 288–289
Satsuma imo to Nihon jin: wasure rareta shoku no ashiato (Itō Shōji), 25
Satsuma imo/Satsuma-imo, 221, 229, 231, 253, 285. *See* sweet potato
Savoring the Past: The French Kitchen and Table from 1300 to 1789 (Barbara Ketcham Wheaton), 119
Scappi, Bartolomeo, 111, 115
Sekaishi o ōkiku ugokashita shokubutsu (Inagaki Hidehiro), 24

Seo Yeongbo, 245, 434n41
Seo Yugu, 161, 163–165, 169, 176, 246, 254–255, 417n32, 430n1
Seville, 79–80, 110, 129
Shakarkand rabdi, 347
shakarkand, 347
Shakespeare, William, 1–2, 29, 49, 98, 124, 387, 406n89
Shanyao/Chinese yam (Dioscorea polystachya), 153, 183–184, 364
Shengji pian, 191
Shi Zhihong, 203–204, 426n140
shichifuku, 294
Shō Nei, 223, 285, 287
Shōchū/shochu, 10, 228–229, 249, 269
Shōgun, 224, 226, 230–231, 241, 243, 245, 284–285
Shomotsu bugyō, 232
Shoushi tongkao (E Ertai, Zhang Tingyu, et al), 173–174, 245
shuliang, 174
Shunsaku Nishikawa, 254, 435
shushoku, 227, 249, 263, 287
Siegmund, Felix, xi-xii, 244, 246, 261, 430n1, 433n37, n39, 434n40, n42, n43, 435n65, 436n86, 437n101
Siku quanshu (Ji Yun, et al), 160
Silhak, 244, 246
Sillitoe, Paul, 299–300, 310, 441n56, n57, 443n85
Sim, Alison, 105, 109, 407n114, n116
Simmons, Amelia, 134, 410n195
Simoons, Frederick, 218, 429n194
Slow Boat Model, 45
Smith, Adam, 27–28, 393n73
Smith, Andrew F, 4, 22, 90, 95, 388n11, n14, n17, 392n61, 393n73, 394n78, 404n48, n59, 405n67, n74
Smith, Eliza, 131–134, 410n185
Soenarjo, Roberto, 337, 447n17
soju, 10, 269, 380

Solanum tuberosum. See potato/white potato
Song Yingxing, 193
sorghum, 13, 177, 185, 189, 203, 209, 212, 214, 218, 365
Squanto. *See* Tisquantum
Statguide for Farmers, 335
Staunton, George, 187–190, 290, 422n97, n101, 440n36
Stevenson, Christopher, 61, 399n75
Stokes, John F. G, 62, 399n76, 446n126
Story of Corn, The (Betty Fussell), 22
Studies on the Population of China, 1368–1953 (Ping-ti Ho), 6, 195
Sugar: A Global History (Andrew F. Smith), 95
suito poteto, 268, 437n104. *See* sweet potato
suribachi, 288
Suyama Don'ō, 240–242, 432n34, 433n36
Suzuki Teiichi, 261
sweet potato (Ipomoea batatas), ix–xii, 2, 4–14, 16, 21, 23–26, 28–32, 59–69, 248–254, 267–269, 330, 371–380; and Africa, 349–370; and African American foodway, 136–145; and Latin America, 146–154; and North America, 129–135; as aphrodisiac, 93, 95–96, 98–103; as famine food, 225–235, 237–241; as staple, 165–175, 199, 271–299, 349–370; as superfood, 381–385; Columbus's encounter of, 70–74; and Hawaii, 311–329; impact on China's population, 180, 182–186, 189, 193; introduction to China, 155–164; introduction to Europe, 75–88, 91–92, 104–105, 107–124, 126–128; introduction to Korea, 242–247; introduction to Japan, 225–228; introduction to Ryukyu/Okinawa, 220–223; and Middle East, 341–348; and Oceania (other than Hawaii and Papua New Guinea), 46–58; Origins of 36–42; and Papua New Guinea, 300–310; and South Asia, 341–349; and Southeast Asia, 331–341; and Taiwan, 271–282; and Thanksgiving, 136–139; and World War II, 255–266
Sweet Potato and Oceania, The (D. E. Yen), 39, 60, 68
Sweet Potato Culture for Profit (R. H. Price), 135
Sweet Potato History in Kawagoe, 376
Sweet Potato Production Guide (Department of Agriculture, Phillippines), 338
Sweet Potato Research and Development for Small Farmers (Kenneth Mackay, Manuel Palomar, and Rolinda Sanico), 25
Sweetness and Power (Sydney Mintz), 91, 93
sweetpotato, 8, 26, 335, 350, 353, 356, 365–366, 369–370
Syllacio, Nicolò, 72, 74–75

taanmu, 292
Taiwan fanshu wenhuazhi (Cai Chenghao and Yang Yunping), 25, 274, 277
tama, 147–148
Tanegashima Hisamoto, 228–229
Tardif-Douglin, David Gregory, 361–362, 369, 451–452
taro (Colocasia esculenta), 29, 31, 71, 286, 288, 291–292; and Americas, 147, 153; and China, 156, 184, 220; and Hawaii, 312–324, 326–327; and Japan and Korea, 227, 241, 243, 251, 253; and Oceania (other than Hawaii and Papua New Guinea), 47, 58–59, 64–66, 69–70; and Papua New Guinea, 296, 301, 304–311; and

taro (*continued*)
  South and Southeast Asia, 333,
    339–340, 364; and Taiwan, 282–283
Taste of Country Cooking, The (Edna
  Lewis), 141
Taste of War, The (Lizze Collingham), 259
tata, 351
Terrón, Eloy, 2, 84, 99, 111, 113–114,
    387n5, 403n31, 406n91, 408n133
Terry, Edward, 342–343, 345, 448n28
Tewe, O.O, 352–354, 449n50, 450n52, n55
Thanksgiving Dinner (Kathryn and
  Anthony Blue), 145
Things Japanese (Basil Hall
  Chamberlain), 253
Thorsby, Erik, 46, 397n41
ti (Cordyline fruticosa), 59,
tiandi you fa, 193
Tiangong kaiwu (Song Yingxing), 193
tinabudlo, 336
Tisquantum, 136
tlapalcamotli, 83
Todorov, Tzvetan, 19–20, 391n52
tōimo, 288
Tōimoden, 229
Tokugawa Ieyasu, 224–225, 240
Tokugawa Yoshimune, 230–231, 240
Tombstone, 216
Torquemada, Juan de, 63
Toyotomi Hideyoshi, 224, 240
Travels in China, 1966–1971 (Rewi Alley),
  208
True History of Chocolate, The (Sophie
  and Michael Coe), 97
tsubo yaki-imo, 266–267, 377–378
Tsuboi Shōgorō, 250, 302, 337, 434n52,
    441n61
*Tudor Kitchen: What the Tudors Ate and
  Drank, The* (Terry Breverton), 118
Twitty, Michael W, 141, 411n209,
    412n217

uala/'uala, 62, 312, 317–319
Uchibayashi Masao, 41, 45, 56, 59,
    396n19, n21, 397n39, 399n65
udon, 269
ufi, 60, 399n72
uma, 333
umaa, 62
umar', 62
umara, 62–63
Umugiraneza, Regis, 363
undhiyu, 346–347
United States Agency for International
  Development, 358
United States Sweetpotato Council, 8
unka, 229, 232, 237
*Unleashing the Potential of Sweetpotato in
  Sub-Saharan Africa* (Maria Andrade,
  et al), 356
upan, 147–148
Upma, 347
uwala, 62, 312

Varenne, François Pierre de la, 119,
    408n148
Vaughan, William, 104
Vega, Garcilaso de la, 114
Venner, Tobias, 120–121, 409n152
Verschuer, Charlotte von, 250–251,
    253–254, 434n55, 435n61
Vespucci, Amerigo, 17
Via Recta ad Vitam Longam (Tobias
  Venner), 120
viandas, 147
Vikings of the Sunrise (Peter Buck), 55
Villamayor, F.G. Jr., 334–335, 446n8,
    447n10, n13
Villareal, Ruben L, 7, 329–330, 337,
    389n20, 390n36, n42, 392n71,
    400n97, 438n8, n14, 440n29,
    441n58, 446n1, 447n17, 450n58, n67,
    456n34

Virginia Housewife, The (Mary Randolph), 129, 139, 143
Völkerwanderung, 208

Wallin, Paul, 61, 399n75, 400n86
Waltner, Ann, 197–198, 424n23
Wan Guoding, 171–172, 416n18, 419n52, n54
Wang Dayuan, 58, 399n69
Wang Feng, 27, 393n74, 424n122, 425n128
Wang Guei-ling, 216
Wang Ji, 289
Wang Jiaqi, 153–154, 156, 163, 414n3
Wang Shiduo, 194–197, 423n110
Wang Siming, 199, 203–204, 425n130, 426n141, n142, n144
Wang Yangming, 244–245
Warman, Arturo, 5, 22, 388n13, 391n60
Warren, Eliza, 130, 410n181
Warren, Rachel Meltzer, 382, 455n26, 456n31
Washington, Booker T, 140
Washington, George, 6, 129–130, 136, 411n201
Washington, Martha, 130, 381, 410n181
Watson, James B, 302–305, 311, 441n62, 442n67, n68, 443n81
Watt, George, 344–346, 448n31, n32, n33
Wenwu, 153
Wheaton, Barbara K, 119, 408n149
white potato (*Solanum tuberosum*), ix, 2–6, 8–9, 11–13, 16, 22–23, 25–29, 31–32, 130–133, 135, 381–382, 384–385; and Africa, 356, 364, 366; and China, 151, 174, 176–177, 185–186, 188–189, 191, 202, 212–213, 155–220, 371–375; and Europe, 85, 87, 90, 93, 104, 106, 108–109, 114–116, 118–119, 121, 123, 125, 128; and Japan and Korea, 223, 245–246, 252, 254; and Latin America, 146, 148–149; and Oceania, 53–54, 70; and South Asia, 342, 344–346, 348–349; and Southeast Asia, 338, 340
white potato stapleization, 371–375, 382
White, Gilbert, 133, 410n192
Whole Body of Cookery Dissected, The (William Rabisha), 172
Winsor, Justin, 18
wiy, 300
wokou, 168
Wong, R, Bin, 27, 196, 206, 393n74, 424n121, 427n151
Woolfe, Jennifer, 8, 13, 150, 301, 376, 383, 389n22, n23, n29, n31, n33, n34, 390n39, n40, n41, n42, n43, 394n77, 396n18, 403n40, n41, 413n242, 414n251, 417n33, 429n192, 432n35, 436n94, 437n107, 441n58, n60, 442n69, 443n80, 449n42, 451n68, n74, n75, n77, 452n87, n88, n94, 454n14, 456n32, n34; on sweet potato in Africa, 349, 359–362, 366–367, 370; on sweet potato in Asia, 165, 242, 264, 269; on sweet potato in Euro-America, 87, 148
World Vegetable Center, 329
Worlds of Christopher Columbus, The (William and Carla Rahn Phillips), 20
Wright, Clarissa Dickinson, 109, 128, 407n122, 408n150, 409n175
Wu Deduo, 153, 160, 414n3, 416n19
Wu Zhenfang, 170–171, 419n49
Wugu shihua (Wan Guoding), 171

Xi Jinping, 372
Xia Nai, 153, 414n3
xiaochi, 281
xochicamotli, 82–83

Xu Guangqi ji, 161
Xu Guangqi, 201, 232, 242, 287, 387, 417n28, n29, n30, 420n72; concerns of population growth, 185, 193, 196; promotion of the sweet potato, 1, 4, 7, 160, 165, 168–169, 171–172, 174–176; and *Nongzheng quanshu*, 161–163, 226–227, 229, 231, 242, 243, 245

yaki-imo, 265–266, 377–378
yam (Dioscorea), 9, 14, 29, 153, 334, 339–341; and Africa, 351–352, 356–357, 361, 364; and Americas, 70–72, 79, 139, 141–142, 145, 147; and China, 156, 163, 183–184, 220; and Japan, 231, 241, 243, 251, 258; and Hawaii, 313, 316; and Oceania (other than Hawaii and Papua New Guinea), 58–60, 64, 66–67; and Papua New Guinea, 300–301, 305–306, 308–309; and Ryukyu/Okinawa, 288, 296
Yamada Shōzi (Syoji), 228, 237, 248, 265, 285, 287, 430n5, n6, n7, 432n11, n12, n13, n14, n30, 434n49, 436n96, 439n23, 440n30, n33
Yamakawa Osamu, 263, 269, 293, 312, 436n91, 437n106, n109, 440n41, n45
Yamamoto Norio, 24, 392n65
yamanoimo, 227
Yang Baolin, 157, 416n16
Yang Jisheng, 216, 429n186

Yang Yunping, 25, 274, 277, 282, 392n69, 438n5, n6, n7, n10, n11, n13, n15
Yang, Martin, 209–210, 428
Yen, D.E, 39–40, 47–48, 57, 60–61, 64, 66, 68, 85, 301, 312, 323, 395n4, 396n16, n20, n23, n24, n25, n28, 397n43, n45, 398n47, n54, 399n74, n76, n77, 400n82, n88, n89, n92, n94, n95, n96, 403n32, n35, 441n59, 443n87, n88, 445n116, 448n24
yeshi, 281
Yi Gwangryeo, 242–245
Yi Ruolan, 197–198, 425n127
yōkan, 269
Yokoo Terukazu, 265–266
Yu Xinrong, 371
yucca, 75–78
yukatchu, 289

zakkoku, 251
zaliang, 174, 212
Zhang Jian, 23, 392n64, 415n12
Zheng Chenggong, 276–277
Zhiping pian (Hong Liangji), 192
Zhongzhi hongshu faze shiertiao (Li Wei), 172, 176
zhou (rice porridge), 217
Zhou Lianggong, 168, 418n42
Zhou Xun, 214–215, 428n173, n174, n175, n176, n179
Zhou Yuanhe, 154–155, 414n5, n8
zongzi, 292
Zuckerman, Larry, 3, 90, 388n9, 404n49

GPSR Authorized Representative: Easy Access System Europe, Mustamäe tee 50, 10621 Tallinn, Estonia, gpsr.requests@easproject.com

www.ingramcontent.com/pod-product-compliance
Lightning Source LLC
Chambersburg PA
CBHW022024290426
44109CB00014B/731